Sponge Cities: Emerging Approaches, Challenges and Opportunities

Sponge Cities: Emerging Approaches, Challenges and Opportunities

Special Issue Editors

Chris Zevenbergen
Dafang Fu
Assela Pathirana

MDPI • Basel • Beijing • Wuhan • Barcelona • Belgrade

MDPI

Special Issue Editors

Chris Zevenbergen
IHE Delft Institute for Water Education (IHE Delft)
and Delft University of Technology (TuDelft)
The Netherlands

Dafang Fu
Southeast University (SEU)
China

Assela Pathirana
Flood Resilience Group, IHE Delft Institute for Water Education
The Netherlands

Editorial Office
MDPI
St. Alban-Anlage 66
Basel, Switzerland

This is a reprint of articles from the Special Issue published online in the open access journal *Water* (ISSN 2073-4441) from 2017 to 2018 (available at: https://www.mdpi.com/journal/water/special_issues/Sponge-Cities)

For citation purposes, cite each article independently as indicated on the article page online and as indicated below:

LastName, A.A.; LastName, B.B.; LastName, C.C. Article Title. *Journal Name* **Year**, *Article Number*, Page Range.

ISBN 978-3-03897-272-3 (Pbk)
ISBN 978-3-03897-273-0 (PDF)

Cover image courtesy of Chris Zevenbergen.

Contents

About the Special Issue Editors

Chris Zevenbergen is (full-time) professor at the Water Engineering Department of IHE Delft Institute for Water Education (IHE Delft) and at Delft University of Technology (TuDelft), The Netherlands and visiting professor at the Southeast University (SEU) and at Norh China University of Water Resources and Electric Power (NCWU), China. He leads the Flood Resilience Chair Group at IHE Delft. In the past 20 years, he has accumulated extensive national and international experience with integrated approaches to manage floods in the urban environment. His research interests and teaching are specifically on innovative concepts to mitigate urban flood risks, on flood proofing building designs and technologies, and on decision-support tool development in urban planning with practical applications in urban flood management. He has published/edited five books and more than 130 scientific publications in the field of environmental engineering and urban flood management.

Dafang Fu is a (full-time) Professor and Assistant-Dean at the School of Civil Engineering, Southeast University, Nanjing, China. Currently, he is the Chair of Southeast University-Monash University Joint Research Centre for Future Cities. Prof. Dafang is also the Vice-President of Water Environment & Eco-Restoration Research Centre, Jiangsu Province, China. He is actively engaged in research including sustainable urban water system planning & design, low-impact development, ecological improvement for urban road and public space, real-time control of urban water supply and drainage systems, climate change, urban water environment and water ecosystem restoration, and rural ecological protection and restoration. He is also leading three big projects funded by MOST and NSFC, China. Prof. Dafang has published more than 100 research papers in reputed international journals and more than 20 patents since 1990. Prof. Dafang is the Associate Editor of *Journal of Hazardous, Toxic, and Radioactive Waste* since 2014, and served as guest editor in various international journals. He is a professional engineer and expert on Sponge Cities in China.

Assela Pathirana is the Associate Professor of Integrated Urban Water Cycle Management in the Department of Water Science and Engineering in the core group of Flood Resilience at IHE Delft and leads the research programme on Water Sensitive Cities. He has published some 55 indexed journal publications that have been collectively cited more than 1500 times. Assela has 18 years of experience in teaching at post-doctoral-level and currently teaches subjects of urban drainage, water-sensitive cities, asset management of urban water systems, climate change adaptation, urban flood management, modelling water transport and distribution systems, numerical modelling, and computer programming in the context of water and environment. Currently, he manages a research group consisting of around 12 researchers including PhDs, Post-Docs, and Masters thesis researchers. Assela is an associate editor of the *Hydrological Research Letters Journal* and has been a reviewer of more than 15 different journals.

Preface to "Sponge Cities: Emerging Approaches, Challenges and Opportunities"

Primarily as a response to increasing flood impacts, the Chinese Central Government called for widespread uptake of the Sponge Cities Approach across China in 2013 and provided financial support to foster the implementation of this approach in a selection of pilot cities. At present, the Sponge Cities Approach is gaining ground and becoming more and more accepted by city governments. The best practices of Chinese cities are being shared, and international exchange activities between research institutions and cities are providing guidance to the design and implementation of new concepts and technologies. However, there are still many challenges ahead that hamper uptake by the selected pilot cities and up-scaling to the remainder of the 600-plus cities in China. City governments at all institutional levels in China have to support the implementation of the Sponge Cities Approach in new built-up areas of city districts, industrial parks, and development zones. Moreover, in existing urban areas, retrofitting of neighborhoods, refurbishment of existing buildings and infrastructure, and other rebuild activities of old city areas should comply with the Sponge Cities Approach.

A Sponge City is a city that has the capacity to mainstream urban water management into urban planning policies and designs. It should have the appropriate planning and legal frameworks and tools in place to implement, maintain, and adapt the infrastructure systems to collect, store, and treat (excess) rainwater. In addition, a Sponge City will not only be able to deal with "too much water" but will also be able to reuse rain water to help to mitigate the impacts of "too little" and "too dirty" water.

This books consists of 27 chapters bringing together emerging approaches, challenges, and opportunities related to Sponge Cities, with the ultimate aim to foster upscaling and widespread uptake. While the Sponge Cities Concept is new, the approaches and technologies involved in it have been tried out in many different parts of the globe under the guise of terminologies like Water Sensitive Cities, Sustainable Drainage Systems, Low-Impact Development, ABC waters, etc. This book aims to draw on both Chinese and worldwide experiences. Therefore, the emphasis of this book is on Chinese cities, but case studies from Singapore, India, UK, USA, Vietnam, Uruguay, Norway, and The Netherlands are also presented.

Chris Zevenbergen, Dafang Fu, Assela Pathirana

Special Issue Editors

water MDPI

Editorial

Transitioning to Sponge Cities: Challenges and Opportunities to Address Urban Water Problems in China

Chris Zevenbergen [1,*], Dafang Fu [2,3] and Assela Pathirana [1]

[1] Department of Water Science Engineering, IHE Delft Institute for Water Education, 2611 AX Delft, The Netherlands; assela@pathirana.net
[2] School of Civil Engineering, Southeast University, Nanjing 210096, China; fdf@seu.edu.cn
[3] Southeast University-Monash University Joint Research Centre for Water Sensitive Cities, Nanjing 210096, China
* Correspondence: c.zevenbergen@un-ihe.org; Tel.: +31-653-599-654

Received: 16 August 2018; Accepted: 6 September 2018; Published: 12 September 2018

Abstract: At present, the Sponge City Concept (SCC) is gaining ground, Sponge Cities technologies are becoming more and more accepted by Chinese city governments, and the first best practices are being shared. However, there are still many challenges ahead which hamper effective implementation and upscaling. This paper presents an overview of some opportunities and constraints for the take up of this approach and has drawn upon international experiences. In China at the national level, the State Council has set a progressive target for the SCC initiative to be achieved in 2030. This target seems to be ambitious as the time needed for integrative planning and design and implementation is much longer than traditional sectoral approaches often omitting to address social well-being, the (local) economy, and ecosystem health. This particularly holds true for the existing building stock. Transforming the existing building stock requires a long-term planning horizon, with urban restoration, regeneration, and modernization being key drivers for adapting the city to become a sponge city. A key challenge will be to align the sponge city initiative (SCI) projects with infrastructure and urban renovation portfolios. Moreover, substantial investment needs and a lack of reliable financing schemes and experience also provide a huge challenge for China. This calls for an integrative opportunistic strategy that creates enabling conditions for linking the SCI investment agenda with those from other sectors. These transformations cannot be made overnight: completing the transformation process will typically take a life time of one generation. The progress in sustainable urban water management is also impacted by innovations in technologies as well as in management strategies. These technological innovations create fertile ground for businesses to adapt state-of-the-art developments from around the world and contextualize them into fit-for-purpose products. China is well-placed to play a leading role in this process in the coming decade.

Keywords: sponge city; water sensitive city; urban water cycle; nature-based solutions; resilience; urban flooding; eco-restoration; stormwater management; low-impact development; sustainable drainage systems

---

1. Introduction

Cities around the world are facing a dire need to take action to manage water-related risks that are exacerbated by forces of change (e.g., climate, urban growth). At the same time, they encounter the challenge to become more sustainable and adaptive to cope with resource restrictions, environmental concerns, and the urgency and uncertainties posed by anticipated future change. Consequently, they have to capture the synergies between sustainability planning and urban water management and mainstream adaptation into all aspects of urban development, service provision, and emergency management [1–3]. Particularly in relation to extreme weather events, there is a call for adaptation of infrastructure and the spatial layout of built-up areas to reduce risk [4–6].

China is a country with severe water problems, both in terms of water scarcity and flooding and water quality. Due to the rapid progress of industrialization and urbanization as well as to increasing frequencies of extreme weather events, urban water problems have become very prominent in the socio-political discourse in the country during the last two decades [7]. The damages caused by floods are exponentially increasing. Amongst these problems, flooding ranks on the top of the most destructive natural hazards in China. These damages are concentrated in cities and are to a large extent a result of heavy summer rainfalls. Their frequencies and intensities have significantly increased in the past few decades [8].

As a response to those increasing flood impacts the Chinese Central Government called in 2013 for widespread uptake of the "Sponge City" approach across China and launched a program to provide financial support to foster implementation of this approach in a selection of pilot cities [9,10]. The Sponge City approach aims to enhance infiltration, evapotranspiration, and capturing and reuse of stormwater in the urban environment.

At present, the Sponge City approach is gaining ground, Sponge Cities technologies are becoming more and more accepted by city governments, and the first best practices are being shared. However, there are still many challenges ahead which hamper effective implementation of Sponge Cities technologies in the selected pilot cities and adoption of the approach as a steering concept in urban planning by the remaining 600 (1 million plus citizens) cities in China. In the coming decade, innovations in research, education, and technology are urgently required to design, engineer, and construct Sponge Cities across the whole of China [7].

An important requirement for implementing the Sponge City program is to learn from similar experiences from around the world. While the term "Sponge City" originates from China and aims to mimic natural hydrological and ecological processes, the last three decades have seen numerous developments in using nature-based solutions to address urban water problems particularly in the U.S., E.U., and Australia [11–14]. Practical experiences with concepts such as low-impact development, sustainable (urban) drainage systems, water sensitive cities, and green infrastructure, while not identical, are more and more generating useful lessons and inspiration for the design and implementation of the Sponge Cities approach and technologies [14,15].

The objective of this paper is to give a brief overview of the recent developments and challenges in the area of urban water management from a global perspective relevant for the Sponge City Program. The first section of this paper presents some key features of the Sponge City Program and the challenges it is currently facing, followed by the second section that provides recent international experiences in in this domain captured by the term Integrated Urban Water Management (IUWM). Finally, based on the previous sections, the last section will give an outlook on what the future holds for the Chinese context and beyond.

2. Sponge Cities Challenges

China has experienced unprecedented urban expansion and growth in wealth since the 1980s. The urban population has increased a 6-fold in number since 1980, which is about 750 million people. A similar relative increase in urban population in Europe took place over a time period of more than 120 years. At present, the urban population in China is more than 54% of the total population and the total urbanized area is about 44.5×10^3 km^2.

During this process of urbanization, the design, implementation, and maintenance of the underground infrastructure, such as the drainage system, of built-up areas could not keep pace with the aboveground urban development processes and were not part of an integrated planning strategy. The design of these urban drainage systems is often based on a single design storm and thereby does not consider a range of plausible ways in which flood risk may shift in the future due to, for instance, land-use change. For example, the capital city of China, Beijing, has witnessed more than a doubling of the total land coverage in the past 10 years and several devastating flood events in that same period. The affected city areas of Beijing have been constructed in the last two decades. Various studies have indicated that one of the major causes of Beijing's recent flood events is a lack of surplus capacity of the drainage system to cope with extreme weather events [16]. Addition, these studies revealed that these urban areas did not have sufficient retention capacity to allow for infiltration and retention of stormwater during heavy downpours. Another cause of increasing urban flooding in China is that many cities have not preserved their indigenous water infrastructure systems, such as city lakes and ponds, canals, and peri-urban wetlands. In most cases, urban expansion has resulted in a staggering loss of unpaved, green, or open areas and loss of water bodies in the peri-urban areas as well as within the old city centres. As the process of urbanization will continue to rise in China for the next two decades and probably beyond, it follows from the above that if no proper responses will be taken in the near future (to move cities away from the business as usual scenario), these cities are predisposed to experience more frequent flooding with increasing consequences.

Among the current 654 cities in China, 641 of them are exposed to frequent flooding [8]. In 2015, more than 150 cities were affected by serious flooding causing a total direct flood damage of about 160 billion RMB (or 22 billion Euro) [8]. Over the past 10 years, 2010 ranks number one in terms of the number of affected cities by flooding and the total direct flood damage. More than 250 cities were affected with a total direct damage of more than 350 billion RMB (or 46.9 billion Euro) in that year.

It should be noted here that information publicly available on the indirect damage of flooding occurring in Chinese cities is relatively scarce compared to those reported on the direct consequences of flood disasters to infrastructure and property (damage to buildings). One of the possible reasons for this is that the indirect consequences (for instance on urban transport, property value) are generally spread over a much wider area and therefore more difficult to assess than localized direct flood impacts. Hence, flooding may indirectly affect the services in cities. If disrupted or destroyed, these would have a serious impact on the well-being of citizens and the operation of an organization, sector, region, or government. Based on international experiences, indirect losses are generally larger than direct flood losses in the case of extreme flood events. By contrast, the direct losses of more frequent flooding seem to outweigh the indirect economic losses. Indirect economic losses, however, depend on contextual factors, such as flood exposure and sensitivity of critical infrastructure comprising, amongst other things, energy supply systems, transport services, water supply, and information and communication services.

2.1. The Sponge City Concept (SCC)

The term "sponge city", used in the context of urban water, did probably not originate from China. For example, Rooijen, et al. [17] used the term to describe the potential of stormwater runoff from the city of Hydrabad in India to offset the impact of water demand of the city on the surrounding agricultural water supply. Shannon [18] describes an urban design project for Vinh city, Vietnam proposed as a "City as Sponge", which embeds a system of alternating low-land and high-land strips that could allow for seasonal floods of the Lam River and Vinh River to penetrate the territory, yet not destroy urbanity. In the project, the existing waterways are developed to be a completely open and interconnected network in order to maintain their irrigation and drainage functions as well as to become a local transportation system [19].

The Sponge City (pilot) program was launched at the end of 2014 under the direct guidance and support of the Ministry of Housing and Rural-Urban Development (MOHURD), the Ministry of Finance, and the Ministry of Water Resources (MWR). The general objectives of the concept entail "restoring" the city's capacity to absorb, infiltrate, store, purify, drain, and manage rainwater and "regulating" the water cycle as much as possible to mimic the natural hydrological cycle. Hence, a "Sponge City" is a city that has the capacity to mainstream urban flood risk management into its urban planning policies and designs. It should have the appropriate planning and legal frameworks and tools in place to implement, maintain, and adapt the infrastructure systems to collect, store, and purify (excess) rainwater. In addition, a "sponge city" will not only be able to deal with "too much water", but also reuse rain water to help to mitigate the impacts of "droughts". The anticipated benefits of a Sponge City are: (i) a reduction of the economic losses (due to flooding), (ii) an enhancement of the livability of cities, and (iii) the establishment of an environment where investment opportunities in infrastructure upgrading and engineering products and new technologies are created and fostered. The State Council issued a guideline on 16 October 2015, referred to as the *"Directive on promoting Sponge City Construction"*, which sets out the target that 20% of the urban areas of Chinese cities should absorb, retain, and reuse 70% of the rainwater by 2020. By 2030, this percentage should increase up to 80%.

To meet these targets, provincial governments in China at all institutional levels have to support the implementation of Sponge City Construction in new built-up areas of city districts, industrial parks, and development zones. In existing urban areas, retrofitting of neighborhoods, refurbishment of existing buildings and infrastructure, and other rebuilding activities of old city areas should comply with Sponge Cities Construction. These autonomous dynamics should be exploited to adapt cities to better manage the water cycle. Hence, these activities should be aligned with interventions required to transform these areas into areas that meet the targets of Sponge Cities.

The public sector (national and city governments) is coordinating and managing the implementation of Sponge Cities Construction. The private sector will have to play an important role in allocating additional resources. Various fund-raising methods, including public–private partnership and franchising, will be promoted and required to meet the targets of the guideline. The central government will support the implementation of Sponge Cities Construction with an allocated central budget, governments at provincial levels should increase investments, and governments at city levels should prioritize Sponge Cities projects in mid-term financial budgets and annual construction plans. Banks will also be given incentives to provide mid- and long-term loans.

The Ministry of Housing and Urban-Rural Development (MOHURD), which is responsible for the implementation of Sponge City Construction, has estimated that the required investments will be about 100 to 150 million RMB per km^2 (or 13.4 to 21.1 million Euro per km^2).

2.2. Pilot Sponge Cities

As of 2015, MOHURD has invested about 6.9 billion RMB (or 0.5 billion Euro) per year in the first batch of 16 pilot projects. This subsidy varies between 600 and 400 million RMB (or 54 million Euro) per city. The total investment required to implement the Sponge City approach in these 16 pilot cities is estimated at 86.5 billion RMB (or 11 billion Euro). With a total surface area of 450 km^2, the average investment will be around 190 million RMB/km^2 (or 23 million Euro/km^2).

It follows from the above that the subsidy from the central government is far from enough to finance the implementation of the projects. The majority of the financial resources need to come from the cities themselves and the private sector. Local governments are generally very keen to become a national pilot of the Sponge City Program as in times of economic slowdown they expect that the pilot status will bring in capital to the city, attract private investments, and enhance the investment climate. However, the private sector is still reluctant to engage as the business opportunities are not self-evident. It is assumed that the implementation of Sponge City Construction should be a public responsibility and not a private matter.

In 2016, a second batch of 14 pilot cities was selected by MOHURD. These cities include, amongst others, Beijing, Tianjin, Shanghai, and Shenzhen. The selection of these cities has been made on the basis of an application document pertaining to the commitment of the cities to invest in the implementation of Sponge City Construction.

Based on the information publicly available, it can be discerned that in 2016 about 20 cities have invested 267 billion RMB (about 36 billion Euro), which on average is about 13.4 billion RMB (or 1.8 billion Euro per city).

There is limited insight into the costs of available Sponge City technologies both in terms of direct costs and maintenance costs. They are dependent on the local context, which may vary considerably. A fair cost comparison should also consider the additional benefits of a sponge city for its residents, such as recreational areas, a healthier environment, increased biodiversity, and cooler temperatures. It is likely that the economic case for such solutions will become clearer as China is moving ahead with the implementation of the Sponge City pilot projects.

3. IUWM: International Experiences

It is increasingly recognized that floods, droughts, and all sorts of uses of water have to be managed in an integrative way requiring an approach that acknowledges the importance of the water cycle as a coherent and many-faceted system [1,20–23]. Integrated approaches to urban water management have been around for decades. Their popularity has grown rapidly since the early 1990s. Probably the best example of such an integrative approach is the concept of integrated urban water management (IUWM) [20]. This concept has multiple definitions and interpretations, but what they share is that they aim to integrate the management of different sub-systems of the urban water cycle, namely, water supply, groundwater, wastewater, and stormwater [24]. The SWITCH project was one of the first research projects which has developed, applied, and demonstrated IUWM concepts and approaches in 12 cities across the world [25].

Looking at the way different cities and regions historically approached IUWM shows the pattern that, depending on the acute problem they faced, they had chosen the "point-of-departure" for IUWM. For example, in Japan the ideas of IUWM originated from addressing the peak runoffs by infiltration technologies [26]. In the United States, it was largely the need for stormwater quality management, and in the Netherlands (and arguably in Western Europe) the point-of-departure was the intention of limiting the stormwater peaks entering the combined sewer systems (disconnection). The primary driving forces behind SCC in China are the rising flood damages caused by rainstorms as described above. However, it should be noted that Chinese cities vary in terms of hydrological, socio-economic, and historical conditions and thus may have very different acute water issues ranging from urban/river flooding to a lack of water resources to water quality issues. Some "non-water" issues, such as urban air quality, are also contributing drivers in many cases.

Nowadays, IUWM is sometimes seen as a strategy for achieving the goals of Water-Sensitive Urban Design (WSUD) [24]. Along similar lines as the Sponge Cities Concept (SCC), WSUD seeks to change the impact of urban development on the natural water cycle based on the premise that, by managing the urban water cycle as a whole, "a more efficient use of resources can be achieved providing not only economic benefits, but also improved social and environmental outcomes" [12,27]. WSUD brings together "water-sensitive" and "urban design". The term *Water-Sensitive* refers to a state in which "residents, community organisations, businesses and land developers, and governmental organisations value water as a finite and vulnerable resource that is critical to the liveability of livelihoods, and that this is reflected in their behaviours related to dealing with water resources" [28]. *Urban Design* refers to the process of shaping neighbourhoods, districts, and entire cities with the aim of creating functional, attractive, and sustainable urban areas. Hence, it is connected to planning and architectural design. Ideally, WSUD is the process that brings about Water-Sensitive Cities (WSC) as the outcome [11,29]. The Australian National Water Initiative defines WSUD as "*the integration of urban planning with the management, protection, and conservation of the urban water cycle that ensures that urban water management is sensitive to natural hydrological and ecological processes*" [29]. It follows from the above that WSUD and SCC are concepts that embrace water sustainability and environmental protection, which are closely linked with urban design and explicitly aim to harness all of the opportunities in the water cycle. Sustainable urban drainage systems (SUDS) are technologies which aim to offset the excess runoff created as a result of urbanization by promoting infiltration, detention, retention, and treatment as close as possible to the generating source. SUDS (Sustainable Drainage Systems, U.K.) and its equivalent BMPs (Best Management Practices, U.S.), in contrast to WSUD, typically take one component of the urban water cycle, stormwater, as the point of departure, albeit ideally in a multifunctional manner [30]. SUDS rely on natural processes, such as evapotranspiration, infiltration, and natural treatment, and often complement traditional "grey" infrastructure. *Green infrastructure* (GI) is often defined as a planned approach to the delivery of nature in the city in order to provide benefits to residents [31,32]. GI uses natural or semi-natural systems, such as Nature-Based Solutions (NBS), to provide water resources management options with benefits to providing important regulatory services, such as pollution filtration, water treatment, flood risk reduction, and modulation of high urban temperatures. The challenge of maximising benefits and value from GI investment is receiving much attention by both the scientific community and the finance sector as the awareness is growing that the broader value of GI needs to be considered when dealing with any aspect of water systems [32–34].

In 2016, the International Water Association launched the Principles for Water Wise Cities [34]. The IWA has developed the Principles for Water Wise Cities to catalyze a shift in the current water management paradigm towards a more sustainable state in order to make cities more resilient and livable. The Principles for Water Wise Cities outline a framework to assist urban leaders and professionals to develop and implement their vision for sustainable urban water and resilient planning and design in their cities. Through the Principles, urban leaders, water managers, citizens, and other stakeholders are inspired to collaboratively find solutions on urban water management challenges and implement water wise management strategies through a shared vision that will enable the development of flexible and adaptable cities.

In summary, IUWM, SCC, WSUD, SUDs, and GI are terms that, while each being unique, have considerable overlap as illustrated in Figure 1.

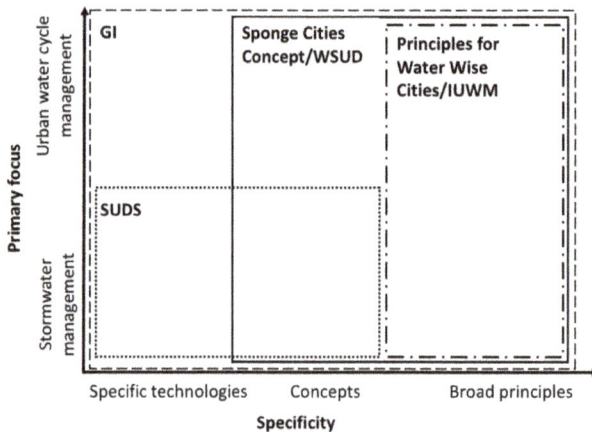

Figure 1. A classification of terminology related to the domain of integrated urban water management (IUWM) according to the terms' specificity and their primary focus (modified from [24]). GI, green infrastructure; SUDS, sustainable urban drainage systems; WSUD, water-sensitive urban design.

4. SPC China's Experience

Since the launching of the SPC programme in 2014, China has started a number of specific SPC projects in different parts of the country. In April 2015, the first 16 pilot cities were selected and this was expanded with another 14 new cities in 2016. The pilot sites cover a range of different climatic regions, though almost no cities are included from the western part of the country. Ding, et al. [35] observe that the range of SPC technologies considered in the SPC programme is limited. This is likely due to the fact that most pilot cities are located in the central and southeastern regions, where annual precipitation ranges from 410 to 1830 mm.

In each of these pilot cities, an area greater than 15 km^2 was designated as pilot. Most of these pilot areas are green-field developments [35]. Urban retrofitting opportunities are largely unidentified in the programme. Li et al. [10] describes case studies in several pilot cities. Jinan city in Shandong Province has cultural heritage value due to its famous natural springs. The problems that are addressed by SPC here include spring protection, flooding, water shortage, and water pollution and the technologies used consist mainly of stormwater detention, storage, and infiltration with possible reuse. Baicheng city in Jilin Province on the other hand has harsh winter conditions. The main issues faced by Baicheng during the Low Impact Development (LID) system construction process are the soil freeze–thaw process, environmental pollution of deicing salts, and restoring alkaline lands.

During the period of the launch of SPC in 2014 up to the present time, China has implemented a number of SCP projects and some comparative and analytical studies were done by researchers on the overall experience. Dai, et al. [36], using Wuhan City as a case study, observed that the government in China plays a significant role and that civil society is merely the recipient. They recommend that the government should slow down the implementation of this ambitious programme and help society and the private sector to catch up and effectively participate. Ma, et al. [37] agree, stating that inadequate supervision and extensive construction are the critical reasons that hinder a sponge city from developing healthily and sustainably. Researchers observe that there are considerable knowledge gaps in the areas of finance and cost of sustaining a sponge city program in larger areas in cities, the co-ordination across bureaus, the public perception and support, the monitoring and evaluation of the effectiveness of the sponge city technologies and program, etc., which remain a challenge [38]. They recommend that China should strengthen their international cooperation and collaboration so that they are able to gain from the experience of institutions around the world.

5. How to Get There?

Cities are adaptive, dynamic systems. They face changing environments and adapt to these changes [39]. Over a longer historical period, cities have always successfully adapted to changing environmental conditions. However, what is being observed today is that rapid urbanization, although leading to an increase of economic and social wealth, has also increased the number of people and assets exposed to water-related disasters. Water-related disasters, such as floods, have become more devastating than in the past due to an excessive concentration of population and wealth [40]. The "temporized" developments of cities in the past allowed them to learn and adapt to changing environmental conditions and increasing risks. However, today, due to the unprecedented rate of urbanization, cities do not have sufficient time to learn from mistakes to build adaptive capacity. This lack of adaptive capacity is exacerbating the trend of increasing water-related risks of cities due to a combination of the following "imperfections", which have been nested in current urban development processes [41]:

(i)   Encroachment and expansion of urban developments onto flood plains and lowlands resulting in a loss of the natural water retention capacity of peri-urban areas;

(ii) Redevelopment and densification of built-up areas through "infill" of remaining open (green/blue) spaces, leading to an overall density increase and subsequent increase of surface sealing and disruption of the natural drainage channels; and

(iii) Increasing interdependency on more diversified infrastructure and a reduction in safety margins due to ageing and deferred maintenance.

It becomes increasingly important that cities have to build capacity to adapt proactively to the rapid changes they are currently facing and to anticipate and deal with disturbances and shocks, such as extreme weather events. They have to adopt flexible, adaptable, and distributed solutions for urban water management, which are embedded in a long-term integrative and adaptive strategy [2,42]. Monitoring and evaluation will be required to assess the effectiveness of interventions and to allow for timely adjustment of the strategy.

In China at the national level, the State Council has set a progressive target: the SCC initiative is to be achieved in 2030 (see previous section). This target seems to be ambitious as the time needed for integrative planning and design and implementation, given the technical, governance, and financial challenges to adopt a broad view on how to manage the present and future water-related risks, is much longer than traditional sectoral approaches often omitting to address social well-being, the (local) economy, and ecosystem health. For the existing urban fabric, the management debt from the past in achieving the standards and goals of flood protection and drainage system upgrading across cities needs to be resolved first [43]. Transforming the existing building stock requires a long-term, adaptive, and opportunistic strategy, with urban restoration, regeneration, and modernization being key drivers for adapting the city while offering opportunities to align the sponge-city initiative (SCI) projects with infrastructure and urban renovation portfolios [2,44,45]. The substantial investment needs and a lack of reliable financing schemes and experience also provide huge opportunities for China as they call for a strategy that creates enabling conditions for linking the SCI investment agenda with those from other sectors. These transformations cannot be made overnight: completing the transformation process will typically take a life time of one generation.

From the viewpoint of urban development, typically three distinct zones can be discerned (see Figure 2). The first is the historical city center (zone 1), which has the oldest building stock and sometimes monuments or buildings of historical value. Surrounding this historical center is an established urban zone (zone 2), which is rapidly undergoing densification. In the periphery (zone 3), there is an ever-expanding urban fringe that consists mainly of newly built areas. Each of these zones require a distinctly different strategy under SCC. In the historical centre (zone 1), the preferred strategy, referred to as "accept and accommodate", is to prevent floodwater coming from the surrounding areas to enter this zone during periods of heavy rainfall. The historical buildings will be protected against floodwater and floodwater and excessive rainwater in exceptional circumstances will be allowed to enter into or around the sites in a controlled and predetermined manner. In zone 2, the strategy is to "adapt and retrofit" the existing urban fabric and aligns with autonomous urban regeneration processes to restore the health of waterways, mitigate flood risk, and to create public spaces that harvest, clean, and recycle water. This transformation process is slow and will likely take a few decades to reach full transformation. Finally, in zone 3, the strategy aims to create "water-sensitive developments" from scratch (tabula rasa) and newly developed buildings and public spaces will fully comply with the SCC standards (such as to allow for green roofs, underground storage in the floodplain, and ground measures, such as swales and rain gardens in the floodplain, where rainwater is collected within a communal harvesting system).

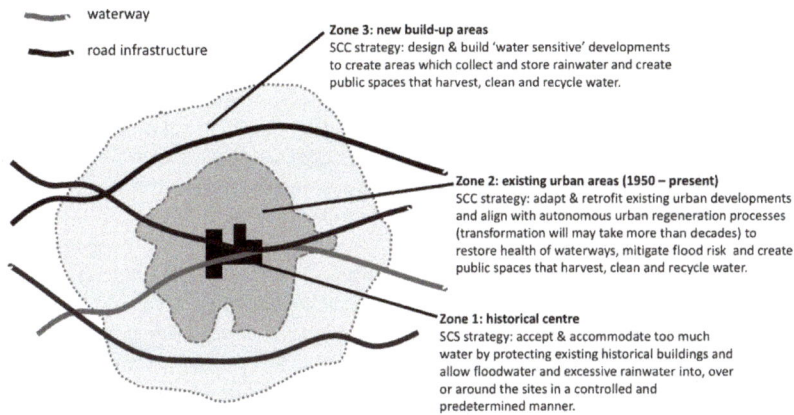

Figure 2. Schematic presentation of a typical layout of a typical Chinese city and associated strategies related to the different zones. SCC, Sponge City Construction.

It is difficult to assume that Chinese cities have the capability to leapfrog directly into a sustainable future while avoiding the negative consequences associated with the "traditional" transition pathway to a water-sensitive city and thus circumvent the evolutionary step of managing the water cycle in a segmented manner. This will require customized strategies as depicted above placing water more centrally within the urban (re)development process and taking into account the opportunities and constraints imposed by the local context utilizing synergies with the management of other urban systems and services. The SCC sets out targets for the take up of such an approach, but the translation of this approach into feasible and practical customized strategies is still a challenge. The concept of "leapfrogging" has drawn the attention of both developing and developed countries as well as international organizations to create a strategy that responds to the water challenges that cities are facing [46]. The literature on leapfrogging concepts in sustainability transitions is scarce [47,48] and very few studies in this field related to the Chinese context have been conducted. Among them, the most relevant are low carbon energy technology and new energy and wastewater treatment technologies [49,50]. These experiences reveal that technological innovations are necessary to leapfrogging; however, that itself is not sufficient. Leapfrogging entails not only technological innovation but innovations in governance that support the development of customized strategies capturing a portfolio of local management policies and measures to be implemented at various spatial scales ranging from the very local scale to the scale of an entire city and even beyond. This will challenge existing organizational structures, governance mechanisms, beliefs, and traditions [30]. Hence, leapfrogging towards SCC includes the adoption of new governance approaches to support SCC transformation [51]. The pilot cities now emerging are generating practical experiences and tailoring existing concepts to a local context and some of them are experimenting with innovative technologies. It goes without saying that the capability of a city to leapfrog towards SCC also depends on the extent to which these experiences and outcomes are being effectively shared between the pilot cities as well as among other Chinese cities. This requires a well-designed city-to-city peer learning process that supports learning from experiments and full-scale projects between pilot cities and beyond [24,52,53] That is why the capacity to learn and to engage in city-to-city networks of Chinese cities will play an imminent role in the process of leapfrogging and upscaling.

6. Outlook

China has the potential to play a leading role internationally in sustainable urban water management in the coming decade. There are several reasons that support this assertion. One is that existing urban environmental problems in China (such as water, but there are many others, such as air quality) are so pressing that they provide immediate-term incentives to invest resources to manage these problems. The Sponge Cities (pilot) program is a good example of such an immediate-term incentive. Another is that China has the ambition to play a leading role in science and technology globally. Therefore, it has dramatically increased its budget for innovation and strengthened its engagement in international research, such as in the Chinese European Water partnership (CEWP). It is to be expected that in the near future a combination of home-grown innovations, joint ventures with foreign companies, and modifications to existing technology will change significantly the way in which the water challenges of Chinese cities will be managed.

One type of innovation which is currently pushing developments in this area is the wide-spread use of crowdsourced data (CSD) in the context of participatory monitoring [54]. Early warning, real-time control systems and smart-phone-based sensors now emerging will disclose distributed monitoring capabilities at a fraction of the cost and effort compared to that of traditional monitoring systems [55–59]. These capabilities in conjunction with innovations in data management and participation related to the internet (such as device-generated content (the Internet of Things, IOT)), are providing unique opportunities to connect a large number of information generation endpoints (be it devices or humans) to the process of gathering relevant information. Particularly in the water management domain, these developments, increasingly referred to as "Web 3.0" [60,61] will likely change the way our water systems will be managed in the near future. For example, they will allow us to have a much more detailed, accurate, and (due to its decentralized nature) *fault-tolerant* flood forecasting/warning system in place, which builds upon multidirectional risk communication and massively distributed sources of information. It will also increase our capability to assess the performance (and quality) of urban infrastructure systems, such as urban drainage systems, road infrastructure, and pumps through real-time field monitoring using IOT devices. However, successfully applying crowdsourced data to environmental monitoring and forecasting (e.g., flood forecasting) comes with its own challenges [62]. For example, traditionally collected data (e.g., flow gauge measurements) and CSD (flood levels reported by citizens using their smartphones) have very different error characteristics. They cannot be used in conjunction without applying novel data-assimilation techniques. Here, urban water managers can learn from other fields; for example, atmospheric science [63].

Such innovations will not only create opportunities for cities in developed countries, such as China, to leapfrog into a more resilient and sustainable state, but also for cities in rapidly growing developing countries, particularly those with poorly developed water management systems. These innovations create fertile ground for businesses to adapt state-of-the-art developments from around the world and contextualize them into fit-for-purpose products. China is well-placed to play a leading role in this process.

Author Contributions: Conceptualization, C.Z., D.F. and A.P. Resources, C.Z., D.F. and A.P., Writing—Original Draft Preparation, C.Z., D.F. and A.P.; Writing—Review & Editing, C.Z., D.F. and A.P.

Funding: This research received no external funding.

Conflicts of Interest: The authors declare no conflict of interest.

References

1. Harremoes, P. Integrated urban drainage, status and perspectives. *Water Sci. Technol.* **2002**, *45*, 1–10. [CrossRef] [PubMed]
2. Gersonius, B.; Nasruddin, F.; Ashley, R.; Jeuken, A.; Pathirana, A.; Zevenbergen, C. Developing the evidence base for mainstreaming adaptation of stormwater systems to climate change. *Water Res.* **2012**, *46*, 6824–6835. [CrossRef] [PubMed]
3. Kirshen, P.; Caputo, L.; Vogel, R.; Mathisen, P.; Rosner, A.; Renaud, T. Adapting Urban Infrastructure to Climate Change: A Drainage Case Study. *J. Water Res. Plan. Manag.* **2015**, *141*, 04014064. [CrossRef]

4. Dickson, E.; Baker, J.; Hoornweg, D. *Urban Risk Assessments: Understanding Disaster and Climate Risk in Cities*; World Bank Publications: Washington, DC, USA, 2012.
5. CIRIA. *C635 Designing for Exceedance in Urban Drainage—Good Practice*; CIRIA: London, UK, 2014.
6. Georgiadis, T.; Iglesias, A.; Iglesias, P. City Resilience to Climate Change. In *Rooftop Urban Agriculture*; Springer: Cham, Switzerland, 2017; pp. 253–262.
7. Jiang, Y.; Zevenbergen, C.; Fu, D. Understanding the challenges for the governance of China's "sponge cities" initiative to sustainably manage urban stormwater and flooding. *Nat. Hazards* **2017**, *89*, 521–529. [CrossRef]
8. Xiaotao, C. Flooding in China. Presented at the River Flows, St. Louis, MO, USA, 12–15 July 2016.
9. Jia, H.; Yu, S.L.; Qin, H. Low impact development and Sponge City construction for urban stormwater management. *Front. Environ. Sci. Eng.* **2017**, *11*, 20. [CrossRef]
10. Li, X.; Li, J.; Fang, X.; Gong, Y.; Wang, W. Case Studies of the Sponge City Program in China. In Proceedings of the World Environmental and Water Resources Congress 2016, West Palm Beach, FL, USA, 22–26 May 2016; pp. 295–308.
11. Wong, T. Water Sensitive Urban Design—The Journey Thus Far. *Aust. J. Water Resour.* **2006**, *10*, 213–222. [CrossRef]
12. Brown, R.; Keath, N.; Wong, T. Transitioning to Water Sensitive Cities: Historical, Current and Future Transition States. In Proceedings of the 11th International Conference on Urban Drainage, Edinburgh, UK, 31 August–5 September 2008; pp. 1–10.
13. Zhou, Q.; Panduro, T.; Thorsen, B.; Arnbjerg-Nielsen, K. Adaption to Extreme Rainfall with Open Urban Drainage System: An Integrated Hydrological Cost-Benefit Analysis. *Environ. Manag.* **2013**, *51*, 586–601. [CrossRef] [PubMed]
14. Butler, D.; Ward, S.; Sweetapple, C.; Astaraie-Imani, M.; Diao, K.; Farmani, R.; Fu, G. Reliable, resilient and sustainable water management: The Safe & SuRe approach. *Glob. Chall.* **2017**, *1*, 63–77.
15. Arnbjerg-Nielsen, K. Past, present, and future design of urban drainage systems with focus on Danish experiences. *Water Sci. Technol.* **2011**, *63*, 527–535. [CrossRef] [PubMed]
16. Duan, W.; He, B.; Nover, D.; Fan, J.; Yang, G.; Chen, W.; Liu, C. Floods and associated spciioeconomic damages in China over the last century. *Nat. Hazards* **2016**, *82*, 401–413. [CrossRef]
17. Rooijen, D.; van Turral, H.; Wade Biggs, T. Sponge city: Water balance of mega-city water use and wastewater use in Hyderabad, India. *Irrig. Drain.* **2005**, *54*, S81–S91. [CrossRef]
18. Shannon, K. Water urbanism: Hydrological infrastructure as an urban frame in Vietnam. In *Water and Urban Development Paradigms*; CRC Press: Boca Raton, FL, USA, 2008; pp. 73–84.
19. Nguyen, P. Deltaic Urbanism for Living with Flooding in Southern Vietnam. Ph.D. Thesis, Queensland University of Technology, Brisbane, Australia, 2015.
20. Mitchell, V. Applying Integrated Urban Water Management Concepts: A Review of Australian Experience. *Environ. Manag.* **2006**, *37*, 589–605. [CrossRef] [PubMed]
21. Mostert, E. Integrated Water Resources Management in The Netherlands: How Concepts Function. *J. Cont. Water Res. Educ.* **2006**, *135*, 19–27. [CrossRef]
22. Santos, D.; van der Steen, P. *Understanding the IUWM Principles: An Activity Based on Role Play Approach (for Facilitator and Specialists)*; UNESCO-IHE: Delft, The Netherlands, 2011; 66p.
23. Burn, S.; Maheephla, A.S.; Sharma, A. Utilising integrated urban water management to assess the viability of decentralised water solutions. *Water Sci. Technol.* **2012**, *66*, 113–121. [CrossRef] [PubMed]
24. Fletcher, T.; Shuster, W.; Hunt, W.; Ashley, R.; Butler, D.; Arthur, S.; Trowsdale, S.; Barraud, S.; Semadeni-Davies, A.; Bertrand-Krajewski, J.; et al. SUDS, LID, BMPs, WSUD and more—The evolution and application of terminology surrounding urban drainage. *Urban Water J.* **2014**, *12*, 525–542. [CrossRef]
25. SWITCH. SWITCH Approach to Strategic Planning for Integrated Urban Water Management (IUWM). 2010, p. 39. Available online: http://www.switchtraining.eu/fileadmin/template/projects/switch_training/db/event_upload_folder/103/SWITCHtrategyplanningapproach28october.pdf (accessed on 10 July 2018).
26. Fujita, S. Measures to promote stormwater infiltration. *Water Sci. Technol.* **1997**, *1997 36*, 289–293. [CrossRef]
27. Salinas Rodriguez, C.; Ashley, R.; Gersonius, B.; Rijke, J.; Pathirana, A.; Zevenbergen, C. Incorporation and application of resilience in the context of water-sensitive urban design: Linking European and Australian perspectives. *Wiley Interdiscip. Rev. Water* **2014**, *1*, 173–186. [CrossRef]

28. Hoyer, J.; Dickhaut, W.; Kronawitter, L.; Weber, B. *Water Sensitive Urban Design: Principles and Inspiration for Sustainable Stormwater Management in the City of the Future*, 1st ed.; JOVIS Verlag: Hamburg, Germany, 2011; p. 118. ISBN 9783868591064.

29. Brown, R.; Rogers, B.; Werbeloff, L. *Moving toward Water Sensitive Cities a Guidance Manual for Strategists and Policy Makers*; Cooperative Research Centre for Water Sensitive Cities: Clayton, Australia, 2016; ISBN 9781921912351.

30. CIRIA. *The SUDS Manual*; CIRIA: London, UK, 2007; p. 606. ISBN 9780860176978.

31. Benedict, M.; McMahon, E. *Green Infrastructure: Smart Conservation for the 21st Century*; Sprawl Watch Clearinghouse: Washington, DC, USA, 2002; p. 36.

32. Foster, J.; Lowe, A.; Winkelman, S. *The Value of Green Infrastructure for Urban*; Center for Clean Air Policy: Washington, DC, USA, 2011.

33. Wise, S.; Braden, J.; Ghalayini, D.; Grant, J.; Kloss, C.; Macmullan, E.; Morse, S.; Montalto, F.; Nees, D.; Nowak, D.; et al. Integrating Valuation Methods to Recognize Green Infrastructure's Multiple Benefits. *Low Impact Dev.* **2010**, *2010*, 1123–1143. Available online: https://ascelibrary.org/doi/abs/10.1061/41099(367)98 (accessed on 21 July2018).

34. International Water Association (IWA). *Principles for Water Wise Cities. For Urban Stakeholders to Develop a Shared Vision and Act towards Sustainable Urban Water in Resilient and Liveable Cities IWA Report*; The International Water Association: London, UK, 2016; Available online: http://www.iwa-network.org/wp-content/uploads/2016/08/IWA_Principles_Water_Wise_Cities.pdf (accessed on 21 July 2018).

35. Ding, L.Q.; Ren, M.L.; Li, C.Z.; Wang, H. Sponge City construction in China: A survey of the challenges and opportunities. *Water* **2017**, *9*, 594.

36. Dai, L.; van Rijswick, H.F.; Driessen, P.P.; Keessen, A.M. Governance of the Sponge City Programme in China with Wuhan as a case study. *Int. J. Water Resour. Dev.* **2018**, *34*, 578–596. [CrossRef]

37. Ma, T.; Wang, Z.; Ding, J. Governing the Moral Hazard in China's Sponge City Projects: A Managerial Analysis of the Construction in the Non-Public Land. *Sustainability* **2018**, *10*, 3018. [CrossRef]

38. Chan, F.K.; Griffiths, J.A.; Higgitt, D.; Xu, S.; Zhu, F.; Tang, Y.T.; Xu, Y.; Thorne, C.R. "Sponge City" in China—A breakthrough of planning and flood risk management in the urban context. *Land Use Policy* **2018**, *76*, 772–778. [CrossRef]

39. Bai, X.; Surveyer, A.; Elmqvist, T.; Gatzweiler, F.W.; Güneralp, B.; Parnell, S.; Prieur-Richard, A.H.; Shrivastava, P.; Siri, J.G.; Stafford-Smith, M.; et al. Defining and advancing a systems approach for sustainable cities. *Curr. Opin. Environ. Sustain.* **2016**, *23*, 69–78. [CrossRef]

40. Estrada, F.; Botzen, W.J.W.; Tol, R.S.J. A global economic assessment of city policies to reduce climate change impacts. *Nat. Clim. Chang.* **2017**, *7*, 403–406. [CrossRef]

41. Zevenbergen, C.; Cashman, A.; Evelpidou, N.; Pasche, E.; Garvin, S.; Ashley, R. *Urban Flood Management*; CRC Press: Boca Raton, FL, USA, 2010; p. 340. ISBN 139780415559447.

42. Klijn, F.; Kreibich, H.; De Moel, H.; Penning-Rowsell, E. Adaptive flood risk management planning based on a comprehensive flood risk conceptualisation. *Mitig. Adapt. Strateg. Glob. Chang.* **2015**, *20*, 845–864. [CrossRef] [PubMed]

43. Liu, H.; Jia, Y.; Niu, C. "Sponge City" concept helps to solve China's urban water problems. *Environ. Erath Sci.* **2017**, *76*, 473. [CrossRef]

44. Veerbeek, W.; Ashley, R.; Zevenbergen, C.; Rijke, J.S.; Gersonius, B. Building adaptive capacity for flood proofing in urban areas through synergistic interventions. In Proceedings of the ICSU 2010 First International Conference on Sustainable Urbanization, Hong Kong Polytechnic University, Faculty of Construction and Land Use, Hong Kong, China, 15–17 December 2010.

45. Zevenbergen, C.; van Herk, S.; Rijke, J. *Future-Proofing Flood Risk Management: Setting the Stage for an Integrative, Adaptive, and Synergistic Approach, Public Works Management & Policy*; SAGE Publications: Los Angeles, CA, USA, 2017; Volume 22, pp. 49–54.

46. Clemens, M.; Rijke, J.; Pathirana, A.; Evers, J.; Quan, N. Social learning for adaptation to climate change in developing countries: Insights from Vietnam. *J. Water Clim. Chang.* **2016**, *7*, 365–378. [CrossRef]

47. Perkins, R. Environmental leapfrogging in developing countries: A critical assessment and reconstruction. *Nat. Resour. Forum* **2003**, *27*, 177–188. [CrossRef]

48. Goldenberg, J.; Oreg, S. Laggards in disguise: Resistance to adopt and the leapfrogging effect. *Technol. Forecast. Soc. Chang.* **2007**, *74*, 1272–1281. [CrossRef]

49. Sauter, R.; Watson, J. *Technology Leapfrogging: A Review of the Evidence A Report for DFID*; Sussex Energy Group SPRU (Science and Technology Policy Research), University of Sussex: Brighton, UK, 2008; p. 32.
50. Binz, C.; Truffer, B.; Li, L.; Shi, Y.; Lu, Y. Conceptualizing leapfrogging with spatially coupled innovation systems: The case of onsite wastewater treatment in China. *Technol. Forecast. Soc. Chang.* **2012**, *79*, 155–171. [CrossRef]
51. Rijke, J.; Farrelly, M.; Brown, R.; Zevenbergen, C. Configuring transformative governance to enhance resilient urban water systems. *Environ. Sci. Policy* **2013**, *25*, 62–72. [CrossRef]
52. Seymoar, N.-K.; Mullard, Z.; Winstanley, M. *City-to-City Learning*; Sustainable Cities: Vancouver, BC, Canada, 2009; Available online: http://sustainablecities.net/city-to-city-learning/ (accessed on 20 January 2015).
53. Lundby, L.; Sjöberg, H. *An Evaluation of City-to-City Learning in the Campaign Making Cities Resilient-Matching Criteria and Implementation*; Report: 5420; Lund University: Lund, Sweden, 2013; ISSN: 1402-3504; ISRN: LUTVDG/TVBB-5420-SE.
54. Turreira-García, N.; Lund, J.; Domínguez, P.; Carrillo-Anglés, E.; Brummer, M.; Duenn, P.; Reyes-García, V. What's in a name? Unpacking "participatory" environmental monitoring. *Ecol. Soc.* **2018**, *23*, 24. [CrossRef]
55. Fienen, M.; Lowry, C. Social. Water—A crowdsourcing tool for environmental data acquisition. *Comput. Geosci.* **2012**, *49*, 164–169. [CrossRef]
56. Overeem, A.; Robinson, J.; Leijnse, H.; Steeneveld, G.P.; Horn, B.; Uijlenhoet, R. Crowdsourcing urban air temperatures from smartphone battery temperatures. *Geophys. Res. Lett.* **2013**, *40*, 4081–4085. [CrossRef]
57. Certoma, C.; Corsini, F.; Rizzi, F. Crowdsourcing urban sustainability. Data, people and technologies in participatory governance. *Futures* **2015**, *74*, 93–106. [CrossRef]
58. Eggimann, S.; Mutzner, L.; Wani, O.; Schneider, M.; Spuhler, D.; Moy de Vitry, M.; Beutler, P.; Maurer, M. The potential of knowing more: A review of data-driven urban water management. *Environ. Sci. Technol.* **2017**, *51*, 2538–2553. [CrossRef] [PubMed]
59. Wang, R.; Mao, H.; Wang, Y.; Rae, C.; Shaw, W. Hyper-resolution monitoring of urban flooding with social media and crowdsourcing data. *Comput. Geosci.* **2018**, *111*, 139–147. [CrossRef]
60. Hendler, J. Web 3.0 Emerging. *Computer* **2009**, *42*, 111–113. [CrossRef]
61. Garrigos-Simon, F.; Lapiedra Alcami, R.; Barbera Ribera, T. Social networks and Web 3.0: Their impact on the management and marketing of organizations. *Manag. Decis.* **2012**, *50*, 1880–1890. [CrossRef]
62. Viero, D. Comment on "Can assimilation of crowdsourced data in hydrological modelling improve flood prediction?" by Mazzoleni et al. (2017). *Hydrol. Earth Syst. Sci.* **2018**, *22*, 171. [CrossRef]
63. Dee, D.; Uppala, S.; Simmons, A.; Berrisford, P.; Poli, P.; Kobayashi, S.; Andrae, U.; Balmaseda, M.; Balsamo, G.; Bauer, D.; et al. The ERA-Interim reanalysis: Configuration and performance of the data assimilation system. *Q. J. R. Meteorol. Soc.* **2011**, *137*, 553–597. [CrossRef]

water

MDPI

Article

Performance of Earthworm-Enhanced Horizontal Sub-Surface Flow Filter and Constructed Wetland

Rajendra Prasad Singh [1,2,*]**, Dafang Fu** [1,2]**, Jing Jia** [1,2] **and Jiaguo Wu** [1,2]

1   School of Civil Engineering, Southeast University, Sipai Lou 2#, Nanjing 210096, China;
    fdf@seu.edu.cn (D.F.); 108209024@seu.edu.cn (J.J.); wujiaguo2017@163.com (J.W.)
2   Southeast University-Monash University Joint Research Centre for Water Sensitive Cities,
    Nanjing 210096, China
*   Correspondence: rajupsc@seu.edu.cn; Tel.: +86-131-6005-2265

Received: 15 June 2018; Accepted: 17 September 2018; Published: 22 September 2018

Abstract: In this study, the performance of the horizontal sub-surface flow filter (HSSFF) and constructed wetland (HSSFCW) experimental units enhanced with earthworms was investigated for the treatment of construction camp sewage wastewater. All the experimental units (filter and constructed wetland) were filled with the same filler except *Eisenia foetida* earthworms and *Lolium perenne Linn* plants. The performance of the earthworm-enhanced filter (EEF) and the earthworm-enhanced constructed wetland (EECW) was compared to that of the blank filter (BF) units. The results revealed that the removal efficiencies for chemical oxygen demand (COD), ammonium-nitrogen (NH_4^+-N), total nitrogen (TN) and total phosphorus (TP) in EEF were higher than the BF unit. In order to optimize the operating conditions, the experiments were conducted in three different water levels. The results revealed that the removal efficiencies of EEF for these pollutants are the highest in experimental conditions no. 2 (water level ~30 cm; HRT ~3 days; hydraulic load ~4.05 cm/day; and Inflow discharge ~0.27 L/h). Compared to the EEF and BF units, the EECW has higher removal efficiency for COD and TN and has more stable performance than the filters. This work will aid the design and improvement of filters and CWs for treatment of effluent wastewater from construction camps. The selection of appropriate hydraulic parameters and experimental conditions could be very beneficial in achieving the goal of implantation of low impact development (LID).

Keywords: filters; horizontal sub-surface flow CW; earthworms; LID; removal efficiency

---

1. Introduction

With the increase in the investment and scale of China's construction industry, the environmental impact, especially the impact of sewage discharged from construction camps, is attracting more and more attention. Since many construction camps are very far away from the municipal pipe network, it is very expensive to discharge sewage from construction camps into the municipal pipe network. Most of the sewage is discharged on the spot after simple treatment, which ultimately causes serious pollution to the surrounding water bodies. In general, the amount of sewage in construction camps varies with human activities and seasons. Its main pollutant index includes suspended solids (SS), COD and ammonium-nitrogen (NH_4^+-N). A filter is a biological treatment structure, which can be employed to remove pollutants from air and water [1]. It consists of a filler and biofilm on the surface of the filler, which integrates solid–liquid separation technology and biological treatment technology. The mechanism of the pollutant removal process in filters involves microbial metabolism filtration, adsorption and sedimentation of the matrix [2]. The filter system can be divided into two categories according to the flow style, including horizontal sub-surface flow filters and vertical sub-surface flow filters. It is usually used for the secondary treatment or advanced treatment of municipal sewage [3].

The removal efficiency of the filter depends on the growth and maintenance of microorganisms that are attached to the filler surface [1]. In addition, the type of filler also affects the efficiency. Various substrates, such as sand, soil, bark, charcoal and granular activated charcoal (CAC), have been used as the filler, which forms the biofilm attachment surface [4]. In the research of Kong et al., the performance of a filler mixed with crushed stone, zeolite and ceramsite was better than that of limestone, with the former removal efficiency of COD, NH_4^+-N, TN and TP being 74.10%, 94.14%, 73.57% and 69.53%, respectively [5]. Simultaneously, the effect is influenced by other factors, such as filtration rate, initial water quality, temperature and oxygen content in the system [6]. However, filters still have some limitations. First, if the back washing system is lacking, the matrix channel can be easily blocked by the organic matter that has not decomposed [7]. Secondly, because the medium chemisorption is the main mechanism for dephosphorization, filters may lose their phosphorus removal capability after working for 1 or 2 years as the adsorption capacity gradually depletes during operation [8,9]. Thus, how to strengthen the effect of filters has become an important research area.

In some earlier studies, researchers planted decontamination plants in filters to improve its efficiency, which creates an environment that is similar to constructed wetlands (CWs). The constructed wetland system is a unique ecological environment that consists of a packed bed, plants, organisms and microorganisms in the substrate, which has a certain length to width ratio and a ground line gradient. When the polluted water flows through the wetland, it can be purified efficiently through filtration, adsorption, sedimentation, ion exchange, plant uptake and microbial decomposition, etc. [10]. Kenatu et al. constructed three HSSFCWs (planted with *Vetiveria zizanioide* and *Phragmites karka* and one without plants) to investigate the performance of contaminant removal and their results showed that the removal rates in planted CWs were higher for all parameters than that in unplanted CW [11]. In order to further improve the sewage treatment effect, researchers introduced earthworms into the filter. It was reported that adding earthworms to the sludge treatment wetlands had a positive effect on sludge characteristics [12,13]. Xu et al. reported that earthworms can increase the above-ground biomass, enzyme activity, nitrification potentials and total number of bacteria [14]. In addition, it was reported that earthworms can effectively improve the matrix channel clogging conditions [15], thereby improving the hypoxic environment in the wetland system [16–18]. Vermicompost can improve soil structure, increase soil fertilizer, promote plant growth and promote the absorption of nitrogen, phosphorus and other organics [9,19–21]. However, earthworms in filters have yet been used to treat construction camp sewage wastewater.

Therefore, the objectives of the current study were to investigate the growth of earthworms in horizontal sub-surface flow filter (HSSFF) and the performance of earthworm-enhanced filter (EEF) for treating the wastewater released from construction camp under local conditions. In addition, the effect of various influencing factors, such as temperature, HRT and water level, were also investigated in this current study.

2. Material and Methods

2.1. Experimental Units

Experimental CW and filter units, which are referred as HSSFCW and HSSFF, were constructed in our Laboratory at "Southeast University-Monash University Joint Research Centre for Water Sensitive Cities," located in Nanjing, China (Figure 1). The experimental device consists of a high water storage tank, water inlet, filter and water outlet. The experimental equipment was made of organic glass and the water pipes were made of polyvinyl chloride (PVC).

The size of the experimental device was as follows: 100 cm length, 40 cm width and 65 cm height. Filter units were divided into three parts: inlet area (10 cm), treatment area (80 cm) and outlet area (10 cm) in the current study. Therefore, the total length was 100 cm long, while the treatment area was 80 cm long. The device was divided longitudinally into two identical parts by an impermeable baffle. It can be considered that the two filters are mutually independent. The purpose of the baffle design was just to save lab space. Furthermore, the size of the experimental units for each side was as follows: 100 cm length, 20 cm width and 65 cm height. The treatment area of the experimental unit (matrix) was a surface soil layer with a height of 20 cm. The gravel layer had a height of 30 cm (gravel size approximately 1–2 cm diameter), which was placed below the soil layer. The device inlet and outlet were 10 cm long, which was filled with gravel (gravel size approximately 2–4 cm diameter). The CWs units were the same as filter units apart from being planted with *Lolium perenne Linn* plants (Figure 1b). According to the experimental measurements, the porosity ε of the matrix was 40%. The porosity ε was calculated by the following formula [22]:

$$\varepsilon = \frac{V_0 - V}{V_0} \times 100\% \tag{1}$$

where V_0 represents the apparent volume (m³), the gravel volume plus the interspace volume; and V represents the absolute volume (m³), the gravel volume.

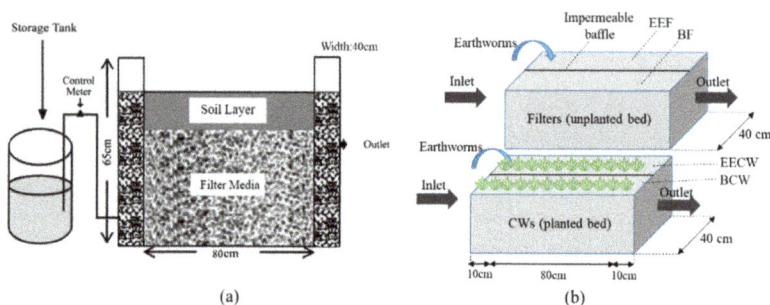

Figure 1. Experimental setup of experimental units: (**a**) Sectional view; and (**b**) Schematic view of experimental setup of the CW and Filters systems (Earthworm-enhanced filter (EEF); Blank-filter (BF); Earthworm-enhanced constructed wetland (EECW); Blank constructed wetland (BCW)).

2.2. Earthworm Selection and Placement

The earthworm species *Eisenia foetida* was selected for this current study (Figure 2). These earthworms were purchased from an earthworm farm located in Changlu town, Liuhe district, Nanjing, China. A total of 20 "*Eisenia foetida*" earthworms were placed in one HSSFF system, while the other system was a BF to explore the differences in the performance of ordinary filters and EEF. Similarly, 20 "*Eisenia foetida*" earthworms were placed in one HSSFCW system, while the other system was a BCW.

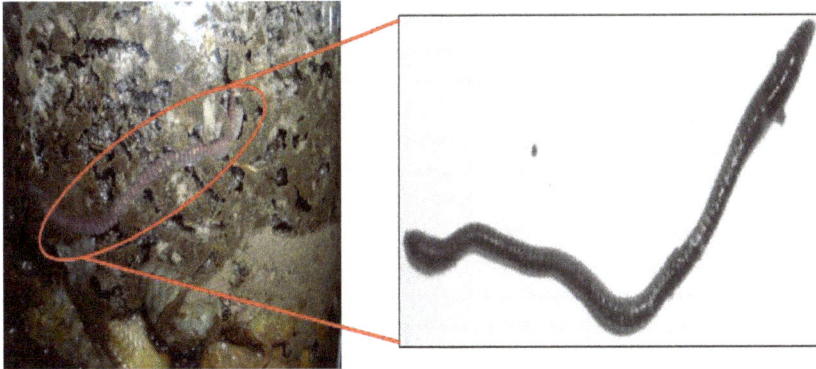

Figure 2. Presence of *Eisenia foetida* earthworms in experimental filter units.

2.3. Chemicals and Experimental Water Quality

In this current study, synthetic water was used to simulate the sewage. The quality of wastewater released from construction camps was investigated before starting the experiment. The chemicals included analytical reagent (AR), glucose, starch, peptone, carbamide, $MgSO_4 \cdot 7H_2O$, $NaHCO_3$, $CaCl_2$, $FeSO_4 \cdot 7H_2O$, KH_2PO_4 and $MnSO_4 \cdot 7H_2O$. These were all acquired from Sinopharm Chemical Reagent Beijing Co., Ltd. (Beijing, China) and were used for the synthesis of wastewater. The water distribution experimental scheme is listed in Table 1. In order to simulate the real situation, fluctuating concentrations of pollutants were set, which are listed in Table 2.

Table 1. Composition of synthetic water used in current study.

Composition	Mean Concentration (mg/L)	Composition	Mean Concentration (mg/L)
Glucose	278	$NaHCO_3$	111
Starch	278	$CaCl_2$	6
Peptone	28	$FeSO_4 \cdot 7H_2O$	0.549
Carbamide	167	KH_2PO_4	26.4
$MgSO_4 \cdot 7H_2O$	66	$MnSO_4 \cdot 7H_2O$	6

Table 2. Concentration of various pollutants in synthetic water used in current study.

Index	COD (mg/L)	NH_4^+-N (mg/L)	TN (mg/L)	TP (mg/L)	pH
Range	360~516	40.2~57.9	56.3~79.2	2.27~4.13	7~8

2.4. System Operation and Sampling

The experimental units of filter and CW systems were fed continuously with synthetic wastewater at the rate (0.27 L/h) that was equivalent to a 3-day hydraulic retention time (HRT). The HRT was calculated by the following formula [22]:

$$HRT = \frac{\varepsilon \times L \times W \times D}{Q}$$

(2)

where ε represents the effective porosity of media (% as a decimal); L represents the length of bed (cm); W represents the width of bed (cm); D represents the average depth of water in bed (cm); and Q represents the inflow discharge (cm^3/day).

Water levels are an important consideration in the operation of filter units, which have significant effects on the activity of earthworms and the performance of pollution removal. In order to investigate the best conditions for pollutant removal, three water levels were set in the current experiment, which are listed in Table 3. The initial commissioning phase for filter units lasted for 60 days, before the performance of the system was monitored in a period of 105 days (summer of 2017). In the first 35 days, the filter system was operated in experimental condition 1. After this, it changed to experimental condition 2 in the second period (approximately days 36–70). During approximately days 71–105, the system was operated in experimental condition 3 and the matrix was not replaced during the process. The CW units were planted with *Lolium perenne Linn* for 150 days so that plants can develop their root system where the biofilm can also develop very well. The experiments were started after the plants grows up to 10 cm above the ground and lasted for 20 days.

The water samples were collected with at a sampling interval of seven days to analyze following water quality parameters: temperature, pH, COD, NH_4^+-N, TN and TP. Prior to transferring earthworms into the filter, their initial mean weight was measured. Therefore, the initial mean weight of earthworm was the same. After 7 days, the weight of earthworms in each layer were measured respectively, before the final mean weight in each layer was also calculated more than 3 times in a simulation experimental column. However, in the filters, we just measured once and collected soil layer without dismantling the system to find out the earthworms that were present in the system.

Table 3. Parameters in different experimental conditions.

Technological Parameter	Experimental Condition 1	Experimental Condition 2	Experimental Condition 3
Water level (cm)	35	30	25
HRT (day)	3	3	3
Hydraulic load (cm/day)	4.65	4.05	3.30
Inflow discharge (L/h)	0.31	0.27	0.22

2.5. Sample Analysis

Temperature and pH were measured by the mercury thermometer, INESA, Shanghai, China), pH meter (HANNA, Padova, Italy) and portable dissolved oxygen (DO) analyzer (INESA, Shanghai, China), respectively. COD was measured by the standard procedure described in APHA [23]. Pollutants, such as NH_4^+-N, TN and TP, were determined by UV–visible spectrophotometer (SHIMADZU, Kyoto, Japan).

2.6. Characterization of Pollutant Removal Effect

The removal efficiency was used to describe the pollutant removal effect of the filter. It can express the differences in the performance of the filters and the CWs. The removal efficiency (η) was calculated by the following formula.

$$\eta = \frac{C_{in} - C_{out}}{C_{in}} \times 100\% \tag{3}$$

where η is the removal efficiency (%); C_{in} is the concentration of the influent (mg/L); C_{out} is the concentration of the effluent (mg/L). For a specific water quality parameter, a higher η value means better removal performance.

3. Results and Discussion

3.1. Growth and Distribution of Earthworms

The growth and distribution of earthworms are important prerequisite knowledge in the current study because earthworm activity directly affects the operating status of filters. The weight change of earthworms was used to evaluate their growth status. Earthworms were placed in the device in March. The related experimental results were observed 7 days later and recorded in Table 4.

Table 4. Weight and distribution of earthworms in the filter.

Depth		Earthworm Number	Initial Mean Weight (g)	Final Mean Weight (g)	Growth Rate (%)	Earthworm Color
Soil	0~5 cm	2	0.48	1.14	137.5	Dark
	5~10 cm	5	0.48	1.28	166.7	Dark
	10~15 cm	6	0.48	1.62	237.5	Light
	15~20 cm	7	0.48	2.07	331.3	Light
Gravel	20~50 cm	0	— —	— —	— —	— —

A total of 20 *Eisenia foetida* earthworms were introduced only in the soil, not in the gravel. Hence, it is clear that earthworms were mainly distributed in the soil layer close to the gravel. It was reported in earlier research that due to their need for food and air, most earthworms live in the boundary between the soil and gravel [15]. The soil near the interface between the soil and the gravel layer adsorbs more organic matter, with these soils with adsorbed organic matter being suitable for the growth of earthworms. On the other hand, the activity of earthworms is greatly affected by light and they live in the deep soil (15–20 cm) for most of the time. They are less affected by temperature. The results revealed that the mean weight of earthworms increased in the four layers of soil, from which the growth rate was calculated. It can be seen that the growth rate increases with the depth of the soil, which reflects the fact that earthworms experience better growth when they are close to gravel and water level. It can be seen the earthworm colors were different in the different soil layers from Figure 3. The color was dark in 0~10 cm soil, while it was light in 10~20 cm soil. In the rearing of earthworms, when they had a better metabolism, their body color is bright and shiny, while they have strong activity. On the contrary, when the growth of earthworms was slow, their body color was dark and they move slowly [24]. Therefore, the body color of earthworms also proved that it was suitable to survive in the soil near the gravel.

(a)   (b)

Figure 3. The earthworm color in different depths of the soil layer ((**a**) earthworm in 0~10 cm soil layer; and (**b**) earthworm in 10~20 cm soil layer).

3.2. Pollutants Removal Effect of EEF and EECW

The performance of the EEF and BF were monitored for 16 weeks. The analysis results were listed in Figure 4. It can be seen that the pollutant removal effect of EEF and BF fluctuated constantly. The average EEF removal efficiency of COD, NH_4^+-N, TN and TP were 53.8%, 65.9%, 34.3% and 73.6%, respectively. In contrast, the average BF η of COD, NH_4^+-N, TN and TP were 44.2%, 55.3%, 24.3% and 55.3%, respectively. The area load of COD, NH_4^+-N, TN and TP were 16.2 g/m^2/day, 2.03 g/m^2/day, 2.84 g/m^2/day and 0.12 g/m^2/day, respectively. These results differ from the findings of Kenatu et al. who used unplanted HSSF CW with the hydraulic retention time of six days and obtained η[BOD] of 73%, η[NH_4^+-N] of 61.0%, η[NO_3^--N] of 55.5% and η[PO_4^{3-}] of 67.6% as well as an areal load of BOD of 5.5–6.16 g/m^2/day. [11]. Reza et al. found that for compost leachate, with the hydraulic retention time of five days, η of COD, NH_4^+-N and TN were 26.2%, 17.1% and 35.0%, respectively, while the areal loads of COD, NH_4^+-N and TN were 785.6 g/m^2/day, 1.64 g/m^2/day and 29.67 g/m^2/day, respectively [25]. The removal effects in this current study were found to be lower compared with the findings of the above-mentioned study [25]. They have also reported that the percentage reductions of COD, NH_4^+-N and TN increases with the increasing HRT [25]. In addition, these removal results depend on the type of matrix and certain environmental conditions, such as pH, redox potential, temperature, organic matter content and the concentrations of certain nutrients or even the presence of heavy metals [26,27].

Figure 4 shows that the effect of EEF was better than that of BF unit. The activity of earthworms could leave many tiny holes in the soil, which can increase the oxygen content in the micro-environment. It is reported earlier that the behavior of earthworms can enhance the activity of microorganisms [28]. Earlier studies have also revealed that the total bacterial count, the numbers of ammonifier, ammonia-oxidizing bacteria and nitrite-oxidizing bacteria as well as urease and protease activities in the upper layer were positively related to earthworms [29–31].

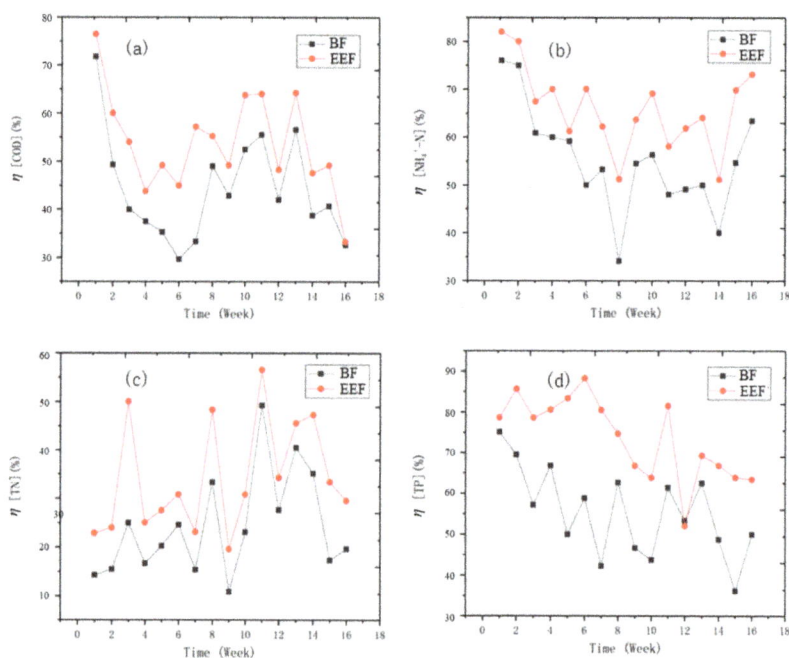

Figure 4. Performance contrast in terms of pollutant removal between the EEF and BF (HRT ~3 days, water level ~30 cm): (**a**) removal of COD; (**b**) removal of NH_4^+-N; (**c**) removal of TN; (**d**) removal of TP.

To draw a much more solid conclusion, the performance of 2 planted CWs were also investigated in the current study. These CWs were planted with *Lolium perenne Linn* for 150 days. Observation started after plants were grown up to 10 cm above the ground and lasted for 20 days. The performance contrast in terms of the pollutant removal between the earthworm-enhanced constructed wetland (EECW) and blank constructed wetland (BCW) is shown in Figure 5. It reflected a similar result as the performance of EECW was better than that of BCW. Compared to the unplanted filters (EEF and BF), the EECW has higher efficiency in the removal of COD and TN. It is worth noting that the CW (planted unit) has more stable removal efficiency than the filters (unplanted units). The results revealed that plants can increase the impact resistance of the device, while more stable influent quality leads to more stable removal efficiency.

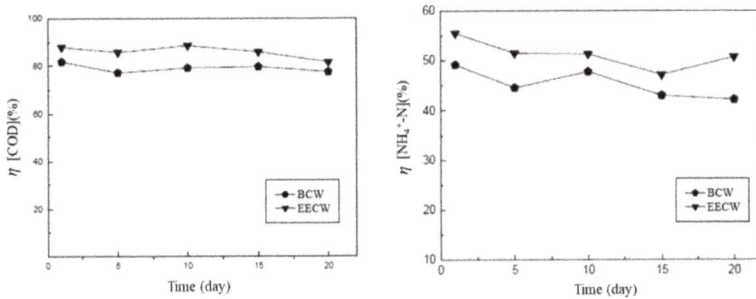

Figure 5. *Cont.*

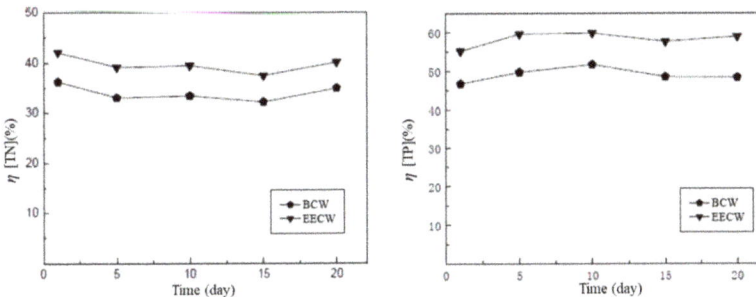

Figure 5. Performance contrast in terms of pollutant removal between the EECW and BCW (HRT ~3 days, water level ~30 cm).

3.3. Influence of Water Level on EEF Performance

The pollutant removal efficiency of EEF was studied under three different water levels (25 cm, 30 cm and 35 cm, respectively), before the η box charts for the monitor sample points of four index (COD, NH$_4^+$-N, TN and TP) were plotted to observe the influence of water level on EEF performance (Figure 6). For COD, it can be seen obviously that the order of η among three water levels is 30 cm > 25 cm > 35 cm. The η[COD] is more than 50% when the water level is 30 cm. In contrast, it is around 30% and 15% when the water level is 25 cm and 35 cm, respectively. In addition, the order of η among three water levels for the three other water quality indices was the same. Because the soil is rich in organic matter and soluble nitrogen and phosphate, when the water level is above the boundary between the soil and gravel, the dissolution of soil increased, which leads to a decrease in the pollutant removal effect. It is clear from Section 3.1 that earthworms exist around the boundary between the soil and gravel as a result, there are more microorganisms around the boundary. Therefore, the EEF had the best performance when the water level was 30 cm. For COD and NH$_4^+$-N, the removal efficiency under an optimized water level (30 cm) is 40% higher than that under adverse conditions (35 cm water level), which demonstrates that it is important to find the best condition for decision making.

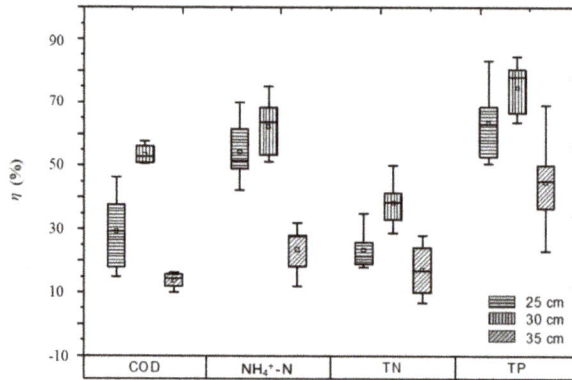

Figure 6. The influence of water level on EEF performance of pollutant removal (6 sample points for per box).

3.4. Influence of Temperature on EEF Performance

In order to study the influence of temperature, 11 individual temperature data and the corresponding η of NH_4^+-N, TN and TP were monitored. The results of correlation analysis are listed in Figure 7. The influent temperature in filter units ranged between 5 to 15 °C. An earlier study reported that low temperatures can have a considerable effect on water quality as a result of a decrease in biological decomposition [32]. Figure 7a,b show that NH_4^+-N and TN removal rate and temperature have the same trend in the EEF. Nitrogen removal is mainly dependent on the microorganisms, which play an important role in nitrification and denitrification process, while microbial activity is greatly affected by temperature [29]. In Figure 7a,b, the p values of the curves were both lower than 0.01, which reflect that temperature has a significant effect on the removal efficiency of NH_4^+-N and TN. Increased temperature leads to increased activity of earthworms and microorganisms, which increases aeration and availability of DO to micro-organisms. Being different from NH_4^+-N and TN, the removal efficiency of TP was less affected by temperature. It can be seen clearly in Figure 7c that the p value was far more than 0.05, which shows that temperature is not significantly related to TP removal rate. The possible reason could be that the removal of phosphorus in the device mainly depends on the contribution of the matrix [30]. Previous studies have also suggested that the actual effect of temperature on removal efficiency for various pollutants could be effectively analyzed by longer operations [33,34].

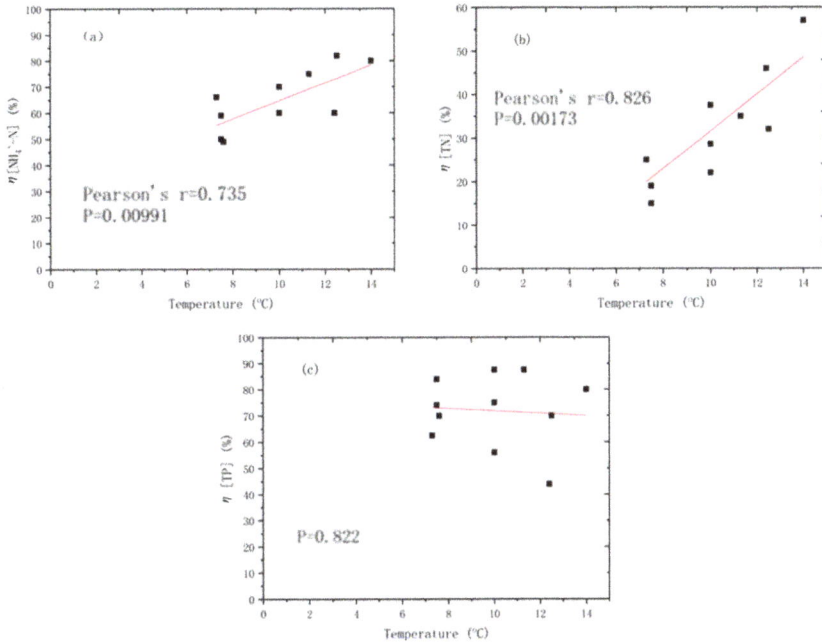

Figure 7. Correlation analysis between pollutant removal efficiency and temperature (result considered significant when $p < 0.05$): (**a**) correlation analysis between NH_4^+-N removal efficiency and temperature; (**b**) correlation analysis between TN removal efficiency and temperature; (**c**) correlation analysis between TP removal efficiency and temperature.

4. Conclusions

The results indicated that the performance of the filter unit with *Eisenia foetida* earthworms was comparatively better than the filter unit without earthworms. Earthworms grew better if they were closer to the gravel and the water level. The results also revealed that removal efficiencies for COD, NH_4^+-N, TN, TP were generally higher when the water level is around the level of the boundary between the soil and gravel. The removal efficiencies for NH_4^+-N and TN were positively correlated with temperature, while the removal efficiency of TP and temperature were not significantly related. The results revealed that compared to the unplanted units (EEF and BF), the EECW has higher efficiency in the removal of COD and TN. It is also worth noting that the CWs (planted units) have more stable removal efficiency than the filters (unplanted units). The results also showed that plants can increase the impact resistance of the device, while more stable influent quality leads to more stable removal efficiency. Overall, the results revealed that performance of filters differs with different hydraulic conditions and presence of *Eisenia foetida*. The findings of the current study could be very useful for the implementation of small-scale HSSFF treatment facilities for the treatment of different types of wastewater in the place where sewage collection and treatment is a big challenge. The evaluation of the performance of filters in treating various types of wastewaters will help us to select the appropriate parameters for better performance of filters under local conditions.

Author Contributions: D.F. and R.P.S. conceived and designed the experiments; J.J. and R.P.S. performed the experiments; J.W. contributed analysis tools and analyzed the data; and R.P.S. wrote the paper.

Funding: This research was supported by the National Natural Science foundation of China NSFC Grants No. 51550110231, 51650410657.

Conflicts of Interest: The authors declare no conflict of interest.

References

1. Durganandа, S.C.; Saravanamuthu, V.; Huu-Hao, N.; Wang, G.S.; Hee, M. Biofilter in water and wastewater treatment. *Korean J. Chem. Eng.* **2003**, *20*, 1054–1065.
2. Wang, C.; Hu, J.; Wang, Y.C.; Ma, Y.F.; Zeng, M.; Li, L. Study on denitrification effect of non-reflux biofilter. *J. Agro-Environ. Sci.* **2018**, *37*, 316–322. (In Chinese)
3. Zhang, H.G. Study on the Effect of Bio-Filter and Constructed Wetlands Purifying Mariculture Effluent. Master's Thesis, Shanghai Ocean University, Shanghai, China, 2011.
4. Starla, G.T.; Kumar, M. Biological filters and their use in potable water filtration systems in spaceflight conditions. *Life Sci. Space Res.* **2018**, *17*, 40–43.
5. Kong, L.W.; Zhang, Y.; Wang, L.; Mei, R.W.; Zhang, Y.; Li, Y. Study on Enhanced treatment of domestic sewage by a novel biological filter-constructed wetland coupling system. *Technol. Water Treat.* **2018**, *44*, 110–114. (In Chinese)
6. Terry, L.G.; Summers, R.S. Biodegradable organic matter and rapid-rate biofilter performance: A review. *Water Res.* **2018**, *128*, 234–245. [CrossRef] [PubMed]
7. Bagherpour, M.B.; Nikazar, M.; Welander, U.; Bonakdarpour, B.; Sanati, M. Effects of irrigation and water content of packings on alpha-pinene vapours biofiltration performance. *Biochem. Eng. J.* **2005**, *24*, 185–193. [CrossRef]
8. Wang, S.; Xu, Z.X.; Li, H.Z. The strengthening method of subsurface constructed wetland treating domestic sewage. *Environ Sci.* **2006**, *27*, 2432–2438. (In Chinese)
9. Ozawa, T.; Risal, C.P.; Yanagimoto, R. Increase in the nitrogen content of soil by the introduction of earthworms into soil. *Soil Sci. Plant Nutr.* **2005**, *51*, 917–920. [CrossRef]
10. Vymazal, J. Constructed wetlands for wastewater treatment: Five decades of experience. *Environ. Sci. Technol.* **2011**, *45*, 61–69. [CrossRef] [PubMed]
11. Kenatu, A.; Seyoum, L.; Worku, M.; Helmut, K.; Erik, M. Organic matter and nutrient removal performance of horizontal subsurface flow constructed wetlands planted with *Phragmites karka* and *Vetiveria zizanioide* for treating municipal wastewater. *Environ. Process.* **2018**, *5*, 115–130.
12. Hu, S.S.; Chen, Z.B. Earthworm effects on biosolids characteristics in sludge treatment wetlands. *Ecol. Engi.* **2018**, *118*, 12–18. [CrossRef]
13. Chen, Z.B.; Hu, S.S.; Hu, C.X.; Huang, L.L.; Liu, H.B.; Vymazal, J. Preliminary investigation on the effect of earthworm and vegetation for sludge treatment in sludge treatment reed beds system. *Environ. Sci. Pollut. Res.* **2016**, *23*, 11957–11963. [CrossRef] [PubMed]
14. Xu, D.; Li, Y.; Fan, X.; Guan, Y.; Fang, H.; Zhao, X. Influence of earthworm *Eisenia fetida* on *Iris pseudacorus*'s photosynthetic characteristics, evapotranspiration losses and purifying capacity in constructed wetland systems. *Water Sci. Technol.* **2013**, *68*, 335–341. [CrossRef] [PubMed]
15. Davison, L.; Headley, T.; Pratt, K. Aspects of design, structure, performance and operation of reed beds—Eight years' experience in Northeastern New South Wales, Australia. *Water Sci. Technol.* **2005**, *51*, 129–138. [CrossRef] [PubMed]
16. Schaefer, M.; Juliane, F. The influence of earthworms and organic additives on the biodegradation of oil contaminated soil. *Appl. Soil Ecol.* **2007**, *36*, 53–62. [CrossRef]
17. Zirbes, L.; Thonart, P.; Haubruge, E. Microscale interactions between earthworms and microorganisms: A review. *Biotechnol. Agron. Soc.* **2012**, *16*, 125–131.
18. Taylor, M.; Clarke, W.P.; Greenfield, P.F. The treatment of domestic wastewater using small-scale vermicompost filter beds. *Ecol. Eng.* **2003**, *21*, 197–203. [CrossRef]
19. Arancon, N.Q.; Edwards, C.A.; Babenko, A.; Cannon, J.; Galvis, P.; Metzger, J.D. Influences of vermi-composts, produced by earthworms and microorganisms from cattle manure, food waste and paper waste, on the germination, growth and flowering of petunias in the greenhouse. *Appl. Soil Ecol.* **2008**, *39*, 91–94. [CrossRef]
20. Chapuis-Lardy, L.; Brossard, M.; Lavelle, P.; Schouller, E. Phosphorus transformations in a ferralsol through ingestion by *Pantoscolex corethrurus*, a geophagous earthworm. *Eur. J. Soil Biol.* **1998**, *34*, 61–67. [CrossRef]
21. Singh, J.; Kaur, A. Vermicompost as a strong buffer and natural adsorbent for reducing transition metals, BOD, COD from industrial effluent. *Ecol. Eng.* **2015**, *74*, 9–13. [CrossRef]

22. *Technical Specification of Constructed Wetlands for Wastewater Treatment Engineering*; Ministry of Environmental Protection of China: Beijing, China, 2011.

23. American Public Health Association (APHA). *Standard Methods for the Examination of Water and Wastewater*, 20th ed.; American Public Health Association, American Water Works Association: Washington, DC, USA, 1998.

24. Xie, Y.M. *The Use and Cultivation of Earthworms*; Wuzhou Press: Taipei, Taiwan, 1989; pp. 25–41.

25. Reza, B.; Nadali, A.; Monireh, M.; Pooya, P. Compost leachate treatment by a pilot-scale subsurface horizontal flow constructed wetland. *Ecol. Eng.* **2017**, *105*, 7–14.

26. Sohair, I.A.; Mohamed, A.E.; Magdy, T.K.; Mohamed, S.H. Factors affecting the performance of horizontal flow constructed treatment wetland vegetated with *Cyperus papyrus* for municipal wastewater treatment. *Int. J. Phytoremediat.* **2017**, *19*, 1023–1028.

27. Yang, J.; Ye, Z. Metal accumulation and tolerance in wetland plant. *Front. Biol.* **2009**, *4*, 282–288. [CrossRef]

28. Tognetti, C.; Mazzarino, M.J.; Laos, F. Improving the quality of municipal organic waste compost. *Bioresour. Technol.* **2007**, *98*, 1067–1076. [CrossRef] [PubMed]

29. Ji, F.Y.; Luo, G.Y.; Zhou, J. Experimental study on earthworm and wastewater treatment. *Chongqing Environ. Sci.* **1998**, *20*, 12–15. (In Chinese)

30. Li, H.Z.; Wang, S.; Ye, J.F.; Xu, Z.X.; Jin, W. A practical method for the restoration of clogged rural vertical subsurface flow constructed wetlands for domestic wastewater treatment using earthworm. *Water Sci. Technol.* **2011**, *63*, 283. [CrossRef] [PubMed]

31. Wu, L.; Li, X.N.; Song, H.L.; Wang, G.F.; Jin, Q.; Xu, X.L.; Gao, Y.C. Enhanced removal of organic matter and nitrogen in a vertical-flow constructed wetland with *Eisenia foetida*. *Desalin. Water Treat.* **2013**, *51*, 7460–7468. [CrossRef]

32. Bulc, T.; Slak, A.S. Performance of constructed wetland for highway runoff treatment. *Water Sci. Technol.* **2003**, *48*, 315–322. [CrossRef] [PubMed]

33. Poe, A.C.; Piehler, M.F.; Thompson, S.P.; Paerl, H.W. Denitrification in a constructed wetland receiving agricultural runoff. *Wetlands* **2003**, *23*, 817–826. [CrossRef]

34. Aldheimer, G.; Bennerstedt, K. Facilities for treatment of stormwater runoff from highways. *Water Sci. Technol.* **2003**, *48*, 113–121. [CrossRef] [PubMed]

water

MDPI

Article

Effectiveness of ABC Waters Design Features for Runoff Quantity Control in Urban Singapore

Wing Ken Yau [1], Mohanasundar Radhakrishnan [2,*], Shie-Yui Liong [3], Chris Zevenbergen [2] and Assela Pathirana [2]

[1]   PUB Singapore's national water agency, Singapore 228231, Singapore; yau_wing_ken@pub.gov.sg
[2]   Water Science and Engineering Department, IHE Delft Institute for Water Education,
       2611 AX Delft, The Netherlands; c.zevenbergen@un-ihe.org (C.Z.); a.pathirana@un-ihe.org (A.P.)
[3]   Tropical Marine Science Institute, National University of Singapore, Singapore 119227, Singapore;
       tmslsy@nus.edu.sg
*   Correspondence: m.radhakrishnan@un-ihe.org; Tel.: +31-647618689

Received: 6 June 2017; Accepted: 31 July 2017; Published: 3 August 2017

Abstract: Active, Beautiful, Clean Waters (ABC Waters) design features—natural systems consisting of plants and soil that detain and treat rainwater runoff—comprise a major part of Sustainable urban Drainage Systems (SuDS) in Singapore. Although it is generally accepted that ABC Waters design features are able to detain runoff and reduce peak flow, their effectiveness in doing so has not been studied or documented locally. This research aims to determine their effectiveness in reducing peak flow based on a newly constructed pilot precinct named Waterway Ridges. Four types of ABC Waters features have been integrated holistically within the development, and designed innovatively to allow the precinct to achieve an effective C-value of 0.55 for the 10-year design storm; the precinct-wide integration and implemented design with the aim of substantially reducing peak flow are firsts in Singapore. The study is based on results from an uncalibrated 1D hydraulic model developed using the Storm Water Management Model (SWMM). Identification of key design elements and performance enhancement of the features via optimisation were also studied. Results show that the features are effective in reducing peak flow for the 10-year design storm, by 33%, and allowed the precinct to achieve an effective C-value of 0.60.

Keywords: ABC Waters; LID; SuDS; WSUD; water quantity control; runoff control; numerical simulation; sustainable storm water management

---

1. Introduction

Cities are being exposed to the threat of increasing flood risk due to climate-change-induced sea level rise and the increased frequency of extreme rain events, as well as increased urbanisation resulting in more impervious areas and runoff [1]. This is especially true of Singapore, a tropical island that has undergone rapid urbanisation over the past few decades despite its land constraints; the city state has an area of about 719 km^2, is the third most densely populated country in the world, and has an urban population of 100% [2,3]. The historical rainfall data from 1980 to 2012 also indicate a trend of higher rainfall intensities (Figure 1). The conventional approach of building wider and deeper drains to quickly collect and channel rainwater runoff away from the urban catchments has been accepted to be no longer adequate or sustainable to cope with the challenge of climate change, limited land and increasing urbanisation in Singapore. There was a paradigm shift on storm water management in 2006 when PUB Singapore's national water agency, launched the Active, Beautiful, Clean Waters (ABC Waters) Programme [4]. The ABC Waters Programme is an initiative to transform Singapore's drains, canals and reservoirs beyond their traditional functions of drainage, flood control and water supply storage into beautiful streams, rivers and lakes that are well integrated with the surrounding landscapes. The programme is part of PUB's strategic objective to bring Singaporeans closer to water so that they can better appreciate and cherish this precious resource, and in turn keep the waters of Singapore clean and free of litter.

Under the programme, storm water runoff is proposed to be managed in a more sustainable manner via the implementation of ABC Waters design features, natural systems consisting of plants and soil that are able to detain and treat rainwater runoff before discharging the cleansed runoff into the downstream drainage system [4]. There are various types of ABC Waters design features including the bioretention basins or rain gardens, bioretention swales, vegetated swales, constructed wetlands (whose types consist of surface flow, floating and subsurface) and sedimentation basins. While ABC Waters design features are similar to Sustainable urban Drainage Systems (SuDS), Water sensitive Urban Design (WSUD), Low Impact Development (LID) and Green Infrastructure (GI) that are prevalent in other countries [5], ABC Waters design features generally have a stronger focus on the cleansing function and represent a subset of the features represented by the other terms; there are features such as green roofs, detention tanks and pervious pavement which are not considered as ABC Waters design features but are part of the features under SuDS/WSUD/LID/GI.

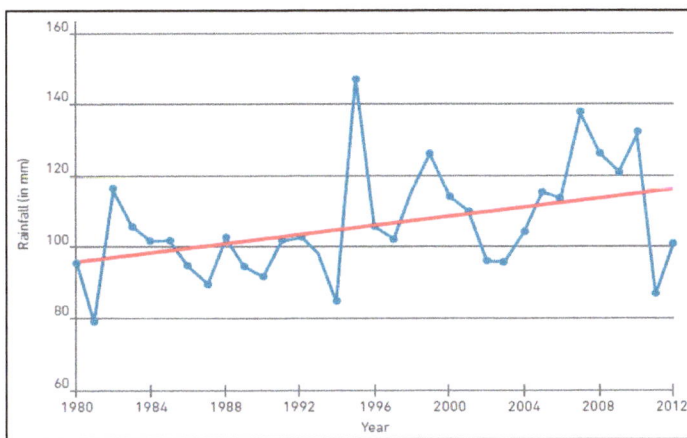

Figure 1. Annual maximum hourly rainfall intensities in Singapore based on records from 28 stations from 1980 to 2012 (blue line/dots), with trend towards higher rainfall intensities (pink line) [6].

The Source-Pathway-Receptor approach in drainage management is also advocated in Singapore and ABC Waters design features are recommended as one of the source control measures to manage storm water runoff [6]. A runoff control legislation, imposed from 2014 onwards via an addendum to the Code of Practice (COP) on Surface Water Drainage, requires "industrial, commercial, institutional and residential developments greater than or equal to 0.2 ha in size" to "control the peak runoff discharged from the development sites" [7]. It is stated that "the maximum allowable peak runoff to be discharged to the public drains will be calculated based on a runoff coefficient of 0.55, and for design storms with a return period of 10 years and for various storm durations of up to 4 h (inclusive)". In addition to detention tanks, ABC Waters design features are also among the measures recommended in the COP on Surface Water Drainage to achieve the peak runoff reduction [7].

Although it is generally accepted that ABC Waters design features are able to detain runoff and help to reduce peak flow, their effectiveness in doing so has not been studied or documented locally in Singapore. Modelling of the runoff control performance of ABC Waters design features is almost non-existent in Singapore and while modelling is carried out, they are done to assess the performance of the features in terms of water quality improvement [8]. This could be inherently due to the fact that ABC Waters design features are not widely used for runoff control in Singapore at the time of writing, as the general perception is that ABC Waters design features are not effective enough to lower peak flow to the standards (C = 0.55) stipulated by the runoff control legislation in Singapore. Instead, the features are normally implemented in Singapore for improving water quality and other ecosystem services such as providing social and recreational spaces for the public or residents in a development [4]. This research thus aims to address the gap of a lack of study on the effectiveness of ABC Waters design features for runoff quantity control in Singapore by performing such a study, and to do so via a demonstration on the use of modelling and optimisation tools for such an assessment.

There are modelling as well as field monitoring studies on the effectiveness of SuDS/LID/WSUD/GI in reducing peak flow within temperate catchments internationally [9–18]. However, the studies on SuDS application in tropical climates are limited. Taking the example of bioretention systems (i.e., vegetated land depressions that are designed to detain and treat storm water runoff via sedimentation, filtration and biological uptake [4], which is a type of ABC Waters design feature), they have been reported to be able to achieve runoff peak flow reduction in the range of 24–99% in the United States of America (USA) [15,17,19–21] and 80–90% in Australia [22,23] based on field monitoring studies. The peak flow reductions depend on the ratio of catchment area to bioretention area, the drainage configuration and the rainfall magnitude, with smaller peak flow reductions being achieved for higher rainfall intensities as well as increased rainfall depths and longer rainfall durations [16–18,21]. Hence there is a need for region specific study due to the large variations in the peak flow reduction effectiveness of the bioretention systems [20].

This research directly addresses the gap of lack of study on the peak flow reduction performance of ABC Waters design features/SuDS/LID/WSUD/GI in the tropical climate setting. In addition, the research aims to identify key design elements that could help to enhance the effectiveness of ABC Waters design features in reducing peak flow. Performance enhancement of the features will also be studied via optimisation.

The relevance of the study results will not be limited to Singapore as the study results will also provide inputs towards bioretention design and practice in many other countries with similar tropical climates, such as countries in South East Asia, Sri Lanka, Uganda, Honduras, El Salvador etc. It is hoped that the study results would help to guide the design and implementation of SuDS/LID/WSUD/GI in the other countries with tropical climates.

2. Details of the Site and ABC Waters Design Features

2.1. The Study Site

The research was based on a newly constructed four hectare pilot project named Waterway Ridges, which is a joint collaboration between PUB and the Housing Development Board of Singapore. The construction of this pioneer precinct began in April 2012 and was completed in April 2017. This public residential development contains four different types of ABC Waters design features that are integrated in the design at precinct level, namely bioretention basins/rain gardens, bioretention lawns, vegetated swales and vegetated swales with gravel layer (termed as "gravel swales" in the rest of this paper). Such integrated application has been adopted at precinct level in Singapore for the first time. The ABC Waters design features are also installed with the aim of substantial peak flow reduction, which is another first in Singapore.

The ABC Waters design features at the site have been innovatively designed to reduce the peak flow of the precinct to correspond to a reduced runoff coefficient of 0.55 for all storm events up to the 10-year return period. The innovative design, which is implemented for the very first time in Singapore, comprises 400–750 mm thick detention gravel storage layers which are incorporated below the rain gardens and gravel swales, and orifice outlets that restrict the outflow from the features. Normally dry, the network of ABC Waters design features (Figure 2) receives runoff from about 60% of the total site area during a rain event [24].

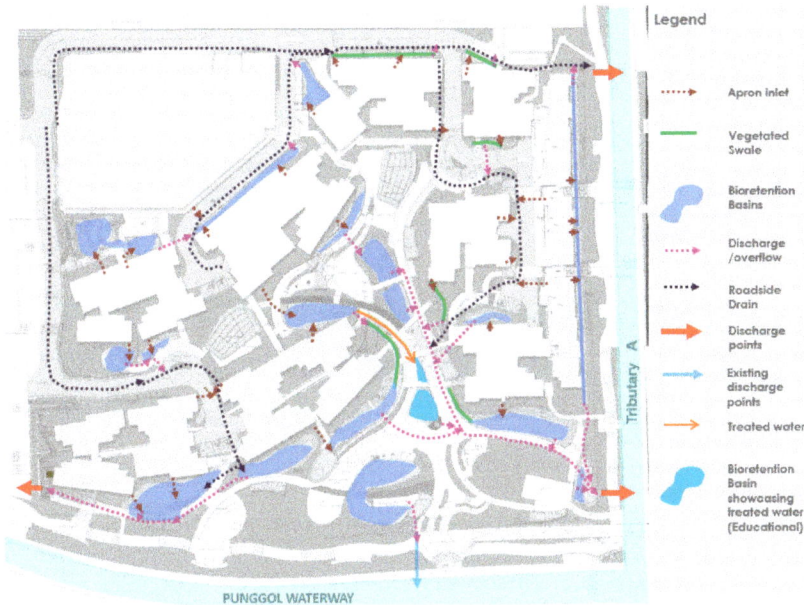

Figure 2. Overview plan and indicative drainage flow paths in Waterway Ridges.

The climate in Singapore is characterised as tropical and coastal. The country experiences high and uniform temperatures (the minimum and maximum temperatures are 23–25 °C and 31–33 °C during the night and day, respectively), high humidity (the mean annual relative humidity is 83.9%), and abundant rainfall throughout the year. The average annual rainfall is 2328.7 mm (based on long-term records from 1869 to 2016) and it rains 178 days a year on average [25]. There are two monsoon seasons in the country, namely the Northeast Monsoon occurring from December to early March, and the Southwest Monsoon from June to September. Though there are no distinct wet or dry seasons in Singapore, there is a monthly variation of rainfall in Singapore. Generally, there is increased rainfall from November to January during the Northeast Monsoon season. On average, the wettest and driest months are December (318.6 mm) and February (112.8 mm), respectively [26].

2.2. Design of ABC Waters Design Features at the Study Site

Four different types of ABC Waters design features are implemented at the precinct. There are 21 rain gardens (with codes FB1 to FB21), four vegetated swales (with codes VS1 to VS4) and two gravel swales (with codes GS1 to GS2) in the precinct.

2.2.1. Bioretention Basins/Rain Gardens

Rain gardens or bioretention basins are vegetated land depressions that are designed to detain and treat storm water runoff via sedimentation, filtration and biological uptake [4]. A typical cross section of the rain gardens implemented at Waterway Ridges is shown in Figure 3; various elements of the design are indicated by the numbers (1) to (5), and will be referred to in the rest of this paragraph. During a rain event, runoff from the subcatchment would flow into the rain garden and be allowed to pond up to a maximum depth of 200 mm on the surface. Concurrently, the runoff percolates through the filter media and is eventually collected by the subsoil perforated pipes (5). Excess runoff greater than the 200 mm surface detention depth will overflow into the overflow manhole (1) and be directed into the underground gravel layer for detention and storage via the subsoil perforated pipes (5). Meanwhile, the amount of flow leaving the feature and entering the discharge pipe (4) will be regulated through the orifice outlet (3), the opening size of which was predetermined through calculations to correspond to C = 0.55 for the 10-year design storm. When the underground gravel layer is full, the water level in the manhole will rise to the overflow standpipe opening (2) and be discharged via the discharge pipe (4) that connects to the roadside drains within the precinct.

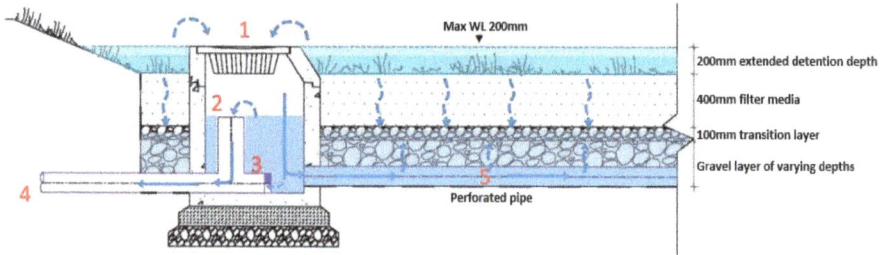

Figure 3. Cross section of a rain garden with overflow manhole in Waterway Ridges.

Unlike the typical rain gardens with drainage layers that are 200–250 mm thick, the rain gardens implemented at Waterway Ridges have a much thicker gravel layer ranging from 400–750 mm, and the orifice outlets are not part of the design of typical rain gardens. The overflow manhole is also fitted with gratings and a "filter basket" to minimise debris from flowing into and choking the gravel storage layers. Literature review suggests that such a design is new not only in Singapore but also internationally. Among the designs found, one that is most similar to the rain gardens at Waterway Ridges is the "Cap-orifice Flow regulator" design [27]. In this design, a cap-orifice is proposed to be installed at the exit of the subsoil perforated pipe for the purpose of regulating the infiltration rate as well as the outflow from the rain garden [27]. However, such a design would not be effective in reducing peak flows in tropical climates with extreme rainfall like Singapore, where a large portion of the subcatchment generated runoff (from storm events greater than the 3-month return period) would overflow into the overflow manholes instead of infiltrating through the filter media of the rain gardens.

2.2.2. Bioretention Lawns

Bioretention lawns have the same design as rain gardens, but have lawns extensively as their planting as compared to shrubs in the rain gardens (Figure 4). When it is not raining, the bioretention lawns serve as recreational spaces for the residents of Waterway Ridges.

Figure 4. Photos of bioretention lawns in Waterway Ridges.

2.2.3. Vegetated Swales

Vegetated swales are essentially earth drains that convey runoff. In Waterway Ridges, they are gravel lined and vegetated with short grass (Figure 5). Due to their increased roughness, the flow of runoff is slower in the swales as compared to concrete-lined drains. The slower runoff velocities help promote sedimentation, thus resulting in cleaner runoff.

Figure 5. Photos of vegetated swales in Waterway Ridges.

2.2.4. Gravel Swales

A typical cross section of the gravel swales implemented at Waterway Ridges is shown in Figure 6; various elements of the design are indicated by the numbers (1) to (4), and will be referred to in the rest of this paragraph. Gravel swales are essentially vegetated swales (1) that are coupled with underground gravel storage layers (2). Runoff flowing through gravel swales would be channelled directly into the gravel layers via the overflow manholes (3) and subsoil perforated pipes (4). The components in the overflow manholes of gravel swales are the same as that of the rain gardens. Similarly to the rain gardens, flow leaving the gravel swales will be regulated through an orifice outlet located within the overflow manhole.

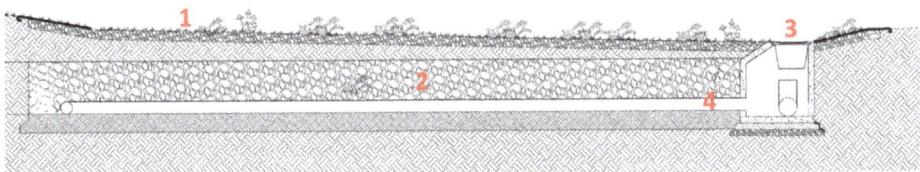

Figure 6. Typical cross section of a gravel swale in Waterway Ridges.

2.3. Monitoring of the Study Site

There are plans for water quality and quantity monitoring to be carried out at the precinct. The plan includes flow monitoring at each of the four precinct discharge outlets as well as at the inlets and outlets of four specific ABC Waters design features (one for each type) to determine the flows entering and leaving each feature type.

At the time of publishing however, the monitoring has yet to commence and as such, the results presented in this manuscript will be based on uncalibrated models. It is acknowledged that the results presented will be preliminary and that model calibration will need to be performed to yield validated results. Nevertheless, the results do provide an indication on the peak flow reduction effectiveness of the ABC Waters design features (and innovative design adopted for the features) at the precinct, and in a tropical climate setting.

3. Research Methodology

The general workflow that was adopted for the research is as shown in Figure 7. More details on the specific research activities that were carried out for each task are as outlined in the following sub-sections.

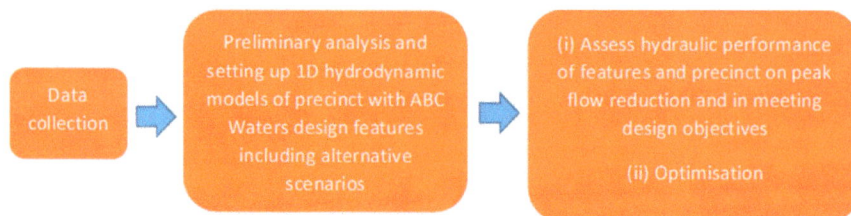

Figure 7. General workflow for the research methodology.

3.1. Data Collection

Prior to setting up the model for the precinct and its suite of ABC Waters design features, design data and general information regarding the precinct and the features were collected. As the final as-built drawings of the precinct and features were not available during the duration of the research, the latest set of construction drawings as obtained from PUB were used for developing the model. Additional information such as rainfall intensity-duration-frequency (IDF) curves and historical rainfall data were also obtained for assessment of the model performance.

3.2. Model Selection

The Storm Water Management Model version 5.1.010 (SWMM v5.1), an open source and freely downloadable hydraulic model, with LID controls developed by the U.S. Environmental Protection Agency (USEPA) was selected and used for the 1D hydrodynamic model study [28]. For Waterway Ridges, backwater effect and reverse flow can occur within the perforated pipes (connecting the gravel storage to the overflow manhole) and orifice outlets of the rain gardens and gravel swales, thus necessitating the use of dynamic-wave hydraulic routing. Due to the small size of the subcatchment and ABC Waters design features (the smallest subcatchment and ABC Waters design feature have areas of 34 m^2 and 18 m^2, respectively, while the average areas for the subcatchments and ABC Waters design features are 632 m^2 and 80 m^2, respectively), sub-hourly time steps in the order of minutes and seconds would also apply for the variation of runoff and associated detention processes in the precinct. These requirements are met by the SWMM model used.

3.3. Preliminary Analysis, Sensitivity Analysis and Model Parameters

Prior to setting up the model for the precinct, preliminary assessment was carried out to determine the best ways to represent the various ABC Waters features in Waterway Ridges. Based on the SWMM Applications Manual [29] as reference, different ways of modelling the subcatchments and features were tested, and the outflow results from the features were compared and analysed. Sensitivity analysis was also performed to determine the sensitivity of runoff from the subcatchments to the various catchment parameters [7,29]. Figure 8 shows a screenshot of the completed 1D SWMM model for the precinct.

3.4. Hydraulic Routing Model and Time Steps

The dynamic-wave flow routing was necessary as backwater effects and reverse flow would take place within the perforated pipes and orifice outlets of the rain gardens and gravel swales. Time step recommendations as listed in the SWMM Applications Manual [29] were adhered to, and model stability was also verified for the time steps used. The time steps for dry weather flow, wet weather flow and routing were set at 1 h, 1 min, and 10 s, respectively. Typically, the wet weather time step and flow routing time step should not exceed the precipitation recording interval and wet weather time step, respectively [29]. The flow routing time step used is less than 30 s due to the dynamic-wave flow routing that is employed in the model, and was adjusted until the stability of the model was satisfactory.

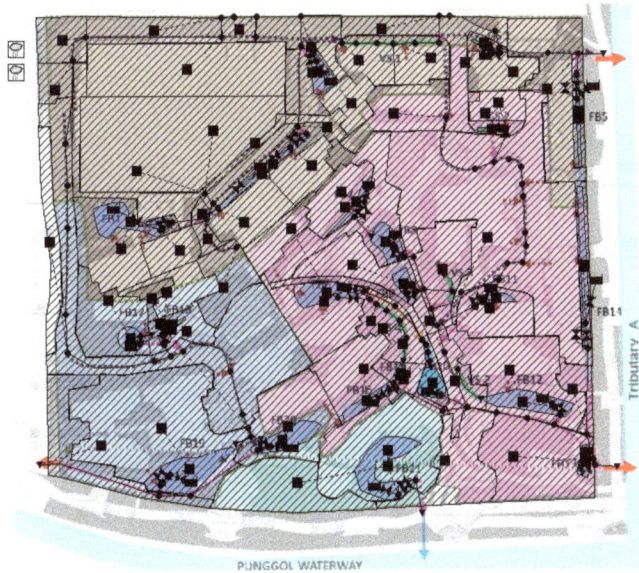

Figure 8. The completed 1D SWMM model for the precinct (in addition to the features whose colour scheme follows that of Figure 2, the other four colours depict the four main subcatchments in the precinct).

3.5. Scenarios

In addition to the base scenario of the precinct model (i.e., base precinct model with its suite of ABC Waters design features that are coupled with gravel storage and orifice outlets), additional scenarios as summarised in Table 1 were also modelled and analysed. The rationale for modelling the various scenarios is also elaborated in the table.

Table 1. Summary of base and alternative scenarios of the precinct which were modelled and analysed.

S/N	Scenario Description	Remarks
1	Precinct without any detention (i.e., conventional design without ABC Waters features, gravel storage or orifice outlets)	To assess performance of the conventional design
2	Precinct without ABC Waters features but with orifice outlets from the respective subcatchments	To assess effectiveness of orifice outlets alone in reducing peak flow
3	Precinct with typical design of ABC Waters features implemented in Singapore (i.e., rain gardens and swales without orifice outlets and gravel storage layer)	To assess performance of the precinct if typical design of ABC Waters features was used
4	Precinct with ABC Waters features coupled with orifice outlets but without gravel storage	To determine importance of gravel storage in reducing peak flow
5	Precinct with ABC Waters features coupled with gravel storage but without orifice outlets	To determine importance of orifice outlets in reducing peak flow
6	Precinct with ABC Waters features coupled with gravel storage and orifice outlets	Base model that corresponds to the actual constructed precinct

3.6. Assessments

3.6.1. Peak Flow Reduction and Compliance with Peak Flow Reduction Objectives

Symmetrical 4 h 10-year and 3-month design rainfall hyetographs in 5-min intervals (Figure 9) derived from Singapore's rainfall IDF curves were routed through each of the models for the respective scenarios. The peak flow from the precinct for each of the scenarios were compared and the corresponding runoff coefficients or C-values were back-calculated using Rational Formula (Equation (1)) [30]. Specifically for the base precinct model, the C-value was compared against the design objective of having an effective runoff coefficient of 0.55. As there are four discharge outlets from the precinct, the peak flow from the precinct was obtained by taking the maximum of the sum of the time series flow results from each of the outlets.

$$Q_r = \frac{1}{360}CIA \tag{1}$$

where

Q_r = peak runoff (m^3/s)
C = runoff coefficient (dimensionless)
I = average rainfall intensity (mm/h)
A = catchment area (ha).

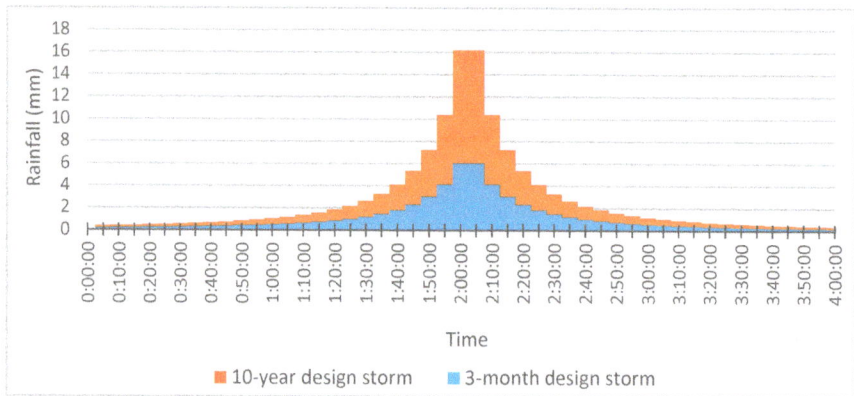

Figure 9. Rainfall hyetographs for the 10-year and 3-month design storms in Singapore [7,31].

3.6.2. Effectiveness of Various ABC Waters Design Feature Types in Reducing Peak Flow

The maximum flows entering and leaving each feature were obtained from the base precinct model for the 10-year design storm, and the corresponding peak flow reductions were calculated using Equation (2). Averages of the peak flow reductions for the various design feature types were then compared to determine if any of the feature types are more effective in reducing peak flow as compared to others.

$$Peak\ flow\ reduction\ (\%) = \frac{maximum\ inflow - maximum\ outflow}{maximum\ inflow} \times 100\% \tag{2}$$

However, it was also noted that the peak flow reduction alone would not be a fair comparison across the different features and even within the same type, as the features have different catchment areas and storage volumes. As such, the peak flow reductions were adjusted using the ratio of the storage volume (both surface detention and gravel storage for the rain gardens, and gravel storage alone for the gravel swales which do not have any surface detention) to the effective impervious catchment area of each feature (using Equations (3)–(5), and then normalised (using Equation (6)). Averages of the normalised peak flow reductions for the various feature types were compared to assess their peak flow reduction effectiveness.

$$
\begin{aligned}
Effective\ catchment\ area \\
= 1 \times impervious\ catchment\ area + 0.45 \\
\times pervious\ catchment\ area
\end{aligned}
\tag{3}
$$

$$
Adjustment\ ratio\ (m) = \frac{Total\ storage\ volume\ of\ feature\ (m^3)}{Effective\ catchment\ area\ of\ feature\ (m^2)}
\tag{4}
$$

$$
\begin{aligned}
Adjusted\ peak\ flow\ reduction\ for\ i^{th}\ feature\ (\%) \\
= \frac{Largest\ adjustment\ ratio\ among\ all\ features}{Adjustment\ ratio\ for\ i^{th}\ feature} \\
\times Peak\ flow\ reduction\ for\ i^{th}\ feature
\end{aligned}
\tag{5}
$$

$$
\begin{aligned}
Normalised\ peak\ flow\ reduction\ for\ i^{th}\ feature\ (\%) \\
= \frac{Adjusted\ peak\ flow\ reduction\ for\ i^{th}\ feature - Smallest\ adjusted\ peak\ flow\ reduction}{Largest\ adjusted\ peak\ flow\ reduction - Smallest\ adjusted\ peak\ flow\ reduction}
\end{aligned}
\tag{6}
$$

4. Results

4.1. Peak Flow Reduction for the 10-Year Design Storm

The outflows and effective C-values (of the base precinct and the various alternative scenarios) for the 10-year design storm are as shown in Figure 10 and Table 2, respectively. From the results, we can see that the best C-value of 0.60 with a corresponding reduction of 32.6% in peak runoff (compared to Scenario 1 where there are no detention or treatment present in the precinct) is obtained in Scenario 6, i.e., when the base precinct with ABC Waters features are coupled with gravel storage and orifice outlets. Although the reduced C-value and peak runoff achieved do not meet the runoff control requirements of Singapore's COP on Surface Water Drainage, the results show that ABC Waters design features are effective in reducing peak runoff when coupled with storage and orifice outlets, and designed appropriately. It is also noted that the peak flow reduction obtained for the base precinct is generally lower as compared to the reductions obtained by the various bioretention systems that have been monitored in the USA [15,17,19–21] and Australia [22,23].

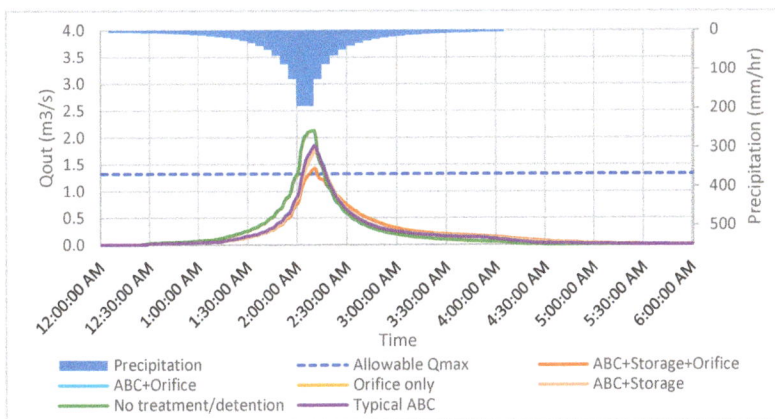

Figure 10. Outflow results of the precinct and alternative scenarios for the 10-year design storm.

Table 2. Outflow and effective C-value results of the precinct and alternative scenarios for the 10-year design storm.

S/N	Label	Scenario	Qmax (m³/s)	Reduction * in Qmax	Effective C-Value
1	No treatment/detention	Precinct without any treatment or detention	2.13	0%	0.89
2	Orifice only	Precinct with orifice outlets only	2.13	0%	0.89
3	Typical ABC	Precinct with typical design of ABC Waters features implemented in Singapore	1.85	13.0%	0.77
4	ABC + Orifice	Precinct with ABC Waters features and orifice outlets but without gravel storage	1.86	12.6%	0.77
5	ABC + Storage	Precinct with ABC Waters features and gravel storage but without orifice outlets	1.75	17.7%	0.73
6	ABC + Storage + Orifice	Precinct with ABC Waters features coupled with gravel storage and orifice outlet	1.44	32.6%	0.60

Note: * Relative to Scenario 1 of the precinct without any treatment or detention.

Meanwhile, it can be observed that the performance of Scenario 2 (precinct with orifice outlets only) is as poor as that of Scenario 1, indicating that there is no benefit in implementing orifice outlets that are not coupled with storage or any ABC Waters design feature.

While the performance of Scenario 3 (precinct with typical design of ABC Waters features implemented in Singapore) is not as good as that of Scenario 6, it still yielded a C-value of 0.77 and a peak runoff reduction of 13% (as compared to Scenario 1). This observation shows that the typical ABC Waters features implemented in Singapore (i.e., without storage and orifice outlets) could still contribute in helping to reduce peak runoff in a development even though they are not able to meet the runoff control requirements of the COP on Surface Water Drainage and will need to be implemented together with other features.

4.2. Effectiveness of the Various ABC Waters Design Feature Types in Reducing Peak Flow for the 10-Year Design Storm

Table 3 summarises the performance of the various feature types in reducing peak flow. From the results, we can see that the rain garden feature type is most effective in reducing peak flow (with average and normalised reductions of 39% and 47%, respectively), followed by the gravel swales (with average and normalised reductions of 9% and 25%, respectively). This result could be attributed to the presence of the 0.4 m thick filter media and extended surface detention in rain gardens but not the gravel swales. The results also show that the vegetated swales do not offer any reduction in peak flow, which is attributed to the lack of gravel storage or any form of surface detention due to the absence of flow regulators (such as weirs) along the vegetated swales.

Table 3. Peak flow reduction performance of the various feature types for the 10-year design storm.

Design Feature Type	Average Peak Flow Reduction (%)	Average Normalised Peak Flow Reduction (%)
Rain gardens *	39%	47%
Gravel swales	9%	25%
Vegetated swales	0%	0%

Note: * Excluding rain gardens without orifice outlets (i.e., rain gardens FB6, FB10, FB15 and FB17).

However, this does not conclude that vegetated swales should not be implemented. While peak flow reduction is the focus of this study, it is noted that ABC Waters design features including vegetated swales also offer other ecosystem services such as improving runoff water quality, enhancing biodiversity [4] and regulating micro-climate via the urban heat island mitigation effect [32].

4.3. Identification of Key Design Elements

It can be observed from the peak flow reduction results from Section 4.1 that the performance of Scenario 5 (precinct with ABC Waters features and gravel storage but without orifice outlets) with a C-value of 0.73 (and peak runoff reduction of 17.7%) is slightly better than that of Scenario 4 (precinct with ABC Waters features and orifice outlets but without gravel storage) with a C-value of 0.77 (and peak runoff reduction of 12.6%). When the orifice outlets and gravel storage are coupled together however, the performance improves to yield a C-value of 0.60, which is an improvement of 18.2% and 22.9% for Scenarios 5 and 4, respectively. It can be concluded from this observation that the orifice outlets and gravel storage are important components of the ABC Waters features (specifically the rain gardens and gravel swales) that will dictate their performance in reducing peak runoff effectively.

4.4. Peak Flow Reduction for the 3-Month Design Storm

The outflows and effective C-values (of the base precinct and the various alternative scenarios) for the 3-month design storm are as shown in Figure 11 and Table 4, respectively. Table 4 also includes the results of the 10-year design storm for comparison. Similar to the results of the 10-year design storm, the constructed base precinct with ABC Waters features that are coupled with gravel storage and orifice outlets achieved the lowest C-value of 0.47. The other trends as earlier observed for the 10-year design storm also apply for the 3-month design storm. In addition (and as expected), the peak flow for all the scenarios are below that of the maximum allowable peak runoff (which is calculated based on the 10-year design storm).

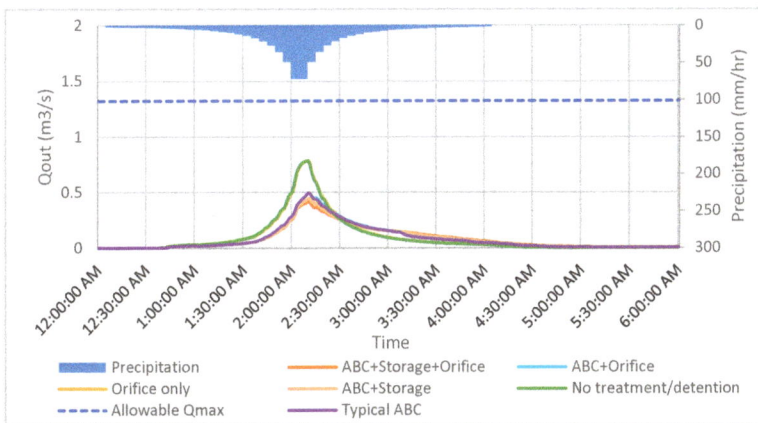

Figure 11. Outflow results of the precinct and alternative scenarios for the 3-month design storm.

Further comparisons of the peak flow difference between the worst and best performing scenario (i.e., Scenarios 1 and 6) were also made for the 3-month and 10-year design storms in the zoomed-in plot shown in Figure 12. From the figure, we can see that the peak flow difference is larger for the 3-month design storm (at 46.7%) as compared to the 10-year design storm (at 32.6%), suggesting that the ABC Waters design features implemented are more effective in reducing peak flow for smaller storm events as compared to larger ones. Similar observations were also reported in other bioretention studies [16–18,22]. This is generally expected for typical rain gardens (since smaller storm events would result in less surface overflow as compared to larger storm events) and is hence a somewhat surprising observation considering the design of the rain gardens and gravel swales in the precinct which are coupled with underground gravel storage and orifice outlets.

Table 4. Comparison of the outflow and effective C-value results of the precinct and alternative scenarios for the 3-month and 10-year design storms.

S/N	Scenario	3-Month Design Storm			10-Year Design Storm		
		Qmax (m³/s)	Reduction * in Qmax (%)	C-Value	Qmax (m³/s)	Reduction * in Qmax (%)	C-Value
1	No treatment/detention	0.78	0%	0.89	2.13	0%	0.89
2	Orifice only	0.78	0%	0.89	2.13	0%	0.89
3	Typical ABC	0.49	36.8%	0.56	1.85	13.0%	0.77
4	ABC + Orifice	0.49	37.3%	0.56	1.86	12.6%	0.77
5	ABC + Storage	0.44	43.2%	0.51	1.75	17.7%	0.73
6	ABC + Storage + orifice	0.42	46.7%	0.47	1.44	32.6%	0.60

Note: * Relative to Scenario 1 of the precinct without any treatment or detention.

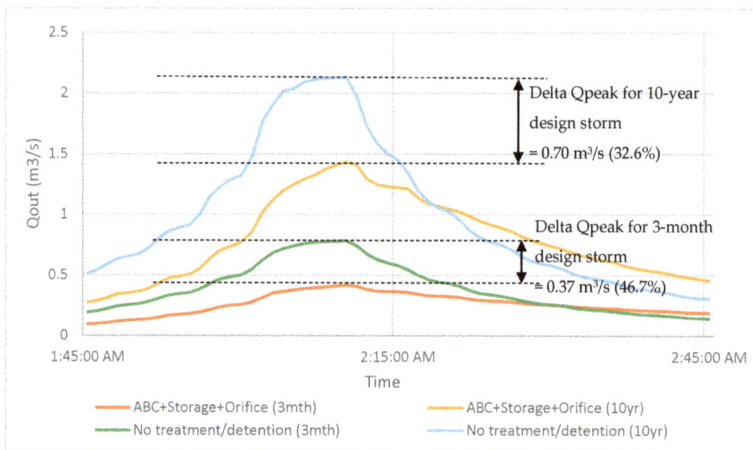

Figure 12. Peak flow difference between the best and worst performing scenario for the 3-month and 10-year design storms.

4.5. Performance Enhancement through Optimisation

The original base precinct has an effective C-value of 0.60 and does not meet the requirements of the runoff control legislation in Singapore. This performance can be attributed to the non-optimal design of the orifice outlets in the precinct. To verify this, optimisation was performed for two scenarios, namely (1) Single-objective optimisation of the resultant C-value of the precinct by varying the diameter of the orifice outlets and (2) Multi-objective optimisation of the C-value and cost of gravel layer cum orifice outlets by varying the gravel storage layer depths and diameter of the orifice outlets. More information on the optimisation can be found in Appendix A.

The single objective optimisation showed that the effective C-value of the precinct could be improved to 0.53. The optimised C-value is 11.7% less than the non-optimised C-value of 0.60 with no increase in cost (since the gravel storage layer depths are the same), and is lower than the 0.55 that is stipulated by the runoff control legislation in Singapore. The multi-objective optimisation also resulted in a similar C-value of 0.53 at an additional cost of S$1610 (~$1180).

A separate optimisation study was also performed on an arbitrary precinct model to determine if there is an optimum rain garden area and treated catchment area (in terms of percentage of total site area) that would allow catchments to achieve C = 0.55 for the 10-year design storm. A catchment area of 4 ha was adopted for the study and three land use types were considered; residential, commercial and mixed (50% residential–50% commercial). Runoff from treated parts of the catchment is channelled into a lumped rain garden with gravel storage and orifice outlets based on the design adopted at Waterway Ridges. The study showed that there is an optimum treated catchment area that would result in the lowest possible C-value for each rain garden size, and beyond which the peak flow reduction performance would deteriorate (Figure 13). It was found that a minimum rain garden size of 6% the total site area with treated catchment areas of 45.2–47.2% (depending on land use type) was needed to meet the runoff control legislation in Singapore. More information on the study is provided in Appendix B.

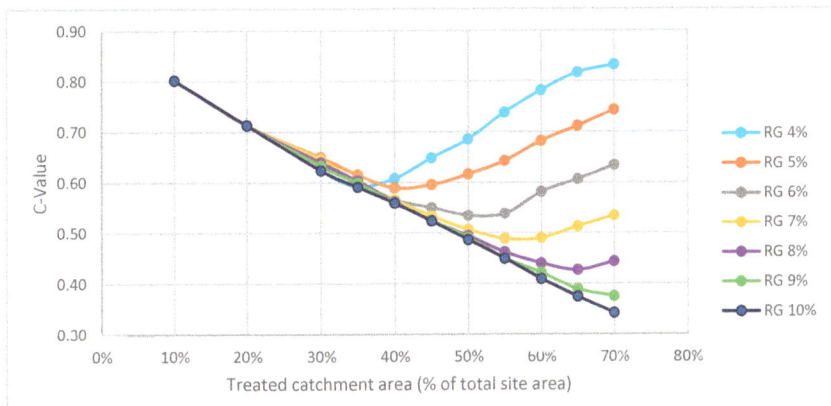

Figure 13. Residential precinct results of effective C-value versus treated catchment area for various rain garden sizes for the separate optimisation study.

5. Discussion

The results prove that the innovative design implemented for the rain gardens and gravel swales at Waterway Ridges, whereby the features are coupled with underground gravel layers for storage and orifice outlets for restricting flow, is indeed workable. These two key design elements (i.e., the gravel storage and orifice outlets) significantly improve the runoff control effectiveness of rain gardens and gravel swales, as can be seen when the precinct scenario with the typical design of ABC Waters design features implemented in Singapore (i.e., without gravel storage and orifice outlets) was only able to achieve an effective C-value of 0.77. By being able to meet the runoff control requirements in Singapore, the new design allows ABC Waters design features to be considered as another effective option in reducing peak flow, in addition to other structural measures such as underground detention tanks.

Upon comparing the performance of ABC Waters design features between smaller and bigger storm events, it was found that ABC Waters design features were more effective in reducing peak flow for the smaller storm events. This was substantiated by peak flow reduction comparisons of the precinct for the 3-month and 10-year design storms, which showed that there is a larger peak flow reduction for the 3-month design storm (at 46.7%) as compared to the 10-year design storm (at 32.6%). While this was expected for typical rain gardens (since smaller storm events would result in less surface overflow from rain gardens as compared to larger storm events), it was a somewhat surprising observation considering the design of the rain gardens and gravel swales in the precinct which are coupled with underground gravel storage and orifice outlets.

Meanwhile, the precinct optimisation results show that the effectiveness of the ABC Waters design features in reducing peak flow and in complying with the runoff control requirements for Singapore can be enhanced. The non-optimised ABC Waters design features in Waterway Ridges were able to reduce the effective C-value of the precinct to 0.60 for the 10-year design storm, a reduction of 32.6% in peak flow when compared to the scenario of the precinct without any treatment or detention features. The peak flow reduction performance of the precinct improved considerably when the orifice outlet sizes were optimised using evolutionary algorithm, resulting in an effective C-value of 0.53 which is 11.7% better than the non-optimised C-value of 0.60 obtained from the base precinct model, at no increase in cost. More importantly, this optimised C-value is lower than the 0.55 stipulated by the runoff control legislation in Singapore. The effective C-value can also be lowered slightly below 0.53 when the gravel storage layer depths were optimised together with the orifice outlet sizes, albeit at an additional cost of S$1610 (~$1180) or 3.1% more (as compared to the cost of gravel layer and orifice outlets in the original base precinct).

The study on optimum rain garden area and treated catchment area (that would allow catchments to achieve C = 0.55 for the 10-year design storm) found that a minimum rain garden size of 6% the total site area, with treated catchment areas of 45.2–47.2% depending on land use, would be required. The recommended rain garden size is only slightly higher than the 4–5% (of equivalent impervious catchment area) which is recommended by PUB [8] in order to meet the storm water quality objectives for Singapore [4]. The 6% is also comparable to the results of Waterway Ridges with optimised orifice outlets, whereby a C-value of 0.53 was obtained with an ABC Waters design feature area (excluding batter slopes for the rain gardens) of 5.4% the total precinct area.

Unlike underground detention tanks, ABC Waters design features (which are natural and sustainable) provide other ecosystem services such as the enhancement of biodiversity and improvements in runoff water quality [4]. The combined design of rain gardens and vegetated swales with underground gravel storage layers also reduces the risk of drowning and mosquito nuisance which are present in open detention systems. The incorporation of green infrastructure in urban areas further adds monetary value to a precinct as there is a willingness by the public to pay more for properties in precincts with green infrastructure; for example, a survey of Australian households reveal that buyers are willing to pay an additional $36,000–54,000 per property in streets with rain gardens [33].

The innovative design of the rain gardens and gravel swales (which are coupled with underground gravel layers for storage and orifice outlets to restrict flow) can also be modified to effectively reduce peak flow in countries beyond Singapore. This would apply in particular to countries with the same tropical climate, many of which are still developing. This design is also relevant for temperate regions in Europe, where the usage of Sustainable urban Drainage Systems (SuDS) is prevalent. The findings of this paper can be used to highlight ABC Waters design features as a viable sustainable option to governments of countries where the extensive use of such features have previously not been considered for storm water management.

6. Conclusions

This research was aimed at determining the effectiveness of Singapore's ABC Waters design features in reducing peak flow during storm events. Hydraulic performance assessment of the innovative design of ABC Waters design features was performed using a 1D SWMM model setup of the Waterway Ridges pilot precinct. The analysis revealed that ABC Waters design features are effective in reducing peak flow and runoff coefficient of the precinct during storm events. The effectiveness is attributed to an innovative design, which comprises the coupling of rain gardens and vegetated swales with underground gravel layers for storage and orifice outlets to restrict outflow. The reduction in peak flow (and effective runoff coefficient i.e., C-value) is 33% and 47% for the 10-year and 3-month design storms, respectively, when compared to the scenario whereby no treatment or detention is performed. When compared with typical designs adopted for the ABC Waters design features, there is also an improvement in peak flow reduction by 23% and 16%, respectively, for the 10-year and 3-month design storms.

The performance of the hybrid ABC Waters design features in terms of reducing the effective C-value of the precinct can also be optimized through design parameters such as orifice outlet diameter and gravel layer thickness; such an optimisation was performed on the precinct which resulted in a further reduction of the C-value by 12%. The reduction of effective C-value via the use of ABC Waters design features can also be improved by optimising the treated catchment area channelled to the feature. In addition to the decrease in peak flow and reduction in runoff coefficient, the innovative design reduces the risk of drowning as well as mosquito breeding when compared against structural detention measures and open detention systems, and also exhibits the other benefits of ABC Waters design features such as in providing ecosystem services and enhancing biodiversity.

It is acknowledged that the results obtained are preliminary and that calibration of the models will need to be carried out to provide validated results; this will be performed when monitoring results become available. Nevertheless, the preliminary results are very encouraging in promoting the use of the modified ABC Waters design features for peak flow and C-value reduction in Singapore and other tropical countries, as well as in the temperate regions of the world where usage of SuDS/LID/WSUD are prevalent.

Acknowledgments: The authors would like to thank IHE Delft Institute for Water Education for covering the costs to publish in open access. The authors would also like to thank SURF Cooperative for providing the opportunity for part of the research to be carried out in a significantly more efficient manner on the Dutch national e-infrastructure.

Author Contributions: The paper was written by Wing Ken Yau based on the results obtained from his research, which was done for the partial fulfilment of requirements for the Master of Science degree at the IHE Delft Institute for Water Education, under the close supervision of his mentors Assela Pathirana and Shie-Yui Liong, as well as his supervisor Chris Zevenbergen. Mohanasundar Radhakrishnan provided timely feedback on the research throughout the research study. All co-authors also helped to review and improve the paper.

Conflicts of Interest: The authors declare no conflict of interest.

Appendix A. Optimisation Study on Waterway Ridges

Appendix A.1. Methodology of Optimization Study

The SWMM5-EA software that applies evolutionary algorithms to optimize drainage/sewerage networks setup in SWMM v5 [34] was used for the optimisation study of the 1D base precinct model.

From the 1D hydrodynamic model, it had been determined that the key design elements affecting the peak flow reduction performance of the precinct are the orifice outlet diameters and the gravel storage layer depths.

With this knowledge, the optimisation was performed for two scenarios, namely (1) Single objective optimisation of the resultant C-value of the precinct by varying the diameter of the orifice outlets; and (2) Multi-objective optimisation of the C-value and cost of gravel layer cum orifice outlets by varying the gravel storage layer depths and diameter of the orifice outlets. Both scenarios are minimisation problems with the constraint that there should be no flooding in the precinct. For Scenario 2, it was assumed that the invert levels of the downstream conduits and drainage channels would be able to match with those optimised gravel layer depths which are deeper than their original depths. In addition, excavation costs were assumed to be constant, and the cost of orifice outlets was assumed to be the same regardless of the orifice size.

To achieve the desired optimisation, the feature of SWMM5-EA allowing modifications to be made to the software (via the introduction of a Python script file named 'swmm5ec_custom.py' in the 'customcode' folder under the SWMM5-EA installation directory) was evoked [34]. Python is a free and open source programming language developed by Python Software Foundation, USA. In particular, customisation had to be made to the "getFitness" function (in the Python script file 'swmm5ec.py'), which defines how the fitness for each individual or phenotype is computed.

For both cases, a population size of $10 \times D$ as recommended by Storn [35], where D is the number of variables being optimised, was adopted. The values ranges of the orifice outlets and gravel storage depths to be optimised are as shown in Table A1. The flooding constraint was also imposed on both scenarios via a flood cost function. For Scenario 1, an arbitrary flood cost function of the total flood volume multiplied by 10^{12} was adopted as the penalty function, and the flooding cost was added to the C-value to define the fitness for each phenotype. Meanwhile, the flooding penalty was imposed on the calculated C-value for the respective phenotypes in Scenario 2; a value of 1 would be assigned to the C-value (instead of the actual calculated value) if flooding is present. Table A2 summarises the optimisation parameters adopted in SWMM5-EA for both scenarios.

Table A1. Value ranges for the optimisation of the orifice outlet and gravel storage depth.

Variable	Value Ranges	
	Minimum	Maximum
Orifice outlet	0.01 m	Diameter of outlet conduit
Gravel storage depth	0.25 m	0.8 m

Table A2. Optimisation parameters for the optimisation of (A) orifice outlet diameter and (B) gravel storage depth and orifice outlet diameter.

Parameter/Case	(A)	(B)
Optimisation type	Single-objective optimisation of effective C-value	Multi-objective optimisation of effective C-value and cost of gravel storage layer cum orifice outlets
Variable to be optimised	Orifice outlet diameters	Gravel storage depth and orifice outlet diameters
Number of variables	18	40
Flood cost function	10^{12} × total flood volume	Imposed on C-value, which becomes 1 when there is flooding
Function for cost of gravel layer [1] and orifice outlets [2]	N.A.	S$55 × (total storage volume)/0.4 × 0.6 + 10 × (number of orifice)
Population size	180	400
Crossover rate	0.9	0.9
Mutation rate	0.2	0.2
Number of evaluations	18,000	200,000
Number of generations	100	500

Notes: [1] Cost of S$55/m^3 (~$40/m^3) for the gravel layer was referenced from the Construction Supplementary Schedule of Rates for the Waterway Ridges project. The design void ratio of 0.4 for the gravel layer was assumed in the cost calculations. [2] Cost of S$10/orifice outlet (~$7/orifice outlet) was assumed as no cost breakdown information was available.

Appendix A.2. Results of Orifice Outlet Optimisation

Results of the single objective optimisation of minimising the C-value of the precinct by varying the orifice outlet sizes with flooding as a constraint are as shown in Figure A1. The lowest C-value obtained after 100 iterations was 0.5298. The optimised C-value is about 11.7% better than the non-optimised C-value of 0.60 with no increase in cost (since the gravel storage layer depths are the same) but more significantly, the optimised C-value is lower than the 0.55 that is stipulated by the runoff control legislation in Singapore. This shows that ABC Waters design features that are coupled with gravel storage and orifice outlets can indeed be effective in reducing peak flow and in meeting the runoff control requirements of Singapore.

The comparisons of the gravel storage utilisation (of the various rain gardens and gravel swales) for the optimised model and the non-optimised base precinct model are as summarised in Table A3. It can be seen that the number of gravel storage units with 100% utilisation has more than doubled from five to 12 when compared to the original precinct model, with the average maximum utilisation also increasing from 65.5% to 78.1%. These results prove the earlier deduction in Section 3.4 that the gravel storage units were sized adequately but were not fully utilised due to non-optimal design of the orifice outlets.

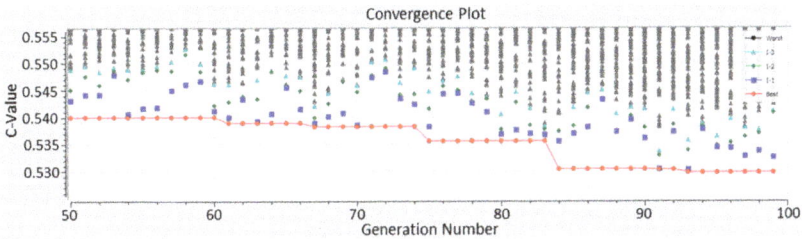

Figure A1. Results of the single objective optimisation of minimising C-value of the precinct by varying the orifice outlet sizes.

Table A3. Comparison of gravel storage utilisations for the optimised precinct model and base precinct model.

Scenario	No. of Gravel Storage Units With 100% Utilisation	Average Maximum Utilization for Gravel Storage Units
Non-optimised base precinct model	5	65.5%
Precinct model with optimised orifice outlets	12	78.1%

Appendix A.3. Results of Orifice Outlet and Gravel Storage Optimisation

The pareto plot for the multi-objective optimisation of the cost of gravel layer cum orifice outlets and effective runoff coefficient of the precinct are as presented in Figure A2. Also included in the plot are the results of the original base precinct, and the precinct with optimised orifice outlets (which also lie on the pareto layer as the single-objective optimisation is a subset of the multi-objective one).

From the results, it can be seen that at an increased low cost of 3.1% (S$1610) (~$1180), the effective C-value can be lowered to 0.5276 which is an improvement of 12.1% as compared to the original precinct. It was noted that the lowest C-value (of 0.5276) obtained for the multi-objective optimisation (by varying both the gravel storage layer depths and orifice outlets) is only 0.4% better than the lowest C-value (of 0.5298) obtained for the single-objective optimisation (by varying the orifice outlets alone). This comparison reveals to us that the gravel storage units are more or less optimised (in terms of their sizing and cost) and reaffirms our earlier deduction that the orifice outlets in the original precinct are not sized optimally. Nevertheless, the optimisation results reveal to us that the efficiency and effectiveness of the original precinct can be very much improved to yield more cost-effective designs with lower effective runoff coefficients.

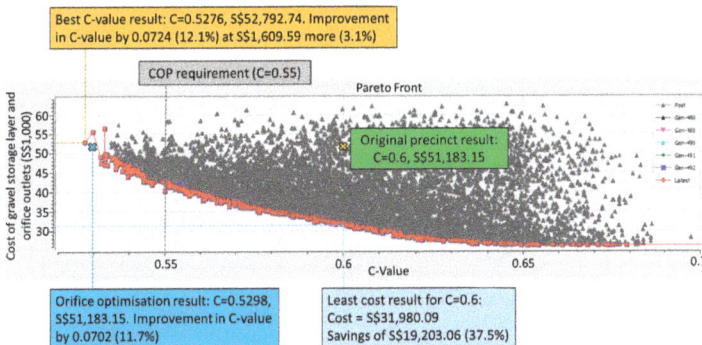

Figure A2. Pareto plot of the multi-objective optimisation of the cost of gravel layer cum orifice outlets and effective C-value of the precinct.

Appendix B. Study of Optimum Rain Garden Area and Treated Catchment Area

Appendix B.1. Scope of Study

The aim of the study was to determine if there is an optimum rain garden area and treated catchment area (in terms of percentage of total site area) that would allow catchments to achieve C = 0.55 for the 10-year design storm. As the study was intended to provide design recommendations pertaining to precinct scale catchments, a catchment area of 4 ha was adopted for the study and three land use types were considered; residential, commercial and mixed (50% residential–50% commercial).

Appendix B.2. Model Setup

An arbitrary model comprising of the treated and untreated parts of a 4-ha catchment was set up using SWMM v5.1. As illustrated in Figure A3, runoff from the treated parts of the catchment will be channelled into a lumped rain garden, whose filtrate and overflow are then channelled into the gravel storage and overflow manhole respectively. Similarly to the design adopted in Waterway Ridges, the gravel storage is connected to the overflow manhole via a perforated pipe (represented by an orifice) and there are two outlets from the overflow manhole, namely the overflow standpipe and the orifice outlet (both of which are also represented as orifices). In the model, the treated parts of the catchment are also split into four equal parts, each of which has its own rain garden and gravel storage areas that are sized equally.

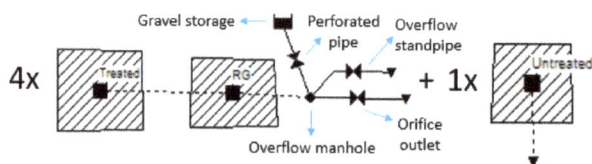

Figure A3. Arbitrary SWMM model setup to study the optimum rain garden area and treated catchment area to achieve C = 0.55.

The runoff coefficients adopted for the various land uses were based on a study on the runoff coefficients for Singapore, which found that runoff coefficients are sensitive to the slope of a catchment as well as the land use [36]. The study by Goh et al. [36] improved upon the runoff coefficient recommendations for various land use types in Singapore's COP on Surface Water Drainage [7] and developed runoff coefficients for different land use types that also took into account the slope of the catchment. The runoff coefficients and percentage imperviousness adopted for the various land uses in our study are as summarised in Table A4. It was noted that the slope of 3% adopted for the catchments in our study is in the "rolling (2–7%)" category by the Florida Department of Transportation, with the two other categories being "flat" (0–2%) and "steep" (>7%) [37].

Table A4. Runoff coefficients and percentage imperviousness adopted for the various land uses.

Land Use Type	Catchment Slope	C-Value	Percentage Imperviousness [1]
Residential		0.88	78%
Commercial	3%	0.97	95%
Mixed (50% residential–50% commercial)		0.925	86%

Notes: [1] Back-calculated using the Rational Formula (Equation (1)) with C-values of 1 and 0.45 being assumed for the impervious and pervious areas, respectively.

Meanwhile, Table A5 summarises the areas of the treated and untreated parts of the catchment that were studied. The areas of the rain gardens and the corresponding gravel storage volume adopted for the study are included in Table A6. An assumption made was that the percentage imperviousness of the treated catchment remains the same regardless of rain garden size. It was felt that this is a reasonable assumption given that the rain garden (which is 100% pervious) is likely to take up areas which were previously occupied by landscaping.

Table A5. Areas of the treated and untreated parts of the catchment that were studied.

Treated Area (% of Total Site Area)	10%	20%	30%	35%	40%	45%	50%	55%	60%	65%	70%
Total treated area (ha)	0.4	0.8	1.2	1.4	1.6	1.8	2.0	2.2	2.4	2.6	2.8
Treated area for each of the four catchments (ha)	0.10	0.20	0.30	0.35	0.40	0.45	0.50	0.55	0.60	0.65	0.70
Untreated area (ha)	3.6	3.2	2.8	2.6	2.4	2.2	2.0	1.8	1.6	1.4	1.2

Table A6. Areas of the rain gardens and the corresponding gravel storage volumes adopted for the study.

Rain Garden Area (% of Total Site Area)	4%	5%	6%	7%	8%	9%	10%
Total rain garden area (ha)	0.16	0.20	0.24	0.28	0.32	0.36	0.40
Total gravel layer storage volume (m^3)	320	400	480	560	640	720	800
Area for each of the four rain gardens (ha)	0.04	0.05	0.06	0.07	0.08	0.09	0.1
Gravel layer storage volume for each of the four rain gardens (m^3)	80	100	120	140	160	180	200
Storage curve area for each of the four rain gardens (m^2)	160	200	240	280	320	360	400

Appendix B.3. Optimisation of Orifice Outlets and Assessment

As the optimum sizes of the orifice outlets (resulting in the lowest peak runoff from the catchments) would vary depending on the size of the treated and untreated parts of the catchment as well as the rain garden, SWMM5-EA was used to optimise the diameter of the orifice outlets under the 10-year design storm for Singapore, which was also used to assess the resultant C-value of each case. A single objective optimisation of the C-value was performed for each case with the optimisation parameters as detailed in Table A7. Similarly to the optimisation study in Appendix A, customisation had to be made to the 'getFitness' function in the Python script file 'swmm5ec.py' in SWMM5-EA [34].

Table A7. Optimisation parameters adopted for the sizing of the orifice outlets for the study.

Parameter	Value
Variable to be optimised	Orifice outlet diameter
Number of variables	1
Flood cost function	10^{12} × total flood volume
Population size	10
Crossover rate	0.9
Mutation rate	0.2
Number of evaluations	300
Number of generations	30

Appendix B.4. Results

Appendix B.4.1. Residential Precinct (C = 0.88)

The study results for the residential precinct are as summarised in Figure A4, whereby the effective C-values are plotted against the treated catchment area for the various rain garden sizes. It can be observed that there is an optimum treated catchment area that would result in the lowest possible C-value for each rain garden size; the optimum has been attained by rain garden sizes of 4–8% but not 9–10% (within the treated catchment area range of 10–70% studied). This optimum treated area and corresponding C-value increases and decreases, respectively, as the rain garden area is increased; the optimum treated area and corresponding C-value for the various rain garden areas are as summarised in Table A8.

While increasing the treated area up to the optimum for each rain garden size would result in better peak flow reduction performance (i.e., lowering of C-values), the performance was found to deteriorate (i.e., C-value increases) when treated areas are increased beyond the optimum. From this, we can conclude that routing more runoff to a rain garden (i.e., having bigger treated catchment areas) may not always result in improved performance. This is especially true for the smaller rain garden sizes of 4–5% whose optimum treated areas are as low as 35–40%.

From the results, it can also be seen that at low treated catchment areas of 35% or less, increasing the rain garden size has either little or no effect on the C-values. For treated catchment areas of 30–35%, the C-values increase slightly before decreasing back to similar values as the rain garden areas are increased from 4% to 10%. These results are substantiated by an alternate representation of Figure A4, whereby C-values are plotted against rain garden sizes for the various treated catchment areas of 10–35%, as shown in Figure A5. Thus, for limited treated catchment areas of 10–35%, a small rain garden can perform as well as a bigger one and should be adopted if peak flow reduction is the only objective to be achieved by the rain garden.

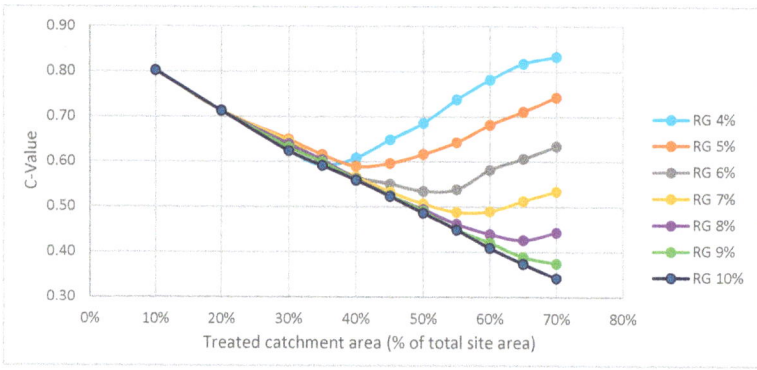

Figure A4. Residential precinct results of effective C-value versus treated catchment area for various rain garden sizes.

Table A8. Residential precinct results of optimum treated areas and C-values for various rain garden sizes.

Rain Garden Area (% of Total Site Area)	4%	5%	6%	7%	8%
Optimum treated area (% of total site area)	35%	40%	50%	57.5%	65%
Corresponding optimum effective C-value	0.592	0.589	0.535	0.488	0.427

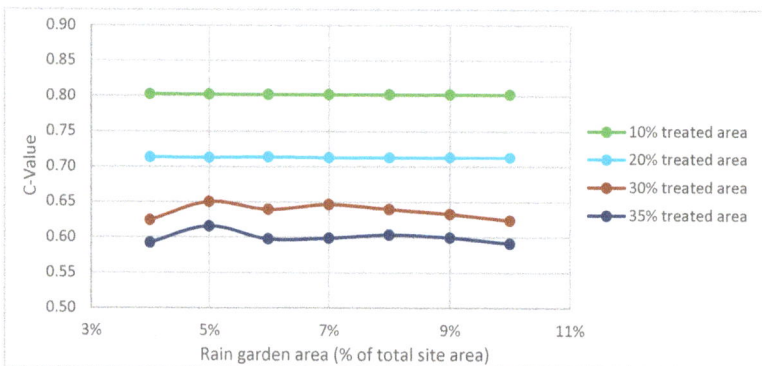

Figure A5. Residential precinct plot of effective runoff coefficient versus rain garden size for treated catchment areas of 10–35%.

Appendix B.4.2. Commercial Precinct (C = 0.97)

The study results for the commercial precinct are as shown in Figures A6 and A7, and Table A9. From the results, it can be observed that the trends as earlier seen for the residential precinct also apply to the commercial precinct. Generally, the effective C-values for the commercial precinct for the various treated catchment areas and rain garden sizes are also larger than the corresponding C-values for the residential precinct, which is expected given that the commercial precinct has a higher runoff coefficient of 0.93, as compared to 0.88 for the residential precinct.

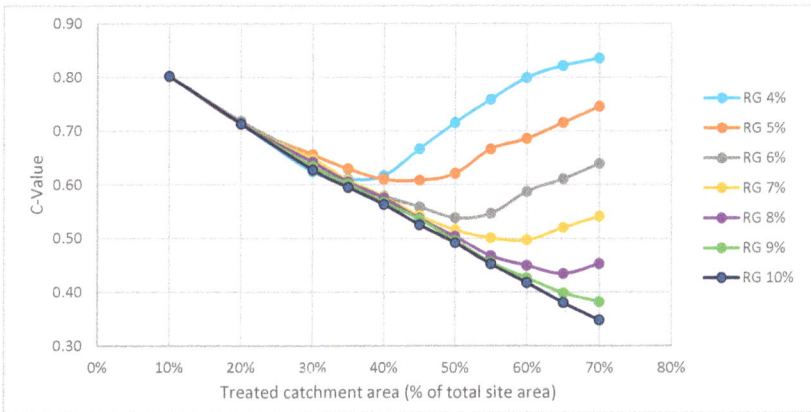

Figure A6. Commercial precinct results of effective C-value versus treated catchment area for various rain garden sizes.

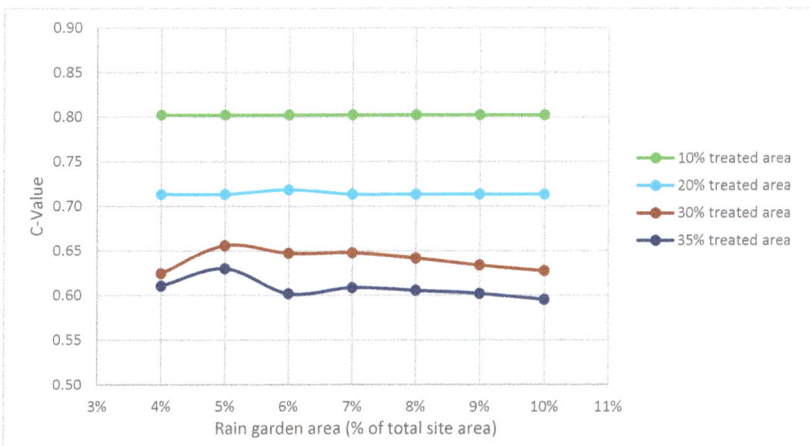

Figure A7. Commercial precinct plot of effective runoff coefficient versus rain garden size for treated catchment areas of 10–35%.

Table A9. Commercial precinct results of optimum treated areas and C-values for various rain garden sizes.

Rain Garden Area (% of Total Site Area)	4%	5%	6%	7%	8%
Optimum treated area (% of total site area)	35%	45%	50%	60%	65%
Corresponding optimum effective C-value	0.610	0.608	0.539	0.497	0.434
Increase in optimum effective C-value compared to residential precinct	0.018	0.019	0.004	0.009	0.008

Appendix B.4.3. Mixed Residential–Commercial Precinct (C = 0.925)

The study results for the mixed precinct with a runoff coefficient of 0.925, which is in between that of the residential and commercial precincts, are as shown in Figures A8 and A9, and Table A10. The same trends as earlier observed for the residential and commercial precincts also apply to the mixed precinct. Generally, the effective C-values for the mixed precinct for the various treated catchment areas and rain garden sizes are also larger than the corresponding C-values of the residential precinct but smaller than those of the commercial precinct.

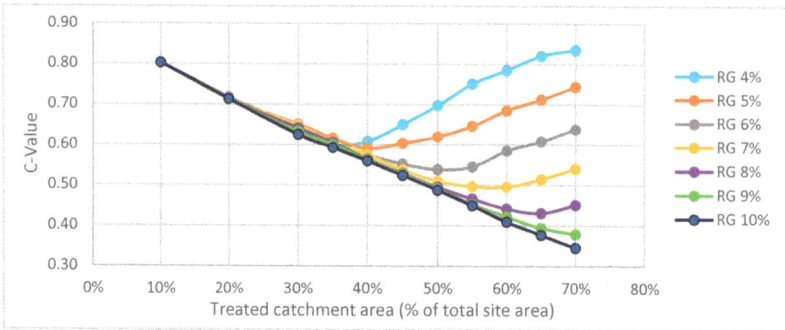

Figure A8. Mixed precinct results of effective C-value versus treated catchment area for various rain garden sizes.

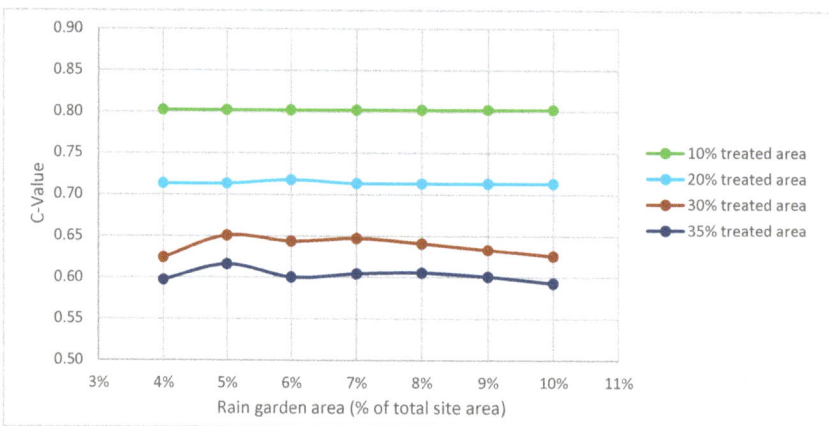

Figure A9. Mixed precinct plot of effective runoff coefficient versus rain garden size for treated catchment areas of 10–35%.

Table A10. Mixed precinct results of optimum treated areas and C-values for various rain garden sizes.

Rain Garden Area (% of Total Site Area)	4%	5%	6%	7%	8%
Optimum treated area (% of total site area)	35%	40%	50%	60%	65%
Corresponding optimum effective C-value	0.597	0.590	0.539	0.496	0.431
Increase in optimum effective C-value compared to residential precinct	0.005	0.001	0.004	0.007	0.004

Appendix B.4.4. Recommended Treated Catchment Areas and Rain Garden Sizes for Achieving Effective C = 0.55

The optimum treated catchment areas and rain garden sizes for meeting the runoff control requirements of Singapore's COP of Surface Water Drainage are as summarised in Table A11 for the three land use types studied.

Table A11. Required treated catchment areas and rain garden sizes to achieve C = 0.55 for the various land use types (percentages in brackets represents the upper limits beyond which the C-value would exceed 0.55 due to deteriorating performance of the rain garden).

Land Use Type	Rain Garden Area (% of Total Site Area)						
	4%	5%	6%	7%	8%	9%	10%
Residential (C = 0.88)	N.A.	N.A.	45.2% (56.4%)	42.4% (74.4%)	41.6% (87.8%)	41.5% (87.8%)	41.3% (87.8%)
Mixed residential–commercial (C = 0.925)	N.A.	N.A.	45.8% (55.0%)	43.3% (72.2%)	41.9% (87.4%)	41.8% (87.4%)	41.4% (87.4%)
Commercial (C = 0.97)	N.A.	N.A.	47.2% (55.0%)	43.9% (72.1%)	43.3% (85.5%)	42.8% (85.5%)	41.8% (85.5%)

From the results, it can be seen that a minimum rain garden size of 6% (of the total site area) is needed for all three land use types, with the corresponding required (lower bound) treated catchment areas of 45.2%, 45.8% and 47.2% for the residential, mixed and commercial land use types, respectively. There is thus an increase in the required treated catchment area (for a particular rain garden size) when the runoff coefficient of the land use type increases. The required treated catchment area also decreases with increasing rain garden size for all three land use types. In other words, less runoff needs to be routed to a bigger rain garden as compared to a smaller one to achieve the effective C-value of 0.55.

If there is any reason the treated catchment area needs to be increased beyond the lower bound percentage areas recommended (e.g., for runoff water quality treatment purposes), it needs to be noted that increasing the area beyond the upper limit as specified in brackets in Table A11 would cause the effective C-value to exceed 0.55 due to deteriorating runoff reduction performance; these upper limits are beyond the optimum treated catchment area for the various rain gardens. The upper limits for rain garden sizes of 7–8% were estimated using linear interpolation as they exceeded the treated catchment area range being studied. Also, the upper limits for rain garden sizes of 9–10% were not able to be determined as their optimum treated catchment area were not reached within the range of treated catchment areas studied, and were set to be equal to the upper limits of rain garden size 8% to be on the cautious side. Nevertheless, it is noted that the treated catchment area upper limits for rain garden sizes of 7% and beyond are quite high, and should typically not be exceeded by developments.

From this study, the rain garden sizes which are recommended to comply with the runoff control legislation are also deemed to be reasonable and practical. The recommended minimum rain garden size of 6% is only slightly higher than the 4–5% (of equivalent impervious catchment area) which is recommended by PUB's Engineering Procedures for ABC Waters Design Features in order to meet the storm water quality objectives for Singapore [4]. The 6% is also comparable to the results of Waterway Ridges with optimised orifice outlets, whereby a C-value of 0.5298 was obtained with an ABC Waters design feature area (excluding batter slopes for the rain gardens) of 5.4% the total precinct area.

It is noted that the recommended treated catchment areas ranging from 41.3% to 47.2% (depending on the rain garden size and land use type) are slightly higher than the 35% (of total site area) that will yield the maximum points for the criteria on "incorporation of ABC Waters design features to treat surface runoff from site" under PUB's ABC Water's Certification Scheme [38], a scheme that serves to "provide recognition to public agencies and private developers who embrace the ABC Waters concept and incorporate ABC Waters design features in their developments" [39].

References

1. Inter Governmental Panel on Climate Change. Summary for policymakers. In *Climate Change 2013: The Physical Science Basis*; Stocker, T.F., Qin, D., Plattner, G.-K., Tignor, M., Allen, S.K., Boschung, J., Nauels, A., Xia, Y., Bex, V., Midgley, P.M., Eds.; Contribution of Working Group I to the Fifth Assessment Report of the Intergovernmental Panel on Climate Change; IPCC: Geneva, Switzerland, 2013.
2. World Bank Group. Climate Change Knowledge Portal. Available online: http://sdwebx.worldbank.org/climateportal/index.cfm?page=country_historical_climate&ThisRegion=Asia&ThisCCode=SGP (accessed on 6 May 2017).
3. United Nations. *The World's Cities in 2016—Data Booklet*; ST/ESA/SER.A/392; Department of Economic and Social Affairs, Population Division: New York, NY, USA, 2016.
4. PUB Singapore's National Water Agency. *Active, Beautiful, Clean Waters Design Guidelines*; PUB Singapore's National Water Agency: Singapore, 2014.
5. Fletcher, T.D.; Shuster, W.; Hunt, W.F.; Ashley, R.; Butler, D.; Arthur, S.; Trowsdale, S.; Barraud, S.; Semadeni-Davies, A.; Bertrand-Krajewski, J.-L.; et al. SUDS, LID, BMPS, WSUD and more—The evolution and application of terminology surrounding urban drainage. *Urban Water J.* **2015**, *12*, 525–542. [CrossRef]
6. PUB Singapore's National Water Agency. *Managing Urban Runoff—Drainage Handbook*, 1st ed.; Singapore's National Water Agency and Institution of Engineers: Singapore, 2013.
7. PUB Singapore's National Water Agency. *Code of Practice on Surface Water Drainage*, 6th ed.; Addendum No. 1: Jun 2013; PUB Singapore's National Water Agency: Singapore, 2011. Available online: http://www.pub.gov.sg/general/Documents/CP2013/COP_Final.pdf (accessed on 11 September 2014).
8. PUB Singapore's National Water Agency. *Engineering Procedures for ABC Waters Design Features*; PUB Singapore's National Water Agency: Singapore, 2014.
9. Zhou, Q.; Panduro, T.E.; Thorsen, B.J.; Arnbjerg-Nielsen, K. Adaption to Extreme Rainfall with Open Urban Drainage System: An Integrated Hydrological Cost-Benefit Analysis. *Environ. Manag.* **2013**, *51*, 586–601. [CrossRef] [PubMed]
10. Wang, W.; Zhao, Z.; Qin, H. Hydrological effect assessment of low impact development for urbanized area based on SWMM. *Acta Sci. Nat. Univ. Pekin.* **2012**, *48*, 303–309.
11. Jia, H.; Lu, Y.; Shaw, L.Y.; Chen, Y. Planning of LID–BMPs for urban runoff control: The case of Beijing Olympic Village. *Sep. Purif. Technol.* **2012**, *84*, 112–119. [CrossRef]
12. Tang, Y. Study of urban stormwater runoff BMPs Planning with support of SUSTAIN system. In *School of Environment*; Tsinghua University: Beijing, China, 2010.
13. Mugume, S.N.; Melville-Shreeve, P.; Gomez, D.; Butler, D. Multifunctional urban flood resilience enhancement strategies. *Proc. Inst. Civ. Eng. Water Manag.* **2016**, *169*, 115–127. [CrossRef]
14. Jato-Espino, D.; Charlesworth, S.; Bayon, J.; Warwick, F. Rainfall–Runoff Simulations to Assess the Potential of SuDS for Mitigating Flooding in Highly Urbanized Catchments. *Int. J. Environ. Res. Public Health* **2016**, *13*, 149. [CrossRef] [PubMed]
15. Dietz, M.E.; Clausen, J.C. A Field Evaluation of Rain Garden Flow and Pollutant Treatment. *Water Air Soil Pollut.* **2005**, *167*, 123–138. [CrossRef]
16. Davis, A.P. Field Performance of Bioretention: Hydrology Impacts. *J. Hydrol. Eng.* **2008**, *13*, 90–95. [CrossRef]
17. Hunt, W.F.; Smith, J.T.; Jadlocki, S.J.; Hathaway, J.M.; Eubanks, P.R. Pollutant Removal and Peak Flow Mitigation by a Bioretention Cell in Urban Charlotte, N.C. *J. Environ. Eng.* **2008**, *134*, 403–408. [CrossRef]
18. Winston, R.J.; Dorsey, J.D.; Hunt, W.F. Quantifying volume reduction and peak flow mitigation for three bioretention cells in clay soils in northeast Ohio. *Sci. Total Environ.* **2016**, *553*, 83–95. [CrossRef] [PubMed]
19. Davis, A.P.; Hunt, W.F.; Traver, R.G.; Clar, M. Bioretention Technology: Overview of Current Practice and Future Needs. *J. Environ. Eng.* **2009**, *135*, 109–117. [CrossRef]

20. Texas A&M Agrilife Extension. Texas A&M AgriLife Ecological Engineering Program. Available online: http://agrilife.org/lid/projects/lid/rain-garden-and-detention-pond (accessed on 2 July 2017).

21. Liu, J.; Sample, D.; Bell, C.; Guan, Y. Review and Research Needs of Bioretention Used for the Treatment of Urban Stormwater. *Water* **2014**, *6*, 1069–1099. [CrossRef]

22. Hatt, B.E.; Fletcher, T.D.; Deletic, A. Hydrologic and pollutant removal performance of stormwater biofiltration systems at the field scale. *J. Hydrol.* **2009**, *365*, 310–321. [CrossRef]

23. Lucke, T.; Nichols, P.W.B. The pollution removal and stormwater reduction performance of street-side bioretention basins after ten years in operation. *Sci. Total Environ.* **2015**, *536*, 784–792. [CrossRef] [PubMed]

24. Atelier Dreiseitl Asia Pte Ltd. *Hydraulic Calculation Report for Pilot ABC Waters Project at Punggol East c39 Housing Precinct and Common Green*; PUB Singapore's National Water Agency: Singapore, 2015.

25. Meteorological Service Singapore. Climate of Singapore. Available online: http://www.weather.gov.sg/climate-climate-of-singapore (accessed on 3 July 2017).

26. National Environment Agency. Weather Statistics. Available online: http://www.nea.gov.sg/weather-climate./climate/weather-statistics (accessed on 3 July 2017).

27. Guo, J.C.Y. Cap-Orifice as a Flow Regulator for Rain Garden Design. *J. Irrig. Drain. Eng.* **2012**, *138*, 198–202. [CrossRef]

28. United States Environmental Protection Agency. Storm Water Management Model (SWMM). 2014. Available online: http://www2.epa.gov/water-research/storm-water-management-model-swmm (accessed on 3 March 2015).

29. Gironás, J.; Roesner, L.A.; Davis, J.; Rossman, L.A.; Supply, W. *Storm Water Management Model Applications Manual*; National Risk Management Research Laboratory, Office of Research and Development, US Environmental Protection Agency: Cincinnati, OH, USA, 2009.

30. Aron, G.; Kibler, D.F. Pond Sizing for Rational Formula Hydrographs. *J. Am. Water Res. Assoc.* **1990**, *26*, 255–258. [CrossRef]

31. Shuy, E.B. *Derivation of Hydrological Curves for ABC Waters Design Features*; PUB Singapore's National Water Agency: Singapore, 2012.

32. Goh, X.P. Broader environmental impact of runoff control using SuDS in the context of urban development in Singapore. In *Water Science and Engineering*; UNESCO-IHE Institute for Water Education: Delft, The Netherlands, 2015.

33. Australian Government. *Stakeholder Annual Report*; Cooperative Research Centre for Water Sensitive Cities Ltd.: Clayton, Australia, 2017.

34. Pathirana, A. SWMM5-EA—A tool for learning optimization of urban drainage and sewerage systems with genetic algorithms. In Proceedings of the 11th Innternational Conference on Hydroinformatics, New York, NY, USA, 17–21 August 2014; CUNY Academic Works: New York, NY, USA, 2014.

35. Storn, R. On the usage of differential evolution for function optimization. Presented at the North American Fuzzy Information Processing Society, Berkeley, CA, USA, 19–22 June 1996.

36. Goh, X.P.; Radhakrishnan, M.; Pathirana, A.; Zevenbergen, C. Effectiveness of runoff control legislation and Active, Beautiful, Clean (ABC) Waters design features in Singapore. *Water* **2017**. submitted.

37. Poullain, J. *Estimating Storm Water Runoff*; PDH online; PDH Centre: Fairfax, VA, USA, 2012.

38. PUB Singapore's National Water Agency. Certification Critieria. 2017. Available online: https://www.pub.gov.sg/abcwaters/certification/criteria (accessed on 6 June 2017).

39. PUB Singapore's National Water Agency. Certification. 2017. Available online: https://www.pub.gov.sg/abcwaters/certification (accessed on 6 June 2017).

water

MDPI

Article

Effectiveness of Runoff Control Legislation and Active, Beautiful, Clean (ABC) Waters Design Features in Singapore

Xue Ping Goh [1], Mohanasundar Radhakrishnan [2,*], Chris Zevenbergen [2] and Assela Pathirana [2]

[1] PUB, Singapore's national water agency, 40 Scotts Road, Environment Building, #07-01, Singapore 228231, Singapore; adelinegohxp@gmail.com
[2] Water Science and Engineering Department, IHE Delft Institute for Water Education, 2611 AX Delft, The Netherlands; c.zevenbergen@un-ihe.org (C.Z.); a.pathirana@un-ihe.org (A.P.)
* Correspondence: m.radhakrishnan@un-ihe.org; Tel.: +31-647-618-689

Received: 9 June 2017; Accepted: 10 August 2017; Published: 22 August 2017

Abstract: Storm water management in Singapore has always been a challenge due to intense rainfall in a flat, low-lying and urbanised catchment. PUB's (Singapore's National Water Agency) recent runoff control regulation limits the runoff coefficient to 0.55 for developments larger than or equal to 0.2 ha. The use of Active, Beautiful, Clean (ABC) Waters design features are encouraged to attain peak runoff reduction. Hence the paper focuses on (i) determining the actual hydrological response regime of Singapore using the relationship between runoff coefficient (C), land use and slope; and (ii) investigating the effectiveness of ABC Waters design features in delaying and reducing peak runoff using a modelling approach. Based on a Storm Water Management Model (SWMM) model and using elevation, land use and soil data as inputs, the peak C-values were obtained for 50 m × 50 m grid cells. The results show that for the same land use, the one with steeper slope resulted in a higher runoff coefficient. Simulations were carried out in two study areas, Green Walk District and Tengah Subcatchment, where ABC Waters design features (such as porous pavements, green roofs, rain gardens) and detention tanks were incorporated to reduce C-values. Results showed that peak C-values can be reduced to less than 0.55 after increasing the green areas and constructing detention facilities. Reduction in peak discharge (22% to 63%) and a delay in peak discharge by up to 30 min were also observed. Hence, it is recommended to consider the relationship between slope and land use while determining runoff coefficients; and to incorporate ABC Waters design features in urban design to reduce the peak flow and runoff coefficient (C).

Keywords: ABC Waters; LID; SuDS; WSUD; water quantity control; runoff control; numerical simulation; sustainable storm water management

---

1. Introduction

Urbanisation results in hydrological changes such as increased overland runoff due to increased imperviousness [1]. The increase in runoff exacerbates the flood risk, which is expected to increase further due to climatic, land use and further changes [2]. A range of measures are being put in place to mitigate the effects of flooding or adapt to flooding [3]. Cities across the world are adapting to urban flooding using various adaptation measures comprising innovative drainage design approaches, policy measures, legislative controls, and so forth [4–6]. The risk of flooding is likely to affect the liveability and sustainability in densely populated cities like Singapore, which is highly urbanised but with a small land area of 718 km^2 [7]. Storm water management in Singapore is a challenge due to its geographical location, climatic regime and topography. As an island state, Singapore is surrounded by the sea and parts of the drainage system are influenced by tidal conditions. Located at the equator, Singapore receives abundant rainfall throughout the year. Solar radiation coupling with sea breeze convergence bring about short but very intense bursts of rainfall. With continued urbanisation, the water balance of the island is highly modified. Increasing the conveyance capacity of the storm water network by enlarging drains is difficult due to space constraints. Therefore Singapore's National Water Agency, PUB, adopts the multi-pronged source-pathway-receptor approach as its storm water management strategy.

Drainage design requirements and considerations in Singapore are specified in PUB's Code of Practice (COP) on Surface Water Drainage, where the rational formula has been recommended to compute peak runoff [8]. The runoff coefficient (C) is dependent on the degree and type of development within the catchment. Classification of catchments is based on their expected general characteristics when fully developed. The C values used in the COP consider only land use of catchment and are shown in Table 1.

Table 1. C-values indicated in PUB's Code of Practice on Surface Water Drainage [8].

Characteristics of Catchment When Fully Developed	Runoff Coefficient (C)
Roads, highways, airport runways, paved areas	1.00
Urban areas fully and closely built up	0.90
Residential/industrial areas densely built up	0.80
Residential/industrial areas not densely built up	0.65
Rural areas with fish ponds and vegetable gardens	0.45

Note: For catchments with composite land use or surface characteristics, a weighted value of C may be adopted. Singapore's National Water Agency (PUB).

In 2006, PUB launched the Active, Beautiful, Clean (ABC) Waters programme [9]. The concept of ABC Waters design features is similar to sustainable drainage systems (SuDS), water-sensitive urban drainage systems (WSUD), low impact developments (LID), best management practices (BMPs) and sponge cities that are prevalent in other countries [10]. Through holistic integration of the waterways and reservoirs with the surrounding environment, the ABC Waters programme aims to improve water quality, enhance the aesthetics of living environment and control the storm water runoff.

In addition to runoff control, some of these features provide a host of additional benefits such as providing recreational space, regulating ambient temperature through reducing urban heat island (UHI) [9]. Apart from enhancing liveability by moderating urban temperatures, reducing UHI may also minimise possible increases in convective thunderstorm activities which would otherwise result in increased rainfall [11].

Through the revised requirements in COP (Clause 7.1.5), PUB adopted the source control approach to reduce and delay runoff in the catchment. In 2014, it was enforced that "industrial, commercial, institutional and residential developments greater than or equal to 0.2 hectares in size are required to control the peak runoff discharged from the development sites. The maximum allowable peak runoff to be discharged to the public drains will be calculated based on a runoff coefficient of 0.55, and for design storms with a return period of 10 years and for various storm durations of up to 4 h (inclusive)" [8]. The peak runoff reduction is achieved through the implementation of ABC Waters Design features (e.g., porous pavements, wetlands, green roofs, planter boxes, bio retention swales, bio retention basins or rain gardens); and through structural detention/retention features (e.g., detention tanks, retention/sedimentation ponds) [9].

In addition to land use, the slope of the catchment plays an important role in determining the runoff. Ascertaining the hydrological response regime only with the land use without considering the slope is not ideal. This might lead to ineffective or inefficient design of drainage systems such as oversizing or under sizing of ABC Waters design features. The expert panel on drainage design and flood protection in Singapore also recommends—recommendation 4B-further research on C value [12]. Hence, this paper focuses on determining the actual hydrological response regime of Singapore using the relationship between coefficient value, slope and land use.

The combined effect and relationship between the land use and slope in Singapore are being ascertained through a hydrological response regime map. Furthermore, this paper also explores the effectiveness of using ABC Waters design features to reduce C values in Singapore.

2. Methodology

2.1. Hydrological Response Regime Map for Singapore

The presented research is divided into two stages. In the first stage, summarised in Figure 1, a hydrological response regime map for Singapore was determined. In the second stage the effectiveness of using ABC Waters design features to reduce C values was explored.

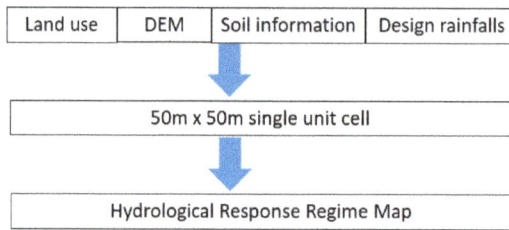

Land use	DEM	Soil information	Design rainfalls

50m x 50m single unit cell

Hydrological Response Regime Map

Figure 1. Methodology for determining the hydrological response regime. Digital elevation model (DEM).

As seen in Figure 1, the high resolution (0.25 m × 0.25 m) digital elevation model (DEM) was resampled to a 50 × 50 m percent slope map. The resampled resolution of 50 m × 50 m was selected to be consistent with the 0.2 ha development size indicated in PUB's regulation. The 32 land uses defined by Singapore's Master Plan 2014 was reclassified into 10 broader categories, such as agriculture, beach, cemetery, port/airport, road, water body, greens, business, residential and special reserve sites. Land use parameters for each land use type were identified based on literature review as well as the author's understanding of Singapore [13–16]. Soil properties in Singapore was assumed to be homogenous and belonged to hydrological soil group (HSG) D due to its small land area [16].

Based on the Singapore's rainfall intensity-duration-frequency (IDF) curves [17] and Chicago hyetograph method, 10 design rainfall time series (for durations 15, 20, 30, 60, 90, 120, 150, 180, 210, 240 min) of a 10-year return period were obtained. The accuracy of the results can be improved by adopting a shorter rainfall duration such as 2 or 5 min. As Singapore's COP allowed for a 15 min rainfall duration, the least rainfall duration was fixed at 15 min [8]. The hydrological modelling aspect of this research was carried out using version 5.0 of Storm Water Management Model (SWMM), an open source hydraulic model, developed and freely distributed online, by USEPA [18]. SWMM conceptualizes the subcatchment as a rectangular surface with uniform slope and calculates the flow based on non-linear reservoir routing method, with resistance formulated by manning's equation [19]. An example of inputs necessary to compute the runoff in a 50 m × 50 m single unit subcatchment on a SWMM interface is shown in Figure 2.The percent slope of a single unit subcatchment (indicated by the green dotted box in Figure 2) was computed from the DEM using ArcGIS 10.1, a commercial GIS software developed and supplied by ESRI, US. Land use parameters of the subcatchment—such as percent imperviousness, Manning's n of overland flow, depression storage depth and curve number (indicated by the red dotted boxes in Figure 2)—were determined based on the dominating land use of the single unit subcatchment. In addition to using the typical C values suggested by SWMM 5.0, design practices from different countries were also reviewed to identify the most representative C value for each land use parameter in Singapore. The following design practices were reviewed: (i) recommended percentage imperviousness values [13]; (ii) overland texture factor N [14], and; (iii) runoff curve numbers for urban areas [15,16]. With the 10 rainfall time series as upper boundary conditions, each single unit subcatchment produced 10 runoff values. The highest of the 10 runoff values, that is, the peak runoff

(m^3/s), was selected. The corresponding rainfall intensity at which this peak runoff occurred was termed the peak rainfall intensity (mm/h).

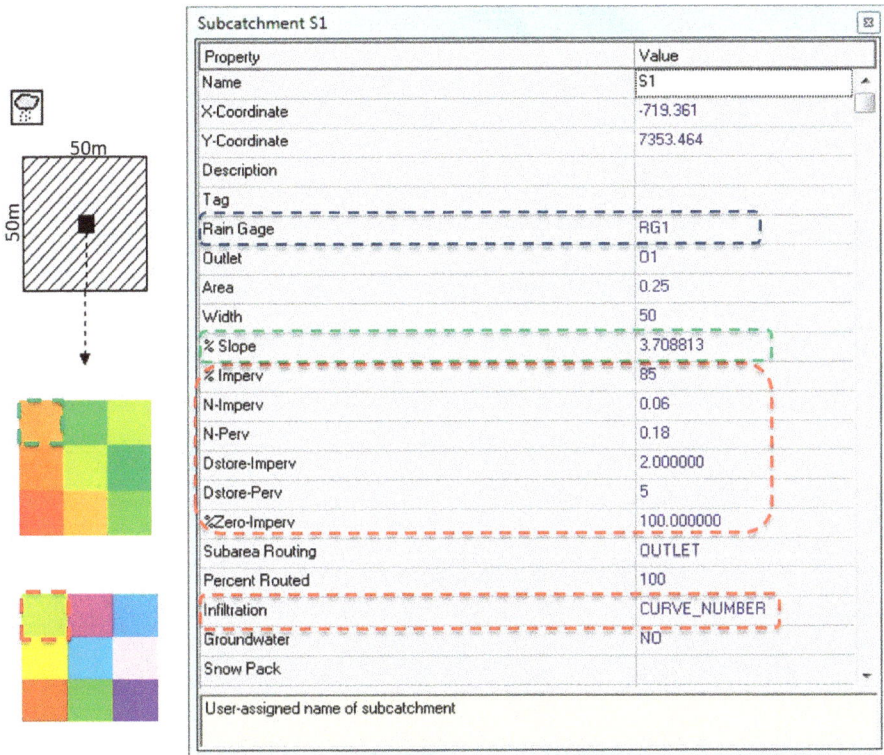

Figure 2. Inputs for a single unit subcatchment in the Storm Water Management Model (SWMM).

To calculate the peak C value of each single unit subcatchment, the following equations were used.

$$Peak\ runoff\ rate\ (mm/s) = \frac{Peak\ runoff\ \left(\frac{m3}{s}\right)}{Area\ of\ sub\ catchment\ (m2)} \times 1000 \qquad (1)$$

$$Peak\ rainfall\ intensity\ \left(\frac{mm}{s}\right) = \frac{Peak\ rainfall\ inetnsity\ \left(\frac{mm}{hr}\right)}{360} \qquad (2)$$

$$Peak\ runoff\ coefficient = \frac{Peak\ runoff\ rate\ \left(\frac{mm}{s}\right)}{Peak\ rainfall\ intensity\ \left(\frac{mm}{s}\right)} \qquad (3)$$

Every 50 × 50 m grid cell was then simulated for 10-year design rainstorms with varying durations. The maximum C value from these simulations were selected for further processing. In total, 554,441 single unit subcatchment cells are modelled to obtain peak runoff coefficients of the various land uses in entire Singapore's main island.

2.2. Effectiveness of ABC Features in Reducing C Value

Residential areas form the largest land use in Singapore and the urban town planners' emphasis for more green spaces housing estates. Hence, residential areas are ideal locations for testing the extensive use of ABC Waters design features as runoff control measures. Two hypothetical study areas,

Green Walk District (GWD) in Tampines North New Town and Tengah Subcatchment (TSC) in Tengah New Town, were identified for Stage 2 of the study. Figure 3 summarises the methodology for Stage 2, which aimed at determining the amount of ABC Waters design features needed to reduce C values to the stipulated 0.55.

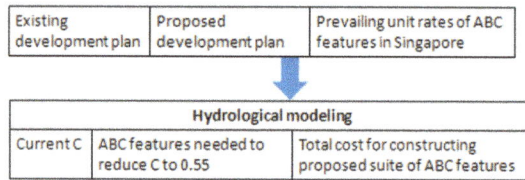

Existing development plan	Proposed development plan	Prevailing unit rates of ABC features in Singapore

Hydrological modeling		
Current C	ABC features needed to reduce C to 0.55	Total cost for constructing proposed suite of ABC features

Figure 3. Methodology for assessing the effectiveness of Active, Beautiful, Clean (ABC) features.

In Tampines North New Town a 'leaf concept' design has been planned, where a comprehensive network of green spaces in the estate are designed using features such as green canyons, elevated roof garden links, extensive landscaped decks and community living rooms [20,21]. However, this, study focussed only on GWD, which is adjacent Sungei Api Api channel (Figure 4).

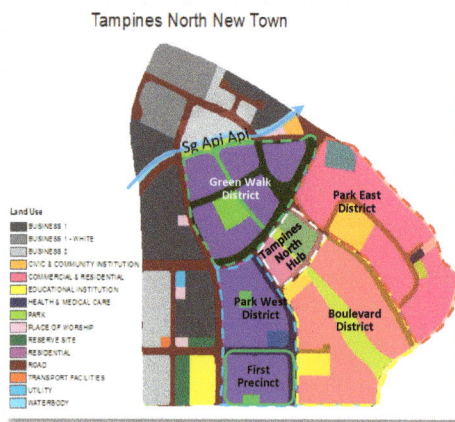

Figure 4. Proposed development of Tampines North New Town.

The second study area identified for application of ABC Waters design features in this research was a 10 ha subcatchment in Tengah New Town (TSC-Tengah Subcatchment), which is currently undeveloped but has the potential to become a high-income residential area. Hence an increase in paved areas is expected in TSC. In addition to the usual civic amenities—such as schools, residential and commercial building—urban farming has been planned in Tengah New Town [22].

In Stage 2 of the study, three ABC Waters design features (green roofs, rain gardens and porous pavements) and detention tanks were used to reduce the C values. Green roof is a low-maintenance vegetated roof system that uses a lightweight plant growing medium with shallow drainage/storage layer to store storm water. The shallow storage layer of the green roof helps in reducing and delaying the peak storm water runoff as compared to conventional rooftops. Rain garden is a vegetated land depression designed to detain and treat storm water runoff. Typically, subsoil pipes are installed in the drainage layer to discharge the filtered water to a nearby drain, thus reducing flow velocities and increasing the catchment's time of concentration. Due to similar design features such as storage layer and overflow drains, green infrastructure and green roofs were modelled as shallow bio retention cells in SWMM. Porous pavement allows storm water to pass through the voids of the paved surfaces and infiltrate into the subbase. The infiltrated water can be drained via full exfiltration, partial exfiltration or tanked systems. This helps to increase infiltration and reduce peak discharge into the public drainage system. Detention tanks are structural measures that temporarily store surface overflow during a heavy rainfall event and discharge the runoff via pumping after the rain event.

As the focus of this research was on overland runoff from subcatchments (including roads), the first priority of the conveyance system in the model was to drain the overland flow from the subcatchment through an outfall using open concrete channels. Actual site constraints were not considered when sizing the channels. ABC Waters design features were modelled using the low impact development (LID) controls module in SWMM (Figure 5). The ABC Waters design features were added into the model in the following sequence: green roofs, rain gardens, porous pavements and, lastly, detention tanks until each subcatchment's C value was reduced to 0.55 or lower. Detention tanks were the last to be added as the preference was to have more green features as there is a preference for green space compared to hard concrete structures. The coverage of each feature was maximised based on the permissible site condition and this varied by subcatchment. Additional details about the representation of ABC Waters design features as LID modules in SWMM is presented in Figures A1–A3 in Appendix A. The simulations were run for a 240 min duration rainfall with a peak intensity of 108.7 mm/h. This storm event was assumed as the worst-case scenario, where the prolonged rain would have saturated the soil, hence generating the maximum overland runoff.

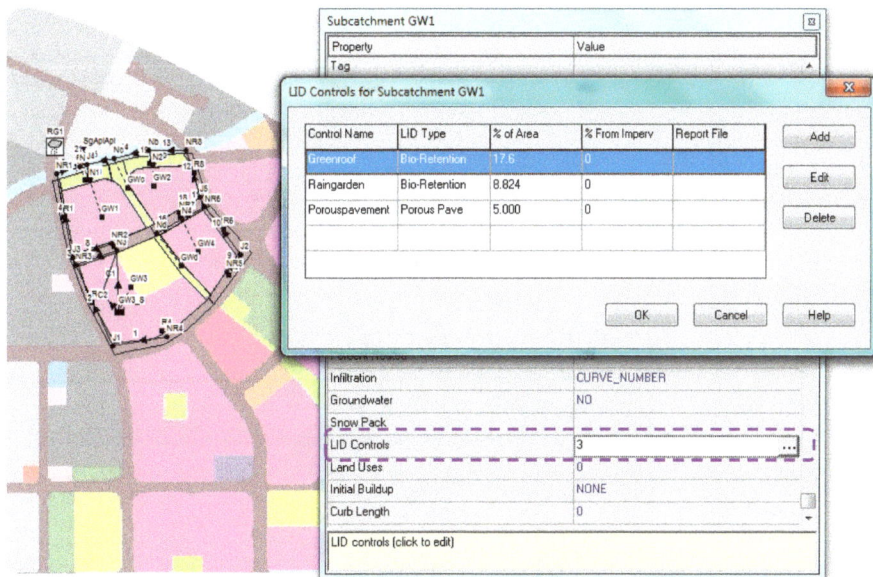

Figure 5. Modelling of low impact development (LID) in SWMM.

Assuming that GWD would be a traditional urban development, that has a lesser green area, an average peak C value of 0.78 was assumed to establish a baseline for comparison. Based on the leaf concept design (Figure 6), green roofs were included at all residential blocks and in low-storey car parks. Also the green canyons, elevated roof garden links and extensive landscape decks suggested in leaf concept were represented as rain gardens in the SWMM model of GWD catchment. New parks (green patches in Figure 6) separate the subcatchments in GWD catchment. Walkways between blocks, playground turf and open parts of the communal areas are designed as porous pavements to enhance runoff control. Also the presence of an educational institution in subcatchment GW3 made it an ideal location for an underground storage tank to be built beneath the school field.

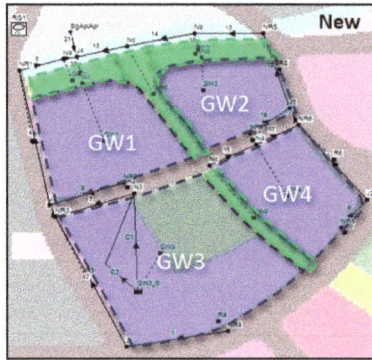

Figure 6. SWMM setup of the leaf concept design in Green Walk District (GWD).

Based on traditional urban development scenario, which is heavily paved, the average peak C value of TSC, the study area, was fixed at 0.9. The planned shopping mall in TSC provided opportunity for including rain gardens as well as a shallow water roof in the design. The water roof, in addition to marginal runoff detention, is expected to improve the aesthetics and cool the building. Decentralised detention tanks are proposed beneath the low-storey car parks and the mall. The 23 m trapezoidal concrete channel known as Sungei Peng Siang, which runs through TSC, was modelled as a naturalised channel similar to Singapore's Kallang River at Bishan-Ang Mo Kio Park ABC Waters project. Similar to GWD, the ABC Waters design features—comprising green roofs, green canyons, elevated roof garden links and extensive landscape decks—are represented in the form of raingardens in SWMM and the porous pavements were implemented in TSC (Figure 7). Also, a community garden and an open community space (represented as rain gardens in SWMM) were included.

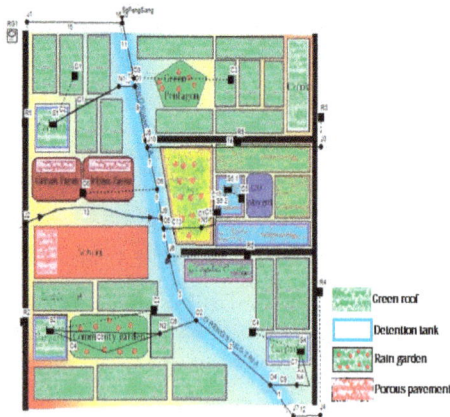

Figure 7. SWMM set up of Tengah Subcatchment (TSC) with ABC Waters design features incorporated.

TSC was modelled as six smaller subcatchments (C1 to C6) in SWMM as shown in Figure 8. Based on the leaf concept design, a total of 3.9 ha of ABC Waters design features were incorporated, which constituted to about 39% increase in ABC Waters design features.

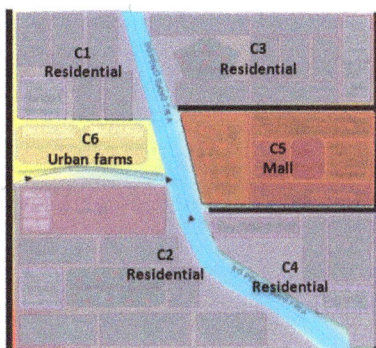

Figure 8. TSC Subcatchments in SWMM.

3. Results

The hydrological response regime map (Figure 9) based on the peak C value of each cell modelled shows that the 70% of Singapore's land surface has C value above 0.55. The areas with zero C values in Figure 9 are the water bodies such as reservoirs and sea.

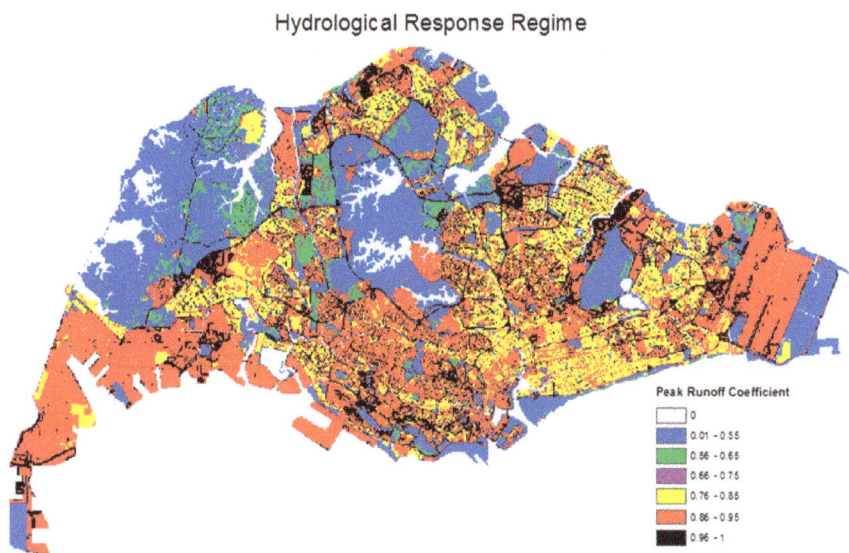

Figure 9. Hydrological response regime of Singapore (Peak C values) for current land use without any ABC Waters design features.

Through the simulations, it was noted that C values were affected not just by land use, but also had an intuitive relationship with the slope of terrain. While steeper terrains generated more runoff, gentler terrains allowed for more infiltration which resulted in lower runoff coefficient [23]. To gauge the impact of slopes on C values, sensitivity analyses were carried out for every land use type across a range of slopes from 0 to 100%. Though rare, steeps slopes of 100% exist in Singapore's urbanised catchments, as there are hilly nature reserves within the city. Figure 10 showed the C value vs. slope percent plots for different land uses.

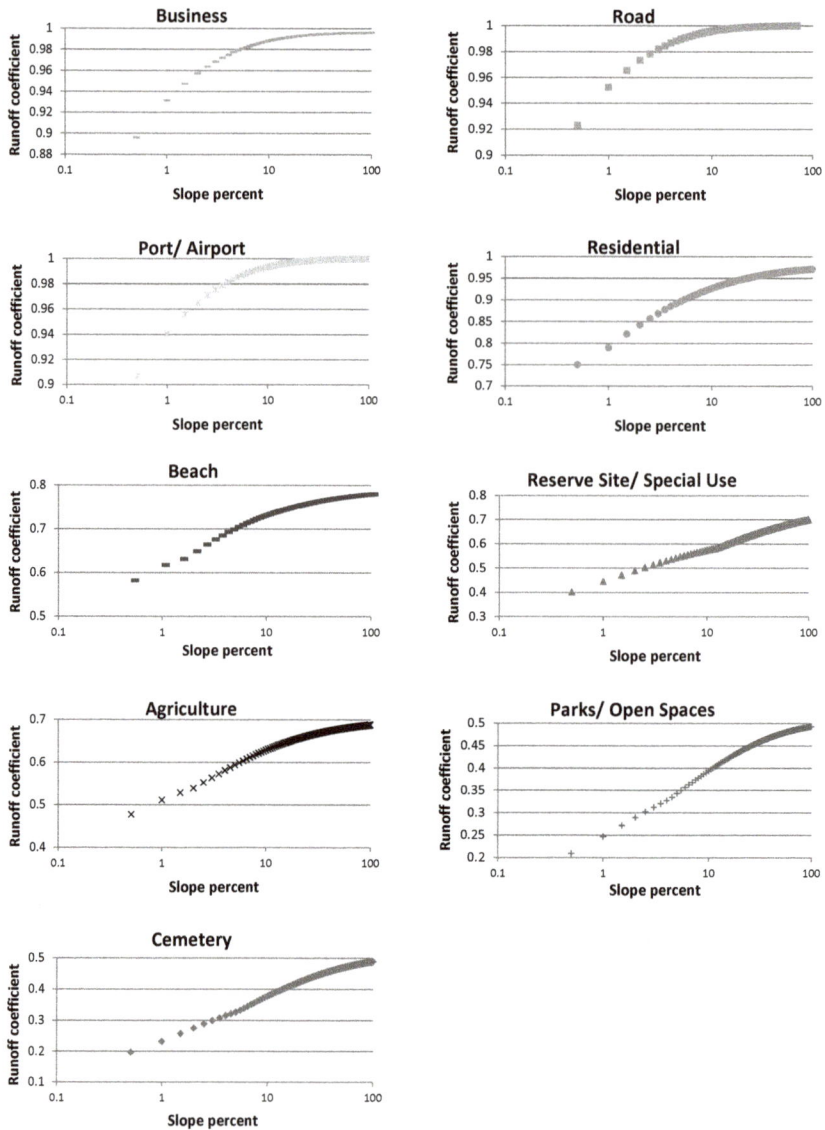

Figure 10. Runoff coefficient vs. percent slope plots for different land uses in Singapore.

The plots in Figure 10 depict the magnitude of variation in C values for the same land use with different slopes. It can be observed that steeper the slope higher the C value. It can also be observed that for all land uses, C value is more sensitive to slope changes at gentler terrain but with increasing steepness, the sensitivity is reduced. This shows that the slope in a catchment need to be considered when determining the C value for drainage design. The C values obtained from the hydrological modelling in this research are presented in Table 2. Compared to the C values indicated in PUB's existing COP [8], the C values presented in Table 2 are true representation of ground reality as it is based on both land use and slope. Hence the computation of discharge (Q) with these C values using Rational Formula will be more accurate than using the C values in Singapore's code of practice.

The area-wise implementation of ABC Waters design features and detention facilities in GWD and TSC are summarised in Figures 11 and 12, respectively. In GWD, 5.7 ha (i.e., 26% of GWD's total area) of ABC Waters design features and a 1200 m^3 detention tank reduced the C value from 0.78 to below 0.55. In TSC, 3.9 ha (i.e., 39% of TSC's total area) of ABC Waters design features and a 1330 m^3 detention facility reduced the C values from 0.9 to below 0.55.

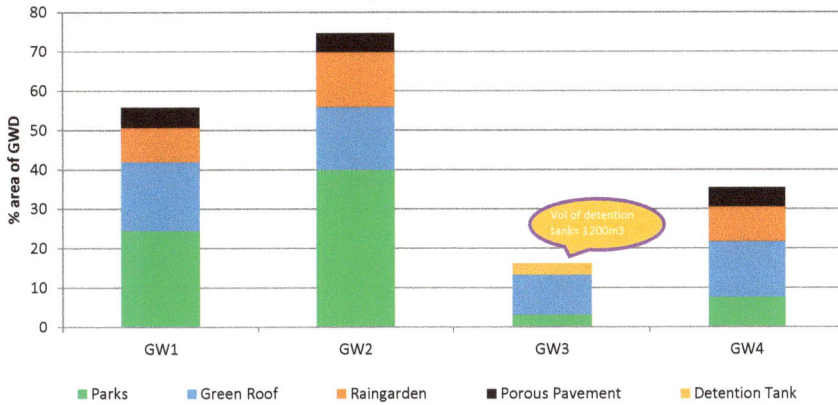

Figure 11. ABC Waters design features and detention facilities implemented in GWD.

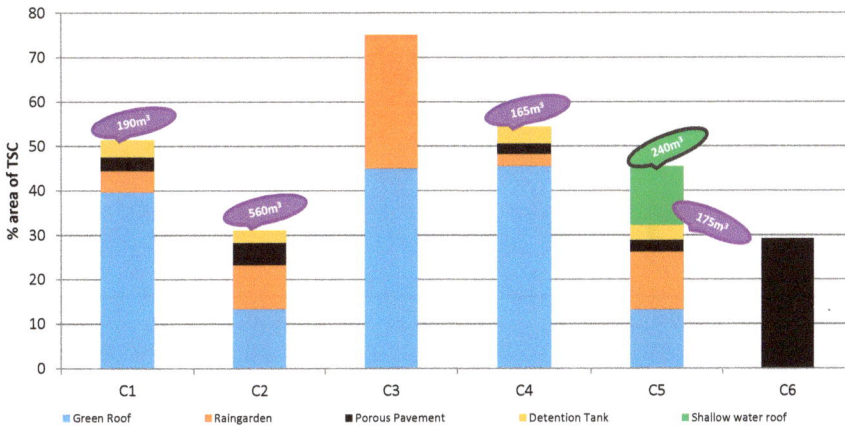

Figure 12. ABC Waters design features and detention facilities implemented in TSC.

Table 2. C values for different land uses and slopes in Singapore.

Road		Port/Airport		Business		Residential		Beach		Agriculture		Reserve Site/Special Use		Parks/Open Spaces		Cemetery	
Slope %	C	Slope %	C	Slope %	C	Slope %	C	Slope %	C	Slope %	C	Slope %	C	Slope %	C	Slope %	C
0 to 0.5	0.92	0 to 0.5	0.91	0 to 0.5	0.90	0 to 0.5	0.75	0 to 0.5	0.58	0 to 0.5	0.48	0 to 0.5	0.40	0 to 0.5	0.21	0 to 0.5	0.20
0.5 to 1	0.95	0.5 to 1	0.94	0.5 to 1	0.93	0.5 to 1	0.79	0.5 to 1	0.62	0.5 to 1	0.51	0.5 to 1	0.45	0.5 to 1	0.25	0.5 to 1	0.23
1 to 2	0.97	1 to 1.5	0.96	1 to 1.5	0.95	1 to 1.5	0.82	1 to 1.5	0.63	1 to 1.5	0.53	1 to 1.5	0.47	1 to 1.5	0.27	1 to 1.5	0.26
2 to 3.5	0.98	1.5 to 2.5	0.97	1.5 to 2.5	0.96	1.5 to 2	0.84	1.5 to 2	0.65	1.5 to 2	0.54	1.5 to 2	0.49	1.5 to 2	0.29	1.5 to 2	0.27
3.5 to 8	0.99	2.5 to 4.5	0.98	2.5 to 4	0.97	2 to 2.5	0.86	2 to 2.5	0.66	2 to 2.5	0.55	2 to 2.5	0.50	2 to 2.5	0.30	2 to 2.5	0.29
8 to 100	1.00	4.5 to 10	0.99	4 to 7	0.98	2.5 to 3	0.87	2.5 to 3.5	0.68	2.5 to 3	0.56	2.5 to 3	0.51	2.5 to 3	0.31	2.5 to 3	0.30
		10 to 100	1	7 to 35	0.99	3 to 3.5	0.88	3.5 to 4	0.69	3 to 3.5	0.57	3 to 3.5	0.52	3 to 3.5	0.32	3 to 4	0.31
				35 to 100	1	3.5 to 4.5	0.89	4 to 5	0.70	3.5 to 4	0.58	3.5 to 4	0.53	3.5 to 4.5	0.33	4 to 4.5	0.32
						4.5 to 5	0.90	5 to 6	0.71	4 to 5	0.59	4 to 5	0.54	4.5 to 5	0.34	4.5 to 5	0.33
						5 to 7	0.91	6 to 7	0.72	5 to 6	0.60	5 to 6	0.55	5 to 6	0.36	5 to 6	0.34
						7 to 9	0.92	7 to 10	0.73	6 to 7	0.61	6 to 7	0.56	6 to 7	0.37	6 to 7	0.35
						9 to 10	0.93	10 to 20	0.75	7 to 9	0.62	7 to 10	0.57	7 to 8	0.38	7 to 8	0.36
						10 to 15	0.94	20 to 30	0.76	9 to 10	0.63	10 to 15	0.60	8 to 10	0.39	8 to 9	0.37
						15 to 25	0.95	30 to 50	0.77	10 to 15	0.65	15 to 20	0.62	10 to 15	0.42	9 to 10	0.38
						25 to 50	0.96	50 to 100	0.78	15 to 25	0.66	20 to 25	0.63	15 to 20	0.44	10 to 15	0.41
						50 to 100	0.97			25 to 40	0.67	25 to 30	0.64	20 to 25	0.45	15 to 20	0.42
										40 to 50	0.68	30 to 35	0.65	25 to 35	0.46	20 to 25	0.44
										50 to 100	0.69	35 to 40	0.66	35 to 40	0.47	25 to 35	0.45
												40 to 50	0.67	40 to 50	0.48	35 to 40	0.46
												50 to 100	0.70	50 to 100	0.49	40 to 50	0.47
																50 to 100	0.49

In addition to reduction in C values, the incorporation of ABC Waters design features and detention facilities in GWD also caused four minutes of delay in peak discharge occurrence and a reduction of peak discharge (22% reduction) in Sungei Api Api (Figure 13). For TSC, there was a significant reduction in peak discharge (63% reduction) at Sungei Peng Siang, and about 30 min delay in the occurrence of the peak discharge as shown in Figure 14.

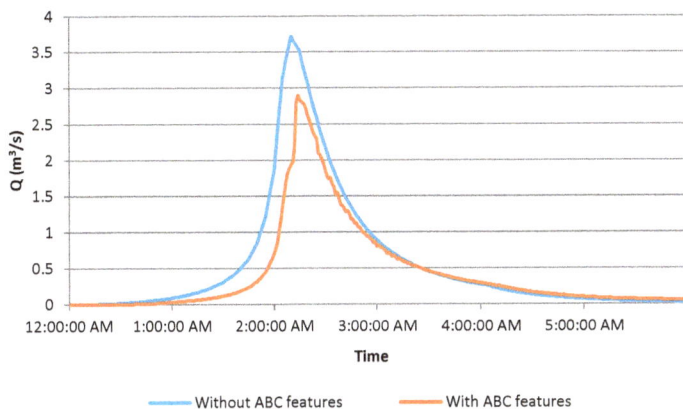

Figure 13. Discharge at Sungei Api Api outfall.

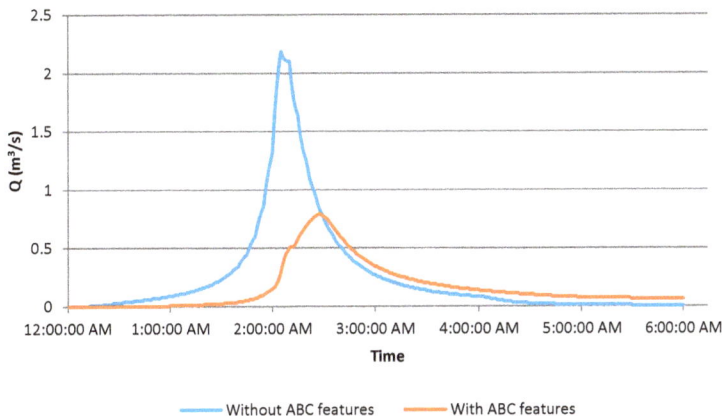

Figure 14. Discharge at Sungei Peng Siang outfall.

A sensitivity analysis conducted using the simulation results from TSC gives an indication on the effectiveness of individual ABC Waters design (and detention) feature in reducing C values. Three set of values were extracted from the simulation results of TSC: (i) the reduction in C values by each ABC Waters design feature; (ii) the area occupied by each ABC Waters design feature, and; (iii) the volume of the detention facility included. The reduction in C values reduced per m^2 of ABC Waters design feature and reduction per m^3 of detention facility is presented in Table 3. In the TSC catchment, 45% of the total area of the catchments is assumed to be covered by a green roof, an ambitious aim which is yet be achieved in Singapore. However, with the ongoing efforts of the Housing and Development Board (HDB) to intensify greening in Singapore's residential estates, there is a possibility for such an extensive green roof coverage in residential land use [24–27].

Table 3. Effectiveness of ABC Waters features and detention facilities in reducing C values in Green Walk District and Tengah Subcatchment.

Subcatchment ID	Area of Subcatchment (ha)	Initial C Value	Area of Green Roof (% Area of Subcatchment)	% Reduction in C Values (Green Roofs)	Area of Rain Gardens (% Area of Subcatchment)	% Reduction in C Values (Rain Gardens)	Area of Porous Pavement (% Area of Subcatchment)	% Reduction in C Values (Porous Pavements)	Area of Detention Tanks (% Area of Subcatchment)	% Reduction in C Values (Detention Tanks)
GW1	3.4	0.67	17.7	3	8.8	9.2	5	5.1	-	-
GW2	2.5	0.72	16	4.2	14	13	4.8	6.7	-	-
GW3	6.9	0.60	10.1	0	0	0	0	0	2.9 (1200 m^3)	8.3
GW4	2.83	0.66	14.1	3	8.83	7.8	5	6.8	-	-
C1	1.01	0.9	39.6	8.9	5	8.9	3	3.7	2.8 (190 m^3)	38.9
C2	3	0.9	13.3	17.8	10	10.8	5	4.5	2.7 (560 m^3)	11.1
C3	2	0.9	45	10	30	32	-	-	-	-
C4	0.88	0.9	45.5	7.8	2.8	4.8	2.3	5.1	3.8 (165 m^3)	32
C5	1.52	0.9	13.2	0	13.2	13.8	3.3	3.7	16.5 (175 m^3) (240 m^3—water roof)	30.8
C6	0.65	0.9	-	-	-	-	29.2	46.7	-	-

From Figure 15 it can be concluded that the detention tank is the most effective feature in reducing C values. For every 1% of detention tank (with 1 m depth) C value was reduced by about 80%, whereas the green roof was the least effective feature, as for every 1% of green roof incorporated the C value was reduced by 5%. The performance of rain gardens and porous pavements were similar, reducing C values by 14% and 19%, respectively, for every 1% of feature incorporated. However, it is important to realise that these conclusions were drawn based on the results from a single urban catchment in Tengah Subcatchment, which had its unique conditions and characteristics. These conclusion should not be taken as universal conclusions which is applicable for all kinds of catchments across Singapore.

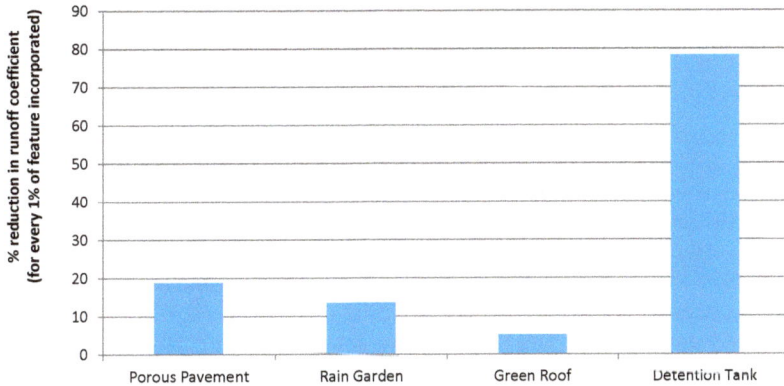

Figure 15. Reduction in runoff coefficient vs. amount of features incorporated.

The effectiveness of each feature was also quantified based on the percentage reduction in C values for every SGD$10,000 of feature incorporated in TSC. As seen from Figure 16, porous pavements fared as the most cost effective feature in terms of C values reduction. The cost reflected here is based solely on construction. Maintenance and operational cost were not included in the current calculations, consideration of which would have resulted in a different outcome.

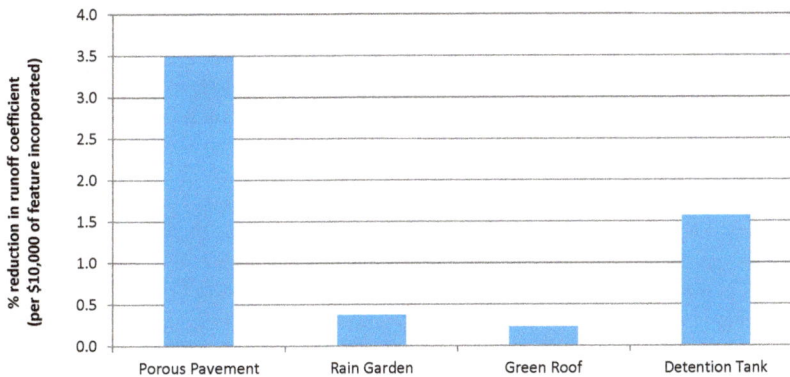

Figure 16. Percentage reduction in C for every SGD$10,000 of features incorporated.

This research highlighted the importance of available storage capacity in reducing C values. This can be seen from detention tanks which (i) have the capacity to store large volume of runoff to be detained; and, (ii) flexibility to control the release of detained water after rainfall event contributed largely to its effectiveness in runoff control. Green roofs, on the other hand, were expected to have the lowest reduction in C values as they have very limited detention capacity. The green roofs modelled in this research have available storage of 70 mm as compared to 500 and 400 mm in porous pavement and rain garden, respectively [28]. The storage layer for porous pavement is in the range of 6 to 36 inches (i.e., 152–914 mm) [29], hence the storage of porous pavement was assumed to be 500 mm in this study. The porous pavements have been incorporated in carparks, walkways, open plazas and playgrounds. While the porous pavements in carparks and walkways are usually shallower, deeper porous pavements can be built in open plazas and playground, hence an average depth of 500 mm was assumed in this study. The under-drain system in porous pavements and rain gardens implies that water stored in these features infiltrated instead of draining in to the conveyance system. It could be inferred that infiltration is an effective mechanism in reducing surface runoff. The runoff reduction potential of raingardens in Singapore is found to increase upon considering the slope of the raingarden.

4. Discussion

This research is a first attempt to determine C values based on land use and slope (Table 2) and quantify the hydrological response regime of Singapore based on its C values (Figure 9). Though computation of C values based on land use and slope is currently not adopted in Singapore, it is common in other cities. For example, the Florida Department of Transportation (FDOT) has adopted a range of C values based on slope in addition to the C values based on land use [30]. Similarly, in Section 7.5.2 (b) of the Storm Water Drainage Manual prepared by the Drainage Services Department (DSD) in Hong Kong, surface characteristics and slope (two categories, "flat" and "steep") of the area are considered in computing C value [31]. Also a recent study by HR Wallingford in UK also states that, the runoff at the site scales are gradient dependent and current practices are inappropriate in predicting the site runoff [32]. Hence there is enough evidence and need to revise the C values table (i.e., Section 7.1.2 of PUB's COP on Surface Water Drainage [8]) to be used in Singapore, which is the motivation behind this research.

Additionally, there is scope to improve the research findings reported here. The first recommendation is with regard to improving the accuracy of Singapore's hydrological response regime map as shown in Figure 9. In this research, land use parameters such as percentage imperviousness, Manning's N and depression storage values, have been determined with reference to guidelines by other countries. Though conscientious efforts have been made to represent each land use parameter as accurately as possible based on the author's understanding of Singapore, calibration against actual rainfall and discharge data have to be carried out at the subcatchment level to adjust these land use parameters values for better representation of the onsite situation. In addition, soil properties across the entire island have been assumed to be homogenous and belong to soil group D in the HSG classification. This aspect can be refined with detailed borehole information from various agencies and complemented with the borehole investigation works for the missing areas. With resources, time and effort invested in this data collection process, a complete soil map can be consolidated and more accurate curve number inputs can be used for the SWMM simulations.

The second recommendation is to improve the hydraulics of the simulations. In this study, emphasis has not been placed on the hydraulics effects of the channels and drainage networks. The two study areas have been designed based on the general slope direction of the terrain as no internal drainage plans were available. It has been assumed in this research that all overland flow generated form the subcatchments will be channelled to the outflow completely and quickly. This is sufficient for this research which aims to test the hydrological feasibility of the new runoff control regulation. However, for a more reliable hydrological and hydraulics analysis of a catchment, it is better to update the model with actual internal drainage plans that will address practical constraints, if any.

The hydrological response regime map of Singapore if complemented by information such as the expected end of lifespan of buildings, will project the duration needed for all urban pockets in Singapore to be renewed/retrofitted and finally have C values within PUB's limits of 0.55. With this projection, flood hotspots that do not have short term renewal or retrofitting opportunities can be identified. Alternative flood management strategies can be explored to alleviate the situation in these areas. This recommendation will allow prioritization of focus areas and identify mainstreaming opportunities, thus allowing PUB to constantly review the robustness of Singapore's storm water management plans.

The ABC Waters programme may have limited effectiveness in reducing C values. However there are several other attractive benefits such as UHI mitigation, increasing land and property values, psychological and social benefits and encouraging biodiversity [9]. This research has further highlighted that there is no one solution fits all approach when dealing with C values reduction in Singapore. Even though Singapore is small, each piece of development on the island is unique. A feature that is highly cost effective in a development may not be suitable for another due to varying site conditions and operational regimes. The need to consider land availability, construction, maintenance, operational and end of life cycle cost remains prudent when determining the most suitable suite of features for each development site.

With the knowledge gained from the more in-depth researches proposed above, PUB will be able to establish a set of guidelines for consultants and developers to follow when identifying the best combination of features to reduce C values and reap decent cost savings as suggested by the expert panel [12]. Also, this research outcomes can be used by PUB to further explore the possibility of including slope in runoff coefficient table in Section 7.1.2 of PUB's Code of Practice for Surface Water Drainage [8].

5. Conclusions

In Singapore, the policy decision of reduction in runoff has translated into a legislative requirement such as runoff control regulation restricting C value of developments to below 0.55. This paper focused on establishing the actual hydrological response regime of Singapore and the relationship between runoff coefficient, slope and land use. This resulted in an elaborate hydrologic response map for Singapore, which comprises a range of C values for the land use based on varying slopes. This might lead to optimal design of drainage systems such as ABC Waters design features. Additionally, the findings from the two study areas have shown that ABC Waters design features have the capacity to reduce C values. Hence, ABC Waters design features in its many forms can be incorporated into new developments or retrofitted onto existing buildings as a means of complying with the new regulatory requirement, without occupying excessive space. Hence, it is recommended to consider the relationship between slope and land use while determining runoff coefficients, and to incorporate ABC Waters design features in urban design to improve runoff coefficient (C), which can reduce the peak flow and delay the occurrence of peak discharge.

Acknowledgments: The authors would like to thank PUB, Singapore's National Water Agency, IHE Delft Institute for Water Education and CRC for Water Sensitive Cities, Australia for supporting the research.

Author Contributions: The paper was written by Xue Ping Goh based on the results obtained from her research which was done for the partial fulfilment of requirements for the Master of Science degree (2013–2015) at the IHE Delft Institute for Water Education, under the close supervision of Assela Pathirana and Chris Zevenbergen. Mohanasundar Radhakrishnan provided timely feedback on the research throughout the research study and contributed to the development of the model. All co-authors also helped to review and improve the paper.

Conflicts of Interest: The authors declare no conflict of interest.

Appendix A

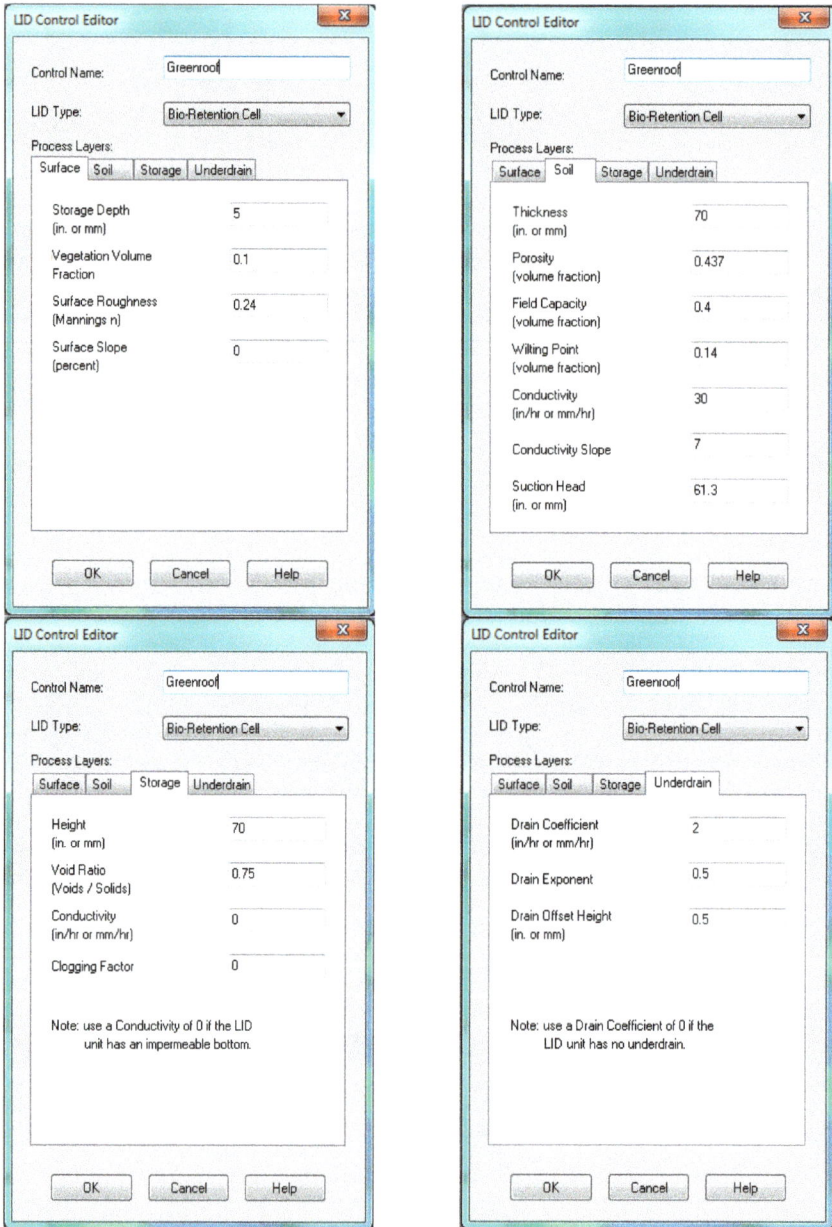

Figure A1. Representation of green roof in SWMM and design parameters used.

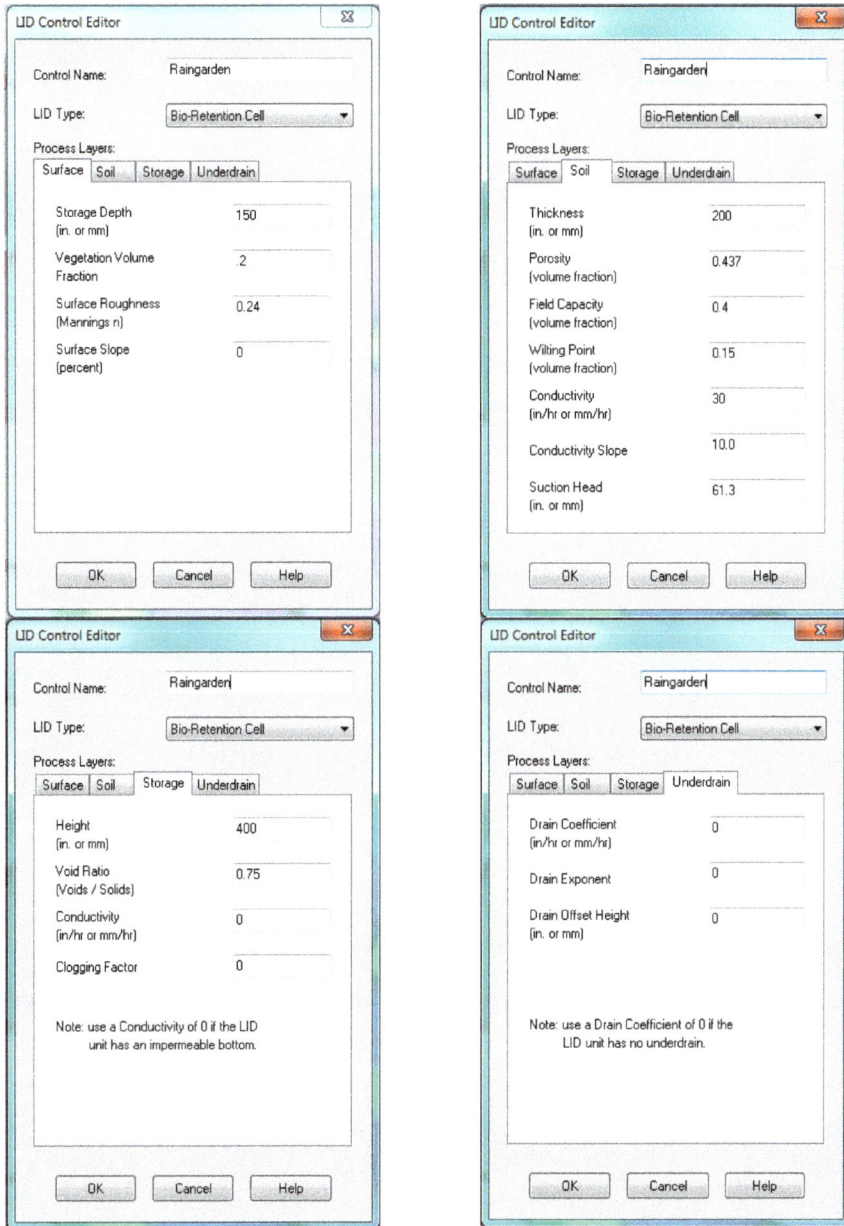

Figure A2. Representation of rain gardens in SWMM and design parameters used.

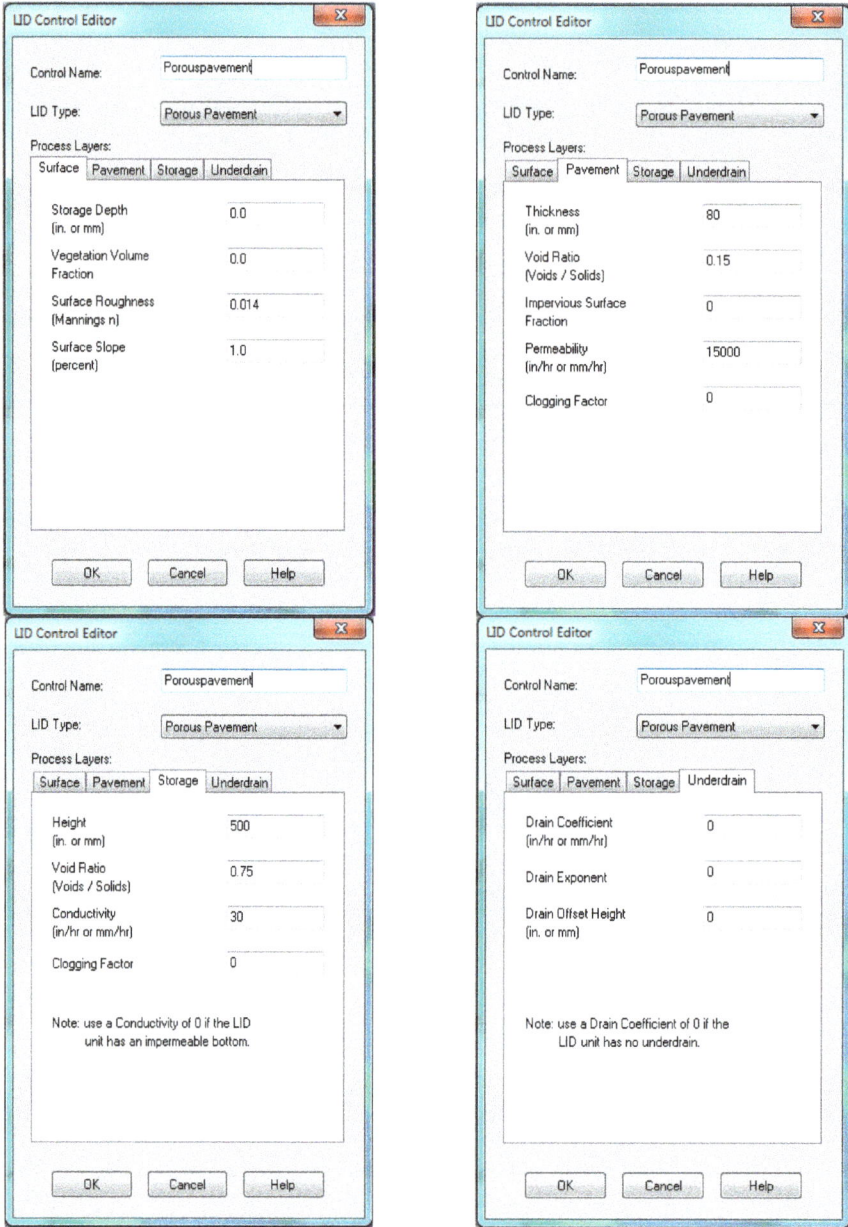

Figure A3. Representation of porous pavement in SWMM and design parameters used.

References

1. Dietz, M.E. Low impact development practices: A review of current research and recommendations for future directions. *Water Air Soil Pollut.* **2007**, *186*, 351–363. [CrossRef]
2. Chinh, D.; Dung, N.; Gain, A.; Kreibich, H. Flood loss models and risk analysis for private households in Can Tho City, Vietnam. *Water* **2017**, *9*, 313. [CrossRef]
3. Radhakrishnan, M.; Pathirana, A.; Ashley, R.; Zevenbergen, C. Structuring climate adaptation through multiple perspectives: Framework and case study on flood risk management. *Water* **2017**, *9*, 129. [CrossRef]
4. European Environment Agency. *Urban Adaptation to Climate Change in Europe: Transforming Cities in a Changing Climate*; European Environment Agency: Copenhagen, Denmark, 2016; p. 135.
5. City of Melbourne. *Resilient Melbourne*; City of Melbourne: Melbourne, Australian, 2016.
6. HM Government. National Flood Resilience Review. Crown Copyright: 2016. Available online: https://www.gov.uk/government/publications/national-flood-resilience-review (accessed on 16 April 2017).
7. Zhang, S.X.; Babovic, V. A real options approach to the design and architecture of water supply systems using innovative water technologies under uncertainty. *J. Hydroinform.* **2012**, *14*, 13–29. [CrossRef]
8. PUB Singapore's National Water Agency. *Code of Practice on Surface Water Drainage (Sixth Edition-December 2011, Addendum No. 1 Junuary 2013)*; Public Utilties Board: Singapore, 7 June 2017. Available online: https://www.pub.gov.sg/Documents/COP_Final.pdf (accessed on 7 June 2017).
9. PUB Singapore's National Water Agency. *Active Beautiful Clean Waters Design Guidelines*; Board, P.U., Ed.; Public Utilties Board: Singapore, 2014.
10. Fletcher, T.D.; Shuster, W.; Hunt, W.F.; Ashley, R.; Butler, D.; Arthur, S.; Trowsdale, S.; Barraud, S.; Semadeni-Davies, A.; Bertrand-Krajewski, J.-L.; et al. SUDS, LID, BMPs, WSUD and more—The evolution and application of terminology surrounding urban drainage. *Urban Water J.* **2015**, *12*, 525–542. [CrossRef]
11. Pathirana, A.; Denekew, H.B.; Veerbeek, W.; Zevenbergen, C.; Banda, A.T. Impact of urban growth-driven landuse change on microclimate and extreme precipitation—A sensitivity study. *Atmos. Res.* **2014**, *138*, 59–72. [CrossRef]
12. PUB Singapore's national water agency. *Report on Key Conclusions and Recommendations of the Expert Panel on Drainage Design and Flood Protection Measures*; PUB Singapore's National Water Agency, Ed.; PUB Singapore's National Water Agency: Singapore, 2012.
13. Urban Drainage and Flood Control District. *Urban Storm Drainage Criteria Manual*; Urban Drainage and Flood Control District, Ed.; Urban Drainage and Flood Control District (UDFCD): Denver, CO, USA, 2016.
14. Nicklow, J.W.; Boulos, P.F.; Muleta, M.K. *Comprehensive Urban Hydrologic Modeling Handbook for Engineers and Planners*; MWH Soft, Inc. Publ.: Pasadena, CA, USA, 2006.
15. Voogt, J.A. Urban heat islands: Hotter cities. *Action Biosci.* **2004**. Available online: http://www.actionbioscience.org/environment/voogt.html?newwindow=true (accessed on 12 August 2017).
16. United States Department of Agriculture. Urban Hydrology for Small Watersheds TR-55. In *US Soil Conservation Service. Technical Release*; Conservation Engineering Division, Natural Resource Conservation Service, United States Department of Agriculture: Washigton, DC, USA, 1986.
17. National Environment Agency. Weather Statistics, 2014. Available online: http://www.nea.gov.sg/weather-climate/climate/historical-daily-records (accessed on 2 October 2014).
18. Environmental Protection Agency, United States. Storm Water Management Model (SWMM), 2014. Available online: http://www2.epa.gov/water-research/storm-water-management-model-swmm (accessed on 3 March 2015).
19. Rossman, L.A.; Huber, W.C. *Stormwater Management model—Reference Manual—Volume I Hydrology (Revised)*; USEPA Agency, Ed.; USEPA: Cincinnati, OH, USA, 2016.
20. Housing and Development Board of Singapore. HDB Homes of the Future: Green & Sustainable, 2014. Available online: http://www20.hdb.gov.sg/fi10/fi10296p.nsf/PressReleases/18994D97FB6D475F48257D4E0023DA3F?OpenDocument (accessed on 30 June 2017).
21. Urban Redevelopment Authority of Singapore. Master Plan—Eastern Region, 2014. Available online: https://www.ura.gov.sg/uol/master-plan/View-Master-Plan/master-plan-2014/master-plan/Regional-highlights/east-region/east-region (accessed on 30 June 2017).

22. Housing and Development Board of Singapore. Unveiling the Masterplan for Tengah: At Home with Nature, 2016. Available online: http://www.hdb.gov.sg/cs/infoweb/press-releases/corporate-pr-unveiling-the-masterplan-for-tengah-08092016 (accessed on 30 June 2017).

23. Liu, Y.B.; Gebremeskel, S.; de Smedt, F.; Hoffmann, L.; Pfister, L. Predicting storm runoff from different land-use classes using a geographical information system-based distributed model. *Hydrol. Process.* **2006**, *20*, 533–548. [CrossRef]

24. Housing and Development Board of Singapore. HDB Greenprint Brings Sustainable & Green Living to 5800 Households in Teck Ghee, 2015. Available online: http://www.hdb.gov.sg/cs/infoweb/press-releases/hdb-greenprint-brings-sustainable-and-green-living (accessed on 30 June 2017).

25. Housing and Development Board of Singapore. Greenery. Undated. Available online: http://www.hdb.gov.sg/cs/infoweb/about-us/our-role/smart-and-sustainable-living/hdb-greenprint/greenery (accessed on 30 June 2017).

26. Housing and Development Board of Singapore. Yuhua Residents First to Benefit from Sustainable Features with Completion of HDB Greenprint, 2015. Available online: http://www.hdb.gov.sg/cs/infoweb/press-release/yuhua-residents-first-to-benefit-from-sustainable-features (accessed on 30 June 2017).

27. Greenroofs.com. Casa Clementi, 2012. Available online: http://www.greenroofs.com/projects/pview.php?id=1711 (accessed on 4 July 2017).

28. Yau, W.K. *Effectiveness of ABC Waters Design Features for Runoff Quantity Control in Urban Singapore, in Water Science and Engineering*; IHE Delft Institute for Water Education: Delft, The Netherlands, 2017.

29. Dickinson, R. Porous Pavement LID Control In #SWMM5, #InfoSWMM Sustain. Blogs about SWMM 5, InfoSWMM, InfoSewer, Stormwater Management Model and (Future) SWMM. Available online: https://swmm5.org/2017/02/01/porous-pavement-lid-control-in-swmm5-infoswmm-sustain/ (accessed on 4 July 2017).

30. Poullain, J. *Estimating Storm Water Runoff*; PDH Centre: Fairfax, VA, USA, 2012.

31. Hong Kong Drainage Services Department. *Stormwater Drainage Manual—Planning Design and Management*; Drainage Services Department, Ed.; Drainage Services Department, Government of the Hong Kong Special Administrative Region: Hong Kong, China, 2013.

32. HR Wallingford. *Greenfield Runoff Rates for Developments—Use of 2D Modelling and the Influence of Site Gradient*; HR Wallingford: Wallingford, UK, 2016.

water

MDPI

Article

Ranking of Storm Water Harvesting Sites Using Heuristic and Non-Heuristic Weighing Approaches

Shray Pathak [1,*], Chandra Shekhar Prasad Ojha [1], Chris Zevenbergen [2] and Rahul Dev Garg [1]

1   Civil Engineering Department, Indian Institute of Technology, Roorkee 247667, Uttarakhand, India;
    cspojha@gmail.com (C.S.P.O.); rdgarg@gmail.com (R.D.G.)
2   Water Science and Engineering Department, IHE Delft Institute for Water Education,
    2611 AX Delft, The Netherlands; c.zevenbergen@un-ihe.org
*   Correspondence: shraypathak@gmail.com; Tel.: +91-8954203033

Received: 31 July 2017; Accepted: 12 September 2017; Published: 16 September 2017

Abstract: Conservation of water is essential as climate change coupled with land use changes influence the distribution of water availability. Stormwater harvesting (SWH) is a widely used conservation measure, which reduces pressure on fresh water resources. However, determining the availability of stormwater and identifying the suitable sites for SWH require consideration of various socio-economic and technical factors. Earlier studies use demand, ratio of runoff to demand and weighted demand distance, as the screening criteria. In this study, a Geographic Information System (GIS) based screening methodology is adopted for identifying potential suitable SWH sites in urban areas as a first pass, and then a detailed study is done by applying suitability criteria. Initially, potential hotspots are identified by a concept of accumulated catchments and later the sites are screened and ranked using various screening parameters namely demand, ratio of runoff to demand and weighted demand distance. During this process, the opinion of experts for finalizing the suitable SWH sites brings subjectivity in the methodology. To obviate this, heuristic (Saaty Analytic hierarchy process (AHP)) and non-heuristic approaches (Entropy weight, and Principal Component Analysis (PCA) weighing techniques) are adapted for allotting weights to the parameters and applied in the ranking of SWH sites in Melbourne, Australia and Dehradun, India. It is observed that heuristic approach is not effective for the study area as it was affected by the subjectivity in the expert opinion. Results obtained by non-heuristic approach come out to be in a good agreement with the sites finalized for SWH by the water planners of the study area. Hence, the proposed ranking methodology has the potential for application in decision making of suitable storm water harvesting sites.

Keywords: stormwater harvesting; rainfall; surface runoff; GIS; suitable sites; Saaty AHP; entropy weight method; PCA; decision making

---

1. Introduction

Scarcity of water is a major concern for the world. There is need to integrate the allocation and management of water supply, wastewater resources and stormwater in order to sustainably manage the scarce resource [1]. A concept is introduced by utilizing stormwater through Storm Water Harvesting (SWH) and treat it as a resource, rather than a problem, to reduce the pressure on fresh water assets [2]. This idea is promoted through public forums [3] and SWH schemes are implemented on international [4].

Although the SWH has the best potential to reduce the pressure on fresh water assets, the approach to determine the potential SWH sites is not yet fully developed [5]. With the knowledge of geospatial technologies, efforts have been made to find the robust technique to shortlist and finalize the suitable sites for SWH. Thus, there is a need for a screening tool that can identify the potential suitable sites for SWH. Different researchers have identified and analyzed different approaches for shortlisting sites.

The decision-making framework (DMF) that is appropriate for SWH scheme is primarily based on technical feasibility and financial costs with a focus on neighborhood-scale development [6]. Recently, the focus of the researchers is to investigate the analogy of urban water supply and its associated energy consumption nexus [7].

Geomatics techniques prove to be the best option to explore the suitability of sites as well as the availability of locations for SWH [8]. Geographic Information System (GIS) facilitates the swift screening of potentially suitable SWH sites in the urban areas. It enables obtaining various parameters in spatial format, which decides the suitability for SWH sites. In India, the methodology to select potential sites for water harvesting were identified by adopting International Mission for Sustainability Development (IMSD) and Indian National Committee on Hydrology (INCOH) guidelines in GIS environment [8,9]. In [9], various parameters, i.e., Geomorphology map, Land Use Land Cover (LULC), road, drainage and lineaments maps were prepared and the knowledge based weights were assigned to all the parameters to compute the ranking of the sites in the GIS environment.

Singh et al. [10] defined some of the criterion for the site selection for various types of storage tanks, i.e., water harvesting structures, check dams, percolation tanks and farm ponds. The criterion suggested for water harvesting structures are that the slope should be less than 15 percent, land use class should be similar to agricultural area and type of soil should be silt loam with low infiltration capacity. In the approach [10], the suitable sites were selected by integrating all the parameters in the GIS environment.

Satellite images, Digital Elevation Model (DEM) and soil map are essential to ascertain and assess the parameters suitable for SWH sites. With GIS techniques, spatial maps of LULC, soil, topography and runoff can be prepared and thus hydrological parameters can be computed and analyzed to cope for the increase in demand of water. The reservoir capacity can be computed by analyzing demand and runoff and subsequently deciding the structure that can be proposed for SWH. The focus should be to design the SWH structures considering the systematic and cost-effective design on a city wide scale. Furthermore, a GIS based screening methodology for identifying suitable SWH sites were developed [11]. Various screening parameters such as Demand, ratio of Runoff to Demand and weighted Demand distance were evaluated for site selection.

Upon identifying the screening parameters, the parameters can be allocated weights based on existing practices such as Analytic hierarchy process [12], Principal Component Analysis [13–17], and entropy weight method [18]. In the AHP method [12], the weights are decided through pairwise comparisons. The method is based upon the opinion from the experts to define the relative scales. Then, the comparison is made between parameters on an absolute scale representing one parameter is more dominant with respect to the other. This method is adopted worldwide for group decision making in the fields such as business, shipbuilding, industry, government, healthcare and education, etc. It represents the decision that best suits the requirements of decision makers and does not a "correct" decision. Thus, it represents a comprehensive and rational framework that evaluates the solutions by representing and quantifying its elements that describes the overall goals.

In this concept, the planners structured their decision problem into a hierarchy that defines the problem in a simpler and systematic manner. The elements involved in the hierarchy can correlate any aspect of decision problem, good or bad, tangible or intangible, well or poorly understood, or anything that applies to the decision at hand. Once the hierarchy is built for the problem, the planners decide the importance of the elements by comparing the elements between each other with respect to their impacts.

However, there are limitations in adopting Saaty AHP methodology for allotting weights to the parameters [19–27]. Some of the major drawbacks are (i) the computations made by the AHP are always guided by the decision maker's experience and may involve the subjective judgments of individuals that constitute an important part in the decision process [27]; (ii) the different methods of designing the hierarchies of the same problem may lead to contrasting results [23,24]; and (iii) the structure of the hierarchy follows the perception of the individual (or the group of individuals) and there is no possible alternative to this problem formulation [25].

The Principal Component Analysis can be defined as a linear combination of optimally-weighted observed variables. This method removes the subjective decisions and totally depends on the data sets. In PCA, the most common used criterion for solving the number of components is to compute eigenvectors and eigenvalues [13–17]. Weights are decided using eigenvalues.

The Entropy Weight method determines the weights associated with the information of values of all the screening parameters. The evaluation through the entropy method for determination of weight is claimed to be a very effective method for evaluating indicators [18,28,29].

The use of different weight allocations methods is yet to be explored in ranking of storm water harvesting sites. It is also not known how these different weighing approaches can help the water planners. Hence, it is decided to explore the potential of these approaches for screening the potential sites for SWH in two case studies, in Melbourne, Australia [11], and in Dehradun, India. The case study in India also includes a step-by-step illustration of ranking methodology. For comparative purposes as well as to assess the merits and demerits of using heuristic versus non-heuristic approaches, the screening parameters are adopted. This paper is organized in following subsections, i.e., methodology used, results, discussion and conclusions.

2. Methodology

The methodology for adopting weight allocation methods to rank SWH sites essentially involves three steps: (i) identification and evaluation of screening parameters; (ii) normalization to a common scale; and (iii) assigning weightage to the normalized parameters by using heuristic and non-heuristic approaches. Details of these three steps are discussed below.

2.1. Identification and Evaluation of Screening Parameters

The methodology used for shortlisting as well as finalizing suitable sites for SWH with the integration of remote sensing and GIS is described [11]. The screening parameters discussed for shortlisting SWH sites are demand, ratio of runoff to demand and weighted demand distance. Variation of these screening parameters with radial distance from each identified hot spot needs to be established as a part of relative ranking of storm water harvesting sites. The process as such involves the following steps:

(i)   Identification of hot spots,
(ii)   Estimation of runoff,
(iii)   Estimation of demand,
(iv)   Weighted demand distance.

In step (i), with the use of DEM, the flow accumulation map is generated and the accumulated catchments are marked on the map. The points/locations of the intersections of the accumulated catchments can be the potential hotspots for the SWH structures where the flow can be trapped.

In step (ii), using precipitation, runoff can be computed using the Natural Resources Conservation Service- Curve Number (NRCS-CN) method [30]. Runoff coefficients for different combinations of pervious-impervious layers and soil type for Indian conditions are described [31]. The surface runoff can be computed by integrating the Land Use Land Cover map with the soil map of the area in the GIS environment [32].

The physical distance from the site to the demand is very critical for considering the economic feasibility.

In step (iii), various demands are considered at different radii of influence, i.e., domestic, irrigation, industrial, commercial and for public uses. All demands are summed up to compute the total demand.

Ratio of runoff to demand assesses the match between the required runoff and the associated demand. It indicates the feasibility of whether the demand can be covered with the available runoff water.

In step (iv), weighted demand distance gives preference to sites close to high demand areas to minimize transport and water infrastructure costs. Thus, the parameter is computed for all the shortlisted potential hotspots.

All of the parameters are computed and analyzed at different radii of influence.

2.2. Normalization to a Common Scale

The inconsistency in the methodology is because of the variability in judging the parameters like high demand, high ratio of runoff to demand and low weighted demand distance, and also the range of the parameters varies differently. The methodology is proposed keeping in view all of these problems. Thus, to address this gap, all of the parameters are transformed to a common range and scale. This will help the water planners to make a quick decision in finalizing the suitable site for SWH.

Equations are proposed for all the parameters as follows:

(a) Demand:

$$D_1 = \alpha \, D_L + \beta, \tag{1}$$

$$D_2 = \alpha \, D_U + \beta, \tag{2}$$

where D_1 is lower value of range, D_2 is upper value of range, D_L is lowest demand of the area, D_U is highest demand of the area, and α and β are constants.

(b) Ratio of Runoff to Demand (RTD):

$$RTD_1 = \gamma \, RTD_L + \delta, \tag{3}$$

$$RTD_2 = \gamma \, RTD_U + \delta, \tag{4}$$

where RTD_1 is lower value of the range, RTD_2 is upper value of the range, RTD_L is lowest value of ratio of runoff to demand of the area, RTD_U is the highest value of ratio of runoff to demand of the area, and γ and δ are constants.

(c) Weighted Demand Distance:

$$WD_1 = \zeta \, WD_L + \eta, \tag{5}$$

$$WD_2 = \zeta \, WD_U + \eta, \tag{6}$$

where WD_1 is lower value of the range, WD_2 is upper value of the range, WD_L is lowest value of inverse weighted demand distance of the area, and WD_U is the highest value of inverse weighted demand distance of the area, ζ and η are constants.

Thus, by solving the above equations (1, 2, ... , 6), α, β, γ, δ, ζ and η constants can be computed. After computing the constants, all the values of parameters of different sites are transformed to a new scale that ranges from D_1 to D_2 for demand, RTD_1 to RTD_2 for ratio of runoff to demand and WD_1 to WD_2 for inverse weighted demand distance by applying the following equations:

(a) For, demand;

$$D_S = \alpha \, D_C + \beta, \tag{7}$$

(b) Ratio of runoff to demand;

$$RTD_S = \gamma \, RTD_C + \delta, \tag{8}$$

(c) Weighted demand distance;

$$WD_S = \zeta \, WD_C + \eta, \tag{9}$$

where D_S is scaled demand, D_C is computed demand for each site, RTD_S is scaled ratio of runoff to demand, RTD_C is computed ratio of runoff to demand for each site, WD_S is scaled inverse weighted distance and WD_C is computed inverse weighted distance for each site.

Thus, by solving the above equations, each value of different parameters for all the shortlisted sites transforms into a common scale.

2.3. Determination of Weights

2.3.1. Saaty Heuristic Approach

The analytic hierarchy process (AHP) is a representation of complex problems by organizing and analyzing them in a more structured manner. It was developed by Thomas L. Saaty in the 1970s and has been applied worldwide for solving the complex problems [12].

The following three-step procedure provides a good approximation of the synthesized priorities.
Step 1: Sum the values in each column of the pairwise comparison matrix.
Step 2: Divide each element in the pairwise matrix by its column total.
The resulting matrix is referred to as the normalized pairwise comparison matrix.
Step 3: Compute the average of the elements in each row of the normalized matrix.
These averages provide an estimate of the relative priorities of the elements being compared.
Computing the vector of criteria weights

(a)   Creating a pairwise comparison matrix A.

Let A = m × m matrix; m = evaluation criteria; each entry a_{jk} represents the importance of jth criteria with respect to k^{th} criteria.

The relative importance between two elements or criteria is by allotting them weights on a scale from 1 to 9.

(b)   Once the matrix A is built, the normalized pairwise comparison matrix A_{norm} is formed, by making the sum equal to 1 of the all of the entries in the column of the matrix A, i.e., each entry a' of the matrix A_{norm} is computed as

$$a'_{jk} = \frac{a_{jk}}{\sum_{l=1}^{m} a_{lk}}.$$ (10)

(c)   Finally, the criteria weight vector w (that is an m-dimensional column vector) is formed by averaging all the entries along the row of matrix A_{norm}, i.e.,

$$w_j = \frac{\sum_{l=1}^{m} a'_{jl}}{m}.$$ (11)

The AHP converts individual evaluations of relative importance of one parameter over another to numerical values, which can be analyzed over the entire range of the problem [33]. A numerical weight or priority is derived using a matrix of such comparisons between various parameters.

2.3.2. Non-Heuristic Approaches

Principal Component Analysis (PCA) Method

PCA is defined as a linear combination of optimally-weighted observed variables. In PCA, the most common used criterion for solving the number of components is to compute eigenvectors and eigenvalues. To solve the eigenvalue problem, the following steps are followed.

Let A be a n × n matrix and consider the vector equation

$$A\vec{v} = \lambda \vec{v},$$ (12)

where λ represents a scalar value.

Thus, if $\vec{v} = \vec{0}$, it represents a solution for any value of λ. Eigenvalue or characteristics value of matrix A is that value of λ for which the equation has a solution with $\vec{v} \neq \vec{0}$. The corresponding solutions $\vec{v} \neq \vec{0}$ are called eigenvectors or characteristic vectors of A.

(i)   Compute the determinant of $A - \lambda I$

With λ subtracted along the diagonal, this determinant starts with λ^n or $-\lambda^n$. It is a polynomial in λ of degree n.

(ii)   Find the roots of this polynomial

By solving $\det (A - \lambda I) = 0$, the n roots are the n eigenvalues of A. It makes $A - \lambda I$ singular.

(iii) For each eigenvalue λ, solve $(A - \lambda I)x = 0$ to find an eigenvector x.

Eigenvalues are used to decide weights in proportions to total of eigenvalues.

Entropy Weight Method

In this approach, the individual elements or criteria are assigned weights by determining entropy and entropy weight. Based on the principle of information theory, entropy is a measure of lack of information regarding a system. If the information entropy of the indicator is small, the amount of information provided by the indicator will be greater and the higher the weight will be, thereby playing a more important role in the comprehensive evaluation [34]. The steps involved in the entropy weight method are (i) formation of the evaluation matrix; (ii) normalization of the evaluation matrix; and (iii) calculation of the entropy and the entropy weight

Let there be m parameters to be evaluated in a problem, n categories of evaluation criteria, and then the evaluation matrix is $X = (x_{ij})_{mxn}$, where x_{ij} represents the actual value of j-th criteria for the ith parameter. The calculation of entropy weight is as follows.

(i)   Normalize the evaluation matrix, X to obtain $R = (r_{ij})_{mxn}$ where r_{ij} is the jth evaluating object for i^{th} indicator and $r_{ij} \in [0,1]$. This will in turn generate a positive indicator for the variables:

$$r_{ij} = \frac{x_{ij} - \min \{x_{ij}\}}{\max\{x_{ij}\} - \min \{x_{ij}\}}. \tag{13}$$

(ii)   Calculate entropy weight value 'H'. the j-th index value of information entropy is computed as

$$H_j = -K \sum_{i=1}^{m} f_{ij} \ln f_{ij}. \tag{14}$$

Here, $f_{ij} = \frac{r_{ij}}{\sum_{i=1}^{m} r_{ij}}$, $f_{ij} \in [0, 1]$.

where, K is a positive constant, relevant to the number of sampling stations, s of the system. When the samples are completely in disordered state, $K = 1/\ln (s)$.

(iii) Calculate the j-th index weight as,

$$W_i = \frac{1 - H_i}{m - \sum_{i=1}^{m} H_i}. \tag{15}$$

3. Case Study Application and Results

Firstly, the methodology is applied on Melbourne city to check results obtained from heuristic and non-heuristic approaches and then to Dehradun city.
Ranking of the potential sites

The main concern of the study is to rank the suitable shortlisted sites according to the various approaches and thus obtain the final site ranking of sites, which are done according to the various approaches discussed earlier by allotting weights to the parameters.

3.1. Application of Methodology to Melbourne City

The study area is a part of the City of Melbourne (COM), where all the water and waste water services are provided by the City West Water. The study area includes residential, industrial areas, public parks and commercial land and covers an area of about 26 km². Further SWH site specific details about Melbourne are elaborated [11]. The methodology is applied to Melbourne sites shortlisted for SWH. The set of data taken as shown in Table 1 is as follows. The ranking of sites according to high demand, high ratio of runoff to demand and low weighted distance had already been computed in the work.

Table 1. Sites shortlisted for Storm Water Harvesting in Melbourne city [11].

Site ID	Possible Options	Demand (ML)	Ratio of Runoff to Demand	Weighted Distance (m)
	76b	49.07	1.3	300
	43c	6.18	29.4	283
43	43b	5.82	31.2	277
	46d	7.47	14	256
	47d	7.47	9.6	256
	44c	6.43	62.6	255
44	44b	6.18	65.2	250
28	28b	6.18	15.8	243
	47c	6.84	10.5	218
	46c	6.84	15.3	217
12	12b	15.88	14.4	210
46	46b	5.82	18	182
47	47b	5.82	12.4	182
14	14b	125.6	1.8	182
69	69b	11.62	81.6	175
	29d	31.65	4.2	136
	52b	13.7	8.5	134
	17d	53.79	1.3	112
	41d	30.65	2.2	103
26	26b	19.35	2.6	87
39	39b	19.35	1.6	87
	29c	28.92	4.6	80
	78b	13.07	1.5	70
	41c	28.92	2.3	67
52	52a	5.33	21.9	0
76	76a	5.3	11.8	0
29	29a	23.14	5.8	0
78	78a	5.3	3.7	0
77	77a	5.3	3.2	0
17	17a	23.14	3	0
41	41a	23.14	2.9	0
20	20a	23.14	2.8	0
9	9b	28.67	1.3	0

ML: Million Litres.

3.1.1. Saaty Heuristic Approach

The Saaty AHP [12] method is difficult to apply on Melbourne city as no such survey to decide the relative weights of screening parameters from water planners of Melbourne city was reported. For this, a survey was done at the Dehradun site, and it was considered desirable to give equal weightage to all the screening parameters.

For Melbourne city, considering all the parameters with equal weights, the ranking is done in two parts, first with a non-zero value of weighted distance and the other with zero value of weighted distance. In the case of Inamdar et al. [11], two types of sites are reported. For certain sites, the hot spots and the demand clusters are located at the same spot, making the value of parameter weighted demand distance be zero.

Sites ranked for non-zero value of weighted distance.

The ranking corresponding to this approach is shown in Table 2 for the non-zero value of weighted distance.

Table 2. Sites ranked for non-zero value of weighted distance for Melbourne city.

Rank	Possible Options	Scaled Demand 10 Intercept	Scaled Inverse WD 5 Intercept	Scaled RTD 0 Intercept	Centroid (m)
1	69b	5.25	20.81	100.00	42.0
2	41c	19.63	100.22	1.25	40.4
3	14b	100.00	18.92	0.62	39.8
4	29c	19.63	79.31	4.11	34.4
5	78b	6.46	94.70	0.25	33.8
6	17d	40.31	48.52	0.00	29.6
7	44b	0.73	6.03	79.57	28.8
8	26b	11.68	70.64	1.62	28.0
9	39b	11.68	70.64	0.37	27.6
10	44c	0.94	5.36	76.34	27.5
11	41d	21.07	55.25	1.12	25.8
12	29d	21.90	34.94	3.61	20.2
13	52b	6.98	35.89	8.97	17.3
14	43b	0.43	2.67	37.23	13.4
15	46b	0.43	18.92	20.80	13.4

Notes: WD: Weighted Demand Distance, RTD: Ratio of Runoff to Demand.

For the zero value of weighted distance, the ranking is computed by considering two parameters, i.e., demand and ratio of runoff to demand with equal weights, as shown in Table 3.

Table 3. Sites ranked for zero value of weighted distance for Melbourne city.

Rank	Possible Options	Scaled Demand 10 Intercept	Scaled RTD 0 Intercept	Center Point (m)
1	52a	0.02	25.65	12.84
2	29a	14.83	5.6	10.22
3	9b	19.43	0	9.71
4	17a	14.83	2.12	8.47
5	41a	14.83	1.99	8.41

3.1.2. Non-Heuristic Method

Sites Ranked According to the PCA Method

In this approach, the Principal Component Analysis method is applied to compute the eigenvectors and eigenvalues of the data sets available for the potential hotspots. Then, the eigenvalues are used to compute the weights for the respective parameters.

The weights computed for the Melbourne city are 0.150, 0.253 and 0.598 for the parameters demand, inverse weighted distance and ratio of runoff to demand, respectively. The rank corresponding to the computed weights are represented in Table 4.

Table 4. Comparative study of ranks of potential hotspots for Melbourne City.

Rank	Saaty AHP Method	Entropy Weight Method	PCA Method
1	69b	14b	69b
2	41c	69b	44b
3	14b	41c	44c
4	29c	44b	41c
5	78b	44c	29c
6	17d	17d	78b
7	44b	29c	43b
8	26b	78b	43c
9	39b	41d	26b
10	44c	26b	14b
11	41d	39b	39b
12	29d	29d	17d
13	52b	76b	41d
14	43b	52b	46b
15	46b	43b	52b
16	43c	43c	29d
17	12b	12b	12b
18	76b	46b	46c
19	47b	47b	47b
20	46c	46c	28b

Notes: AHP: Analytic hierarchy process, PCA: Principal Component Analysis.

Sites Ranked According to Entropy Weight Method

In this, the weights are computed for the parameters according to the entropy weight method and the sites are ranked accordingly. The weights computed for the Melbourne city are obtained as 0.423, 0.228 and 0.350 for the parameters Demand, Inverse weighted distance and Ratio of runoff to demand, respectively. Thus, sites are ranked with these weights and are represented in Table 4.

3.2. Application of Methodology to Dehradun city

The study area is Dehradun, capital city of state Uttarakhand and is of national importance. District Dehradun is situated in the northwest corner of Uttarakhand state and extends from North Latitude 29°58′ to 31°02′30″ and East Longitude 77°34′45″ to 78°18′30″. Uttarakhand is 86 percent covered with mountains and 65 percent is covered with forests. The state is popular, as its northern part is occupied by glaciers and Himalayan peaks. The two India's largest rivers i.e., the Ganga and Yamuna, emanate from the glaciers of Uttarakhand. These rivers are fed by myriad lakes, glacial melts and streams. The Dehradun district is at an altitude of 640 m above Mean Sea Level (MSL) and covers an area of approx. 3088 km^2.

The above methodology is applied for the study area Dehradun. Firstly, the supervised classification is done to generate the LULC map. For applying the NRCS-CN method on the study area, the LULC and soil maps are merged together to form a reclassified image that interprets the curve number [32]. From the table described [31], the runoff coefficients values are allotted to different combinations of LULC and soil maps for Indian conditions.

The data for the study area is prepared for the areas of different combinations of land use and soil type with the knowledge of monthly rainfall data for 25 years, and monthly runoff (mm) is computed for the study area [32]. By delineating DEM, the accumulated catchments are marked on the LULC map, which is also the interpretation for potential hotspots for SWH. Points A, B, C, … , H are the eight potential hotspots shortlisted for the study area as shown in Figure 1.

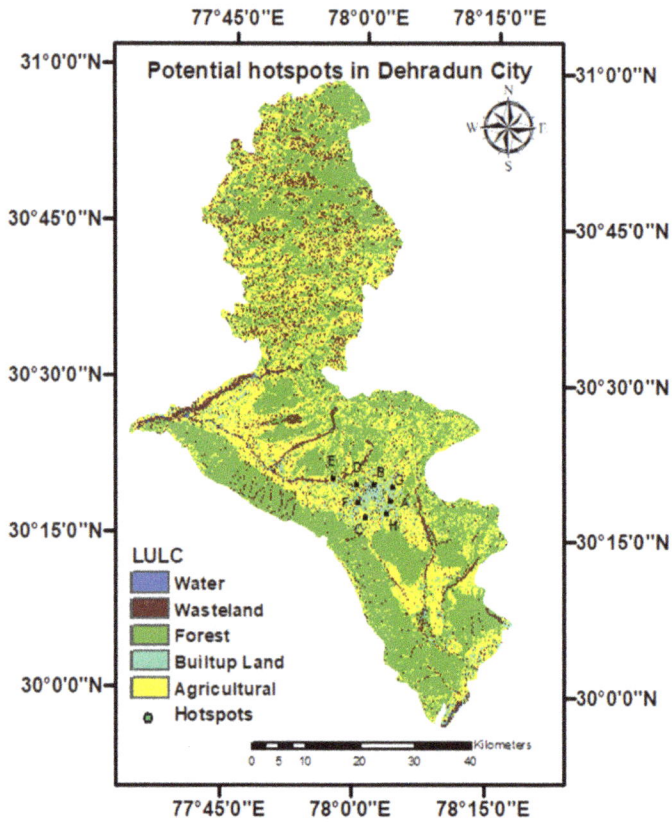

Figure 1. Potential hotspots in Dehradun city.

A radius of influence is drawn for each potential hotspot at a radius of 200 m, 400 m, 600 m, 800 m and 1000 m. The spatial demand is generated for each radius of influence for the sites and the runoff is computed accordingly. Thus, 40 combinations are formed for eight shortlisted sites with different radii of influence suitable for SWH as shown in Table 5.

3.2.1. Saaty Heuristic Approach

Saaty AHP method is applied on the study area as all planners suggested the same weights for all the parameters. Thus, the parameters are allotted the same weight and the ranking is done accordingly as shown in Table 6. By solving all the equations in the proposed methodology, the scaled demand, scaled inverse weighted demand distance and scaled ratio of runoff to demand are calculated for all of the data available for the study area by considering equal weights for all of the parameters.

Table 5. Sites shortlisted for Storm Water Harvesting in Dehradun city.

ID	Radius of Influence (RI) (m)	Total Area (m²)	Urban Area (m²)	Urban Runoff Volume (ML) (Monthly)	Water Demand (ML) (Monthly)	Ratio of Runoff to Demand	Weighted Distance
A	200	125,663.7	93,765	39.3	4.5	8.8	4.2
A	400	502,654.8	313,553	157.4	15	10.5	9.6
A	600	1,130,973	534,297.6	354.1	25.5	13.9	15.9
A	800	2,010,619	956,866.5	629.5	45.7	13.8	19.7
A	1000	3,141,593	1,443,163	983.6	68.9	14.3	25
B	200	125,663.7	4915	39.3	0.2	167.6	8.1
B	400	502,654.8	108,944.4	157.4	5.2	30.2	13.4
B	600	1,130,973	465,995	354.1	22.3	15.9	20.4
B	800	2,010,619	982,995	629.5	46.9	13.4	26.1
B	1000	3,141,593	1,770,438	983.6	84.6	11.6	28.1
C	200	125,663.7	62,829.9	39.3	3	13.1	6.3
C	400	502,654.8	170,539.4	157.4	8.1	19.3	11.6
C	600	1,130,973	309,677	354.1	14.8	23.9	16
C	800	2,010,619	495,541.5	629.5	23.7	26.6	25.5
C	1000	3,141,593	831,424	983.6	39.7	24.8	28.7
D	200	125,663.7	11,413	39.3	0.5	72.2	141.2
D	400	502,654.8	29,059.4	157.4	1.4	113.4	13
D	600	1,130,973	87,418.9	354.1	4.2	84.8	19.4
D	800	2,010,619	251,187.5	629.5	12	52.5	28.9
D	1000	3,141,593	738,246	983.6	35.3	27.9	36.8
E	200	125,663.7	36,676	39.3	1.8	22.5	6.5
E	400	502,654.8	205,243	157.4	9.8	16.1	9.9
E	600	1,130,973	309,245	354.1	14.8	24	11.7
E	800	2,010,619	420,560	629.5	20.1	31.3	18
E	1000	3,141,593	557,655	983.6	26.6	36.9	22.6
F	200	125,663.7	9102	39.3	0.4	90.5	139.5
F	400	502,654.8	180,030	157.4	8.6	18.3	13
F	600	1,130,973	537,586	354.1	25.7	13.8	15
F	800	2,010,619	1,005,158	629.5	48	13.1	20.7
F	1000	3,141,593	1,741,692	983.6	83.2	11.8	29.7
G	200	125,663.7	58,561.5	39.3	2.8	14.1	2.2
G	400	502,654.8	218,306.4	157.4	10.4	15.1	8.9
G	600	1,130,973	619,860	354.1	29.6	12	9.5
G	800	2,010,619	1,192,337	629.5	56.9	11.1	12.9
G	1000	3,141,593	1,932,770	983.6	92.3	10.7	17
H	200	125,663.7	31,584.1	39.3	1.5	26.1	5.8
H	400	502,654.8	245,109.6	157.4	11.7	13.4	8.5
H	600	1,130,973	587,960.8	354.1	28.1	12.6	10.9
H	800	2,010,619	1,026,758	629.5	49	12.8	18.1
H	1000	3,141,593	1,572,652	983.6	75.1	13.1	22.7

Table 6. Comparative study of ranks of potential hotspots for Dehradun city.

Rank	Saaty AHP Method		Entropy Weight Method		PCA Method	
	ID	RI (m)	ID	RI (m)	ID	RI (m)
1	B	200	B	200	B	200
2	G	1000	D	400	D	400
3	G	200	G	200	D	600
4	B	1000	G	1000	G	200
5	F	1000	D	600	D	800
6	H	1000	B	1000	E	1000
7	A	1000	F	1000	H	200
8	D	400	H	1000	G	1000
9	G	800	A	1000	E	800
10	H	800	G	800	E	200
11	F	800	D	800	A	200
12	A	800	E	1000	B	1000
13	D	600	H	800	H	1000
14	B	800	C	1000	F	1000
15	C	1000	A	200	D	1000
16	A	200	H	200	B	400
17	G	600	F	800	C	1000
18	E	1000	A	800	A	1000
19	D	1000	D	1000	E	600
20	H	600	B	800	G	800

3.2.2. Non-Heuristic Method

Sites Ranked According to the PCA Method

In this approach, the Principal Component Analysis method is applied to compute the eigenvectors and eigenvalues of the data sets available for the potential hotspots. Then, the eigenvalues computed corresponds to the weights for the respective parameters.

The weights are computed for the Dehradun city are 0.118, 0.248 and 0.634 for the parameters Demand, Inverse weighted distance and Ratio of runoff to demand respectively. The rank corresponds to the shortlisted sites with corresponding weights are represented in the Table 6.

Sites Ranked According to Entropy Weight Method

In this, the weights are computed for the parameters according to the entropy weight method and the sites are ranked accordingly. The weights are computed for the Dehradun city and these turn out to be 0.297, 0.213 and 0.490 for the parameters Demand, Inverse weighted distance and Ratio of runoff to demand, respectively. Thus, ranks of various shortlisted sites are represented in Table 6.

4. Discussion

The storm water harvesting sites in Melbourne and Dehradun were ranked according to the different possible combinations of parameters with equal weights and then by applying various methods for assigning weights to the parameters for both the sites. All the potential hotspot sites both for Melbourne city and Dehradun city are evaluated using the combination of all three parameters. Tables 4 and 6 show the ranks of different sites in Melbourne city and Dehradun city. In this study, the main focus is to remove subjectivity from the approaches used for ranking water harvesting sites. A definitive approach based on principal components and entropy is introduced to assign the weights. Initially, the sites are shortlisted by using DEM for the study area and applying the concept of accumulated catchments. For the similar sites, different radii of influence are used in order to rank the sites. The use of different heuristics as well non-heuristic approaches is demonstrated using three screening parameters Demand, ratio of Runoff to Demand and weighted Demand distance. The top ranking sites are well captured in Tables 5 and 6 using PCA and entropy based approaches, as evident from the sites selected in the paper [11].

It is noted that the objective here is not to select the best site that has rank one in either of approaches. Instead, the utility of these approaches should be viewed in terms of screening of a few most feasible sites. With the allocation of weights, the ranking is done for the sites in Australia and India, and the suitable sites are thus matched with the sites suggested by Inamdar et al. [11] for the Australian site. For the Australian site, the sequence of finalizing suitable sites by Inamdar et al. [11] is based on ranking of sites using only one attribute at a time. To explain it further, the site having the highest rank for attribute Demand is "14b"; similarly, the site having the highest rank for attribute ratio of runoff to demand is "69b", and likewise for attribute weighted distance, the site is "52a". It is interesting to see that no aggregation of these attributes has been done by Inamdar et al. [11].

The top ranked sites for the Australian site through a non-heuristic approach comes out to be "14b", "69b", "41c", "44b", and "44c". It is interesting to observe that the sites finalized by our approach matches with the sites finalized by Inamdar et al. [11] with the help of planners.

The present approach considers a unified view of all attributes and provides an integrated score. This integrated score forms the basis for selecting a few top sites. This practice of aggregating attributes is widely used in literature [35,36]. It is needless to emphasize that ranking based on "n" attributes will create a pool of "n" options, whereas the present approach will provide only one aggregated score. Thus, the alternatives evolved in the present approach will provide a smaller pool for alternatives to be picked up.

Thus, this methodology reduces a large pool of data sets and also provides the reliable shortlisted sites suitable for stormwater harvesting. Once a pool of the top, say 5 to 10 sites, is identified, more rigorous analysis using financial, social, environmental aspects can be adopted to choose most appropriate sites suitable for a specific location. The results obtained by non- heuristic approaches, i.e., entropy weight method and PCA are in the good agreement with the results obtained by concerning various water planners of Melbourne city and Dehradun city. As a non-heuristic approach does not involve subjectivity and capture the sites of Inamdar et al. [11], certainly there is a merit in using non-heuristic approaches. Utility value of the study is that it provides a rational procedure to rank the sites and this procedure is consistent with a similar application in other disciplines. The work is significant as it removes the ad hoc approach of selecting the sites based on isolated attributes and subjectivity in ranking is also avoided using this approach.

Thus, this methodology reduces time and subjectivity in creating a set of few suitable storm water harvesting sites uses from which planners can take a quick and efficient decision in finalizing suitable sites for Storm Water Harvesting at a specific location.

5. Conclusions

The study focuses on screening a few suitable SWH sites within a region of interest using a GIS based robust methodology, which utilizes Demand, ratio of runoff to demand and weighted demand distance as screening parameters. A suitable site should fulfill the criteria of high demand, high ratio of runoff to demand and a low weighted demand distance. It was observed that, while allotting the ranks to the shortlisted sites, the same sites have obtained different ranks for different parameters and are also influenced by subjectivity in decision-making. This makes it difficult for the water planners to make a quick decision and hence completing the process of site selection for SWH. A new methodology is adopted that transforms the weight of all the parameters for the heuristic and non-heuristic approaches. Saaty AHP, PCA and entropy weight methods are applied for allotting weights to the parameters. The methodology is applied both for Melbourne city and Dehradun city. The results obtained by non-heuristic approaches, i.e., PCA and entropy weight method were good for Melbourne city as well as for Dehradun city. Thus, the proposed methodology has the potential of application in decision-making of suitable storm water harvesting sites. Change in climate and LULC may affect the precipitation patterns and runoff generated. This coupled with dynamic changes in the demand pattern may lead to re-evaluation of existing storm water harvesting sites. Furthermore, changes may also take place in the screening parameters. Under these conditions, the use of non-heuristic approaches can work as a potential tool to screen out several alternate options or augmentation of existing sites. It is also recommended to conduct further studies on sensitivity analysis of the ranking parameters and applications in different rainfall contexts such as in arid regions to strengthen the storm water harvesting site selection process.

Acknowledgments: The authors are thankful to Ashok K. Sharma, Associate Professor, Victoria University for sending us relevant papers and helping us to grasp the concept of stormwater harvesting. The authors are also thankful to Ravi Pandey, Uttarakhand Jal Vidyut Nigam Ltd., Dehradun, India (UJVNL), water irrigation officer for providing their expertise and opinions for the ranking of stormwater harvesting sites for Dehradun city.

Author Contributions: The paper was written by Shray Pathak based on the results obtained from his research which was done under the close supervision of Chandra Shekhar Prasad Ojha, Chris Zevenbergen and Rahul Dev Garg. All co-authors helped to review and improve the paper.

Conflicts of Interest: The authors declare no conflict of interest.

References

1. Brown, R. Impediments to integrated urban stormwater management: The need for institutional reform. *Environ. Manag.* **2005**, *36*, 455–468. [CrossRef] [PubMed]
2. Sörensen, J.; Persson, A.; Sternudd, C.; Aspegren, H.; Nilsson, J.; Nordström, J.; Larsson, R. Re-Thinking Urban Flood Management—Time for a Regime Shift. *Water* **2016**, *8*, 332. [CrossRef]
3. Hamdan, S.M. A literature based study of stormwater harvesting as a new water resource. *Water Sci. Technol.* **2009**, *60*, 1327–1339. [CrossRef] [PubMed]

4. Hatt, B.E.; Deletic, A.; Fletcher, T.D. Integrated treatment and recycling of stormwater: A review of Australian practice. *J. Environ. Manag.* **2006**, *79*, 102–113. [CrossRef] [PubMed]

5. Akram, F.; Rasul, M.G.; Khan, M.M.K.; Amir, M.S.I. A Review on Stormwater Harvesting and Reuse. World Academy of Science, Engineering and Technology. *Int. J. Environ. Chem. Ecol. Geol. Geophys. Eng.* **2014**, *8*, 188–197.

6. Goonrey, C.M.; Perera, B.J.C.; Lechte, P.; Maheepala, S.; Mitchell, V.G. A technical decision-making framework: Stormwater as an alternative supply source. *Urban Water J.* **2009**, *6*, 417–429. [CrossRef]

7. Plappally, A.K.; Lienhard, V.J.J. Energy requirement for water production, treatment, end use, reclamation, and disposal. *Renew. Sustain. Energy Rev.* **2012**, *16*, 4818–4848. [CrossRef]

8. Jato-Espino, D.; Sillanpää, N.; Charlesworth, S.M.; Andrés-Doménech, I. Coupling GIS with Stormwater Modelling for the Location Prioritization and Hydrological Simulation of Permeable Pavements in Urban Catchments. *Water* **2016**, *8*, 451. [CrossRef]

9. Kumar, M.; Agarwal, A.; Bali, R. Delineation of potential sites for water harvesting structures using remote sensing and GIS. *J. Indian Soc. Remote Sens.* **2008**, *36*, 323–334. [CrossRef]

10. Singh, J.; Singh, D.; Litoria, P. Selection of suitable sites for water harvesting structures in Soankhad watershed, Punjab using remote sensing and geographical information system (RS and GIS) approach- A case study. *J. Indian Soc. Remote Sens.* **2009**, *37*, 21–35. [CrossRef]

11. Inamdar, P.M.; Cook, S.; Sharma, A.K.; Corby, N.; O'Connor, J.; Perera, B.J.C. A GIS based screening tool for locating and ranking of suitable stormwater harvesting sites in urban areas. *J. Environ. Manag.* **2013**, *128*, 363–370. [CrossRef] [PubMed]

12. Saaty, T.L. How to make a decision: the analytic hierarchy process. *Eur. J. Oper. Res.* **1990**, *48*, 9–26. [CrossRef]

13. Kim, J.O.; Mueller, C.W. *Introduction to Factor Analysis: What It Is and How to Do It*; Sage Publications: New York, NY, USA, 1978.

14. Kim, J.O.; Mueller, C.W. *Factor Analysis: Statistical Methods and Practical Issues*; Sage Publications: New York, NY, USA, 1978.

15. Rummel, R.J. *Applied Factor Analysis*; Northwestern University Press: Evanston, IL, USA, 1970; pp. 1–616. ISBN 978-0-8101-0824-0.

16. Stevens, J.P. *Applied Multivariate Statistics for the Social Sciences*; Lawrence Erlbaum Associates: Hillsdale, NJ, USA, 1986; pp. 1–650. ISBN 0-898-59568-1.

17. Abdi, H.; Williams, L.J.; Valentin, D. Multiple factor analysis: Principal component analysis for multitable and multiblock data sets. *Wiley Interdiscip. Rev. Comput. Stat.* **2013**, *5*, 149–179. [CrossRef]

18. Luan, W.; Lu, L.; Li, X.; Ma, C. Weight Determination of Sustainable Development Indicators Using a Global Sensitivity Analysis Method. *Sustainability* **2017**, *9*, 303. [CrossRef]

19. Tscheikner-Gratl, F.; Egger, P.; Rauch, W.; Kleidorfer, M. Comparison of Multi-Criteria Decision Support Methods for Integrated Rehabilitation Prioritization. *Water* **2017**, *9*, 68. [CrossRef]

20. Dyer, J.S. A Clarification of "Remarks on the Analytic Hierarchy Process". *Manag. Sci.* **1990**, *36*, 274–275. [CrossRef]

21. Harkar, P.T.; Vargas, L.G. Reply to "Remarks on the Analytic Hierarchy Process" by J.S. Dyer. *Manag. Sci.* **1990**, *36*, 269–273. [CrossRef]

22. Ouma, Y.O.; Tateishi, R. Urban flood vulnerability and risk mapping using integrated multi-parametric AHP and GIS: methodological overview and case study assessment. *Water* **2014**, *6*, 1515–1545. [CrossRef]

23. Adelmann, L.; Sticha, P.J.; Donnell, M.L. An Experimental Investigation of the Relative Effectiveness of Multiattribute Weighting Techniques. *Organ. Behav. Hum. Perform.* **1986**, *33*, 243–262. [CrossRef]

24. Stillwell, W.G.; Winterfeld, D.; John, R.S. Comparing Hierarchical and Nonhierarchical Weighting Methods for Eliciting Multiattribute Value Models. *Manag. Sci.* **1987**, *33*, 442–450. [CrossRef]

25. Adham, A.; Riksen, M.; Ouessar, M.; Ritsema, C.J. A methodology to assess and evaluate rainwater harvesting techniques in (semi-) arid regions. *Water* **2016**, *8*, 198. [CrossRef]

26. Arrington, C.E.; Jensen, R.E.; Tokutani, M. Scaling of corporate multivariate performance criteria subjective composition versus the analytic hierarchy process. *J. Account. Public Policy* **1982**, *1*, 95–123. [CrossRef]

27. Lockett, G.; Hetherington, B.; Yallup, P.; Stratford, M.; Cox, B. Modelling a research portfolio using AHP: A group decision process. *R & D Manag.* **1986**, *16*, 151–160. [CrossRef]

28. Zou, Z.H.; Yi, Y.; Sun, J.N. Entropy method for determination of weight of evaluating indicators in fuzzy synthetic evaluation for water quality assessment. *J. Environ. Sci.* **2006**, *18*, 1020–1023. [CrossRef]

29. Singh, V.P. *Entropy Theory and Its Application in Environmental and Water Engineering*; John Wiley and Sons: Hoboken, NJ, USA, 2013; pp. 1–661. ISBN 978-1-119-97656-1.

30. Ojha, C.S.P.; Bhunya, P.; Berndtsson, R. *Engineering Hydrology*, 1st ed.; Oxford University Press: Oxford, UK, 2008; pp. 1–511. ISBN 9780195694611.

31. Mishra, S.K.; Jain, M.K.; Pandey, R.P.; Singh, V.P. Catchment area based evaluation of the AMC-dependent SCS-CN-inspired rainfall-runoff models. *Hydrol. Process* **2005**, *19*, 546–565. [CrossRef]

32. Pathak, S.; Ojha, C.S.P.; Zevenbergen, C.; Garg, R.D. Assessing Stormwater Harvesting Potential in Dehradun city Using Geospatial Technology. In *Development of Water Resources in India*; Garg, V., Singh, V.P., Raj, V., Eds.; Springer: Cham, Switzerland, 2017; pp. 47–60. ISBN 978-3-319-55124-1.

33. Kułakowski, K. Heuristic rating estimation approach to the pairwise comparisons method. *Fundam. Inform.* **2014**, *133*, 367–386. [CrossRef]

34. Ding, X.; Chong, X.; Bao, Z.; Xue, Y.; Zhang, S. Fuzzy Comprehensive Assessment Method Based on the Entropy Weight Method and Its Application in the Water Environmental Safety Evaluation of the Heshangshan Drinking Water Source Area, Three Gorges Reservoir Area, China. *Water* **2017**, *9*, 329. [CrossRef]

35. Sargaonkar, A.; Deshpande, V. Development of an overall index of pollution for 573 surface water based on a general classification scheme in Indian context. *Environ. Monit. Assess.* **2003**, *89*, 43–67. [CrossRef] [PubMed]

36. Ojha, C.S.P.; Goyal, M.K.; Kumar, S. Applying fuzzy logic and point count system to select landfill sites. *Environ. Monit. Assess.* **2007**, *135*, 99–106. [CrossRef] [PubMed]

water

MDPI

Article

Study on Mercury Distribution and Speciation in Urban Road Runoff in Nanjing City, China

Rajendra Prasad Singh [1,2,*], Jiaguo Wu [1,2], Alagarasan Jagadeesh Kumar [1,2] and Dafang Fu [1,*]

[1]   School of Civil Engineering, Southeast University, Sipai Lou 2#, Nanjing 210096, China;
      wujiaguo2017@163.com (J.W.); jaga.jagadeesh1987@gmail.com (A.J.K.)
[2]   Southeast University-Monash University Joint Research Centre for Water Sensitive Cities,
      Nanjing 210096, China
*   Correspondence: rajupsc@seu.edu.cn (R.P.S.); fdf@seu.edu.cn (D.F.); Tel.: +86-13301580003 (D.F.)

Received: 7 September 2017; Accepted: 2 October 2017; Published: 12 October 2017

Abstract: The current study was aimed to investigate the mercury pollution in urban road runoff. A total of 34 rainfall events were monitored on 5 independent road catchments from 2015 to 2016 in Nanjing city, China. Events mean concentrations of mercury and the impact factors of mercury pollution in urban road runoff were also carried out in the current study. Results revealed that the concentration of various mercury species was very high. Total mercury, dissolved mercury and particulate mercury were found to be in the range of 0.173–8.254 µg/L, 0.069–6.823 µg/L, and 0.086–2.485 µg/L, respectively. The order of total mercury concentration among the five catchments was as follows: Longpan road > Xinjiekou > Jiulonghu > Zhujiang road > Maqun area. Results revealed the existence of different dominant species of mercury in different urban areas. Particularly, mercury in urban road runoff mainly existed in particulate form in Maqun area, and the concentrations of inactive mercury (0.250–2.821 µg/L) were far more than that of volatile mercury (0.023–0.215 µg/L) and active mercury (0.026–0.359 µg/L). The order of impact factors of rainfall characteristics on Hg pollution in runoff was dry periods > runoff time > duration of rainfall > storm intensity > rainfall. Analysis based on the first flush effect showed that the first flush phenomenon of mercury was not significant.

Keywords: mercury species; urban road runoff; events mean concentrations (EMCs); water pollution; first flush

1. Introduction:

Mercury (Hg) is a global pollutant which is accumulated in the long term through natural and anthropogenic activities in the environment in organic or inorganic form [1–3]. As a kind of non-biological metabolism of toxic heavy metals, Hg in low concentrations is still seriously harmful to the natural environment and human health [4]. There are two categories of the main sources of Hg, natural sources and anthropogenic sources. Natural sources mainly include volcanic, forest fire, soil and water Hg release [5], whereas anthropogenic sources include traffic activities, ore smelting, garbage incineration and fossil fuel combustion, etc. [6]. Along with the advancement of urbanization, industrialization and human activities intensifying, Hg pollution prevention is essential to reduce more serious pollution of mercury in the urban environment.

With acceleration of the urbanization process, polluted runoff as one of the major causes of water quality impairment in urban area has attracted more and more attention. In recent decades, researchers have investigated various contaminants in road runoff such as suspended solids, nutrients and common heavy metals like Cu, Zn, Pb and Cr [7–11], but there are few studies about Hg pollution in urban road runoff [12]. Mercury has drawn global attention due to its ability to contaminate entire water bodies from remote non-point source trace level inputs that bio-accumulate through the food chain [13]. It is reported that rivers flowing through urban areas have higher Hg concentrations compared to the rural areas [14,15], probably caused by pollution from expressway runoff. Hg in urban runoff mainly comes from direct anthropogenic activities such as mining, or indirectly through dry and wet deposition such as atmospheric deposition, vehicle sources, and the road surface wear. There are a variety of different forms of Hg in the environment which can be classified as particulate Hg (PHg) and dissolved Hg (DHg), according to their solubility. Furthermore, Hg can also be classified as organic mercury and inorganic mercury [12,16], in which the methyl Hg (MeHg) is best known due to its high toxicity and bioavailability. Because of the different hazards of various Hg species, investigation of each species in urban road runoff is highly significant to assess the pollution level of runoff and to determine the treatment process of Hg pollution.

In order to have a relatively accurate assessment of Hg pollution in urban road runoff in Nanjing city, the current study aimed to investigate the event mean concentrations (EMCs) of different Hg species during various rainfall events. The impact factors of the Hg pollution and the first flush in urban road runoff were also analyzed. Furthermore, the current study also aimed to assess the level of Hg pollution in urban road runoff, and meanwhile, to provide basic data and theoretical support for the treatment of Hg pollution and the reuse of stormwater.

2. Methods

2.1. Site Description

Nanjing is located in the middle and lower reaches of the Yangtze River Delta region in East China (Figure 1), with an average temperature of 15.3 °C. Annual average rainfall in Nanjing is 1106.05 mm, with most of the rainfall events occurring in summer time. Five urban road catchments divided according to the surrounding land use in Nanjing were selected for monitoring during the period of April 2015–May 2016 (Table 1).

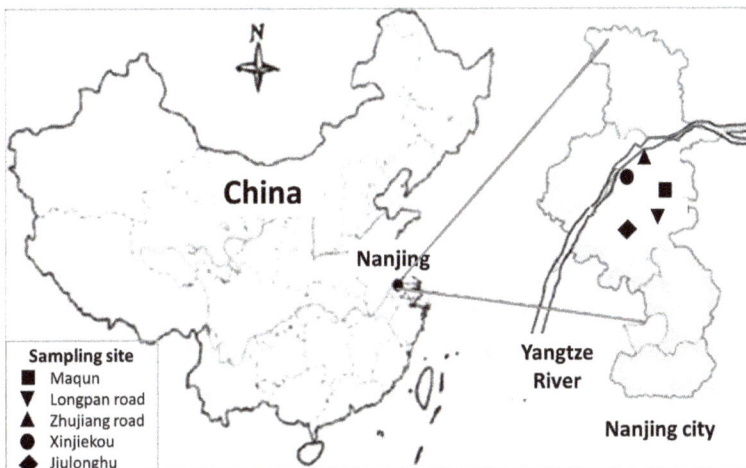

Figure 1. Sampling sites.

Table 1. Characteristics of studied catchments.

Catchment Identification	Area (m^2)	Surrounding Land Use	Road Material	Vehicle Flow Rate (/Day)
Maqun	1100	Transports	Pitch	38,400
Longpan road	1200	Transports	Pitch	39,200
Zhujiang road	1300	Electronic	Pitch, cement	30,000
Xinjiekou	1500	Commercial	Pitch, terrazzo	47,800
Jiulonghu	900	University campus, thermal power plant	Pitch	2000

The land use type in surroundings of the study area is mainly transportation. Maqun area of Nanjing city circle expressway is one of the main expressways into the city center situated on the east of Nanjing. Longpan road area has a vehicle flow rate of 39,200 vehicles/day, which was selected to represent the traffic load. Zhujiang road area is the largest distribution center for electronic products in East China. Xinjiekou area is a commercial area located in the center of Nanjing. It is a high-density population area. Jiulonghu area has Southeast University Jiulong lake campus, as well as a thermal power plant also located very close.

2.2. Rainfall Characteristics

A total of 34 effective rainfall events were monitored from April 2015 to May 2016. Rainfall runoff events data are presented in Table 2.

Samples of each rainfall event were manually collected from running water flowing out of the rainwater collection pipe (made of polyvinyl chloride material) below the pavement, for the purpose of investigating the level of Hg pollution in urban road runoff. 1 L polyethylene bottles were used to collect the runoff samples. In the first 30 min of the runoff formation, samples were collected at 5–10 min intervals. Samples were collected at 10–15 min intervals during 30 to 60 min. Following that, samples were collected at 30–60 min intervals to the end of the rainfall event. Samples were collected with precautions and immediately sent to the laboratory. Prior to analysis, a portion of each sample was extracted and filtered with a 0.45 μm filter membrane for dissolved mercury analysis. Then all of the samples except samples for total suspended solid (TSS) analysis were pretreated by adding HCl to adjust pH to less than 1 and then 0.5g K_2CrO_4 to keep the orange color of water, followed by shaking. All of the samples were kept at 0–4 °C temperature to minimize the loss of Hg. JS22 Siphon-hyetometer (Tianjin Meteorological Instrument Industry, China) can record the attributes of rainfall such as volume and intensity, which was placed at 1.5 m above the ground to collect the rainfall data.

Table 2. Rainfall characteristics in study areas.

Site	Date	Rainfall (mm)	Duration of Rainfall (min)	Runoff Time (min)	Max Storm Intensity (mm/min)	Average Storm Intensity (mm/min)	Min Storm Intensity (mm/min)	Dry Periods (h)
Maqun	12 June 2015	1.96	185	37	0.0238	0.0106	0.0060	32.0
	20 June 2015	7.15	450	65	0.0350	0.0159	0.0017	180.5
	24 June 2015	1.27	100	85	0.0267	0.0127	0.0011	83.8
	29 June 2015	22.01	385	15	0.2908	0.0572	0.0017	47.6
	14 July 2015	6.13	438	213	0.0189	0.0140	0.0012	240.5
	19 July 2015	4.40	450	26	0.0500	0.0098	0.0026	98.3
	8 August 2015	14.30	288	38	0.3497	0.0497	0.0031	120.7
	21 August 2015	9.40	158	8	0.4000	0.0595	0.0019	23.0
	25 August 2015	2.35	138	42	0.0400	0.0013	0.0124	59.0
	2 May 2016	4.80	267	21	0.0800	0.0400	0.0188	72.4
	7 May 2016	0.91	121	52	0.0200	0.0075	0.0031	88.3

Table 2. *Cont.*

Site	Date	Rainfall (mm)	Duration of Rainfall (min)	Runoff Time (min)	Max Storm Intensity (mm/min)	Average Storm Intensity (mm/min)	Min Storm Intensity (mm/min)	Dry Periods (h)
Longpan road	21 April 2015	12.00	480	58	0.2500	0.0921	0.0042	192.0
	29 April 2015	58.00	90	30	3.7000	0.6443	0.0073	84.1
	7 May 2015	100.00	180	47	2.0000	0.5884	0.0023	156.0
	14 May 2015	23.50	204	29	0.2000	0.1966	0.0038	27.2
	18 May 2015	18.00	210	64	0.3900	0.1210	0.0083	82.3
	2 May 2016	4.80	267	21	0.0800	0.0400	0.0188	72.4
	7 May 2016	0.91	121	52	0.0200	0.0075	0.0031	88.3
Zhujiang road	29 April 2015	58.00	90	30	3.7000	0.6443	0.0073	84.1
	7 May 2015	100.00	180	47	2.0000	0.5884	0.0023	156.0
	14 May 2015	23.50	204	29	0.2000	0.1966	0.0038	27.2
	18 May 2015	18.00	210	64	0.3900	0.1210	0.0083	82.3
	2 May 2016	4.80	267	21	0.0800	0.0400	0.0188	72.4
	7 May 2016	0.91	121	52	0.0200	0.0075	0.0031	88.3
Xinjiekou	7 May 2015	100.00	180	47	2.0000	0.5884	0.0023	156.0
	14 May 2015	23.50	204	29	0.2000	0.1966	0.0038	27.2
	18 May 2015	18.00	210	64	0.3900	0.1210	0.0083	82.3
	2 May 2016	4.80	267	21	0.0800	0.0400	0.0188	72.4
	7 May 2016	0.91	121	52	0.0200	0.0075	0.0031	88.3
Jiulonghu	7 May 2015	100.00	180	47	2.0000	0.5884	0.0023	156.0
	14 May 2015	23.50	204	29	0.2000	0.1966	0.0038	27.2
	18 May 2015	18.00	210	64	0.3900	0.1210	0.0083	82.3
	2 May 2016	4.80	267	21	0.0800	0.0400	0.0188	72.4
	7 May 2016	0.91	121	52	0.0200	0.0075	0.0031	88.3

2.3. Hg Analysis

Parameters such as total suspended solid (TSS), total Hg (THg), dissolved Hg (DHg), particulate Hg (PHg), volatile Hg (Hg^0, also often referred to as dissolved gaseous mercury), active Hg (which can be considered to mainly correspond to Hg^{2+} [17]), and inactive Hg (Hg^{re}, which is also referred to as residual Hg) were analyzed in the current study.

Total Hg and DHg were determined by Hydra II A Mercury Vapourmeter (Teledyne Leeman Labs, Hudson, NH, USA), while Hg at various valence states was detected by Hydra II C Mercury Vapourmeter (Teledyne Leeman Labs, Hudson, NH, USA). Experimental methods for analyzing various Hg forms are presented in Table 3.

Table 3. Analytical methods of different Hg species.

Hg	Methods
THg	50 µL stormwater, taken by pipette, was placed in a nickel boat, and Hydra II A mercury analyzer was utilized to measure the absolute values of Hg concentrations directly.
DHg	Stormwater samples were filtered through a 0.45 µm filter membrane, and 50 µL sample was analyzed by the method of THg.
PHg	PHg = THg-DHg [18,19]
Hg^0	Concentrated sulfuric acid by sub-boiling distillation was used to acidify 100 mL stormwater; then, N_2 was used to blow the samples at a rate of 350 to 400 mL/min for 30 min, and Hg^0 was captured onto the gold tube. Finally, samples were analyzed by Hydra IIC mercury analyzer.
Hg^{2+}	Stormwater samples, which have been measured for Hg^0, continue to be used to analysis Hg^{2+}. Hg^{2+} in water samples was reduced to Hg^0 by 5 mL of 20% $SnCl_2$. N_2 was used to blow the samples at a rate of 350 to 400 mL/min for 30 min, and Hg^0 as a redox product was captured onto the gold tube. Finally, samples were analyzed by Hydra IIC mercury analyzer.
Hg^{re}	$Hg^{re} = THg-Hg^{0-}-Hg^{2+}$

2.4. Events Mean Concentrations (EMCs)

Event mean concentration is a generally accepted index to assess the pollution levels in road runoff, which was represented by the ratio of total pollution loads and the total volume of runoff (USEPA) [20]. The formulation of EMCs is as follows:

$$\text{EMC} = \bar{C} = \frac{M}{V} = \frac{\int C(t)Q(t)dt}{\int Q(t)dt} \approx \frac{\sum\limits_{t=1}^{t=T} C(t)Q(t)}{\sum\limits_{t=1}^{t=T} Q(t)} \tag{1}$$

M: total mass of the contaminant;

V: total volume of runoff;

C(t): concentrations of the contaminant at different times;

Q(t): flow of runoff;

t: runoff time.

The total runoff volume was not monitored in the current work due to the limitation of equipment, which was alternated by rainfalls. Similarly, the flow of runoff for a certain period was alternated by the product of rainfall intensity in that period and catchment area, for which the total mass of contaminant can be calculated. The beginning of rainfall and the formation time of runoff was monitored to reduce errors. In addition, the evaporation and infiltration of runoff during the rainfall were limited, which could be neglected.

2.5. Measurement of First Flush

First flush (FF) is the phenomenon in which the concentration of pollutants is substantially higher in stormwater runoff in the initial period of of a storm event compared to those obtained during the later stages [21–24]. The phenomenon is described as the relationship of dimensionless cumulative pollutant mass and dimensionless cumulative runoff volume which is calculated by the following formulas.

$$m'(t) = \frac{m(t)}{M} = \frac{\int_0^t c(t)q(t)dt}{\int_0^{t_r} c(t)q(t)dt} \tag{2}$$

$$v'(t) = \frac{v(t)}{V} = \frac{\int_0^t q(t)dt}{\int_0^{t_r} q(t)dt} \tag{3}$$

M: total mass of the contaminant;

V: total volume of runoff;

m(t): Cumulative mass of the contaminant at time t;

v(t): Cumulative volume of runoff at time t;

c(t): concentrations of the contaminant at different times;

q(t): flow of runoff;

t: runoff time;

t_r: runoff total duration.

There are differences in the determination of the first flush phenomenon in different studies. Earlier findings indicated that a first flush occurs at time t if the $m'(t)$ exceeds the $v'(t)$ at all instances during the storm events [25]. Bertrand-Karajewski et al. believed that first flush phenomenon occurs when at least 80% of the pollution load is transferred in the first 30% of the runoff volume [26].

When we define FF_n as the quotient of $m'(t)$ and $v'(t)$, the previous two standards are equivalent to $FF_n > 1$ and $FF_{30} \geq 2.7$.

$$FF_n = \frac{m'(t)}{v'(t)} \tag{4}$$

where n: the proportion of the runoff volume that has been generated to the total runoff volume.

In the current study, FF_{30} was calculated to determine if the first flush phenomenon exists in the various Hg species in urban road runoff.

3. Results and Discussion

3.1. EMCs of Various Hg Species

Event mean concentrations of Hg in stormwater runoff events in 5 sampling point are presented in Table 4. The data revealed that the concentrations of Hg in different forms varied greatly over 34 rainfall events, ranging from 0.173 to 8.254 µg/L for THg, from 0.069 to 6.823 µg/L for DHg, and from 0.086 to 2.485 µg/L for PHg, respectively. The EMCs of THg in 22 stormwater runoff events and also the average value of all rainfall events far exceeded 1.0 µg/L. The relationship of THg mean concentration among five regions was as follows: Longpan road (4.243 µg/L) > Xinjiekou (2.332 µg/L) > Jiulonghu (1.686 µg/L) > Zhujiang road (1.185 µg/L) > Maqun (1.120 µg/L). This phenomenon indicated that Hg concentration in urban road runoff is independent of the traffic flow. The high concentrations of Hg in urban road runoff, therefore, would cause severe pollution once entering water bodies.

Table 4. Event mean concentrations (EMCs) of Hg pollution in urban storm water runoff.

Site	Date	Hg (µg/L)			DHg/PHg	log K_d	TSS (mg/L)
		THg	DHg	PHg			
Maqun	12 June 2015	0.335	0.085	0.250	0.34	4.49	95
	20 June 2015	2.760	0.786	1.974	0.40	4.16	174
	24 June 2015	0.437	0.175	0.262	0.67	4.49	48
	29 June 2015	0.337	0.102	0.235	0.43	4.44	84
	14 July 2015	3.347	0.862	2.485	0.35	4.05	256
	19 July 2015	0.667	0.236	0.431	0.55	4.30	92
	8 August 2015	3.001	0.778	2.223	0.35	3.77	484
	21 August 2015	0.439	0.215	0.224	0.96	4.51	32
	25 August 2015	0.173	0.069	0.104	0.66	4.20	96
	2 May 2016	0.373	0.127	0.246	0.52	4.45	68
	7 May 2016	0.455	0.136	0.319	0.43	4.28	123
	Mean	1.120	0.234	0.796	0.515	4.38	141
	Median	0.439	0.136	0.262	0.43	4.31	95
	Range	0.173~3.347	0.069~0.862	0.104~2.485	0.34~0.96	3.77~4.51	32~484
Longpan road	21 April 2015	4.311	3.300	1.011	3.26	—	—
	29 April 2015	8.254	6.823	1.431	4.77	—	—
	7 May 2015	5.241	4.872	0.369	13.20	—	—
	14 May 2015	2.892	0.491	2.401	0.20	—	—
	18 May 2015	0.519	0.267	0.252	1.06	—	—
	2 May 2016	3.256	2.431	0.825	2.95	—	—
	7 May 2016	5.230	3.767	1.463	2.57	—	—
	Mean	4.243	3.151	1.092	4.00	—	—
	Median	4.311	3.300	1.011	2.95	—	—
	Range	0.519~8.254	0.267~6.823	0.252~2.401	0.20~13.20	—	—
Zhujiang road	29 April 2015	0.813	0.318	0.495	0.64	—	—
	7 May 2015	1.332	0.953	0.379	2.51	—	—
	14 May 2015	1.882	0.745	1.137	0.66	—	—
	18 May 2015	0.711	0.498	0.214	2.33	—	—
	2 May 2016	1.345	0.917	0.428	2.14	—	—
	7 May 2016	1.025	0.782	0.243	3.22	—	—
	Mean	1.185	0.702	0.483	1.92	—	—
	Median	1.179	0.622	0.355	2.24	—	—
	Range	0.711~1.882	0.318~0.953	0.214~1.137	0.64~3.22	—	—

Table 4. *Cont.*

| Site | Date | Hg (µg/L) | | | DHg/PHg | log K_d | TSS (mg/L) |
		THg	DHg	PHg			
Xinjiekou	7 May 2015	3.862	2.171	1.691	1.28	—	—
	14 May 2015	1.892	0.873	1.019	0.86	—	—
	18 May 2015	1.243	0.531	0.712	0.75	—	—
	2 May 2016	2.659	1.528	1.131	1.35	—	—
	7 May 2016	2.005	1.229	0.776	1.58	—	—
	Mean	2.332	1.351	1.432	1.16	—	—
	Median	2.005	0.531	1.691	1.28	—	—
	Range	1.243~3.862	0.531~2.171	0.712~1.691	0.75~1.58	—	—
Jiulonghu	7 May 2015	2.027	1.464	0.563	2.60	—	—
	14 May 2015	0.986	0.663	0.323	2.05	—	—
	18 May 2015	2.045	1.096	0.949	1.15	—	—
	2 May 2016	1.962	1.593	0.369	4.32	—	—
	7 May 2016	1.410	0.972	0.438	2.22	—	—
	Mean	1.686	1.074	0.612	2.47	—	—
	Median	1.962	1.096	0.623	2.22	—	—
	Range	0.986~2.045	0.663~1.593	0.323~0.949	1.15~4.32	—	—

Table 5 shows the relevant studies carried out by other researchers around the world. Data revealed that Hg pollution in urban road runoff in Nanjing has a higher level than other cities. Hg in urban road runoff mainly comes from the wet and dry deposition processes which include the discharge of automobile, the wear degree of the road materials, and some other human activities. It is hypothesized that land uses and weather conditions caused the differences.

Various forms of Hg can be reflected by the proportion of PHg and DHg (Table 4). In Maqun area, the average EMCs of PHg was 0.796 µg/L, which is much higher than the concentration of DHg. The DHg/PHg ratio of runoff in the 11 rainfall events was less than 1 (Table 3), illustrating that Hg concentrations were predominantly in particulate form. The partition coefficient K_d was calculated to explain the distribution for Hg between dissolved and particulate phases (K_d = [ng of Hg (kg of sediment)$^{-1}$]/[µg of Hg (L of rain water)$^{-1}$]) [27,28]. The log K_d values ranged from 3.77 to 4.51, showing that the Hg is associated with the particulate phase. While the DHg/PHg ratio of runoff in Xinjiekou fluctuated at 1, which shows that two Hg forms existed at an equal level. Therefore, it can be concluded that Hg in the urban road runoff of Longpan road, Zhujiang road and Jiulonghu areas was mainly in the dissolved state.

Table 5. EMCs of Hg pollution in different regions (average values).

Sites		Land Using Type	EMCs (µg/L) THg	Reference
near Ontario lake in Canada		Urban	0.015	[12]
Beijing, China		Urban	0.1075	[16]
Shanghai, China		Urban	0.510	[29]
Tianjin, China	City road A	Urban	0.520 (0.412–2.76)	[30]
	City road B	Urban	0.730 (0.174–1.223)	
	Industrial district	Industry	0.660 (0.104–1.182)	
Nanjing, China		Urban	2.036 (0.173–8.254)	Current study

An earlier study by Eckleya and Branfireuna [12] has reported that PHg accounted for 84% of total Hg. Dissolved Hg can be easily absorbed by aquatic organisms, followed by accumulation in the human body through the food chain; while PHg would be adsorbed by sediments over a long time, and transfer to DHg in suitable conditions. The current study revealed different results because the samples analyzed were all collected from urban road runoff, where the biological absorption and accumulation are not obvious. This is the reason for the occurrence of different dominant Hg species in different regions.

3.2. EMCs of Hg in Different Valence States

It is well known that the toxicity of Hg in different states of Hg is different. In addition to methyl mercury (MeHg), highly active divalent mercury is one of the most toxic pollutants. Investigation of different Hg species in urban road runoff is highly significant to assess the Hg pollution level. The concentrations of Hg in different valence states (Hg^0, Hg^{2+}, and Hg^{re}) in urban road runoff in Maqun area were monitored for 6 rainfall events, and the results are provided in Figure 2.

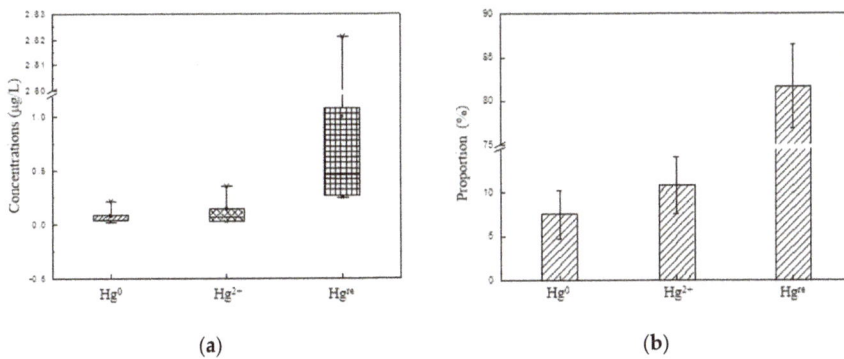

(a)   (b)

Figure 2. Distribution of Hg in different valence states: (a) Concentrations of Hg^0, Hg^{2+} and Hg^{re}, and (b) proportions of Hg^0, Hg^{2+} and Hg^{re}.

The concentrations of EMCs of Hg^0, Hg^{2+}, and Hg^{re} changed greatly, with 0.023–0.215 µg/L, 0.026-0.359 µg/L, and 0.250–2.821 µg/L, respectively (Figure 2a). The average concentration of Hg^{re} was 1.080 µg/L, which was much higher than of the concentrations of Hg^0 (0.090 µg/L) and Hg^{2+} (0.150 µg/L). Wang et al. [16] investigated the EMCs of Hg in different species in a farmland near Beijing and found the EMCs of Hg^0, Hg^{2+} and Hg^{re} were 0.011 µg/L, 0.0429 µg/L and 0.0536 µg/L, respectively, which were all lower than the results of the current study, indicating that Hg pollution of urban road runoff was much more serious, and also demonstrating different occurrence regularity in different regions.

Figure 2 shows the event mean concentrations (EMCs) (Figure 2a) and proportions (Figure 2b) of different Hg species in urban road runoff. The average percentage of Hg^{re} was 81.66%, which is more than 10 times of Hg^0 (7.51%). Hg^{re} is relatively stable and has the least hazard compared to other species. The proportion of Hg^{2+} was 10.85%, which is relatively activated and transforms easily to MeHg.

Earlier findings of a study carried out in Beijing by Liu et al. revealed that the concentration of Hg^{re} (0.171 µg·L^{-1}) was much higher than the concentrations of Hg^0 (0.039 µg·L^{-1}) and Hg^{2+} (0.066 µg·L^{-1}) [31], which is similar to the current study. Whereas, Zhang et al. reported that the concentration of Hg^{re} was at the same level with Hg^{2+}, both of which were about 4 to 5 times higher than the concentration of Hg^0 in Shanghai [29]. Therefore, it is assumed that the presence of Hg species in urban road runoff greatly depends on the sources of Hg in the environment.

3.3. Relationship between EMCs of Hg and TSS

Pearson correlation coefficient data of various Hg species and TSS is presented in Table 6. Results revealed that all of the Hg species are positively correlated with TSS ($p < 0.01$) except Hg^0 ($p < 0.05$ levels) in the current study and the Pearson correlation coefficients ranged from 0.768 to 0.954. For Hg in different occurrence forms in urban road runoff, the Pearson correlation coefficient of PHg was highest, which can also illustrate that Hg mainly existed in particulate form in Maqun area. It was reported that there was a significant TSS/THg relationship ($r^2 = 0.67$, $p < 0.01$) in Toronto near Lake Ontario [12], which is similar with the current study ($r^2 = 0.657$, Figure 3a). The Pearson correlation coefficient of Hg^{re} was highest for Hg in various valence states. Furthermore, the concentrations of Hg^{re} and TSS have liner correlation ($r^2 = 0.9105$, Figure 3b), illustrating that most of Hg^{re} adsorbed on particles in urban road runoff [32].

Table 6. Pearson correlation coefficients of different Hg species and total suspended solid (TSS).

Hg	TSS	Significance Level
THg	0.807	0.01
DHg	0.768	0.01
PHg	0.819	0.01
Hg°	0.850	0.05
Hg^{2+}	0.923	0.01
Hg^{re}	0.954	0.01

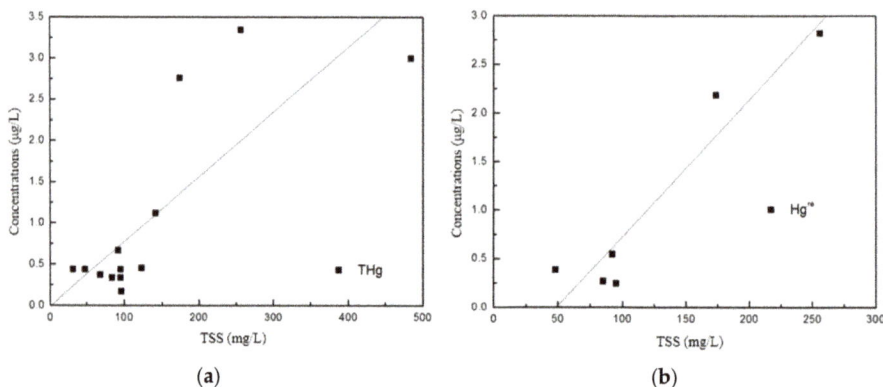

Figure 3. Correlation between TSS and Hg: (**a**) total Hg vs. TSS; and (**b**) residual Hg vs. TSS.

3.4. Influence of Rainfall Characteristics

Cluster analysis is a general method to reveal the relationship among multiple variables and adopted to reveal the different characteristics of rainfalls effect to EMCs of Hg species [33]. The similarity between different variables in tree diagrams of cluster analysis was depicted by Mini tab to represent the impact levels. The length and the correlation are inversely proportional; therefore it can be taken to assess the influence levels of characteristics of rainfall to Hg pollution in urban road runoff.

Nanjing city expressway was chosen to find out that some impact factors, including road materials, land use types, vehicle flow rate, atmospheric sedimentation, and the methods of road cleaning were relatively stable. Dry periods, runoff time, rainfall, rainfall duration, and rainfall intensity (including max. rainfall intensity and average rainfall intensity) were considered as the main influence factors. The cluster analysis of these factors and Hg in different occurrences is shown in Figure 4. The influence factors of rainfall characteristics on different Hg species were found to be similar. Results revealed a notable correlation between dry periods and concentration of all forms of Hg, indicating that Hg pollution in road runoff mainly comes from the accumulation of particle contaminants in dry periods. Results are consistent with an earlier study which reported that pollutants accumulate on urban surfaces mainly from dry atmospheric deposition as well as from vehicle sources during dry periods [34]. The order for impact factors on Hg deposition is as follows: dry periods > runoff time > rainfall duration > rainfall intensity > rainfall.

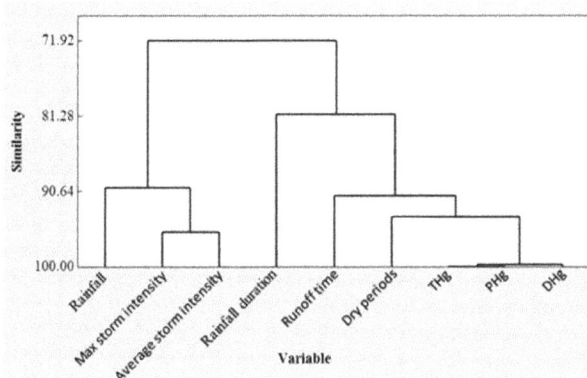

Figure 4. Cluster analysis of different Hg species and characteristics of rainfalls.

3.5. First Flush Effect (FF$_{30}$)

The FF$_{30}$ of various forms of Hg in urban road runoff at 4 different sites are presented in Table 7. Findings revealed that in over 23 rainfall events, the FF$_{30}$ of Hg in different occurrence forms ranges from 0.132 to 0.570 for THg, from 0.036 to 0.793 for DHg, and from 0.177 to 0.907 for PHg, respectively. All of the FF$_{30}$ values were lower than 2.7, means that none of the rainfall events were fulfilling the criteria of 80% of pollution load in the first 30% of the volume. These results are consistent if the relatively weak standard (FF$_n$ > 1) is adopted. Therefore, there is no first flush of Hg but there is a significant dilution effect in the urban road runoff.

Table 7. First Flush effect (FF$_{30}$) of THg, DHg and PHg in different rainfall events.

Site	Date	FF$_{30}$ of Hg		
		THg	DHg	PHg
Longpan road	21 April 2015	0.479	0.485	0.508
	29 April 2015	0.471	0.489	0.384
	7 May 2015	0.378	0.337	0.907
	14 May 2015	0.276	0.636	0.195
	18 May 2015	0.535	0.233	0.372
	2 May 2016	0.523	0.638	0.583
	7 May 2016	0.492	0.448	0.473
Zhujiang road	29 April 2015	0.132	0.793	0.217
	7 May 2015	0.179	0.205	0.181
	14 May 2015	0.496	0.102	0.332
	18 May 2015	0.319	0.284	0.177
	2 May 2016	0.255	0.639	0.291
	7 May 2016	0.364	0.584	0.195
Xinjiekou	7 May 2015	0.287	0.305	0.287
	14 May 2015	0.483	0.402	0.483
	18 May 2015	0.373	0.356	0.386
	2 May 2016	0.404	0.399	0.329
	7 May 2016	0.378	0.428	0.372
Jiulonghu	7 May 2015	0.570	0.160	0.355
	14 May 2015	0.450	0.036	0.692
	18 May 2015	0.407	0.282	0.482
	2 May 2016	0.529	0.183	0.622
	7 May 2016	0.358	0.157	0.475
	Range	0.132~0.570	0.036~0.793	0.177~0.907

4. Conclusions

Results from this study revealed that the Hg concentrations of different forms in urban road runoff varied greatly and ranged from 0.173 to 8.254 µg/L for THg, 0.069 to 6.823 µg/L for DHg, and 0.086 to 2.485 µg/L for PHg. Results also showed the existence of different dominant forms of Hg in different studied regions. The range of EMCs of Hg in different valence states was 0.023–0.215 µg/L for Hg^0, 0.026–0.359 µg/L for Hg^{2+}, and 0.250–2.821 µg/L for Hg^{re}. The concentration of Hg^{re} was higher than the concentration of Hg^0 and Hg^{2+}. Different Hg species in runoff were positively correlated with TSS, indicating that Hg mainly existed as particle type in Maqun area, and most of the Hg can be removed by precipitation. The order of impact factors on Hg pollution was as follows: dry periods > runoff time > duration of rainfall > storm intensity > rainfall. Current results are highly significant for understanding the Hg concentrations in urban road runoff. Outcomes of the current study will also be helpful in carrying out further research aiming to investigate the fate and transformation behavior of Hg during various treatment processes.

Acknowledgments: This research was supported by the National Natural Science foundation of China NSFC Grants No. 51550110231, 51650410657).

Author Contributions: Dafang Fu and Rajendra Prasad Singh conceived and designed the experiments; Jiaguo Wu performed the experiments and analyzed the data with Rajendra Prasad Singh; Alagarasan Jagadeesh Kumar contributed reagents/materials/analysis tools; and Rajendra Prasad Singh wrote the paper.

Conflicts of Interest: The authors declare no conflict of interest.

References

1. Li, P.; Feng, X.B.; Qiu, G.L.; Shang, L.H.; Li, Z.G. Mercury pollution in Asia: A review of the contaminated sites. *J. Hazard. Mater.* **2009**, *168*, 591–601. [CrossRef] [PubMed]
2. Falandysz, J.; Zhang, J.; Wang, Y.Z.; Krasińska, G.; Kojta, A.; Saba, M.; Shen, T.; Li, T.; Liu, H.G. Evaluation of the mercury contamination in mushrooms of genus Leccinum from two different regions of the world: Accumulation, distribution and probable dietary intake. *Sci. Total Environ.* **2015**, *537*, 470–478. [CrossRef] [PubMed]
3. Fricke, I; Götz, R.; Schleyer, R.; Püttmann, W. Analysis of sources and sinks of mercury in the urban water cycle of Frankfurt am Main, Germany. *Water* **2015**, *7*, 6097–6116. [CrossRef]
4. Liu, Y.R.; Lu, X.; Zhao, L.; An, J.; He, J.Z.; Pierce, E.M.; Johs, A.; Gu, B.H. Effects of cellular sorption on mercury bioavailability and methyl mercury production by desulfovibriodesulfuricans ND132. *Environ. Sci. Technol.* **2016**, *50*, 13335–13341. [CrossRef] [PubMed]
5. Zhang, L.; Wu, Z.; Cheng, I.; Wright, L.P.; Olson, M.L.; Gay, D.A.; Risch, M.R.; Brooks, S.; Castro, M.S.; Conley, G.D.; et al. The Estimated six-year mercury dry deposition across North America. *Environ. Sci. Technol.* **2016**, *50*, 12864–12873. [CrossRef] [PubMed]
6. Chung, S.; Chon, H.T. Assessment of the level of mercury contamination from some anthropogenic sources in Ulaanbaatar, Mongolia. *J. Geochem. Explor.* **2014**, *147*(Part B), 237–244. [CrossRef]
7. Gill, L.W.; Ring, P.; Higgins, N.M.P.; Johnston, P.M. Accumulation of heavy metals in a constructed wetland treating road runoff. *Ecol. Eng.* **2014**, *70*, 133–139. [CrossRef]
8. Sörme, L.; Lagerkvist, R. Sources of heavy metals in urban wastewater in Stockholm. *Sci. Total Environ.* **2002**, *298*, 131–145. [CrossRef]
9. Li, H.; Shi, J.; Shen, G. Characteristics of metals pollution in expressway stormwater runoff. *Environ. Sci.* **2009**, *30*, 1621–1625. (In Chinese)
10. Han, J.C.; Gao, X.L.; Liu, Y.; Wang, H.W.; Chen, Y. Distributions and transport of typical contaminants in different urban stormwater runoff under the effect of drainage systems. *Desalin. Water Treat.* **2014**, *52*, 1455–1461. [CrossRef]
11. Borne, K.E.; Fassman-Beck, E.A.; Tanner, C.C. Floating treatment wetland influences on the fate of metals in road runoff retention ponds. *Water Res.* **2014**, *48*, 430–442. [CrossRef] [PubMed]
12. Eckley, C.S.; Brianfireun, B. Mercury mobilization in urban road runoff. *Sci. Total Environ.* **2008**, *403*, 164–177. [CrossRef] [PubMed]

13. Ratcliffe, H.E.; Swanson, G.M.; Fischer, L.J. Human exposure to mercury: A critical assessment of the evidence of adverse health effects. *J. Toxicol. Environ. Health* **1996**, *49*, 221–230. [CrossRef] [PubMed]

14. Mason, R.P.; Sullivan, K.A. Mercury and methyl mercury transport through an urban watershed. *Water Res.* **1998**, *32*, 321–330. [CrossRef]

15. Lawson, N.M.; Mason, R.P.; Laporte, J.M. The fate and transport of mercury, methyl mercury, and other trace metals in Chesapeake Bay tributaries. *Water Res.* **2001**, *35*, 501–515. [CrossRef]

16. Wang, W.; Liu, J.; Peng, A. Mercury origin in surface runoff caused by precipitation. *Agro-Environ. Prot.* **2001**, *20*, 297–301. (In Chinese)

17. Tomiyasu, T.; Minato, T.; Ruiz, W.L.G.; Kodamatani, H.; Kono, Y.; Hidaka, M.; Oki, K.; Kanzaki, R.; Taniguchi, Y.; Matsuyama, A. Influence of submarine fumaroles on the seasonal changes in mercury species in the waters of Kagoshima Bay, Japan. *Mar. Chem.* **2015**, *177*, 763–771. [CrossRef]

18. Rimondi, V.; Costagliola, P.; Gray, J.E.; Lattanzi, P.; Nannucci, M.; Paolieri, M.; Salvadori, A. Mass loads of dissolved and particulate mercury and other trace elements in the Mt. Amiata mining district, Southern Tuscany (Italy). *Environ. Sci. Pol. Res.* **2014**, *21*, 5575–5585. [CrossRef] [PubMed]

19. Zhao, L.; Guo, Y.; Meng, B.; Yao, H.; Feng, X. Effects of damming on the distribution and methylation of mercury in Wujiang River, Southwest China. *Chemosphere* **2017**, *185*, 780–788. [CrossRef] [PubMed]

20. US Environment Protection Agency (USEPA). *Results of the Nationwide Urban Runoff Program*; US Environment Protection Agency: Washington, DC, USA, 1993.

21. Gupta, K.; Saul, A.J. Specific relationships for the first flush load in combined sewer flows. *Water Res.* **1996**, *30*, 1244–1252. [CrossRef]

22. Schriewer, A.; Horn, H; Helmreich, B. Time focused measurements of roof runoff quality. *Corros. Sci.* **2008**, *50*, 384–391. [CrossRef]

23. Gikas, G.D.; Tsihrintzis, V.A. Assessment of water quality of first-flush roof runoff and harvested rainwater. *J. Hydrol.* **2012**, *466*, 115–126. [CrossRef]

24. Ma, Z.B.; Ni, H.G.; Zeng, H.; Wei, J.B. Function formula for first flush analysis in mixed watersheds: A comparison of power and polynomial methods. *J. Hydrol.* **2011**, *402*, 333–339. [CrossRef]

25. Lee, J.H.; Bang, K.W.; Ketchum, L.H.; Choe, J.S.; Yu, M.J. First flush analysis of urban storm runoff. *Sci. Total Environ.* **2002**, *293*, 163–175. [CrossRef]

26. Bertrand-Krajewski, J.L.; Chebbo, G.; Saget, A. Distribution of pollutant mass vs volume in stormwater discharges and the first flush phenomenon. *Water Res.* **1998**, *32*, 2341–2356. [CrossRef]

27. Nevado, J.J.B.; Rodríguez, R.C.; Moreno, M.J. Mercury speciation in the Valdeazogues River–La Serena Reservoir system: Influence of Almadén (Spain) historic mining activities. *Sci. Total Environ.* **2009**, *407*, 2372–2382. [CrossRef] [PubMed]

28. Benoit, J.M.; Gilmour, C.C.; Mason, R.P.; Heyes, A. Sulfide controls on mercury speciation and bioavailability to methylating bacteria in sediment pore waters. *Environ. Sci. Technol.* **1999**, *33*, 951–957. [CrossRef]

29. Zhang, J.; Bi, C.; Chen, Z. Assessment on the effect of surface dust on mercury and arsenic in rainfall-runoff in the Shanghai urban district. *J. East China Norm. Univ. (Nat. Sci.)* **2011**, *1*, 195–202. (In Chinese)

30. Li, Q.; Li, T.; Zhao, Q. Characteristics of heavy metal pollution in road rainfall-runoff of Tianjin city. *Ecol. Environ. Sci.* **2011**, *20*, 143–148. (In Chinese)

31. Liu, J.; Wang, W.; Peng, A. Mercury Species in Precipitation. *Chin. J. Environ. Sci.* **2000**, *21*, 43–47. (In Chinese).

32. Paraquetti, H.H.M.; Ayres, G.A.; de Almeida, M.D.; Molisani, M.M.; Lacerda, L.D. Mercury distribution, speciation and flux in the Sepetiba Bay tributaries, SE Brazil. *Water Res.* **2004**, *38*, 1439–1448. [CrossRef] [PubMed]

33. León-Borges, J.A.; Lizardi-Jiménez, M.A. Hydrocarbon pollution in underwater sinkholes of the Mexican Caribbean caused by tourism and asphalt: Historical data series and cluster analysis. *Tour. Manag.* **2017**, *63*, 179–186. [CrossRef]

34. Charlesworth, S.; Everett, M.; McCarthy, R. A comparative study of heavy metal concentration and distribution in deposited street dusts in a large and a small urban area: Birmingham and Coventry, West Midlands, UK. *Environ. Int.* **2003**, *29*, 563–573. [CrossRef]

water

MDPI

Article

Study on Storm-Water Management of Grassed Swales and Permeable Pavement Based on SWMM

Jianguang Xie *, Chenghao Wu, Hua Li and Gengtian Chen

Department of Civil Engineering, Nanjing University of Aeronautics and Astronautics, Nanjing 210016, China; Williamchenghao@163.com (C.W.); lihua112358@gmail.com (H.L.); cgt43648426@163.com (G.C.)
* Correspondence: xiejg@nuaa.edu.cn; Tel.: +86-025-848-91754

Received: 28 September 2017; Accepted: 26 October 2017; Published: 31 October 2017

Abstract: Grassed swales and permeable pavement that have greater permeable underlying surface relative to hard-pressing surface can cooperate with the city pipe network on participating in urban storm flood regulation. This paper took Nanshan village in Jiangsu Province as an example, the storm-water management model (SWMM) was used to conceptualize the study area reasonably, and the low-impact development (LID) model and the traditional development model were established in the region. Based on the storm-intensity equation, the simulation scene employed the Chicago hydrograph model to synthesize different rainfall scenes with different rainfall repetition periods, and then contrasted the storm-flood-management effect of the two models under the condition of using LID facilities. The results showed that when the rainfall repetition period ranged from 0.33a to 10a (a refers to the rainfall repetition period), the reduction rate of total runoff in the research area that adopted LID ranged from 100% to 27.5%, while the reduction rate of peak flow ranged from 100% to 15.9%, and when the values of unit area were the same, the combined system (permeable pavement + grassed swales) worked more efficiently than the sum of the individuals in the reduction of total runoff and peak flow throughout. This research can provide technical support and theoretical basis for urban LID design.

Keywords: grassed swale; permeable pavement; SWMM; storm management

1. Introduction

China is in a period of rapid urbanization, with expansion of city scale and concentration of human population. Due to this process, the natural characteristics of the land are changed by various anthropogenic activities. The original forestland, farmland and grassland are gradually transformed into impermeable construction land. These anthropogenic activities are among the most important pollutant sources [1]; numerous pollutants are introduced into the urban environment by vehicular traffic, industrial processes, building construction and commercial activities, and are carried by storm-water. Above all, the impact of urbanization on the water environment includes increased risks in terms of floods, erosion and degradation of stream habitats, and deterioration of water quality [2,3]. In addition, the recharge of shallow groundwater resources has been attenuated with the reduction in potential infiltration and recharge. The storm-water runoff originates from residential and commercial areas having large amounts of impervious area connected to the storm-water system known as directly connected impervious areas (DCIAs) [4]. Therefore, the control and utilization of rainwater is imperative, and rainwater drainage systems should be changed from the traditional rainwater drainage into systems that mainly rely on permeability and combine with storage and drainage [5,6]. The concept of sponge city construction emerges as the times require.

The concept of sponge city refers to the city that can absorb water like a sponge. Sponge cities break through the traditional concept of drainage in urban rainwater management, relying on the sunken lawn, grassed swales, green roof, permeable pavement, bioretention swales and other infrastructure, and take them as carriers, taking full account of the operational safety of urban infrastructure and urban water security in the absence of floods. Reasonable resource utilization of rainwater and maintenance of a good hydrological and ecological environment can be made at the same time as the comprehensive utilization of a variety of ecological technologies including penetration, retention, storage, purification, reuse, efflux and so on, which supplement the groundwater and regulate the water cycle.

Grassed swales are an important part of the sponge city, which is a landscape surface-drainage system planted with vegetation [7,8]. Grassed swales allow surface runoff to be detained, filtered and to penetrate at a lower flow rate [9]. As a new type of ecological measure, grassed swales can be used to collect road-surface runoff, and replace the gully, ditch or part of the rainwater pipe network. Grassed swales can not only absorb rainwater, but also pretreat rainwater with plant roots, which controls the runoff of contaminants from the source [10]. If grassed swales are substituted for a traditional underground drainage system, the problem—that traditional sewage pipes and rainwater pipelines were connected randomly or improperly—can be solved.

Pervious asphalt pavement refers to a kind of asphalt pavement structure that is composed of a highly porous mixture that allows road surface water to enter the subgrade [11]. Compared with the present urban roads that mainly have a dense type of pavement, the pervious asphalt pavement can effectively replenish the groundwater; the urban heat island effect would be alleviated, the peak flow during rainstorms and the pressure of the urban drainage system would be reduced, and the vehicle running noise would also be reduced significantly, while the safety and comfort of driving would be improved effectively [12]. Since pervious asphalt pavement can significantly improve the ecological and environmental efficiency of the road, it has attracted widespread attention at home and abroad.

Since the design and construction of a sponge project are unlikely to occur simultaneously, and it is hard to adjust after implementation, it comes to be an inevitable trend that software models are used for simulating the effect of the sponge facilities in projects [13,14]. Green infrastructure is increasingly being considered for application in urban storm-water management designs. Many municipalities, regulatory agencies and advocacy groups promote the use of low-impact development (LID) to reduce runoff and increase infiltration. To show LID benefits, engineers must quantify the advantages of green versus traditional grey infrastructure [15]. Rainfall-runoff models can be classified into three types: physically based models, conceptual models and empirical models. In this latter class of models, the catchment is considered as a black box, without any reference to the internal processes that control the transformation of rainfall to runoff. In recent years, some models derived from studies on artificial intelligence have found increasing use [16]. Currently, storm-water management model (SWMM) and Hydro CAD (http://www.hydrocad.net/company.htm) are the most-commonly used simulation software for LID, and the former is the most-widely used. A comprehensive hydrological model, like SWMM, has been widely used for rainfall-runoff simulation. In recent years, simple and effective modern modeling techniques have also brought great attention to the prediction of runoff with rainfall input [17]. SWMM is a hydrodynamic rainfall-runoff simulation model developed by the US Environmental Protection Agency that includes a hydrology, hydraulic and water-quality module. It has functions of simulating urban rainfall-runoff processes (including the ground runoff, current in drainage systems, the process of flood regulation and storage), BOD (biochemical oxygen demand), COD (chemical oxygen demand), and the migration and diffusion of total phosphorus, total nitrogen and six other kinds of pollutants. The latest version, SWMM 5.1 (United States Environmental Protection Agency, Washington, DC, USA), set up eight LID technical facilities including grassed swales, roof closure facilities, rain water tanks, permeable pavements, infiltration ditchs, green roofs, rain gardens and bioretention units in the LID module [18].

At present, most of the research on sponge cities in China focus on municipal roads, residential quarters or urban area planning, but are rarely concerned with how to quantitatively analyze the sponge facilities involved in urban rainwater management. It is not difficult to make source reductions for permeable pavement and grassed swales because of the large area and permeability of the underlying surface; the difficulty is how to cooperate with the rainwater pipe network, import exogenous rainwater and perform the functions of halfway transmission, terminal regulation and storage [19]. Nanshan village, Tianwang Town, Jurong, Jiangsu Province, is a demonstration point of the National Key Technology Support Program. In this paper, Nanshan village was selected as the research area, where the permeable pavement and grassed swales were arranged reasonably with local conditions in the overall planning of the village [20]. The traditional development model and the LID model based on SWMM were established to simulate the runoff regulation effect in the study area under various rainfall repetition periods [21]. Finally, comparison of the rainfall regulation effect between permeable pavement and grassed swales was done individually. The results of this study can provide guidance for the selection and arrangement of LID facilities in urban areas, and can also provide some technical support for the construction of sponge cities in China.

2. Research Procedure

2.1. Situation of Research Area

Nanshan village lies in the north of Fushan Cherry Garden, adjacent to S340 provincial highway. The research area was named a national demonstration site in rural characteristic tourism and the design area is about four hectares. The village is located in a subtropical zone with subtropical monsoon climate with four distinctive seasons, and the annual average natural precipitation is 1188 mm, which is mainly concentrated from May to September. The plan of the research area is shown in Figure 1.

Figure 1. Plan of Nanshan village in Jurong.

2.2. Sub-Catchment Generalization

The topography of the region, low in south and high in north, has a roughly ladder-like distribution, and the region is composed of 25% permeable area and 75% impervious area. The generalized area is divided into nine water catchments and one outfall. A generalized sketch map of sub-basin arrangement can be seen in Figure 2, where c1 to c9 refer to catchment and S1 to S9 mean pipe system.

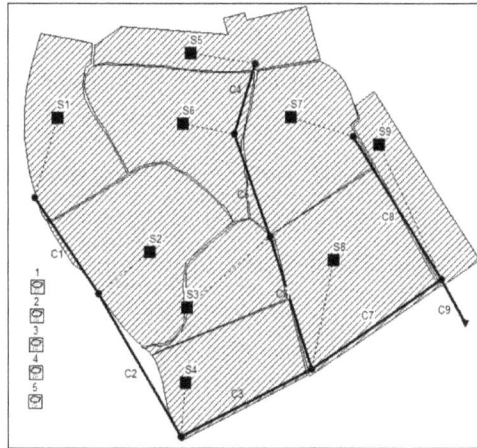

Figure 2. Generalized sketch map of sub-basin arrangement.

In the LID model, permeable pavement and grass swales were arranged in the catchment. The area of grassed swales is 800 square meters, accounting for 4% of the total area, while the area of permeable pavement is 2450 square meters, accounting for 12.25% of the total area. After infiltration and treatment of grassed swale and permeable pavement, rainwater overflows into the municipal rainwater pipe network. The drainage path of runoff in the LID model is shown in Figure 3.

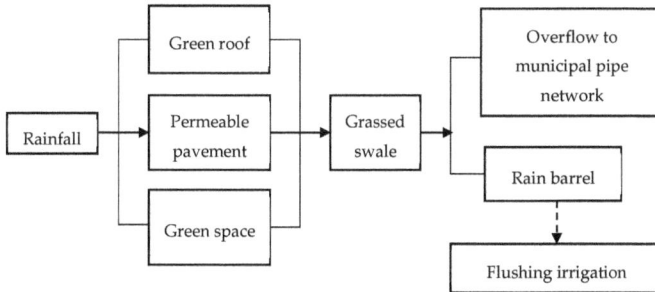

Figure 3. Drainage path of runoff in the LID model.

In the traditional development model, there were no LID facilities. The drainage path of runoff in the traditional development model is shown in Figure 4.

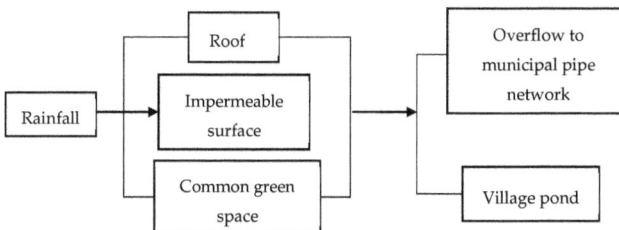

Figure 4. Drainage path of runoff in the traditional development model.

2.3. Selection of Model Parameters

The Horton model, with maximum infiltration rate of 76.2 mm/h, minimum infiltration rate of 3.18 mm/h and attenuation constant of 4.12 per hour according to soil characteristics was adopted to simulate the rainfall infiltration [22]. The nonlinear reservoir model was chosen to calculate the runoff generation at each grid while the equation of kinetic wave was adopted to calculate the pipeline transmission system [23]. In addition, the determination of parameters, which is also determined by the field geotechnical exploration reports and previous settings on the underlying surface, refers to typical values in the user manual and related documents [24]. Parameters of grass swale and permeable pavement are shown in Table 1.

Table 1. Table of parameters of LID controls.

Grassed Swale		Permeable Asphalt					
Surface		Surface		Pavement		Water storage	
Berm height (mm)	200	Berm height (mm)	10	Thickness (mm)	150	Thickness (mm)	300
Vegetation coverage rate	0.2	Manning's n	0.011	Void radio	0.25	Void radio	0.5
Manning's n	0.3	Surface slope (%)	0.3	Permeability (mm/h)	800	Seepage rate (%)	250
Surface slope (%)	0.5			Clogging factor	0	Clogging factor	0
Swale side slope (%)	4						

2.4. Designed Rainfall

Design of rainfall hyetograph refers to the distribution of total rainfall in the designed phase, and it is the boundary condition for delay of simulation based on the hydraulic model; it is also the key module to calculate the designed flow of the drainage pipe network [25].

The Chicago rainfall hydrograph model is a rainfall process based on a rainstorm intensity Equation [26], and Equation (1) has been adopted as a widespread rainstorm intensity equation in China [27]. Equation (1) can reflect the regular pattern of rainfall distribution over time:

$$q = \frac{167A_1(1 + C\lg P)}{(t_d + b)^n}, \tag{1}$$

where q refers to the rainfall intensity of rainfall duration (L/s ha); p refers to the repetition period of designing rainfall (a); t_d refers to the rainfall duration (min); and A_1, C, b and n are local parameters, calculated by a statistical method.

The rainstorm intensity equation of Jurong is Equation (2) [28].

$$q = \frac{167 \times 64.3 \times (1 + 0.836 \log P)}{(t_d + 32.900)^{1.011}}, \tag{2}$$

Peak coefficient r is the only parameter of the Chicago rainfall hydrograph [29], and the hydrograph can be calculated by Equations (3) and (4). Accurate determination of peak coefficient r is the key to fit the curve of designed rainfall:

$$i_{before} = \frac{a[(1-b)\left(\frac{t_b}{r}\right)^b + c]}{[\left(\frac{t_b}{r}\right)^b + c]^2}, \tag{3}$$

$$i_{after} = \frac{a[(1-b)\left(\frac{t_b}{1-r}\right)^b + c]}{[\left(\frac{t_b}{1-r}\right)^b + c]^2}, \tag{4}$$

where i_{before} refers to the rainfall intensity before peak flow (mm/min); i_{after} refers to the rainfall intensity after peak flow (mm/min); t_b refers to the time before peak flow (min); t_a refers to the time after peak flow (min); and r refers to the ratio of peak flow time and rainfall duration (peak coefficient).

In this research, the rainfall duration was set as 3 h and the peak coefficient was set as 0.34 [30]. The distribution with time of rainfall under different rainfall repetition periods is shown in Table 2, and the rainfall hydrograph curve of various repeating rainfall periods when the rainfall peak coefficient is 0.34 is shown in Figure 5.

Table 2. Distribution of 3 h of rainfall under different rainfall repetition periods.

Time Step/min	0.33a *	1a	3a	5a	10a
0	0.025	0.041	0.058	0.066	0.076
10	0.034	0.056	0.079	0.089	0.104
20	0.048	0.081	0.113	0.128	0.149
30	0.075	0.125	0.175	0.198	0.230
40	0.130	0.217	0.304	0.344	0.399
50	0.275	0.461	0.645	0.731	0.847
60	0.914	1.530	2.142	2.426	2.811
70	0.564	0.945	1.322	1.497	1.735
80	0.317	0.531	0.744	0.842	0.976
90	0.203	0.339	0.475	0.538	0.623
100	0.140	0.235	0.328	0.372	0.431
110	0.103	0.172	0.240	0.272	0.315
120	0.078	0.131	0.183	0.207	0.240
130	0.061	0.103	0.144	0.163	0.189
140	0.049	0.083	0.116	0.131	0.152
150	0.041	0.068	0.095	0.108	0.125
160	0.034	0.057	0.080	0.090	0.105
170	0.029	0.048	0.067	0.076	0.089
180	0.025	0.041	0.058	0.066	0.076

Note: * a refers to the rainfall repetition period.

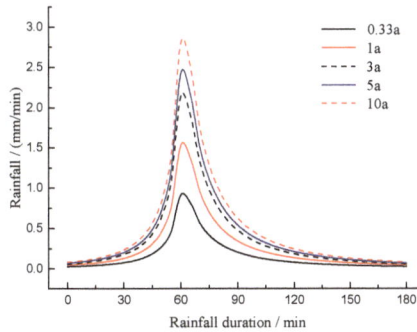

Figure 5. Rainfall hydrograph of various repeating rainfall periods when the rainfall peak coefficient is 0.34.

3. Results and Discussion

3.1. Comparison of LID and Traditional Development

The total runoff of outfall changed with the rainfall intensity under the LID model and the traditional development model, and the simulation results can be seen in Figure 6. The total rainfall here was the product of rainfall and catchment area, so its unit had changed. The rainfall repetition periods were 0.33a, 1a, 3a, 5a and 10a, the rainfall duration was 3 h and the simulation time was 4 h; total runoff, peak flow, runoff-yielding time and peak-flow occurrence time can be seen in Table 3.

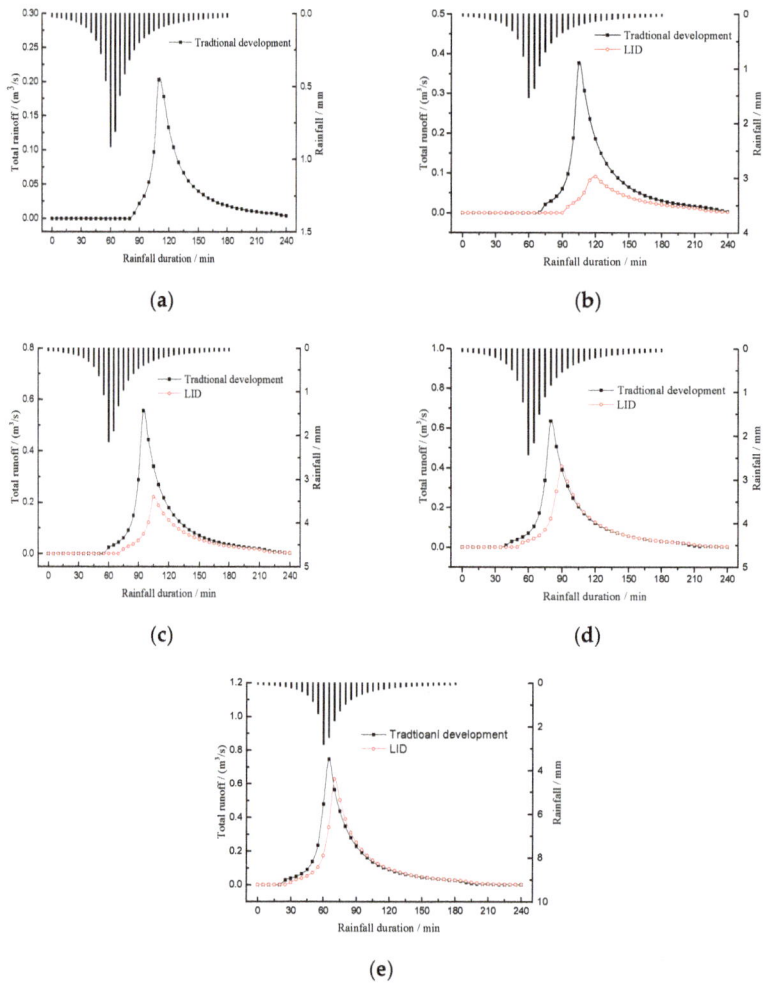

(a)

(b)

(c)

(d)

(e)

Figure 6. Comparison curve of runoff control effect: (**a**) when rainfall repetition period is 0.33a; (**b**) when rainfall repetition period is 1a; (**c**) when rainfall repetition period is 3a; (**d**) when rainfall repetition period is 5a; (**e**) when rainfall repetition period is 10a.

Table 3. Simulated results of the runoff under different rainfall intensities.

Simulation Item	0.33a			1a			3a		
	LID	Traditional Development	Difference Value	LID	Traditional Development	Difference Value	LID	Traditional Development	Difference Value
Total rainfall (m³)		1080			2052			2865	
Total runoff (m³)	0	503	503	323	1174	851	817	2007	1190
Peak flow (m³/s)		0.377	0.377	0.091	0.377	0.286	0.221	0.556	0.335
Peak-flow occurrence time (min)		110		123	106	17	106	94	12
Runoff-yielding time (min)		38		44	31	13	36	25	11

Table 3. *Cont.*

Simulation Item	5a			10a		
	LID	Traditional Development	Difference Value	LID	Traditional Development	Difference Value
Total rainfall (m³)		3247			3765	
Total runoff (m³)	1163	2598	1435	2321	3200	879
Peak flow (m³/s)	0.408	0.635	0.227	0.629	0.748	0.119
Peak-flow occurrence time (min)	91	80	11	70	65	5
Runoff-yielding time (min)	25	17	8	13	8	5

According to Table 3, the total rainfalls over 3 h were 1080, 2052, 2865, 3247 and 3765 m³ when the rainfall repetition periods were 0.33a, 1a, 3a, 5a and 10a, respectively. After the management effect of LID (i.e., permeable pavement + grassed swales), the outfall runoff was decreased in all groups. Especially, when the rainfall repetition period was 0.33a, the outfall had no flow. According to the comparison from Figure 6, it can be seen that the peak flow was reduced by 0.377, 0.286, 0.335, 0.227 and 0.119 m³/s when the rainfall repetition periods were 0.33a, 1a, 3a, 5a and 10a, respectively. With the comparative analysis of sub-catchment runoff control data, it is proposed that the runoff reduction in the traditional development model mainly depends on the infiltration of the surface permeable area, while the runoff reduction in the LID model mainly depends on reduction at the source, infiltration at halfway and regulation on the terminal.

Grassed swale and permeable pavement were conducive to reducing the peak flow under different rainfall repetition periods; however, as the rainfall repetition period and rainfall intensity increased, the runoff reduction efficiency decreased. When the rainfall repetition period was 10a, the difference in values of peak flow, runoff-yielding time and peak-flow occurrence time between the two models was negligible. The result is consistent with the simulation results of the effect of the rainwater garden's regulation and control of the urban rainwater runoff from Jiake Li et al. [31]. Furthermore, in Li's design model, the rainwater garden accounts for 2% of the total land area, while the rainfall repetition periods were 2a, 5a, 10a and 20a, the total runoff reduction was more than 25.69% and the peak flow occurrence time was delayed by 5–7 min.

It can be seen from Figure 7 that when the rainfall intensity rises to a certain extent, the difference value of the total runoff between the two development models tends to be downward, indicating that all LID facilities have entered a saturated state, and the regulation and storage of the LID model are not obvious at this point.

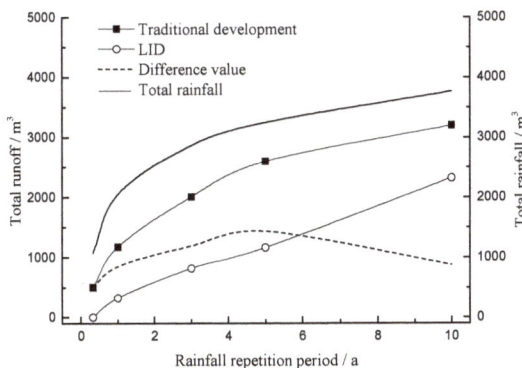

Figure 7. Rainfall curve under different rainfall intensities.

3.2. Comparison of Combined System and Single LID Facility

The rainfall repetition individual LID controls, one 5-year 3 h of rainfall event period was set as 5a while the rainfall duration was set as 3 h, the LID facilities (including grassed swales and permeable pavement) were simulated in the study area individually, and the simulation results were compared with the traditional development model and LID model; the final results can be seen in Table 4.

Table 4. Simulated results of runoff regulation effect under.

Type of LID Facilities	No LID Facilities	Grassed Swale	Permeable Pavement	Combination of LID Facilities
Area ratio		4%	12.25%	16.25%
Total runoff (m^3)	2598	2430	1730	1163
Reduction in total runoff (m^3)		168	868	1435
Reduction in total runoff per ha (m^3)		2100	3542	4416
Peak flow (m^3/s)	0.635	0.583	0.509	0.408
Reduction in peak flow (m^3/s)		0.052	0.126	0.227
Reduction in peak flow per ha (m^3/s)		0.650	0.514	0.698
Runoff-yielding time (min)	17	19	22	25
Reduction in time of runoff-yielding (min)		2	5	8
Reduction in time of runoff-yielding per ha (min)		25	20	25
Peak-flow occurrence time (min)	80	82	87	91
Reduction in time of peak-flow occurrence (min)		2	7	11
Reduction in time of peak flow occurrence per ha (min)		25	29	34

It can be seen from Table 4 that the LID facilities have effect in reducing total runoff and peak flow, delaying the runoff-yielding time and peak-flow occurrence time. Particularly, the combined system (i.e., permeable pavement + grassed swales) worked more efficiently than the sum of the individuals in the reduction of total runoff and peak flow, and the delay of peak-flow occurrence time throughout. The reason is that after runoff from sub-catchment areas infiltrates through the grassed swales and permeable pavement surface, the water in storage layers of permeable pavement can infiltrate into the grassed swales through lateral infiltration. The grassed swales reduce the runoff velocity in the transmission of rainwater runoff, which results in the time delay or elimination in runoff-yielding. The connection of an internal infiltration path between the individual LID facilities increases the water-storage space of the sub-catchment area, which leads to a reduction in peak flow and delay in the occurrence time of peak flow.

When the values of unit area are the same, the efficiency of the LID facilities on the total runoff reduction is ranked as: combined system (i.e., permeable pavement + grassed swales) > permeable pavement > grassed swales. The efficiency of the peak-flow reduction is ranked as: combined system (i.e., permeable pavement + grassed swales) > permeable pavement > grassed swales. The efficiency of delay in peak-flow occurrence time is ranked as: combined system (i.e., permeable pavement + grassed swales) > permeable pavement > grassed swales. The reason is that the permeable pavement is composed of surface course, cushion course and porous base course, so that the runoff on the surface can infiltrate freely from top to bottom in the permeable pavement, and finally recharge groundwater. Per unit area, the permeable pavement has a larger reservoir space than grassed swales, the rainwater in the water-storage formation continues to infiltrate after rainfall, and finally the extra water overflows and discharges. The effect of delay in runoff-yielding time is ranked as: combined system (i.e., permeable pavement + grassed swales) > grassed swales > permeable pavement. The reason is that grassed swales have a larger Manning coefficient on the surface due to plantation coverage (the Manning coefficient on the surface is generally 0.3), which leads to the reduction in runoff velocity, and a slower runoff velocity would increase the transmission time of rainwater runoff from runoff-yield to outfall, resulting in the delay or elimination of runoff-yielding time.

4. Conclusions

This paper presented a research method on storm-water management using SWMM. As a part of this study, the superiority of LID in storm regulation was verified through comparison of the LID model and the traditional development model, and the contrast between the combined system (i.e., permeable pavement + grassed swales) and a single LID facility on the storm-water management effect was further investigated. Based on the results and analysis, the following conclusions were made:

(1) The SWMM model can quantify the management effect of LID facilities on storm-water runoff. According to the nonlinear reservoir theory, the model took the extracted information of land surface as the basis to calculate runoff, hydrodynamic theory was used to compute concentration flow unit by unit, and the computational data can reflect the running effect of sponge facilities. In practical engineering, flow meters and level gauges are arranged in the infall and outfall to monitor actual treatment results, and the actual operating results of LID facilities are to be verified in the follow-up field tests for real-world authentication.

(2) Better results of storm-water management could be obtained through LID, including reduction in the total runoff, decrease in the peak flow, elimination or delay in the runoff-yielding time and delay in the peak-flow occurrence time. When the rainfall repetition period ranged from 0.33a to 10a, the reduction rate of total runoff in the research area where LID facilities were arranged ranged from 100% to 27.5%. The reduction rate of peak flow ranged from 100% to 15.9%. The runoff-yielding time was eliminated or delayed by 13 min to 5 min and the peak-flow occurrence time was eliminated or delayed by 17 min to 5 min after LID treatment.

(3) When the values of unit area are the same, the efficiency of the LID facilities on the total runoff reduction is ranked as: combined system (i.e., permeable pavement + grassed swales) > permeable pavement > grassed swales. The efficiency of the peak-flow reduction is ranked as: combined system (i.e., permeable pavement + grassed swales) > permeable pavement > grassed swales. The effect of delay in runoff-yielding time is ranked as: combined system (i.e., permeable pavement + grassed swales) > grassed swales > permeable pavement. The efficiency of delay in peak-flow occurrence time is ranked as: combined system (i.e., permeable pavement + grassed swales) > permeable pavement > grassed swales. The capacity for storm-water management of permeable pavement and grassed swales was limited, and the combined system was well-operated when the rain was light. However, when the area suffered too much rain in a short time, the management was mostly negligible.

(4) The results can provide reference for the selection and arrangement of LID facilities in urban areas; for example, when the regional space is ample, the united arrangement of LID facilities are recommended in the design of the drainage system, whereas when the area is limited, permeable pavement should be a top priority. Grassed swales and permeable pavement emphasize site-specific recommendations, intensive management, improved efficiency and environmentally sound use of inputs, rather than pursuit of the largest area of sinking space.

Acknowledgments: The authors would like to acknowledge National Key Technology Research and Development Program of the Ministry of Science and Technology of China (2015BAL02B00) and Jiangsu Scientific and Technological Development Program (BE2015349) for its financial support in this project.

Author Contributions: Each author had his own jobs. The first author, Jianguang Xie, had the responsibility of test designs and the construction of SWMM. Chenghao Wu and Hua Li helped to collect the test data and make analysis based on the data. Jianguang Xie wrote this paper and Gengtian Chen had the job of checking the paper on grammar and revision. As the corresponding author, Jianguang Xie was also the person who submitted the paper and kept communication with editors.

Conflicts of Interest: The authors declare no conflict of interest.

References

1. Masamba, W.R.L.; Mazvimavi, D. Impact on water quality of land uses along Tamalakane-Boteti River: An outlet of the Okavango Delta. *Phys. Chem. Earth Parts A/B/C* **2008**, *33*, 687–694. [CrossRef]
2. Beniston, J.W.; Lal, R.; Mercer, K.L. Assessing and managing soil quality for urban agriculture in a degraded vacant lot soil. *Land Degrad. Dev.* **2016**, *27*, 996–1006. [CrossRef]
3. Gorgoglione, A.; Gioia, A.; Iacobellis, V.; Piccinni, A.F.; Ranieri, E. A Rationale for Pollutograph Evaluation in Ungauged Areas, Using Daily Rainfall Patterns: Case Studies of the Apulian Region in Southern Italy. *Appl. Environ. Soil Sci.* **2016**, *2016*, 9327614. [CrossRef]
4. Aad, M.P.A.; Suidan, M.T.; Shuster, W.D. Modeling Techniques of Best Management Practices: Rain Barrels and Rain Gardens Using EPA SWMM-5. *J. Hydrol. Eng.* **2010**, *15*, 434–443. [CrossRef]
5. Ji, G.X. Experimental study on urban rainwater infiltration and rainwater drainage system. *J. Univ. Shanghai Sci. Technol.* **2003**, *25*, 72–76.
6. Yang, B.; Xu, T.; Shi, L. Analysis on sustainable urban development levels and trends in China's cities. *J. Clean. Prod.* **2017**, *141*, 868–880. [CrossRef]
7. Fletcher, T.D.; Peljo, L.; Fielding, J.; Weber, T.R. The Performance of Vegetated Swales for Urban Stormwater Pollution Control. In Proceedings of the Ninth International Conference on Urban Drainage, Portland, OR, USA, 8–13 September 2002; pp. 1–16.
8. Shen, Z.X.; Kan, L.Y.; Che, S.W. Effects of grass swales structure parameters on storage and pollutant removal of rainfall runoff. *J. Shanghai Jiaotong Univ. (Agric. Sci.)* **2015**, *33*, 46–52.
9. Xu, P.; Xi, W.J.; Sun, K.P.; Ren, X.X.; Zhang, Y.J. The Control Effects of Conveyance Grass Swales in Park on Rainfall Runoff under Moderate and High Rainfall Intensity. *Environ. Sci. Technol.* **2016**, *11*, 47–51. (In Chinese)
10. Huang, Y.; Wei, P.; Li, H.Y.; Zhang, Y. Impact of Vegetated Check Dam and Overflow Weir on Operation of Grass Swale Perforated Pipe System. *China Water Wastewater* **2015**, *13*, 99–104. (In Chinese)
11. Jiang, W.; Sha, A.M.; Xiao, J.J.; Pei, J.Z. Water storage-infiltration model for permeable asphalt pavement and its efficiency. *J. Tongji Univ. (Nat. Sci.)* **2013**, *41*, 72–77. (In Chinese)
12. Brattebo, B.O.; Booth, D.B. Long-term storm water quantity and quality performance of permeable pavement systems. *Water Res.* **2003**, *37*, 4369–4376. [CrossRef]
13. Cai, L.H. Introduction of Hydrological and Hydraulic Models for "Sponge City". *Digit. Landsc. Archit.* **2016**, *2*, 33–43. (In Chinese)
14. Park, S.Y.; Lee, K.W.; Park, I.H.; Ha, S.R. Effect of the aggregation level of surface runoff fields and sewer network for a SWMM simulation. *Desalination* **2008**, *226*, 328–337. [CrossRef]
15. Mccutcheon, M.; Wride, D. Shades of Green: Using SWMM LID Controls to Simulate Green Infrastructure. *J. Water Manag. Model.* **2013**, *R246-15*, 289–301.
16. Granata, F.; Gargano, R.; Marinis, G.D. Support Vector Regression for Rainfall-Runoff Modeling in Urban Drainage: A Comparison with the EPA's Storm Water Management Model. *Water* **2016**, *8*, 69. [CrossRef]
17. Wang, K.H.; Altunkaynak, A. Comparative Case Study of Rainfall-Runoff Modeling between SWMM and Fuzzy Logic Approach. *J. Hydrol. Eng.* **2013**, *18*, 283–291. [CrossRef]
18. Song, C.P.; Wang, H.C.; Tang, D.S. Research progress and development trend of storm water management model. *China Water Wastewater* **2015**, *16*, 16–20. (In Chinese)
19. Mao, X.; Jia, H.; Yu, S.L. Assessing the ecological benefits of aggregate LID-BMPs through modelling. *Ecol. Model.* **2017**, *353*, 139–149. [CrossRef]
20. Xing, W.; Li, P.; Cao, S.-B.; Gan, L.-L.; Liu, F.-L.; Zuo, J. Layout effects and optimization of runoff storage and filtration facilities based on SWMM simulation in a demonstration area. *Water Sci. Eng.* **2016**, *9*, 115–124. [CrossRef]
21. Guan, Y.H.; Lu, M.; Wang, C. LID Stormwater Control Effect and Water Quality Simulation Based on SWMM. *China Rural Water Hydropower* **2017**, *1*, 84–87. (In Chinese)
22. Jang, S.; Cho, M.; Yoon, J.; Yoon, Y.; Kim, S.; Kim, G.; Kim, L.; Aksoy, H. Using SWMM as a tool for hydrologic impact assessment. *Desalination* **2007**, *212*, 344–356. [CrossRef]
23. Campbell, C.W.; Sullivan, S.M. Simulating time-varying cave flow and water levels using the Storm Water Management Model. *Eng. Geol.* **2002**, *65*, 133–139. [CrossRef]

24. Kuang, X.; Sansalone, J.; Ying, G.; Ranieri, V. Pore-structure models of hydraulic conductivity for permeable pavement. *J. Hydrol.* **2011**, *399*, 148–157. [CrossRef]
25. Fortunato, A.; Oliveri, E.; Mazzola, M.R. Selection of the Optimal Design Rainfall Return Period of Urban Drainage Systems. *Procedia Eng.* **2014**, *89*, 742–749. [CrossRef]
26. Zhang, D.W.; Zhao, D.Q.; Chen, J.N.; Wang, H.Z.; Wang, H.C. Application of Chicago rainfall hydrograph model in simulation of drainage system. *Geomat. World* **2008**, *34*, 354–357. (In Chinese)
27. Jin, J.M. Formulation and application method of urban rainstorm intensity formula. *China Munic. Eng.* **2010**, *1*, 38–67. (In Chinese)
28. Zhuang, Z.F.; Wang, K.Q.; Yang, J.; Chen, B.; Zhu, H.T. Research on new generation rainstorm intensity formula and design of rainfall hyetograph in Zhenjiang. *J. Meteorol. Sci.* **2015**, *35*, 506–513. (In Chinese)
29. Li, W.T.; Sui, J.; Liu, C.L.; Niu, Y.; Zhou, J.H.; Tan, J.X. Analysis of influence of design rainfall peak coefficient on design flow of drainage pipeline network. *Water Purif. Technol.* **2015**, *5*, 100–103. (In Chinese)
30. Xu, S.R.; Wang, Y. Impact analyze of rainfall intensity and interval under low impact development on stormwater control. *Sci. Technol. Eng.* **2015**, *15*, 219–223. (In Chinese)
31. Li, J.; Li, Y.; Shen, B. Simulation of rain garden effects in urbanized area based on SWMM. *J. Hydroelectr. Eng.* **2014**, *33*, 60–67.

water

MDPI

Article

Enhancing the Economic Value of Large Investments in Sustainable Drainage Systems (SuDS) through Inclusion of Ecosystems Services Benefits

Santiago Urrestarazu Vincent [1,2], Mohanasundar Radhakrishnan [1], Laszlo Hayde [1] and Assela Pathirana [1,*]

[1]   Water Science and Engineering Department, IHE Delft Institute for Water Education,
      2611 AX Delft, The Netherlands; surresta@gmail.com (S.U.V.); m.radhakrishnan@un-ihe.org (M.R.);
      l.hayde@un-ihe.org (L.H.)
[2]   CSI Ingenieros, Soriano 1180, 11100 Montevideo, Uruguay
*   Correspondence: a.pathirana@un-ihe.org; Tel.: +31-15-215-1854

Received: 13 July 2017; Accepted: 16 October 2017; Published: 31 October 2017

Abstract: Although Sustainable Drainage Systems (SuDS) are used in cities across the world as effective flood adaptation responses, their economic viability has frequently been questioned. Inclusion of the monetary value of ecosystem services (ES) provided by SuDS can increase the rate of return on investments made. Hence, this paper aims at reviewing the enhancement of the economic value of large-scale investments in SuDS through inclusion of ecosystem services. This study focuses on the flood reduction capacity and the ES benefits of green roofs and rain barrels in the combined sewerage network of Montevideo Municipality in Uruguay. The methodology comprises a cost–benefit analysis—with and without monetised ES provided by SuDS—of two drainage network configurations comprising: (i) SuDS; and (ii) SuDS and detention storage. The optimal drainage design for both these drainage configurations have been determined using SWMM-EA, a tool which uses multi-objective optimisation based evolutionary algorithm (EA) and the storm water management model (SWMM). In both design configurations, total benefits comprising both flood reduction and ES benefits are always higher than their costs. The use of storage along with SuDS provides greater benefits with a larger reduction in flooding, and thus is more cost-effective than using SuDS alone. The results show that, for both of the drainage configurations, the larger investments are not beneficial unless ES benefits are taken into account. Hence, it can be concluded that the inclusion of ES benefits is necessary to justify large-scale investments in SuDS.

Keywords: Sustainable Drainage Systems; multi-objective optimization; multiple values

1. Introduction

Urban floods driven by climate variability, climate change and rapid urbanisation at various scales cause damage to lives and property [1]. The flood damages are direct economic losses (e.g., property damage) and indirect losses (e.g., health impacts and disruption of transport). A variety of flood mitigation and adaptation measures can be implemented to mitigate or adapt to urban floods [2]. Most conventional measures tend to increase the capacity of drainage infrastructure, for example by increasing conduit sizes. Other measures aim to reduce the demand for conveyance capacity by retaining or detaining stormwater at source thereby reducing the runoff before entry to the drainage system. In parallel, there is an increasing awareness of sustainable development needs [3], which necessitates the use of sustainable materials and sustainable management practices. The range of technology and techniques used to manage stormwater or surface water in a way that is more sustainable than conventional drainage techniques are known in several parts of the world as sustainable drainage systems (SuDS) [4]. For example, SuDS include rain gardens, green roofs, rainwater harvesting systems, and bio-swales.

Many SuDS are a part of Low Impact Development (LID), Water Sensitive Urban Design (WSUD), Green Infrastructure (GI), Active Beautiful and Clean Waters Design Features (ABC Waters Design Features) and Best Management Practices (BMP) [4,5]. SuDS, or their equivalent, are being used extensively in UK, USA, Canada, Australia, Netherlands, Denmark, Malaysia and Singapore [4–7]. In addition to storm water management benefits, SuDS are known to deliver other benefits such as to ecological systems, enhancement of the liveability of urban environments, reduction in ambient temperature, urban agriculture, etc. [8]. SuDS such as green roofs, swales, bio-retention systems and rain gardens may be considered as part of the natural urban ecosystem which provides ecosystem services to urban dwellers and others [4,8,9]. Ecosystem services are the goods or services provided by ecosystems to society [10,11]. The value of ecosystem services can be expressed in monetary units, so that the cost incurred and benefits accrued can be compared [8,10]. One such example is the BeST tool (Benefits of SuDS Tool) developed by CIRIA (Construction Industry Research and Information Association, London, UK), which provides a structured approach to evaluate the wide range of benefits based on overall drainage system performance and monetises many of the benefits [8].

The inclusion of SuDS in the urban environment has highlighted many of the interdependencies between the various systems in the urban environment and has also illustrated the important interactions between the stormwater management components [7]. Hence, the cost effectiveness and benefits of using SuDS needs to be assessed in the framework of these interdependencies and interactions. In places where SuDS have been implemented [4–7], they are not in isolation, but are applied to support or relieve the demand pressure on the existing traditional (grey) infrastructure. Many of the investments made in green and grey drainage infrastructure are large-scale, as these are often implemented alongside major drainage retrofitting [12]. Thus, determining the cost effectiveness of a combination of green and grey drainage infrastructure and determining the most effective mix based on the overall drainage performance are essential for making decisions for investments in stormwater drainage infrastructure. As the overall benefits of SuDS, including ecosystem services (ES), can be monetised using tools such as BeST, there is an opportunity for an increase in cost effectiveness of the large-scale investments made in green-grey drainage systems.

This paper aims at providing insights into enhancing the economic value of large-scale investments in SuDS through the inclusion of ecosystem services benefits in an urban context. The general objective is to incorporate the monetary value of SuDS ecosystem services into the decision making process for selecting storm water management (SWM) measures for an urban catchment affected by regular floods. In so doing, to optimize as far as practicable, the design of the drainage system, in order to examine whether it is possible to financially justify larger-scale investments for these types of mixed systems. SuDS are considered here in combination with underground storage to broaden the range of flood reduction measures and compare these. This study focuses on the flood reduction capacity and the ecosystem services benefits due to the incorporation of green roofs and rain barrels in the combined sewerage network of Montevideo Municipality in Uruguay. The methodology comprises a cost–benefit analysis of the two drainage network configurations, with the first configuration including SuDS and the second with SuDS and detention storage together. The analysis has been carried out using SWMM-EA [13], a tool which combines multi-objective optimisation based on the non-dominated sorting genetic algorithm II (e.g., Deb et al. [14]) with the storm water management model (SWMM) [15].

2. Methodology

The framework for this research is shown in Figure 1. This includes a first stage (I in Figure 1), which comprises the collection of climatic, physical, environmental and socio-economic data.

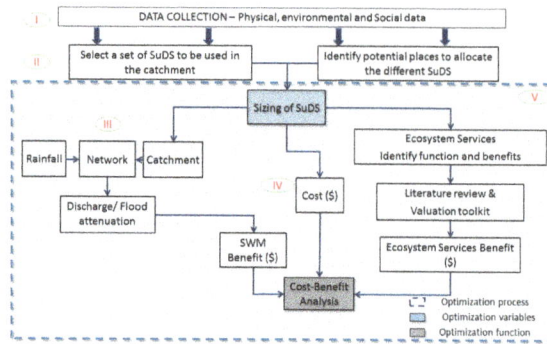

Figure 1. Research framework.

In the second stage (II in Figure 1), the types of SuDS are determined based on the characteristics of the catchment under study (e.g., rain barrels, rain gardens and green roofs). The maximum or potential area (or number of these, depending on the SuDS element under consideration) to be treated by each SuDS element are determined based on the area of the different sub-catchments, their land use and land cover. There could be some dependence between the potential areas in some cases, hence the relation between the potential areas of the different SuDS elements has been established.

In the third stage (III in Figure 1), a one-dimensional rainfall–runoff simulation model is set up with the design rainfall, the sub-catchments and the main conduits (network), and is also calibrated. Subsequently, the general characteristics of the SuDS elements have been incorporated, keeping the area or number of these elements in the model as variables. The variable SuDS parameters are optimized later in stage five.

The fourth stage is (IV in Figure 1) the estimation of the costs and ES benefits of the SuDS, which are usually expressed in monetary terms such as US$/unit or US$/m^2. The accepted approach is to derive the flood damage by using inundation depths calculated by a two-dimensional (or 1D-2D coupled) flood model together with a depth–damage curve. Inundation models are computationally expensive to run because of the tens-of-thousands of iterations needed by optimization algorithms. Thus, an approach using correlation between flood damage and water depth simulated at two observation nodes in the one-dimensional model has been used in this study by using flood damage assessments based on flood depths for various rainfall events across the entire catchment. The resulting depth–damage correlation has then been used to estimate flood damage under different optimization-iterations. The depth–damage correlation represents the total costs for the entire area due to a single flooding event and therefore inputs to the computation of the SWM benefits for the different candidate options in the optimization process.

The fifth stage (V in Figure 1) is the implementation of a multi-objective optimization (MOO) process using an optimization tool that couples hydraulic models such as supply–demand models or rainfall–runoff models with evolutionary algorithm computations [16,17]. Within this process, a layout of the SuDS elements has been defined for each candidate option by setting the number and/or area of each of these over each of the sub-catchments. This definition is represented in the model and the reduction in flooding costs for this specific layout is computed and translated into monetary values (SWM benefit in Figure 1). At the same time, the ecosystem services provided by this option are evaluated and translated into monetary values (ecosystem services benefit in Figure 1). Finally, the costs of the SuDS have been calculated.

The iteration of this process by changing the area covered by each SuDS or the number of SuDS units over the different sub-catchments (within a predetermined range of SuDS parameters and using a MOO process) results in various cost–benefit values. The MOO is set to find solutions that minimize costs and maximize total benefits. The outputs that have a maximum benefit for a given cost will be the optimal solutions, and the rest will be non-optimal since more benefits can be derived for the same cost (by changing the SuDS layout). The "benefit vs. cost" curve formed by the optimal solutions is called the "Pareto front". This optimization process can be repeated as many times as necessary in order to assess different cases.

2.1. Storm Water Runoff Modelling

The modelling of the hydrological process and the conveyance of the storm water in the network has used version 5.0 of the Storm Water Management Model (SWMM 5.0), developed by the United States Environmental Protection Agency (US EPA). To model flooding, the roads have been included in the model, leading to a 1D-1D model, which is a simple approximation of reality. A more accurate approach would use a 1D-2D model. Several rainfall return periods have been considered and modelled to compute the flooding costs (and therefore the SWM benefits), from 2 to 50 years.

2.2. Cost and Benefits Valuation

All costs and benefits have been considered as cash flows that happen either in the present (initial costs and benefits) or in the future (future costs and benefits). To compare different options that may have different value flows during the project lifespan, the Net Present Value (NPV) is used. The NPV is the net sum of the discounted benefits (positive values) and the discounted costs (negative values). The discount rate is based on the prevailing local practices. In addition, private costs and benefits are distinguished from purely social costs and benefits. This study focuses on the benefits to the whole society, adding private to social costs and benefits whenever these apply, avoiding double counting and not considering costs and benefits that are simply money transfers from private to social, or vice versa.

2.2.1. Costs

The costs comprise initial costs and maintenance costs of SuDS during the lifespan of the project. In addition, the social costs such as environmental impacts related to the production of the materials used for the SuDS have to be considered. The costs, usually expressed as unit costs (i.e., US\$/m^2 or US\$/unit), have been computed from local prices for the case study and through literature review. The costs of conventional drainage measures have been adapted from estimates made by the local authorities in the area.

2.2.2. Benefits

The benefits derived from the SuDS can be classified into storm water management (SWM) benefits and ecosystem services (ES) benefits.

Storm Water Management Benefits

For every assessed option, storms with different return periods will produce different floods (if any) and therefore different damage costs. The differences in these damage costs are due to the difference in the number of houses affected and also due to the difference in the depth of water in the flooded houses. The depth of water in a particular house can be translated into flooding costs (for a single house) by applying a depth–damage curve. These combined effects (flooded extent and water depth in houses) have been represented by a depth–damage correlation, assuming that the damage cost for the whole area for one flooding event can be derived based on the water depth at a particular location in the network). The depth–damage curve is different from the depth–damage correlation. The former relates the damage in a single house to flood depth in the house, whereas the latter is a derived relationship between the total flood loss in the model area and the water level at a selected node in the model. Thus, the latter provides a way to estimate the loss due to flooding, which can be simulated using a 1-D model such as SWMM. This is essential to facilitate the tens of thousands of simulations that are necessary to employ the optimization algorithms.

The SWM benefit of each individual (candidate) solution is the difference between the actual damage cost due to regular floods in the area and the damage cost that is expected after the implementation of that particular solution. Expected Annual Damage (EAD), which is the integral of the damage probability function, is determined for every candidate solution [16]. A depth–damage correlation $(D(h_f))$ is determined for two control points in the network. Hedonic pricing for the loss in the market value of a property or household has been used to calculate the losses due to flooding and to derive the depth–damage correlation for control points in the network. The difference in price is assumed to represent, at least in part, the impact of flooding on the market value of a house which is regularly flooded. This difference in the price of a house implicitly includes not only the direct damage on the infrastructure, but also other factors such as distress at being flooded and all the related costs that market prices can reflect. It is not only the cost related to a single flooding event, but also the cost of recurring floods that is important, hence the analysis includes the severity and frequency of the flooding events. For example, in Quitacalzones catchment in Montevideo, it is estimated that the average decrease in price of a regularly flooded property (house) is US$ 30,000. There are 610 houses that are regularly flooded and, thus, a total loss of around US$ 18,300,000 [18,19]. Although roads cover a significant area of the catchment (26%), the damage to the roads has not been included due to a lack of reliable data.

This cost, calculated with the hedonic pricing, is comparable with the present value of all of the EADs, assuming that if the cost is calculated by any of the two methods, the final result should be the same. Based on this assumption a "damage cost vs. depth" curve has been constructed, which represents the cost per house and per flood event. The process by which the "damage cost vs. depth" curve is adjusted and therefore the depth–damage correlation is derived is presented in Figure 2.

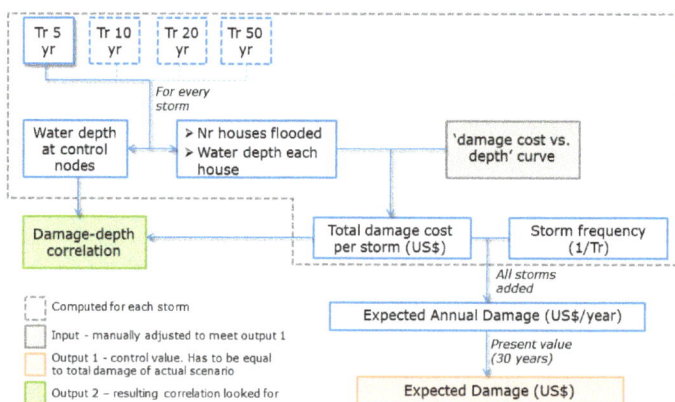

Figure 2. Process for deriving the depth–damage correlation. The depth–damage correlation links the total damage in the catchment to water level in a selected (1-D) model-node.

Ecosystem Services Benefits

The valuation of the benefits from the ecosystem services provided by a SuDS element (and for green infrastructure in general) is presented in Figure 3.

The benefits, value of benefits and the method to quantify the value of benefits have been based on literature on Ecosystem Services. Ecosystem Services Benefits of the SuDS can be computed with the help of the Green Infrastructure Valuation Toolkit Calculator [10], a valuation toolkit developed by Green Infrastructure North West, UK. However, the values can be changed with the help of literature on SuDS [20–27]. Realistic local values have been chosen whenever possible.

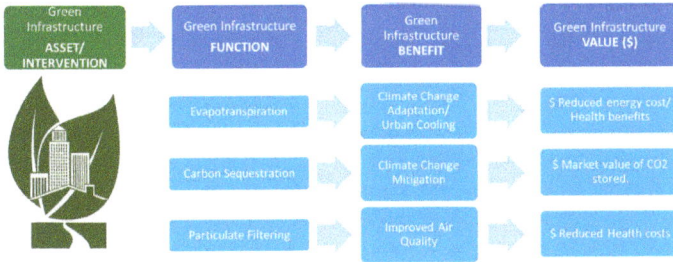

Figure 3. Translating green infrastructure intervention into monetized benefit values. After Ashton et al. 2010 [10].

2.3. Optimization Process

The optimization of the layout of the SuDS over the catchment has been set up as a Multi-Objective Optimization (MOO) problem, where the two objectives to be optimized are the total costs and the total benefits of SuDS, i.e., SWM benefits and ES benefits (Figure 4). When storage is included, the costs and benefits are also part of the objective functions.

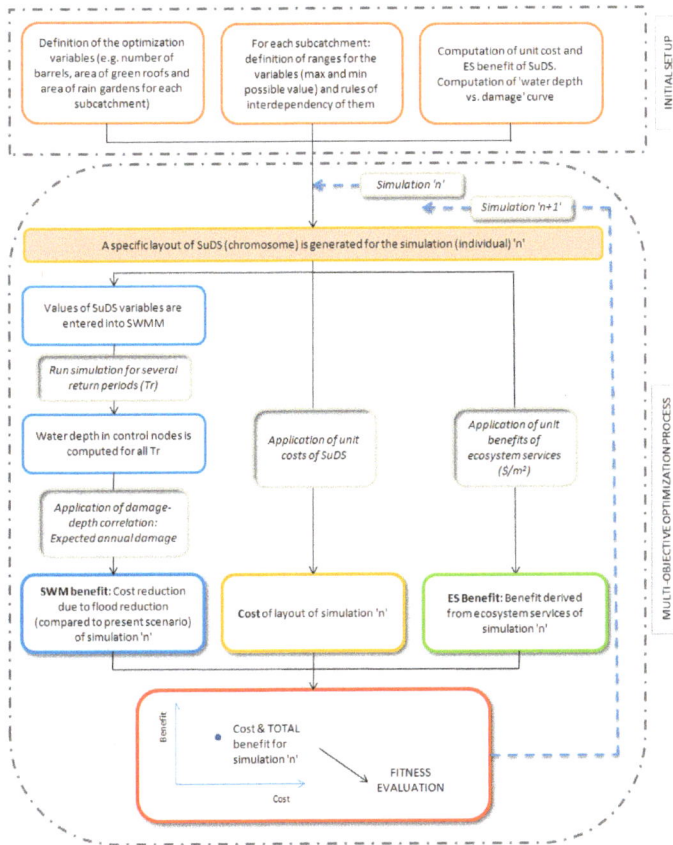

Figure 4. Optimization process framework.

The tool used for optimization was SWMM5-EA [13]. This is a software tool that applies evolutionary algorithms in urban drainage systems. This optimization tool couples the SWMM 5.0 with an evolutionary computing library. The SWMM has here been used as part of the evaluation step of the optimization process in order to assess the SWM benefits of the solutions. SWMM5-EA is a flexible tool that was originally intended for educational use and therefore provides a limited number of ways to express objective functions (e.g., cost as a sum of implementation cost and a penalty for flooding). However, the tool provides a basic Application Programming Interface (API) that facilitates implementation of arbitrary objective functions with a limited amount of coding. In this study, the API was used to implement the objective functions that were required (e.g., including ecosystem service benefits and SWM benefits).

3. Case Study: Montevideo, Uruguay

The case study area is a part of the dense urban area of Montevideo, capital city of Uruguay. Mean annual precipitation in Montevideo is 1100 mm/year. The annual distribution of the rainfall is relatively homogenous, with a minimum of 81 mm/month in June and a maximum of 108 mm/month in October [28]. This study covers 220 ha of the upper part of an urban catchment called "Quitacalzones". The whole Quitacalzones catchment has an area of 600 ha and drains towards the Montevideo bay located to the West. The catchment is entirely urbanized, with dense residential land use. However, buildings with more than three storeys are rarely found. Green spaces, as well as free public areas are very few. From the analysis of aerial imagery it can be said that roughly 26% of the area corresponds to roads and pathways, 64% corresponds to buildings and the rest (10%) to private gardens [29].

The area is served by a combined sewer system, which was constructed between 1920 and 1950, using design criteria that were internationally used at that time [18]. Interceptors, which are located along the coast of the city, collect sewage from most of the catchments of Montevideo, including Quitacalzones. During dry periods, sewage collected from the catchment is brought to Punta Carretas pre-treatment plant through an intermediate pumping station and is pumped out to the sea after treatment. However, these interceptors do not have sufficient capacity to convey storm water flows, except smaller events (i.e., storms of less than one month return period). This causes frequent overflows of combined sewage and storm water to the sea. In Quitacalzones, these overflows are discharged into the bay.

The Municipality of Montevideo (MM) estimates that approximately 610 houses are regularly flooded. For storms of three-year return period, the drainage system's capacity is surpassed and polluted storm water starts to flow into the streets and accumulates in low lying areas for typically less than 3 h. This flooded area is located in the most downstream part of the 220 ha study area. Once the inflows to the drainage system start to decrease, the ponded surface water is drained from the catchment by the drainage system.

Flood water depths can reach more than 1 m in the worst cases and water enters the houses, causing economic and health damage. The flood extents for storms of 5-, 10- and 20-year return periods are shown in Figure 5. Montevideo Municipality has started making adaptation and mitigation plans to overcome flooding problems. Three off-line underground storage tanks, Conservación, Quijote and Liceo, are planned to be implemented as flood detention measures [19]. These storage areas (Figure 5) are designed to avoid flooding during a storm with one in ten-year frequency. Characteristics and costs of these underground storage tanks are presented in Table S1 (Supplementary Materials).

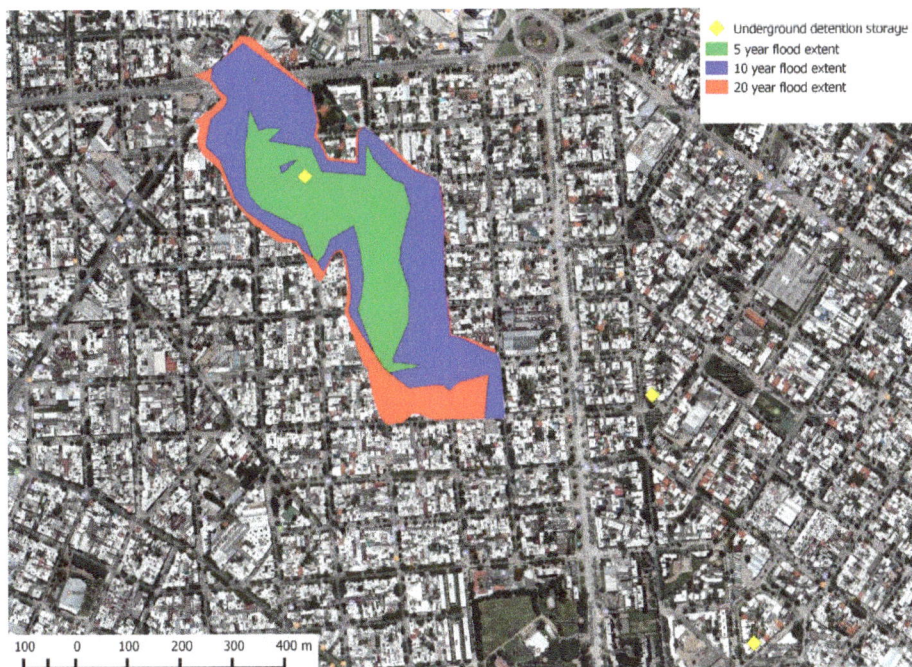

Figure 5. Location of proposed underground detention storage tanks and flooded extents for 5-, 10- and 20-year storm events in Quitacalzones catchment.

4. Application of SWMM-EA in Montevideo

4.1. Set Up of Storm Water Runoff Model

Montevideo Municipality has constructed a SWMM model of the study area, which is already calibrated, where the main network has been defined and the sub-catchments delineated. A number of streets are also included, represented as rectangular channels of 17 m width, to model their conveyance of surface storm water (which cannot be conveyed by the drainage network). This is a simplification, since it considers neither the real cross sections formed by the pavement, or the pathways and the houses, nor the elements that might affect the flow pattern, such as trees. Moreover, the entry of stormwater into the houses is also not considered in detail. However, the hydrology is not modelled in the SWMM model provided by the municipality and has therefore been included in the model as inflows into the different network calculation nodes. Since SuDS are part of the hydrologic components of the SWMM model (i.e., sub-catchments), sub-catchments elements have been incorporated into the model replacing the original inflow hydrographs. A part of the SWMM model of the catchment with sub-catchments is shown in Figure 6.

Extensive green roofs were considered in the selection of the SuDS. Extensive green roofs are shallow, lighter and therefore can be implemented on roofs (even on sloped roofs) that are not originally designed to support green roofs [30]. There is usually no need for additional structural reinforcement. These extensive green roof modules are made of several layers. The bottom layer is a plastic support, which prevents the soil to be in contact with the roof and provides space, which acts as the under-drain for the green roof. Above this support, there is a geotextile layer that prevents the soil being washed out, but allows water to drain. The growing media, with a depth of 90 mm, is on top of the geotextile.

A photograph of green roof modules installed in Montevideo by a local manufacturer (Verde facil, Montevideo, Uruguay) is shown in Figure 7.

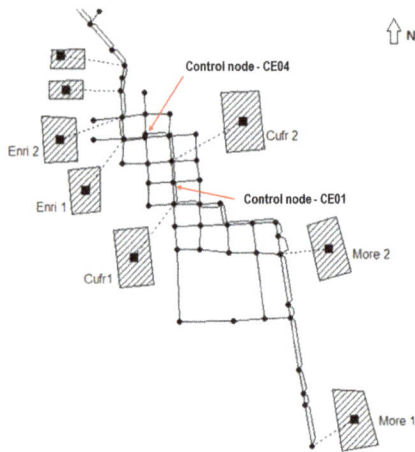

Figure 6. Representation of sub-catchments in Storm Water Management Model (SWMM).

Figure 7. Green roof modules in Montevideo [31].

Rain barrels are storage devices that collect storm water from roof downspouts. In this study, only individual household barrels have been considered. Larger cisterns that collect water from two or more houses were not considered, even though they are cost-effective in some cases [32]. There are cases in which rain barrels of 208 and 284 L (relatively small) are successfully implemented at parcel level in a small sub-urban watershed, e.g., in Cincinnati, OH, USA [33]. Here, the selected rain barrel has a volume of 600 L.

The extensive green roofs and rain barrels were incorporated into the SWMM model that was used for generating the Pareto-optimal front solutions. In the SWMM model, the height and area of the rain barrels, the total impervious area served by the barrels and the initial level of saturation of the impervious area had to be specified. The initial water level was set as 30% of the barrel (rather than assuming them to be completely empty, as they will be used for other purposes such as watering plants), and the impervious area connected by each barrel was set as 40 m². The SWMM modelling parameters were set according to the approximate characteristics of the selected green roof. However, some of the parameters, such as the hydraulic conductivity of the soil, are difficult to estimate and influence the storm water runoff performance of the green roof, which is also sensitive to the surface storage depth or the under-drain's drain coefficient. A value of 38 mm/h was considered as the hydraulic conductivity, which is based on a similar study done by Tang et al. [34]. The initial saturation of the soil was set as 14%, which corresponds to the midpoint between field capacity and wilting point. This is a simplification and it is possible for the soil moisture to be much closer to field capacity if a rainstorm occurs preceded by a rain event (e.g., 2–3 days or less antecedent dry weather period).

The SWMM model was coupled with a multiple objective optimisation evolutionary algorithm, set up in such a way that it received the SuDS/LID input parameters from the EA module. The inputs to the model that vary from one option to the another during the optimization process were:

- Percentage area covered by each SuDS control and the percentage of water treated from the impervious area;
- Width of the sub-catchment: the width of the non-SuDS portion of the sub-catchment, where, if the area occupied by the SuDS changed, the width should also be changed;
- Percentage of impervious area in the sub-catchment: the percentage of impervious area over the non-SuDS portion of the sub-catchment; if the area occupied by the SuDS changed, the impervious area should be changed; and
- Per cent routed: the percentage of the non-SuDS portion of the routed subarea (in this case, the impervious area was routed through the pervious area) that is routed through the corresponding routing subarea.

4.2. Cost and Benefit Valuation

Costs are those related to investment and maintenance of the systems while benefits were the ecosystem services benefits plus the storm water management benefits (flood reduction benefits in this case). A discount rate of 5% was used to calculate the present value of costs and benefits [35]. The lifespan of green roofs was taken as 40 years [21,23] and could be extended up to 55 years [36]. Rain barrels had a lifespan of 50 years [32,37]. A life span of 30 years was considered for both the SuDS options considered in this study. All the costs and benefits were based on the market rates of suppliers of green roofs (e.g., Verdefacil and Maria Pietranera), rain barrels and services associated with these in Montevideo and Buenos Aires.

4.2.1. Costs

The costs of the SuDS corresponded to local prices at the case study location. However, values from literature were also considered—in some cases to compare with the local prices, and in others to complement these. Private and social costs were added to obtain the total costs related to green roof implementation. The net present value (NPV) of the cost of the green roofs was 142 US$/m² and the NPV of the cost of a rainwater barrel was 300 US$/barrel. The unit cost of the detention storage to be used during the optimization process was obtained from the engineering and planning documents of municipality of Montevideo [18,19]. A detailed cost breakdown of the green roofs and rain barrels is provided in Tables S2 and S3.

4.2.2. Benefits

Storm Water Management Benefits

On average, the price of the houses that flood regularly decreases by about US$30,000 and, with 610 houses in the study area, the total loss is about US$18,300,000 [18,19]. The total damage in the entire area was correlated to the water depths at two control nodes CE01 and CE04 and a depth–damage correlation curve obtained (Figure 6). This is a major simplification in the study. Flood depths are typically non-linear functions of runoff and therefore can be difficult to accurately estimate using a relationship with stage at a limited number of locations within a catchment. However, due to the computational burden of a 2D model, it was decided to make this assumption, as optimization typically demands thousands of model runs. The depth–damage curve (Figure 8), which represents the cost per house and per flooding event based on the maximum depth of water, was used to generate the depth–damage correlation (Figure 9). A trend line is also shown for each curve with the respective correlation, which has been used subsequently to compute the flood related costs when assessing the candidate solutions during the optimization process.

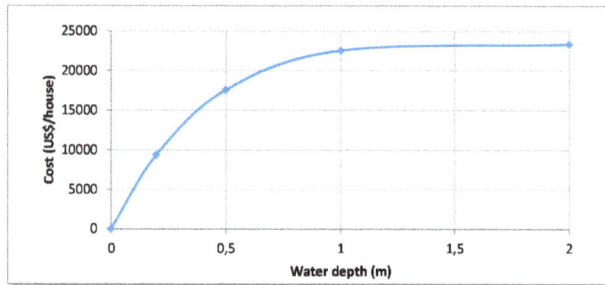

Figure 8. Damage cost per house and per flooding vs. water depth at the house—Quitacalzones.

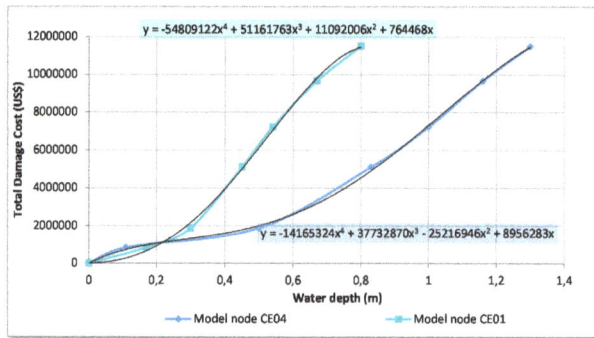

Figure 9. Depth–damage correlation: Damage cost per flooding event for the whole area vs. water depth at the control node. The total damage depends on water level in each house that is flooded. As the depth of water in the control nodes increases, both the flooded area (hence the number of houses flooded) and the depth of flooding in each house increases. Therefore, the shape of the depth–damage correlation function is different to that of the depth–damage curve (Figure 8) for a single house. The depth–damage correlation is contextual as it depends on many factors such as topography, spatial distribution of houses, etc.

Ecosystem Services Benefits

The private benefits from green roofs considered in this study are the reduction in energy consumption for cooling and heating, increase in value of private property, food production and increase in roof longevity. Other benefits such as avoided infrastructure costs for drainage installations are not considered because these green roofs will be installed on to existing buildings. The social benefits from green roofs considered are the avoided carbon emissions from energy savings, energy and carbon emission savings from reduced storm water volumes entering into the combined sewers, avoided costs for air pollution control measures and increased aesthetic value. Other benefits such as CSO (combined sewer overflow) control, habitat creation and job generation are not considered due to lack of reliable local data. Even though it is expected that jobs will increase in relation to maintenance of green roofs, it is also likely that this will lead to a reduction in other jobs such as maintenance of traditional roofs [24]. The total benefit presented in US$/m^2 is the net present value (NPV) of the benefits during the lifespan of the green roof (assumed as 30 years). The total ES benefits were calculated with the help of the GIVTC toolkit [38]. The values were adjusted to the local context as much as possible with local data. The total ES benefits (net present value of benefits during the life time) from green roofs is 132 US$/m^2. Detailed calculations of benefits are presented in the supplement with costs in Tables S4 and S5.

The private benefits from rain barrels considered in this study are the water savings due to the use of rainwater [39] and the reduction on sanitation fees due to the lower water consumption even though this may not lead to changes in sanitary waste discharges. In Uruguay, for every dollar charged in water fee, another 0.6 dollars are charged as sanitation fees [40]. However, sanitation fees are just a transfer of sanitation costs to the final users, the households. Other benefits such as reduction in detergent use due to the reduced hardness of rainwater were not considered. Social benefits of rain barrels include savings on energy and carbon emissions due to the reduction of storm water entering the sewers. Other benefits such as reduction in CSO discharges and employment generation are not considered for the same reasons as already stated for green roofs. The total ES benefits from rain barrels is 125 US$/barrel. Detailed calculation of benefits can be found in the Supplementary Materials.

4.3. Set up of Multi-Objective Optimization Model

Two different drainage configurations are optimized. The first one is a drainage system where only SuDS (green roofs and rain barrels) can be installed within the sub-catchment. In this configuration, the optimization variables are the number of modelling units of the two different SuDS groups in each sub-catchment. In the second configuration, storage tanks are also considered, where the area of each of the three tanks, as well as orifices and pipes dimensions are considered as optimization variables. Further, each of these two configurations has been optimized for two different cases: when computing the ecosystem services benefits of using SuDS and when not considering these. Thus in total, four different cases have been assessed, as presented in Table 1.

Table 1. Combination of drainage configurations with and without ES benefits for optimization.

Configuration	With ES	Without ES
SuDS	Case 1	Case 2
SuDS and storage tanks	Case 3	Case 4

Optimisation based on the population of 40 individuals across 70 generations yielded consistent results. The optimisation runs were performed by restricting the algorithm to find solutions based on total costs. The maximum total cost for the solution was fixed at US$35,000,000, approximately twice the flooding cost. The individuals whose total cost was above this set cost were penalized so that they are less likely to be passed on to the next generation.

5. Results and Discussion

The output of the optimization process for Case 1 is shown in Figure 10. Every assessed option is represented by a point with the total costs and the total benefits as axes. The upper envelope is the Pareto front, formed by the optimal solutions. In the figure it can be seen that the benefits are higher than the costs for the optimal solutions. The Pareto front has a steeper slope for the low-cost solutions than for the more costly options. The net benefits increase with the costs; however, the benefit–cost ratio remains approximately constant for the solutions below US$10 million cost and then decreases for higher investment costs.

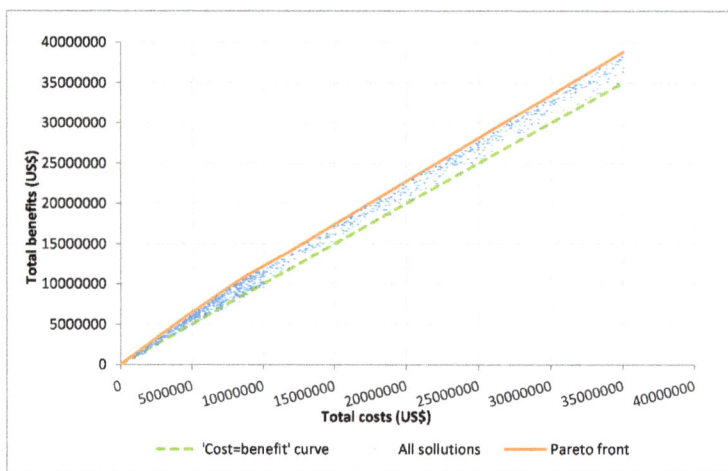

Figure 10. The Pareto-front relating total cost and total benefits for Case 1 (SuDS without underground storage; benefits including ES). The dots represent sub-optimal solutions, which are not a part of the final optimal results.

Each point in Figure 10 is a drainage network configuration comprising rain barrels and green roofs without added storage, each of which has distinct costs and benefits.

The results of the optimization process for the four assessed cases are shown in Figure 11. In Figure 11, we can see that storage tanks combined with SuDS give a cost-effective solution, regardless of the consideration of ES benefits. However, when ES benefits of SuDS are considered, it is possible to justify larger investments as they yield positive net benefits. Conversely, for SuDS only without storage tanks, the inclusion of ES benefits results in a positive benefit–cost ratio. The economic indicators for the assessed cases are presented in Table 2. Only the case with SuDS without ES benefits (case 2) results in negative net benefits for any total cost. Case 3 is the one with higher net benefits, while the one with higher total benefit–cost ratio is Case 4, which is similar to Case 3.

Larger flood reductions are achieved when the system is optimized with the combination of SuDS and storage tanks, and especially when ES benefits are not considered. The results are similar when only SuDS are considered. The SWM benefits of optimal solutions are higher for the case in which ES benefits are not taken into account. This suggests that when considering the ES benefits, the solutions that optimize the total benefits are not those that optimize the SWM benefits. Thus, if the drainage system were to be optimized for maximum SWM benefits instead of maximum total benefits, the results would have been different [41].

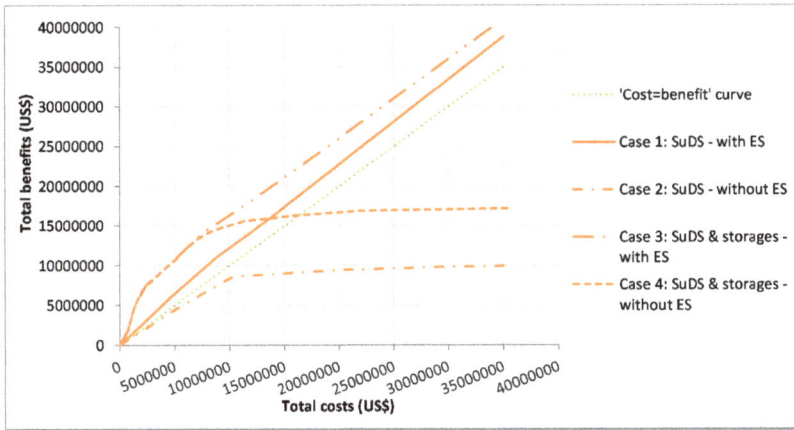

Figure 11. Pareto fronts for the four assessed cases/options.

Table 2. Maximum net benefits and benefit–cost ratios for the assessed cases.

Item	Case 1 (SuDS/with ES)	Case 2 (SuDS/without ES)	Case 3 (SuDS & Storage Tanks/with ES)	Case 4 (SuDS & Storage Tanks/without ES)
Max net benefit (million US$)	4.5 [a]	-	6.3	6
Max total B/C ratio	1.25	0.85	3.3	3.5
Max profitable investment (million US$)	>120	-	>120	16.3

Note: [a] for an investment of approximately US$75 million.

An important consideration when estimating the costs and benefits is the level of uncertainty of the input parameters. The magnitude of the differences between cost and benefits can vary due to the uncertainty of the underlying information. Although it is possible to systematically treat (known) uncertainties in optimization, by approaches such as robust-optimization, it was beyond the scope of this study. The objective of the current study was to demonstrate that it is possible to better justify investments in SuDS when ES benefits are considered alongside SWM benefits.

The analysis presented in this paper does not provide a complete picture of all the measures required when decisions have to be made. Social, legal, institutional and political implications related to the installation of SuDS are beyond the scope of this paper. However, these results can give an important input to the decision making process. Moreover, legal and social analyses can also utilise the results from this type of study, for instance, if the budget for solving flooding problems is relatively low, results have shown that storage tanks are the best option and therefore it would not be necessary to analyse the willingness of homeowners to install these SuDS or to consider the legal and other implications of installing SuDS.

6. Conclusions

This paper aimed at enhancing understanding of the economic value of large investments in SuDS through inclusion of ecosystem services valuation. The study has focused on the flood risk reduction capacity and the ES benefits of green roofs and rain barrels in the combined sewerage network of Montevideo Municipality in Uruguay. The cost and benefits of two drainage network configurations were analysed: (i) comprising SuDS; and (ii) comprising SuDS and detention storage in underground detention tanks. Optimal design configurations of SuDS, based on their costs and benefits, with and without ecosystem service benefits were determined for a wide range of costs and benefits. This provided a Pareto-optimal front of design configurations. In both of the design configurations, total benefits comprising both flood reduction and ES benefits are always higher than their costs. However, the exclusion of ES benefits when using SuDS alone was not found to deliver a positive BCR. Where the drainage configuration utilised both SuDS and storage tanks, declining BCRs subsequently leading to sub-zero values were found when the investments were increased without considering ES benefits. Hence, it can be concluded that larger investments can provide a positive return only when the ES benefits of SuDS are included.

Although none of the solutions eliminated flooding completely, the use of storage tanks resulted in the greatest reduction in flooding and, together with ES benefits, was found to be more cost-effective than when only using SuDS. Rain barrels were found to be more effective than green roofs in reducing storm water runoff and thus more effective in reducing flooding. When ES benefits were ignored, rain barrels were more cost effective when compared with green roofs, due to the higher investment costs for green roof compared with rain barrels for the same roof area.

Supplementary Materials: The following are available online at www.mdpi.com/2073-4441/9/11/841/s1.

Author Contributions: This paper is based on a master's thesis research of S.U.V., L.H. and A.P. together with Charlotte de Fraiture (who is acknowledged) supervised the thesis work. A.P. provided guidance on SWMM5EA tool. M.R., L.H., S.U.V. and A.P. wrote the paper.

Conflicts of Interest: The authors declare no conflict of interest.

References

1. Organisation for Economic Co-Operation and Development (OECD). Financial Management of Flood Risks. In *OECD Report Series*; Organisation for Economic Co-Operation and Development (OECD): Paris, France, 2016.
2. Radhakrishnan, M.; Pathirana, A.; Ashley, R.; Zevenbergen, C. Structuring climate adaptation through multiple perspectives: Framework and case study on flood risk management. *Water* **2017**, *9*, 129. [CrossRef]
3. United Nations. *Transforming Our World: The 2030 Agenda for Sustainable Development*; United Nations, Ed.; United Nations: New York, NY, USA, 2005.
4. Fletcher, T.D.; Shuster, W.; Hunt, W.F.; Ashley, R.; Butler, D.; Arthur, S.; Trowsdale, S.; Barraud, S.; Semadeni-Davies, A.; Bertrand-Krajewski, J.-L.; et al. Suds, Lid, Bmps, Wsud and More-The Evolution and Application of Terminology Surrounding Urban Drainage. *Urban Water J.* **2015**, *12*, 525–542. [CrossRef]
5. PUB Singapore's National Water Agency. *Active, Beautiful, Clean Waters Design Guidelines*; P.S.S.N.W. Agency, Ed.; PUB Singapore's National Water Agency: Singapore, 2014.
6. Charlesworth, S.M.; Mezue, M. Sustainable Drainage Out of the Temperate Zone: The Humid Tropics. In *Sustainable Surface Water Management: A Handbook for SUDS*; Wiley-Blackwell: Hoboken, NJ, US, 2016; p. 301.
7. Hoang, L.; Fenner, R.A. System interactions of stormwater management using sustainable urban drainage systems and green infrastructure. *Urban Water J.* **2016**, *13*, 739–758. [CrossRef]
8. Horton, B.; Digman, C.J.; Ashley, R.M.; Gill, E. *BeST (Benefits of SuDS Tool) W045c BeST-Technical Guidance*; Release Version 3; CIRIA: London, UK, 2016.
9. Ashley, R.; Walker, L.; D'Arcy, B.; Wilson, S.; Illman, S.; Shaffer, P.; Woods-Ballard, B.; Chatfield, P. UK sustainable drainage systems: Past, present and future. *Civ. Eng.* **2015**, *168*, 125–130. [CrossRef]
10. Ashton, R.; Baker, R.; Dean, J.; Golshetti, G.; Jaluzot, A.; Jones, N.; Moss, M.; Steele, M.; Williams, W.; Wilmers, P. *Building Natural Value for Sustainable Economic Development: The Green Infrastructure Valuation Toolkit User Guide*; Green Infrastructure Valuation Network: London, UK, 2011.
11. Costanza, R.; d'Arge, R.; De Groot, R.; Farber, S.; Grasso, M.; Hannon, B.; Limburg, K.; Naeem, S.; O'Neill, R.V.; Paruelo, J. The value of the world's ecosystem services and natural capital. *Nature* **1997**, *387*, 253–260. [CrossRef]
12. European Environment Agency. *Urban Adaptation to Climate Change in Europe: Transforming Cities in a Changing Climate*; European Environment Agency: Copenhagen, Denmark, 2016; p. 135.
13. Pathirana, A. SWMM5-EA-A Tool for Learning Optimization of Urban Drainage and Sewerage Systems with Genetic Algorithms. In *International Conference on Hydroinformatics*; CUNY Academic Works: New York, NY, USA, 2014.

14. Deb, K.; Pratap, A.; Agarwal, S.; Meyarivan, T. A fast and elitist multiobjective genetic algorithm: NSGA-II. *IEEE Trans. Evolut. Comput.* **2002**, *6*, 182–197. [CrossRef]

15. James, W.; Huber, W.; Pitt, R.; Dickinson, R.; Rosener, L.; Aldrich, J.; James, W. *SWMM4 User's Manual (User's Guide to the EPA Stormwater Management Model and to PCSWMM)*; University of Guelph and CHAI: Guelph, ON, Canada, 2002.

16. Delelegn, S.; Pathirana, A.; Gersonius, B.; Adeogun, A.; Vairavamoorthy, K. Multi-objective optimisation of cost-benefit of urban flood management using a 1D2D coupled model. *Water Sci. Technol.* **2011**, *63*, 1054. [CrossRef] [PubMed]

17. Radhakrishnan, M.; Pathirana, A.; Ghebremichael, K.; Amy, G. Modelling formation of disinfection by-products in water distribution: Optimisation using a multi-objective evolutionary algorithm. *J. Water Supply Res. Technol. AQUA* **2012**, *61*, 176–188. [CrossRef]

18. CSI Ingenieros S.A. *Estudio de Alternativas y Proyecto Ejecutivo de las Obras de Drenaje Pluvial en las Cuencas de los Arroyos seco y Quitacalzones—Memoria Descriptiva y Justificativa*; Municipality of Montevideo: Montevideo, Uruguay, 2012. (In Spanish)

19. CSI Ingenieros S.A. *Estudio de Alternativas y Proyecto Ejecutivo de las Obras de Drenaje Pluvial en las Cuencas de los Arroyos seco y Quitacalzones—Evaluación SocioeconóMica de Priorización de Inversiones*; Municipality of Montevideo: Montevideo, Uruguay, 2011. (In Spanish)

20. Bianchini, F.; Hewage, K. How "green" are the green roofs? Lifecycle analysis of green roof materials. *Build. Environ.* **2012**, *48*, 57–65. [CrossRef]

21. Carter, T.; Keeler, A. Life-cycle cost–benefit analysis of extensive vegetated roof systems. *J. Environ. Manag.* **2008**, *87*, 350–363. [CrossRef] [PubMed]

22. Lee, A.; Sailor, D.; Larson, T.; Ogle, R. Developing a web-based tool for assessing green roofs. In *Greening Rooftops for Sustainable Communities, Minneapolis*; Green Roofs for Healthy Cities: Toronto, ON, Canada, 2007.

23. Clark, C.; Adriaens, P.; Talbot, F.B. Green roof valuation: A probabilistic economic analysis of environmental benefits. *Environ. Sci. Technol.* **2008**, *42*, 2155–2161. [CrossRef] [PubMed]

24. Doug, B.; Hitesh, D.; James, L.; Paul, M. *Report on the Environmental Benefits and Costs of Green Roof Technology for the City of Toronto*; Ryerson University: Toronto, ON, Canada, 2005.

25. Kosareo, L.; Ries, R. Comparative environmental life cycle assessment of green roofs. *Build. Environ.* **2007**, *42*, 2606–2613. [CrossRef]

26. Rosenzweig, C.; Gaffin, S.; Parshall, L. *Green Roofs in the New York Metropolitan Region: Research Report*; Columbia University Center for Climate Systems Research and NASA Goddard Institute for Space Studies: New York, NY, USA, 2006; pp. 1–59.

27. Smit, J.; Nasr, J.; Ratta, A. *Urban Agriculture: Food, Jobs and Sustainable Cities*; Urban Agriculture Network, Inc.: New York, NY, USA, 1996; pp. 35–37.

28. Dirección Nacional de Meteorología. Rainfall Data. Available online: http://meteorologia.gub.uy/index.php/estadisticasclimaticas (accessed on 12 January 2011).

29. Googlemaps. Satellite imagery of Montevideo. Available online: http://bit.ly/2qnmyws (accessed on 26 May 2017).

30. Getter, K.L.; Rowe, D.B. The role of extensive green roofs in sustainable development. *HortScience* **2006**, *41*, 1276–1285.

31. Verde Facil. Green Roofs. Available online: www.verdefacil.com (accessed on 23 May 2017).

32. Morales-Pinzón, T.; Lurueña, R.; Rieradevall, J.; Gasol, C.M.; Gabarrell, X. Financial feasibility and environmental analysis of potential rainwater harvesting systems: A case study in Spain. *Resour. Conserv. Recycl.* **2012**, *69*, 130–140. [CrossRef]

33. Thurston, H.W.; Taylor, M.A.; Shuster, W.D.; Roy, A.H.; Morrison, M.A. Using a reverse auction to promote household level stormwater control. *Environ. Sci. Policy* **2010**, *13*, 405–414. [CrossRef]

34. Tang, Y. *Exploring the Response of Urban Storm Sewer System to the Implementation of Green Roofs, in Civil & Environmental Eng*; University of Illinois at Urbana-Champaign: Champaign, IL, USA, 2012.

35. Oficina de Planeamiento y Presupuesto. *Metodología General de Formulación y Evaluación Social de Proyectos de Inversión PúBlica*; Oficina de Planeamiento y Presupuesto, Uruguay: Montevideo, Uruguay, 2012.

36. Bianchini, F.; Hewage, K. Probabilistic social cost-benefit analysis for green roofs: A lifecycle approach. *Build. Environ.* **2012**, *58*, 152–162. [CrossRef]

37. Anand, C.; Apul, D. Economic and environmental analysis of standard, high efficiency, rainwater flushed, and composting toilets. *J. Environ. Manag.* **2011**, *92*, 419–428. [CrossRef] [PubMed]

38. Ashton, R.; Baker, R.; Dean, J.; Golshetti, G.; Jaluzot, A.; Jones, N.; Moss, M.; Steele, M.; Williams, W.; Wilmers, P. *Building Natural Value for Sustainable Economic Development: Green Infrastructure Valuation Toolkit*; Colling, R., Ed.; Genecon LLP: Leeds, UK, 2010.

39. Kuhn, K.; Serrat-Capdevila, A.; Curley, E.F.; Hayde, L.G. Alternative water sources towards increased resilience in the Tucson region: Could we do more? In *Water Bankruptcy in the Land of Plenty*; Poupeau, F., Ed.; CRC Press/Taylor and Francis Group Boca Raton: London, UK; New York, NY, USA; Leiden, The Netherlands, 2016; pp. 337–362.

40. Presidencia. Decretos de Setiembre de 2012. Available online: https://www.presidencia.gub.uy/normativa/2010-2015/decretos/decretos-09-2012 (accessed on 23 May 2017).

41. Ashley, R.M.; Digman, C.J.; Horton, B.; Gersonius, B.; Smith, B.; Shaffer, P.; Baylis, A. Evaluating the Longer Term Benefits of Sustainable Drainage. Available online: https://doi.org/10.1680/jwama.16.00118 (accessed on 17 October 2017).

water

MDPI

Article

Thermal Study on Extensive Green Roof Integrated Irrigation in Northwestern Arid Regions of China

Yajun Wang, Rajendra Prasad Singh *, Dafang Fu *, Junyu Zhang and Fang Zhou

Department of Municipal Engineering, School of Civil Engineering, Southeast University, Nanjing 210096, China; 230149631@seu.edu.cn (Y.W.); 230149068@seu.edu.cn (J.Z.); 230169393@seu.edu.cn (F.Z.)
* Correspondence: rajupsc@seu.edu.cn (R.P.S.); fdf@seu.edu.cn (D.F.); Tel.: +86-25-83790757 (D.F.)

Received: 24 August 2017; Accepted: 19 October 2017; Published: 9 November 2017

Abstract: Selection of xerophils and drought tolerant plants is highly crucial in green roof techniques in the drought prone regions of Northwest China. In this study, the thermal performance under the natural conventional climate in summer was analyzed using a self-made simulation experimental platform through comparison of the internal surface temperature with and without green roofs. The distribution frequency of internal surface temperature was investigated by dividing internal surface temperature into several ranges. Statistical analysis showed that the frequency of internal surface temperature lower than 33 °C for green roofs was 91.8%, about 1.09 times higher than that for non-green roofs, and that the sum of internal surface temperature exceeding 35 °C was about one third of that for non-green roofs. The results proved that green roofs have a significant insulation effect. Moreover, the thermal insulation property of green roofs had a strong positive relation with outside temperature. The thermal insulation characteristic was improved as the outdoor temperature increased, additionally, it had a better insulation effect within two hours after irrigation.

Keywords: thermal insulation; extensive green roof; irrigation; arid regions

1. Introduction

Commonly existing urban infrastructure, such as buildings, roads and parking lots, forms a growing area of impervious surfaces across cities globally. Impervious surfaces have considerable influence on stormwater quantity and quality. The runoff coefficient on impervious surfaces is generally 0.9, translating to 90% of rain falling on an impervious surface leaving as surface runoff rather than infiltrating [1]. Surface runoff leaves a site as discharge to existing drainage infrastructure or receiving water bodies; it represents a missed opportunity for storage and reuse. This missed opportunity is especially impactful in the northwestern arid areas of the Loess Plateau in China where rainwater has become the second most important water source [2–4] due to the lack of available surface water resources in the region.

If collected and treated reliably, captured stormwater has the potential to become an important additional source of freshwater. This can be achieved by employing stormwater management practices which aim to intercept, attenuate, and retain stormwater flows to improve water quality and restore flow regimes to pre-development levels. Such techniques are often referred to as low impact development (LID) systems in the USA, water sensitive urban design (WSUD) in Australia, sustainable urban drainage systems (SUDS) in the UK and sponge city construction in China.

Green roofs are a popular stormwater management practice in urban areas as they fit within the existing developed footprint without the additional space requirements of other practices. Green roof technology is also a popular stormwater management practice in sponge city construction. Green roofs are effective in improving the city environment [5] and absorbing radiation heat. Associated thermal insulation characteristics can reduce the burden on building air conditioning equipment [6] and significantly reduce the temperature of the roof [7].

Installing green roofs is now widely considered as an effective strategy to reduce the thermal load of the building's shell. Several studies have been carried out to evaluate the thermal insulation of vegetation for green roofs by using experimental and infrared techniques. Niachou et al. [8] demonstrated that solar radiation and external temperature are reduced by the vegetative cover of planted green roofs, which agrees with previous research [9]. A recent study by Costanzo et al. [10] found that the green leaf area index has the largest impact on thermal performance. Moreover, Sailor [11] studied the energy balance of green roofs and concluded that the dominant way for green roofs to dissipate absorbed heat was through evapotranspiration.

A recent study by Pianella et al., showed that green roof irrigation would be necessary to maintain plant health and reduce substrate temperature and heat flux [12]; however, costs associated with irrigating green roofs in arid climates can nullify savings in energy demand for air-conditioning [13]. Consequently, rainwater collection and automatic irrigation subsystem research is significantly needed in follow-up studies.

The surface temperature of vegetation is nonhomogeneous due to the complex geometry of plant surfaces. It is also difficult to measure surface temperatures using traditional thermocouple probes as there is considerable error [14]. However, results measured using infrared thermography are visual and accurate [15,16].

Research on green roofs constructed on new buildings has made considerable headway [17], while the amount of research involving existing green roofs has not. Moreover, the majority of published research literature contains little related information on the performance of green roofs in arid regions.

The study aims to investigate extensive multi-species vegetated green roofs in order to answer the following questions:

1. What is the best model to simulate green roofs in arid climates?
2. What plant species have the greatest impact on the temperature inside the buildings during long-term drought?
3. How can integrated irrigation systems be designed for drought resistance?
4. How can the thermal characteristics of green roofs be assessed under drought and strong light area conditions?
5. How can thermal changes be assessed after irrigation?

2. Methods

Reduced-scale models of buildings equipped with vegetated green roofs were constructed for this study. Post-construction, plants were allowed to establish before treatments and monitoring began. Greens roofs were monitored for temperature and plant water requirements. Monitoring of the models began on 1 July 2014, and continued, with some interruptions and changes to the experimental design, until 31 July 2014.

2.1. Study Site

The experiment was conducted at Lanzhou University of Technology. Located at the geometric center of China (36°3′27.10″ N, 103°43′43.75″ E), Lanzhou has an arid to semi-arid climate with hot summers. Temperatures range from −11 to 1 °C in January and 16 to 28 °C in August and average annual precipitation is 322.10 mm, with rainfall occurring entirely between June and August (Figure 1).

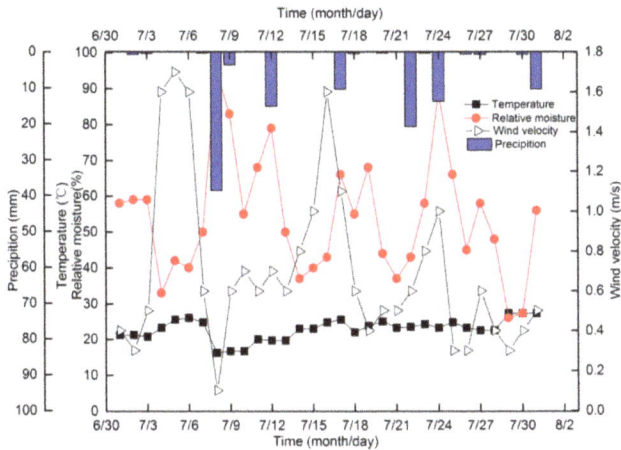

Figure 1. Test value of daily average.

2.2. Selection of Plant Species

Three types of plants with different appearances and adaptation strategies were selected [18–20]. There were six replicates for each plant species. Plants were propagated in greenhouse conditions from seed when possible and from cuttings if seeds were unavailable. Plants were allowed to root and grow for at least four weeks before the experiment began. Mature cuttings propagated from plants were chosen as parent materials. To ensure the same initial condition for all cuttings, the tested cuttings were snipped off at 4 cm from the top of shoot, and the leaves on the two bottom nodes were stripped. The following selected plants represent different classes of potential green roof solutions that would maximize benefit in terms of thermal protection and minimize cost in terms of irrigation:

a   *Sedum lineare* has a dense root system and is a rapidly growing species. It propagates easily and is drought-tolerant. Previous research has shown *Sedum spp.* reliability and dependability to be of crucial importance in green roof implementation [18].
b   *Aptenia cordifolia* is a succulent with bright green leaves and red flowers. It is very efficient in heavy metal uptake for sewage sludge compost recuperation. Earlier studies reveal that it is capable of surviving drought conditions [19].
c   *Tifdwarf bermuda* grass has been one of the most popular choices for putting greens for over 40 years [20]

2.3. Construction of Reduced-Scale Building Models

A simulation experiment platform representing northwest parapet flat-roofed houses was used for the test model. The model was scaled by 1:8 and had similar materials, mechanics, geometry, and had the following dimensions: 1000 mm length, 650 mm width, and 400 mm height (Figure 2). The test model contained two adjacent contrast test rooms each with an area of 0.325 m^2. The steel structural framework consisted of constructional columns and ring beams using #20 square concrete-filled steel tubes. The roof and floor were made of cast-in-place reinforced concrete slab with a thickness of 12.50 mm. Its assembly is shown in Figure 3. Walls were made of fiber concrete slab with an exterior wall thickness of 38 mm and interior wall thickness of 25 mm. The underside of the upper roof space had natural ventilation so that its surface temperature directly reflected the effects of roof insulation. It should be emphasized that an attempt was made to ensure the differential thermal behavior between the test model rooms with and without greening.

131

Figure 2. Model experimental device.

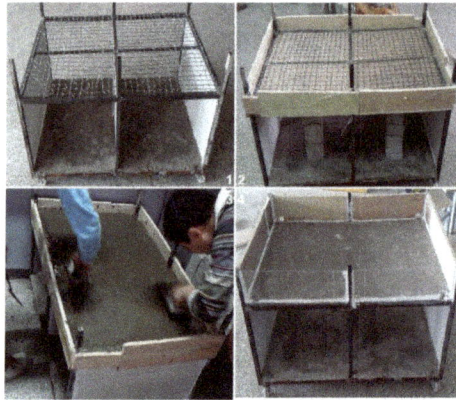

Figure 3. Assembling process of the experimental device.

2.4. Establishment of Irrigation System

The extensive green roof system with light, insulation, and greening layer structure consisted of a rainwater collection subsystem and automatic irrigation subsystem and could implement rainwater collection, storage and irrigation during the dry seasons found in Lanzhou [21] (Figure 4).

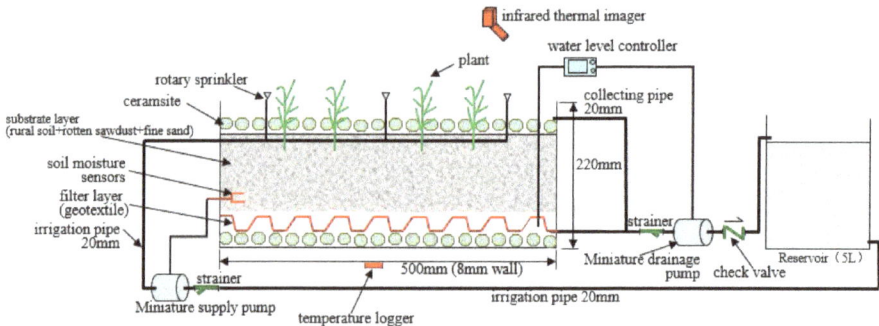

Figure 4. Schematic diagram of the experimental green roof system for integrated rainwater collection and irrigation.

The green roof container assembly consisted of a polypropylene frame (220 mm deep) with a 6 mm watertight high impact polypropylene (HIPP) sheet serving as the tray (Figure 4). Based on technical specifications for green roofs [22], the lightweight media was a mixture of rural soil (30%, sawdust (40%), and fine sand (30%). Additional media characteristics were as follows: 950 kg/m^3 saturated water density, 25% organic content, 65% total porosity, 30% effective moisture, 60 mm/h drainage rate, 60 mm soil layer depth [22]. Both the storage layer and the drainage layer were settled at the bottom of the green roof container. The storage layer can be used to conserve water for vegetation. Simulated rain water was not stored until internal water levels fell to contact the control probe, and was not recharged until soil moisture fell below 20% (m^3 water/m^3 soil).

2.5. Set-up of Monitoring Instrument

Monitoring was conducted during July 2014. Thermodetectors were installed in the internal surface for set-ups with and without the green roof (Figure 4). Data were collected using Elitech Elitech RC-4 temperature data loggers with a multifunction liquid crystal display alarm indicator (Elitech Corp., Milpitas, California, CA, USA) with an internal probe (Elitech Corp., Milpitas, California, CA, USA) temperature range of −30 to 60 °C, external probe (Elitech Corp., Milpitas, California, CA, USA) temperature range of −49 to 85 °C, accuracy of ±1 °C, and resolution of 0.10 °C. Data were recorded at 15 minute intervals and retrieved for processing in Excel spreadsheets. Other climate parameters including humidity, wind speed, and rainfall, were gathered using a WE800 mini-meteorological station (Global Water Corp., College Station, TX, USA). Infrared thermography of sprinkler irrigation and non-sprinkler irrigation areas were measured with a YRH600 infrared thermal imager (Fengte Corp., Guangzhou, China), and were analyzed using SatIrReport software (Fengte Corp., Guangzhou, China).

2.6. Statistical Analysis

All data were analyzed using an ANOVA model by SPSS 13.0 (IBM, Corp., Armonk, NY, USA). Data were checked to ensure ANOVA assumptions were not violated and were transformed appropriately when needed prior to analysis. Figures were drawn using Origin 9.0 (OriginLab Corp., Northampton, MA, USA).

3. Results and Discussion

3.1. Drought Tolerance

The three plant species installed in the water-scarce region may be exposed to stress as a result of prolonged periods without precipitation. To look into the plants' performance under such conditions, limits on irrigation and restrictions on the use of potable water were investigated. Figure 5 shows the effects of the withdrawal of irrigation from previously healthy plants for a period of 40 days. A tray with *Aptenia cordifolia* (Figure 5) survived this extended drought in fairly good health. It is estimated that this plant could have survived for several more weeks. All other tested plant species, including *Sedum lineare*, did not survive the 40-day drought period.

The effect of non-irrigation is illustrated in Figure 5. This result confirmed the conclusion of Fleta-Soriano et al. [19] that *Aptenia cordifolia* is drought tolerant and can survive extended periods without precipitation.

Figure 5. Trays with (1) *Sedum lineare*; (2) *Aptenia cordifolia* and (3) *Tifdwarf bermudagrass* after 40 days without irrigation.

3.2. Comparison of the Effect of Green Roofs on Temperature

Figure 6 shows the changes of the internal temperatures of models with and without a green roof, as influenced by the outside climate. The moisture state of the soil and the plants, which directly affects the heat transfer to the roof, were affected by the change in climate-related parameters, resulting in variable temperature recordings.

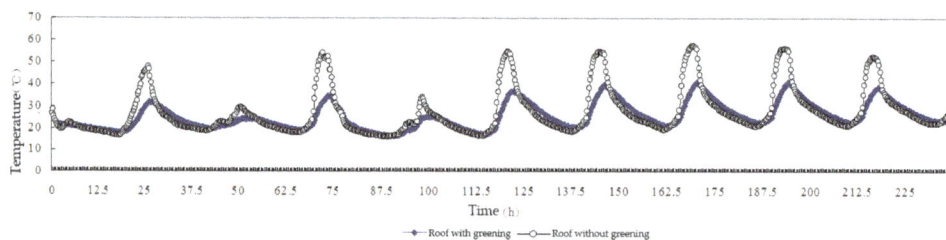

Figure 6. Temperature comparison with and without roof greening.

The internal temperature of the building, a comprehensive natural indoor temperature, is formed from heat transformation at each part of the building envelope. It can be seen that the internal temperatures with green roofs were lower than without green roofs, and had much smaller temperature variations daily. This means that the green roof set-up fully showed the cooling effect on indoor thermal environments. On the contrary, internal temperature on the non-green roof set-up had larger temperature variations. The temperature variations can be directly related to the weather (i.e., the variations on a sunny day will be bigger). The indoor temperature in a non-green roof set-up was higher than that in a green roof set-up on a hot sunny day where the highest temperature can reach up to 57.7 °C, higher than that in the green roof set-up under the same conditions. However, there was no evident difference in internal temperature between non-green roof and green roof set-ups during rainy days.

Therefore, the research's emphasis was to investigate the roof insulation's ability under high temperature conditions. During the experiment, the point data was 1000 times. Using 2 °C as step length, it was divided into 13 time intervals from 33 to 59 °C.

The distribution frequency of each interval is as follows:

$$P(i, i + 1.9) = \frac{\Sigma n}{N + 1} \tag{1}$$

where i = 33, 35, 37, 39, 41, 43, 45, 47, 49, 51, 53, 55, 57, 59 (°C); P is the distribution frequency within $(i,i + 1.9)$ temperature interval (%); N is the test number (1000); and n is the occurrence number in $(i,i + 1.9)$ interval.

The distribution number and frequency of each test point using Equation (1) is shown in Table 1. Table 1 shows that 34 °C of the highest outside temperature distributed in one temperature interval; 37.5 °C of the highest internal temperature with green roof distributed in five temperature intervals; 57.7 °C of the highest internal temperature without green roof distributed in 13 temperature intervals. All results of the exceeding 28 °C temperature frequency show that the outside temperature frequency was 44.10%, the internal temperature frequency with green roof was 60.00%, and the internal temperature frequency without green roof was 69.10%.

Table 1. Frequency distribution of outside temperature and internal surface temperature.

Temperature Interval (°C)	Outside Temperature		Internal Temperature Frequency with Green Roof		Internal Temperature Frequency without Green Roof	
	Number	Frequency (%)	Number	Frequency (%)	Number	Frequency (%)
33~34.9	50	5	30	3	23	2.3
35~36.9	45	4.5	34	3.4	18	1.8
37~38.9	23	2.3	10	1.0	11	1.1
39~40.9	10	1	7	0.7	12	1.2
41~42.9	1	0.1	1	0.1	7	0.7
43~44.9	1	0.1	0	0	11	1.1
45~46.9	0	0	0	0	15	1.5
47~48.9	0	0	0	0	9	0.9
49~50.9	0	0	0	0	12	1.2
51~52.9	0	0	0	0	26	2.6
53~54.9	0	0	0	0	21	2.1
55~56.9	0	0	0	0	15	1.5
57~58.9	0	0	0	0	4	0.4
Σ	130	13	82	8.2	158	15.8
<33 °C	870	87	918	91.8	842	84.2
≥35 °C	80	8	52	5.2	135	13.5

Table 1 shows that the highest distribution frequencies were recorded at the following temperature intervals: 33–34.9 °C interval for outside temperature; 35–36.9 °C interval for the internal temperature with the green roof; and 51–52.9 °C interval for the internal temperature without the green roof. It is worth noting that the frequencies were distributed at each temperature interval for the internal temperature without the green roof.

People feel more comfortable when the surface temperature is less than 33 °C [23]. Table 1 shows that the highest occurrence frequency for temperature recordings less than 33 °C occurred on the recording for internal surface temperature with green roofs at 91.8%, while the occurrence frequency without green roofs was only 84.20%. Obviously, the distribution frequency with green roofs is 0.8 times than that of non-green roofs. However, the distribution frequency of uncomfortably high temperature (i.e., exceeding 35 °C) mainly appeared on the recordings for internal surfaces without green roofs, where the high temperature can reach up to 57.7 °C.

39 °C is a turning point. When the temperature is less than 39 °C, the temperature distribution frequency for both green and non-green roof set-ups were almost higher than the outside temperature. This suggests that the two set-ups have a certain ability to insulate heat. On the other hand, the two set-ups have different tendencies when the temperature is higher than 39 °C. The frequency for the green roof fell sharply and stopped at the 41–42.9 °C temperature range (only 0.1% of the total number), and was lower than the frequency for the outside air temperature. This shows that the green roof set-up had better thermal insulation. However, the frequency without green roofs decreased much

more slowly, its frequency was much higher than the outside temperature and green roofs, especially under the extreme high temperature range (i.e., exceeding 40 °C), where the frequency without green roof emerged 120 times (12% of the total number).

The total frequency at the 15–59 °C temperature range was set to 100%. From this, 91.8% of internal surface temperature with green roofs fell into the low temperature area (i.e., less than 33 °C) while the internal surface temperature with green roofs appeared only 53 times in the high temperature area (i.e., exceeding 35 °C). On the other hand, the internal surface temperature without green roofs appeared 135 times in the high temperature area.

The obtained results confirm that the thermal insulation of green roofs was very effective, which coincides with a series of the previous experimental results, showing that green roofs can reduce the heating loads of residential buildings. For example, Getter et al. [24] quantified that the monthly average values over the course of the year were consistently higher for a gravel-ballasted roof than a green roof, with values up to 20 °C warmer during summer. Qin et al. [25] showed that a green roof test-bed can reduce the internal air temperature by an average value of 0.5 °C if compared with a bare roof.

3.3. Statistical Test and Variance Analysis

Through statistical analysis of the raw data from Table 1, the statistical description of each group (A for outside temperature, B for internal temperature frequency with green roof, C for internal temperature frequency without green roof), statistical tests (Table 2) and variance analysis (Table 3) are given below.

Table 2. Statistical tests of inside surface temperature.

Name	Outside Temperature (A)	Internal Temperature Frequency with Green Roof (B)	Internal Temperature Frequency without Green Roof (C)
	Test Number	Test Number	Test Number
Average value $\bar{x}$	10	6.307692	14.15385
Standard deviation S_x	17.92577	11.863	6.401122
Sample number n	13	13	13
Minimum value x_{min}	0	0	4
Maximum value x_{max}	50	34	26
Range $x_{min} \sim x_{max}$	0~50	0~34	4~26
Confidence interval half width d	0.000714	0.000714	0.000714
Double tail pair comparison t test	0.386006 (A and C)	0.036636 (A and B)	0.026228 (B and C)

Table 3. Variance analysis of inside surface temperature.

Summary				
Group	Observation number	Summation	Average	Variance
A	13	130	10	321.3333
B	13	82	6.307692	140.7308
C	13	184	14.15385	40.97436

Variance Analysis					
Source of variance	SS	df	MS	F	P-value	F crit
Inter-group	400.6154	2	200.3077	1.194587	0.314548	3.259446
Intra-group	6036.462	36	167.6795			
Total	6437.077	38				

Table 2 shows that there is a significant difference between group B and group C (0.026228 < 0.05). It also shows that there is a significant difference between group B and group A—between greening facilities and outside temperature (0.036636 < 0.05). The results are consistent with the results in Table 1.

3.4. Infrared Thermal Imaging of Plants and Analysis

To study the transpiration process of green plants and the optimum conditions of heat insulation in summer under sprinkler irrigation, infrared thermography was performed once every half an hour for two hours.

During the long-term high temperature before irrigation, the temperatures of the plant surface began to rise rapidly, and the temperature field was between 42.5 and 48.5 °C (Figure 7). The main reason for the temperature rise is that the water content on the plant surface evaporated completely, as well as plant transpiration decreasing rapidly, and the leaf entered the dormant state again after it reached the critical state. After further irrigation, the temperature of the leaves decreased obviously, and the temperature field was between 36.9 and 40.9 °C (Figure 8). The leaves received moisture, meanwhile, the surface transpiration took away a lot of heat when the stomata of the leaves were opened, so that the surface temperature dropped again. Then, after half an hour of irrigation, the temperature field was between 34.9 and 37.9 °C (Figure 9), with leaf transpiration occurring continually. After that, after half an hour of irrigation, the temperature field was between 36.2 and 40.2 °C (Figure 10). At this time, the transpiration of leaves entered the most intense period, the average temperature of the leaf surface only increased by 1 °C when the environmental temperature rose by 1.7 °C. Later, after 1.5 h of irrigation, the temperature field was between 37.2 and 41.2 °C (Figure 11). At this time, due to insufficient water supply, the leaf surface temperature increased rapidly with the reduction of leaf transpiration. In the end, after two hours of irrigation, the temperature field was between 38.2 and 41.2 °C (Figure 12). At this time, the leaves of the plant basically entered dormancy, and the leaf surface temperature was close to the environmental temperature.

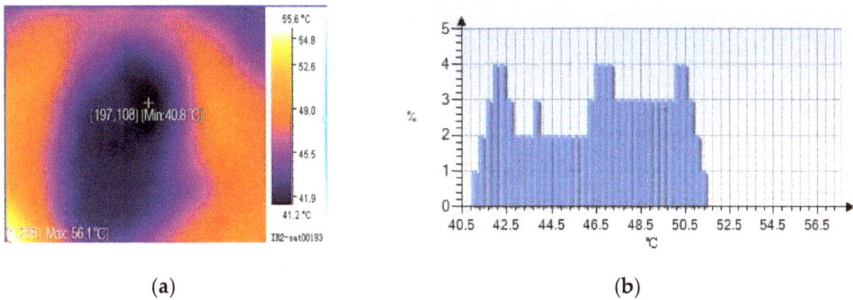

(a)   (b)

Figure 7. Infrared imaging and temperature distribution before irrigation at 10:36 a.m., (**a**) Infrared imaging; (**b**) Temperature distribution diagram.

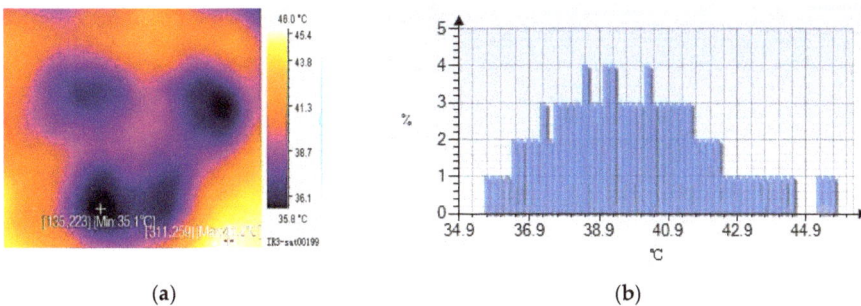

(a)   (b)

Figure 8. Infrared imaging and temperature distribution after irrigation at 10:50 a.m., (**a**) Infrared imaging; (**b**) Temperature distribution diagram.

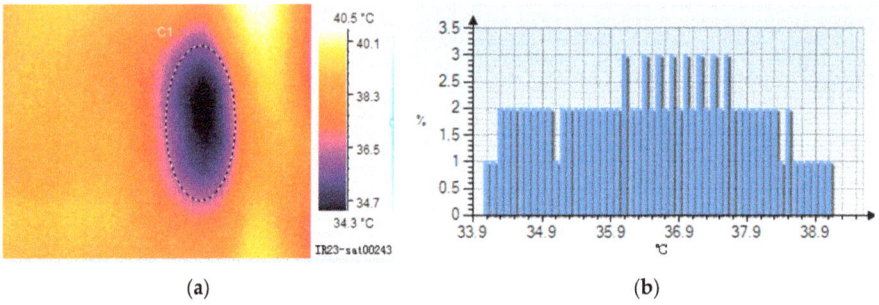

(a) (b)

Figure 9. Infrared imaging and temperature distribution after irrigation at 11:20 a.m., (**a**) Infrared imaging; (**b**) Temperature distribution diagram.

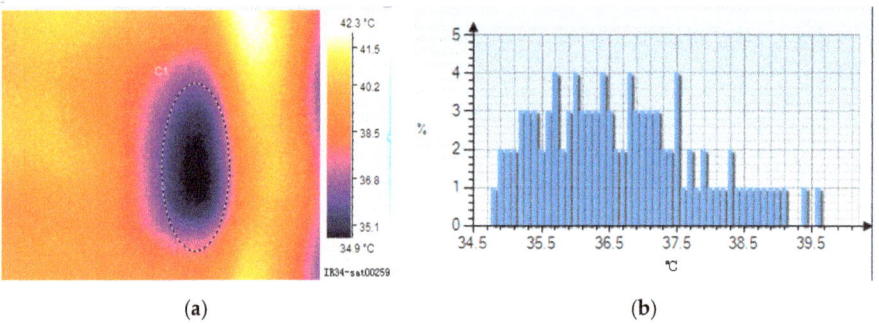

(a) (b)

Figure 10. Infrared imaging and temperature distribution after irrigation at 11:50 a.m., (**a**) Infrared imaging; (**b**) Temperature distribution diagram.

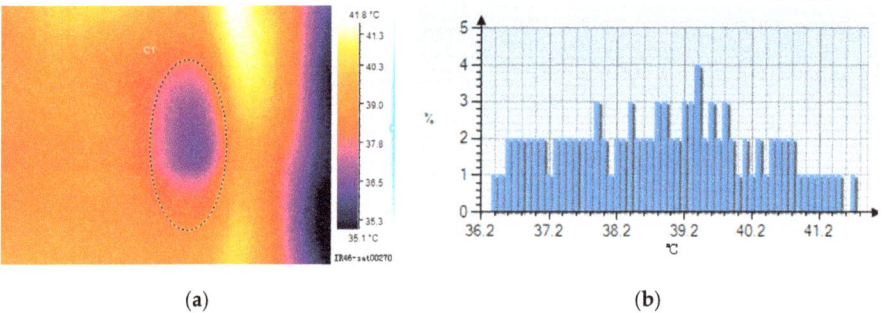

(a) (b)

Figure 11. Infrared imaging and temperature distribution after irrigation at 12:20 p.m., (**a**) Infrared imaging; (**b**) Temperature distribution diagram.

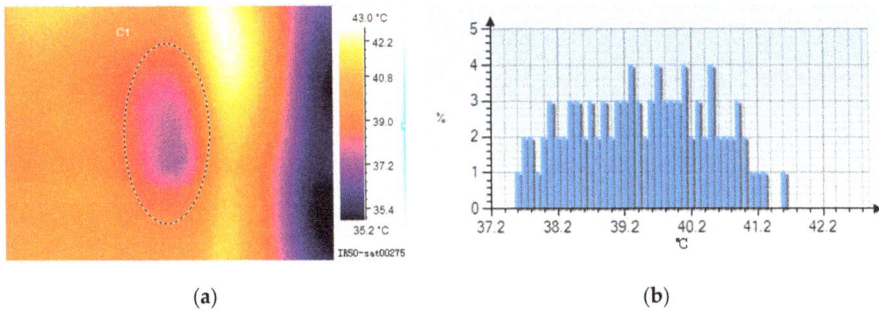

(a)   (b)

Figure 12. Infrared imaging and temperature distribution after irrigation at 12:50 p.m., (**a**) Infrared imaging (**b**) Temperature distribution diagram.

Under the high temperatures in summer, the leaves of the plant enter the dormancy state. When the leaf surface is sprayed with water, the temperature of the leaves drops and then the surface stomata opens, and the transpiration stage begins [26,27]. The leaves enter the dormancy state again, as well as water on the leaf surface evaporating [28]. After irrigation, the transpiration began to weaken with the increase of temperature, and then the leaves became dormant and transpiration disappeared when the temperature reached 40 °C; this entire process took about 1.5 h, thus the greening plants have better insulation within 1.5 h, though the insulation effect dropped dramatically after more than two hours.

4. Conclusions

In climates with extended hot and dry seasons, such as in the northwest arid areas located in the Loess Plateau in China, the installation of green roofs may prove beneficial only if appropriate plants are selected. The proper selection of plants will allow the green roof to survive extended dry periods through rainwater collection and automatic irrigation. Three local plant species were selected for the drought resistance experiments to choose the suitable local greening plants. The study showed that *Aptenia cordifolia* was capable of surviving extended dry periods in fairly good health. An automatic irrigation system was designed to sustain the roofs through an extended dry period.

Through the statistical analysis method, the measured data were used in the statistical description and variance analysis. The results showed that (1) the occurrence frequency of the internal surface temperature below 33 °C for green roofs was 91.8%, and it was 1.09 times higher than that of the internal surface temperature without green roofs; (2) the sum of the internal surface temperature exceeding 35 °C was about one third of that for roofs without greening plants and showed that the green roof showed obvious insulation performance, which became more significant as temperatures rose; (3) the non-green roof set-up had a more sensitive response to outside temperature changes, as exemplified by the rapid increase in the internal surface temperature with the increase of outside temperature. Frequencies were also recorded for each temperature range where the highest temperature could reach up to 57.7 °C; where people could feel uncomfortable in a room without a green roof, which is bound to increase the energy consumption. Also, green roofs have a strong temperature control function as exemplified by the relatively slow internal temperature response. This characteristics also means that the internal surface temperature is relatively stable.

Through monitoring of the plant temperature and heat insulation before and after irrigation, the results showed that timely irrigation can improve insulation performance, which is better within two hours of irrigation.

Although this system might be more sensitive to water scarcity and thermal conduction on reduced-scale models than on full-scale roofs due to spatial heterogeneity effects, it allows us to assess/verify the function of green roofs in dry areas. The automatic irrigation subsystem according to the soil moisture content can make up for the influence of the drought climate on greening plants.

Extensive green roofs are a proven technology in temperate or tropical countries, but there are many barriers to the implementation of such roofs in the arid regions of northwest China with dry climates. Further research is required to solve the vegetation survival in the cold winter, and to quantify the actual benefits, conduction and others. It is clear that green roofs must be designed with great care in accordance with the local climate: they may not, in fact, be an appropriate environmental response in all conditions.

Acknowledgments: This study was co-funded by the National Key Technologies R&D Program (No. 2015BAL02B05); and the Priority Academic Program Development of Jiangsu Higher Education Institutions; the University Scientific Research Project of Gansu Province (2016B-026); the National Natural Science Foundation of China: (No. 51650410657). We thank Jeffrey Johnson (Department of Biological and Agricultural Engineering, North Carolina State University, Raleigh, NC 27695, United States; jpjohnso@ncsu.edu), who carried out the English editing and proofreading.

Author Contributions: Yajun Wang and Dafang Fu conceived and designed the experiments; Fang Zhou performed the experiments; Yajun Wang and Junyu Zhang analyzed the data; Yajun Wang, Dafang Fu and Rajendra Prasad Singh contributed reagents/materials/analysis tools; Yajun Wang and Rajendra Prasad Singh wrote the paper.

Conflicts of Interest: The authors declare no conflict of interest.

References

1. Walsh, C.J.; Fletcher, T.D.; Ladson, A.R. Stream restoration in urban catchments through redesigning stormwater systems: looking to the catchment to save the stream. *J. N. Am. Benthol. Soc.* **2005**, *24*, 690–705. [CrossRef]
2. Wang, G.; Cheng, G. The characteristics of water resources and the changes of the hydrological process and environment in the arid zone of northwest China. *Environ. Geol.* **2000**, *39*, 783–790. [CrossRef]
3. Wang, Y.; Chang, Y. Drinking water technology of rural housing rainwater treatment at arid regions of northwest China. *Chin. J. Environ. Eng.* **2014**, *8*, 1021–1024.
4. Xiao, G.; Wang, J. Research on progress of rainwater harvesting agriculture on the Loess Plateau of China. *Acta Ecol. Sin.* **2003**, *23*, 1003–1011.
5. Payne, E.G.I.; Hatt, B.E.; Deletic, A. *Adoption Guidelines for Stormwater Biofiltration Systems*; Cooperative Research Centre for Water Sensitive Cities: Melbourne, Australia, 2015.
6. Jim, C.Y. Building thermal-insulation effect on ambient and indoor thermal performance of green roofs. *Ecol. Eng.* **2014**, *69*, 265–275. [CrossRef]
7. Teemusk, A.; Ülo, M. Temperature regime of planted roofs compared with conventional roofing systems. *Ecol. Eng.* **2010**, *36*, 91–95. [CrossRef]
8. Niachou, A.; Papakonstantinou, K.; Santamouris, M.; Tsangrassoulis, A.; Mihalakakou, G. Analysis of the green roof thermal properties and investigation of its energy performance. *Energy Build.* **2001**, *33*, 719–729. [CrossRef]
9. Mellor, R.S.; Salisbury, F.B.; Raschke, K. Leaf temperatures in controlled environment. *Planta* **1964**, *61*, 56–72. [CrossRef]
10. Costanzo, V.; Evola, G.; Marletta, L. Energy savings in buildings or UHI mitigation? Comparison between green roofs and cool roofs. *Energy Build.* **2016**, *114*, 247–255. [CrossRef]
11. Sailor, D.J. A green roof model for building energy simulation programs. *Energy Build.* **2008**, *40*, 1466–1478. [CrossRef]
12. Pianella, A.; Lu, A.; Chen, Z.; Williams, N. Substrate Depth, Vegetation and Irrigation Affect Green Roof Thermal Performance in a Mediterranean Type Climate. *Sustainability* **2017**, *9*, 1451–1470. [CrossRef]
13. Ascione, F.; Bianco, N.; Rossi, F.D.; Turni, G.; Vanoli, G.P. Green roofs in European climates. Are effective solutions for the energy savings in air-conditioning? *Appl. Energy* **2013**, *104*, 845–859. [CrossRef]
14. Attivissimo, F.; Di Nisio, A.; Carducci, C.G.; Spadavecchia, M. Fast Thermal Characterization of Thermoelectric Modules Using Infrared Camera. *IEEE Trans. Instrum. Meas.* **2017**, *66*, 305–314. [CrossRef]

15. Santamouris, M.; Pavlou, C.; Doukas, P.; Mihalakakou, G.; Synnefa, A.; Hatzibiros, A.; Patargias, P. Investigating and analysing the energy and environmental performance of an experimental green roof system installed in a nursery school building in Athens, Greece. *Energy* **2007**, *32*, 1781–1788. [CrossRef]

16. Monteiro, C.M.; Calheiros, C.S.; Martins, J.P.; Costa, F.M.; Palha, P.; De Freitas, S.; Ramos, N.M.; Castro, P.M. Substrate influence on aromatic plant growth in extensive green roofs in a Mediterranean climate. *Urban Ecosyst.* **2017**, 1–11. [CrossRef]

17. Berardi, U.; Ghaffarianhoseini, A.H.; Ghaffarian Hoseini, A. State-of-the-art analysis of the environmental benefits of green roofs. *Appl. Energy* **2014**, *115*, 411–428. [CrossRef]

18. Lu, J.; Yuan, J.G.; Yang, J.Z.; Chen, A.K.; Yang, Z.Y. Effect of substrate depth on initial growth and drought tolerance of *sedum lineare*, in extensive green roof system. *Ecol. Eng.* **2015**, *74*, 408–414. [CrossRef]

19. Fleta-Soriano, E.; Pinto-Marijuan, M.; Munne-Bosch, S. Evidence of Drought Stress Memory in the Facultative CAM, Aptenia cordifolia: Possible Role of Phytohormones. *PLoS ONE* **2015**, *10*, 1–12. [CrossRef] [PubMed]

20. Nektarios, P.A.; Ntoulas, N.; Nydrioti, E. Turfgrass use on intensive and extensive green roofs. *Acta Hortic.* **2012**, *938*, 121–128. [CrossRef]

21. Yan, F.X. Analysis and research of utilization status and existing problems of Lanzhou city water resources. *Gansu Water Resour. Hydropower Technol.* **2012**, *48*, 11–13.

22. Ministry of Housing and Urban-Rural Development of the People's Republic of China. *Technical Specification for Green Roof*; China Architecture & Building Press: Beijing, China, 2005. (In Chinese)

23. Watkins, R.; Palmer, J.; Kolokotroni, M. Increased temperature and intensification of the urban heat island: implications for human comfort and urban design. *Built Environ.* **2007**, *33*, 85–96. [CrossRef]

24. Getter, K.L.; Rowe, D.B.; Andresen, J.A.; Wichman, I.S. Seasonal heat flux properties of an extensive green roof in a Midwestern US climate. *Energy Build.* **2011**, *43*, 3548–3557. [CrossRef]

25. Qin, X.; Wu, X.; Chiew, Y.M.; Li, Y. A green roof test bed for stormwater management and reduction of urban heat island effect in Singapore. *Br. J. Environ. Clim. Chang.* **2012**, *2*, 410–420. [CrossRef] [PubMed]

26. Beerling, D.J.; Chaloner, W.G. The Impact of Atmospheric CO_2 and Temperature Changes on Stomatal Density: Observation from *Quercus robur* Lammas Leaves. *Ann. Bot.* **1993**, *71*, 231–235. [CrossRef]

27. Brewer, C.A.; Smith, W.K. Leaf surface wetness and gas exchange in the pond lily nuphar polysepalum (nymphaeaceae). *Am. J. Bot.* **1995**, *82*, 1271–1277. [CrossRef]

28. Wolf, D.; Lundholm, J. Water uptake in green roof microcosms: effects of plant species and water availability. *Ecol. Eng.* **2008**, *33*, 179–186. [CrossRef]

water

MDPI

Article

Detailed Sponge City Planning Based on Hierarchical Fuzzy Decision-Making: A Case Study on Yangchen Lake

Junyu Zhang [1,2], Dafang Fu [1,2], Yajun Wang [2,3] and Rajendra Prasad Singh [1,2,]*

[1]   Joint Research Centre for Water Sensitive Cities, Southeast University-Monash University Joint Graduate School (Suzhou), Southeast University, Suzhou 215123, China; junyu.zhang@monash.edu (J.Z.); fdf@seu.edu.cn (D.F.)

[2]   Department of Civil Engineering, Southeast University, #2Sipailou, Nanjing 210096, China; 230149631@seu.edu.cn

[3]   School of Civil Engineering, Lanzhou University of Technology, 287 Langongping, Lanzhou 730050, China

*   Correspondence: rajupsc@seu.edu.cn

Received: 16 October 2017; Accepted: 17 November 2017; Published: 20 November 2017

Abstract: We proposed a Hierarchical Fuzzy Inference System (HFIS) framework to offer better decision supports with fewer user-defined data (uncertainty). The framework consists two parts: a fuzzified Geographic Information System (GIS) and a HFIS system. The former provides comprehensive information on the criterion unit and the latter helps in making more robust decisions. The HFIS and the traditional Multi-Criteria Decision Making (MCDM) method were applied to a case study and compared. The fuzzified GIS maps maintained a majority of the dominant characteristics of the criterion unit but also revealed some non-significant information according to the surrounding environment. The urban planning map generated by the two methods shares similar strategy choices (6% difference), while the spatial distribution of strategies shares 69.7% in common. The HFIS required fewer subjective decisions than the MCDM (34 user-defined decision rules vs. 141 manual evaluations).

Keywords: Sponge City; urban planning; fuzzy logic; GIS; flooding

1. Introduction

With the tremendous development in China over the past few decades, various problems resulting from rapid urbanization, population growth, and climate change have emerged. The failure of water drainage systems is one of the most common. Due to a low design capacity, a lack of maintenance, and a reduction in natural buffering areas, flooding and waterlogging caused by this failure are in turn causing huge losses in terms of both property and human lives.

The Sponge City concept was proposed in 2012 in China during the Low-Carbon Urban Development and Technology Forum to address the conflict between development and resilience cities face [1]. Similar concepts in urban planning, such as Best Management Practices [2], Low Impact Development [3], and Water Sensitive Urban Design [4] have been successfully practiced, but there is still a long way to go to adapt, improve, and develop proper techniques, strategies, and planning methods to meet local conditions and needs in China.

Followed by a barrage of government-issued policies, several cities in China with different population densities, spatial scales, and climate conditions are currently sponsored to explore the applicative national strategy and practice of Sponge City by the 2020s. Meanwhile, current Sponge City designs and construction plans do not satisfy our expectations [1]. To achieve better performance and more cost-efficiency, it is urged that a proper decision-making method be employed in the planning process. Although many novel, powerful, and accurate models and tools to support decision making have emerged in recent years [5–9], most of them have been developed and used by experienced researchers or developers. It is almost impossible for lay designers and decision makers to correctly and easily apply those tools to their work.

Some widely used methods are usually simple and straightforward. One of the most commonly used decision support method is Multi-Criteria Decision Making (MCDM), which combines quantitative and qualitative criteria to form a single index of evaluation. Implemented in a Geographic Information System(GIS) environment, MCDM has been applied in various studies in such areas as resource management [10], urban planning [11–14], and vulnerability assessment [15–18] over the last few decades. Such spatial-based MCDM involves a set of geographically defined basic units (e.g., polygons, or cells), and a set of evaluation criteria represented as map layers [11]. The criterion maps rank each unit with an overall score according to the attribute values and criteria weights using different analyzing approaches (e.g., Boolean overlay, weighted linear combination, and ordered weighted average) [13]. The Analytical Hierarchy Procedure (AHP) [19] is a method widely used for ranking multi-criteria weights. It calculates the weighting factors using a pairwise comparison matrix where all relevant criteria are compared against each other with reproducible preference factors.

Another decision support method is Fuzzy Decision Making, which is a mathematical method for supporting decision making under uncertain situations with limited information [20]. It consists in an inference structure that enables appropriate human reasoning capabilities. It has been widely applied in studies relating to vulnerability assessment [15,17,21–25] and urban planning [26–30]. The approach sets up a fuzzy inference system (FIS), which consists in user-defined membership functions and decision rules [20]. The value for each criterion is first divided into classes/words, and a membership function is used to identify the range of each class/word. Each class has overlay parts with adjacent classes to represent the fuzziness. The decision rules represent the ambiguous designing principle of the planner (e.g., if the imperviousness is low and the pollution productivity is high, then the vulnerability is high). The criterion maps rank each unit by allocating their input distribution in the membership function and finding out their output distribution according to the rule set and rule strength, which is called the Mamdani method [31].

Realistically, the application of such methods requires a comprehensive understanding of the planning process as well as sufficient data. On one hand, the more data we have, the more comprehensive we can understand the situation and make more reliable decisions. On the other hand, the more data we are dealing with, the more subjective pairwise comparison matrixes (e.g., MCDM) or decision rules (e.g., FIS) we need to establish and therefore the more uncertain we are of the decisions. Nevertheless, the above method usually evaluates criteria units individually (especially polygons) and disregards the surrounding features.

In this study, we developed an easily applicable decision-making framework that applies a hierarchical FIS system [24] on a fuzzified GIS system, in order to offer better decision supports with fewer user-defined data. The hierarchical FIS system aims to reduce the subjective judgement from planners, minimizing uncertainty in the system. The fuzzified GIS system provides comprehensive information on the surrounding environment to support better decisions. The developed framework and the traditional MCDM method were applied on a planning program at Yangchen Lake Resort, Suzhou, Jiangsu, China. The results of both methods were compared so that the pros and cons for each approach could be analyzed.

2. Methods and Data Description

2.1. Study Description

The 61.7 km^2 Yangchen lake resort consists of two peninsulas, over which more than 100 inner rivers are spread (see Figure 1). About 3.5 km^2 of land area is used for various kinds of farming activities (rice, vegetable, fruit, and fishery) and 5 km^2 is used for public landscapes and parks. A new town is gradually developing at the upper end of the left peninsula.

Due to pollution from farming and a lack of maintenance, more than one-third of the inner rivers are blocked. The performance of the drainage system is poor in the town area as the design can be traced back for decades.

The aim of the design is to offer decision support for Sponge City planning (where to introduce new techniques, cost-efficiency on adapting strategies, etc.), such that the retrofit impact in the area of the resort is minimized and the hydrology performance of the resort's water system is more cost-efficient.

There are four candidate strategies: business as usual (BAU), rain tank or green roof (small scale system), rain garden or bioretention cell (large scale system), and re-planning.

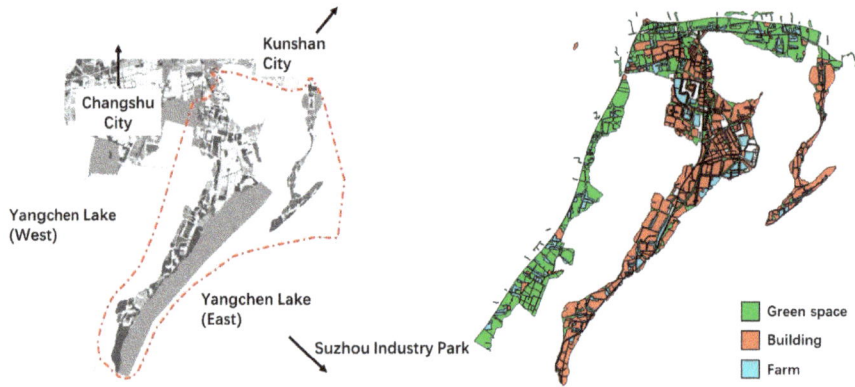

Figure 1. Case Study: Yangchen lake resort.

2.2. Data Collection and Criteria Selection

On the basis of a dwg map from the stakeholders, field investigations were carried out to gather information about land use, the source of pollution, and the environmental status on site. The dwg file was then transformed into a GIS map via ArcGIS, and these data were inputted into each polygon. Together with experienced designers, major criteria pertaining to permeability, pollution productivity, loss from flooding, and retrofit cost, were identified. Due to limitations in data accessibility, the four features are represented by 0, 1, and 2 (indicating low, medium, and high) for every polygon according to the designers' experience (see Table 1).

Table 1. Geographic Information System (GIS) features for the resort.

Land Use	Permeability	Pollutant Productivity	Loss from Flooding	Retrofit Cost
Farm	1	2	1	1
Building	0	1	2	2
Green space	2	0	0	0

Notes: 0: Low; 1: Medium; 2: High.

2.3. Methodology

We applied MCDM and HFIS in this case study according the following procedures (see Figure 2).

2.3.1. Multi-Criteria Decision Making with an Analytical Hierarchy Procedure

The MCDM process requires decision makers to rank the criteria based on pairwise comparisons. In this study, these comparisons were obtained from a survey of 25 experts that included members of the urban planning institute as well as academic experts specialized in urban planning. Each participant was asked to rank the criteria and class by referring to a numerical scale of 1–9, with a score of 1 representing indifference between the two criteria and 9 indicating a great amount of concern [13].

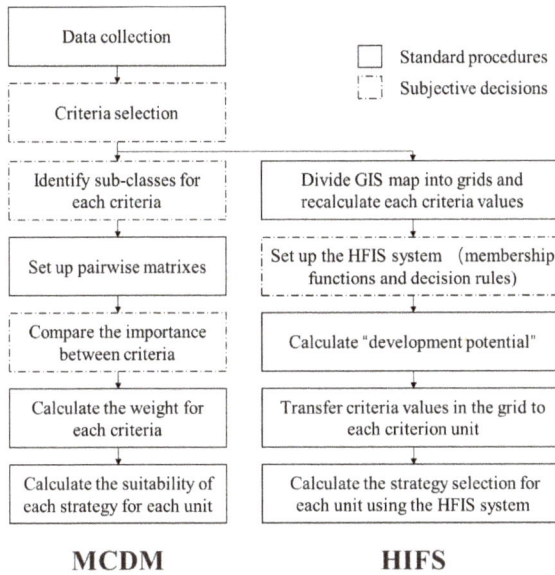

Figure 2. Multi-Criteria Decision Making (MCDM) and Hierarchical Fuzzy Inference System (HFIS) flow chart.

The final 25 pairwise comparisons matrixes were establish based on the mean value of all survey results. The weight for each criteria, class, and strategy were then calculated using the AHP method (see Table 2). The consistency ratio (CR) was calculated to evaluate the consistency of pairwise comparisons. A standard CR threshold value of 0.10 was applied. The pairwise comparisons in this study were consistent with a consistency ratio (CR) of <0.10.

Table 2. Criteria tree and Analytical Hierarchy Procedure (AHP) results.

Goal	Weight	Criteria	Weight	Class	Weight	Strategy Weight			
						Business as Usual	Raintank Green Roof	Rain Garden Bioretention	Re-Planning
		Permea-bility	0.0553	High	0.0754	0.6692	0.1155	0.1155	0.0998
				Medium	0.2290	0.2500	0.2500	0.2500	0.2500
				Low	0.6955	0.0871	0.3854	0.3854	0.1422
		Land use	0.0649	Farm	0.6955	0.0886	0.0952	0.5513	0.2649
				Building	0.2290	0.3812	0.4331	0.1030	0.0828
				Park	0.0754	0.6201	0.0708	0.2166	0.0925
Development	1.0000	Area (ha)	0.0401	1.19–45	0.1140	0.3000	0.3000	0.3000	0.1000
				0.25–1.19	0.4054	0.2500	0.2500	0.2500	0.2500
				0–0.25	0.4806	0.2857	0.2857	0.2857	0.1429
		Flood loss	0.2994	High	0.7514	0.0445	0.1723	0.1958	0.5874
				Medium	0.1782	0.0813	0.3598	0.3598	0.1991
				Low	0.0704	0.3000	0.3000	0.3000	0.1000
		Retrofit cost	0.3286	High	0.0658	0.4167	0.0833	0.0833	0.4167
				Medium	0.2172	0.3000	0.3000	0.3000	0.1000
				Low	0.7171	0.3125	0.3125	0.3125	0.0625
		Pollution	0.2117	High	0.7429	0.0457	0.1451	0.3494	0.4598
				Medium	0.1939	0.0871	0.3854	0.3854	0.1422
				Low	0.0633	0.3313	0.2916	0.2916	0.0855

After the factors, their weights, and all constraints in the decision tree were established for each strategy, the suitability of each strategy was calculated for each unit in the criterion map according to its criteria value (Suitability = Criteria Weight × Class Weight × Strategy Weight). The sponge urban planning map was generated by selecting the most suitable strategy for each unit.

2.3.2. Hierarchical Fuzzy Inference System (HFIS) Decision Making

Step 1   "Fuzzification" of GIS Maps

As discussed before, a major goal of this framework is to provide comprehensive information from the surrounding environment for each criteria unit. In this study, we first divided the resort into 100 m × 100 m (1 ha) grids. The size of the grids was determined considering the mid-value (0.55 ha) and distribution (<1 ha: 63%) of the polygon area in the GIS map. Such grids offered an adequate capacity of embracing characteristics of multiple polygons. Permeability, pollutant productivity, and loss from flooding was calculated for each grid according to the corresponding value in the polygon they intersected (Equation (1)).

$$I_{B,j} = \sum I_{A,j} \frac{a_i'}{A_i'} , \tag{1}$$

where

$I_{B,j}$ is the criteria value of grid j;
$I_{A,i}$ is the value of the corresponding criteria in polygon i;
a_i' is the intersect area of grid j and polygon i;
A_i' is the area of polygon i.

Step 2   Fuzzy Analysis

A Matlab (R2017a) toolbox, the Fuzzy Logic Designer, was used to set up the FIS based on the result from Step 1. The Gaussian Membership Functions are adopted for the following criteria to allow better deviation to these fuzzified values (see Figure 3). "Permeability" and "pollutant productivity" were first analyzed to evaluate the "vulnerability" of each grid. Together with the "loss from flooding", the "vulnerability" went through the FIS again to calculate the "develop potential". The advantage of using a two-layer fuzzy process instead of dealing with three parameters at a time (using permeability, pollutant productivity, and loss for flooding to directly analyze develop potential) is that the former requires fewer inputs. The two-layer fuzzy module requires two rule sets with 25 rules each (for two parameters with 5 value ranges, the minimum amount of rules will be 5 × 5), while the one-layer module would need a set of 125 rules. If fewer rules are designed, more uncertainty will be reduced from the cognitive limitation.

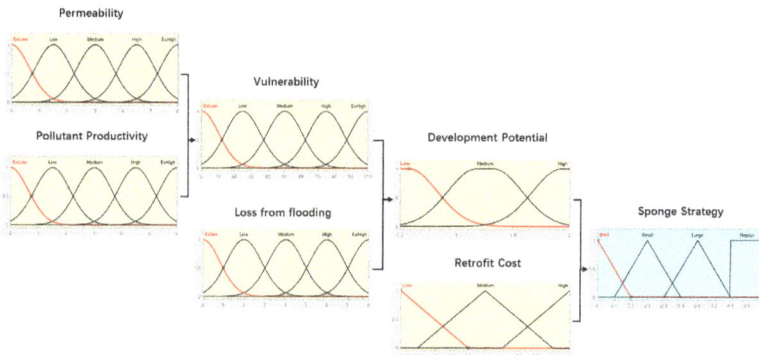

Figure 3. Designed membership functions of the hierarchical fuzzy inference system (FIS).

Step 3   "Defuzzification" of GIS Maps and Strategy Selection

In this stage, information is translated from a grid map to an original GIS map through Equations (2) and (3) so that the need for each land can be understood.

$$D_{A,i} = \sum D_{B,j} \frac{a'_j}{A_i} \tag{2}$$

$$A_i = \sum a'_j \tag{3}$$

where

$D_{A,i}$ is the development potential of polygon i;
$D_{B,i}$ is the development potential of grid j;
a'_j is the intersecting area of grid j and polygon i;
A_i is the area of polygon i.

A third fuzzy process was conducted with the development potential and retrofit cost, in the same way as Step 2, to determine the strategy to be chosen (business as usual, rainwater tank, rain garden, or re-planning). The Trapezoidal Membership Function are adopted in this step to allow crisper decision choices (see Figure 3). The sponge urban design map was then worked out to visualize the strategy choice.

3. Results and Discussion

3.1. Fuzzified GIS Map

Compared to the original GIS maps, the fuzzified maps present the same or less intensive criteria values in most areas (see Figure 4). This indicates that the fuzzification process under this gridding scale can maintain the dominant characteristics of the polygons while considering the surrounding environment and can make reasonable adjustments. Notably, the highlighted Areas A and B initially had the same attribute values but ended up with different fuzzified values. The reason for this difference results from the more intensive land use in Area A. Despite the same land use type, there were more scattered green spaces and inner rivers within Area B. The fuzzified GIS map proved to be efficient in revealing this non-significant information, which influences the final decision.

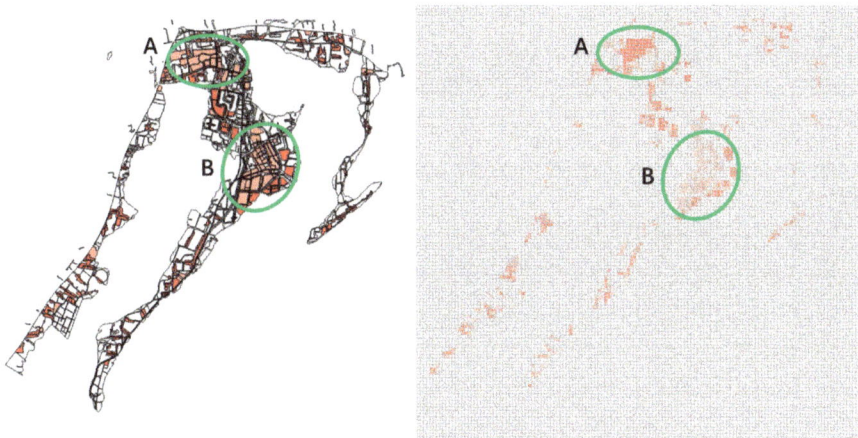

Figure 4. Example of a fuzzified GIS map (pollutant productivity).

3.2. Sponge Urban Plan for Yangchen Lake Resort

The result of the sponge urban planning map is presented in Figure 5. This map suggests that 55.6% (MDCM) or 49.7% (HFIS) of the resort can undertake the business as usual strategy. These areas include the majority of the west and north green spaces of the resort and the residential areas in the middle peninsula. These areas have relatively good permeability (by themselves or by adjacent to the lake), moderate pollutant productivity, or low loss from flood.

It is also suggested that 22.7% (MDCM) or 28.7% (HFIS) of the resort apply rain water tanks or green roofs. These areas consist of high-density buildings whose runoff contributes to the pollution in adjacent areas (such as farming).

The map further suggests that 21.7% (MDCM) or 21.6% (HFIS) of the resort have rain gardens or bioretention cells implemented. Most of these would be located in areas related to farming. The pollution from fish farming is extremely high, so it is advisable to take advantage of the reserved land and to construct a large-scale rain garden to hold and treat the runoff from the farming area.

□ Business as usual
▨ Rain garden or Bioretention cell
■ Rain tank or Green roof
■ Replanning

Figure 5. Sponge urban planning maps for MCDM (**left**) and hierarchical FIS (**right**).

3.3. Comparison of MCDM and HFIS

In this study, the two methods both require selection and identification of criteria and their classes' ranges. To generate a planning map, MCDM methods required the planners to fulfill 25 pairwise comparison matrixes, which consist of 141 manual evaluations for deciding the importance between two criteria to their upper level criteria. The three-level HFIS required to design 34 decision rules to determine which strategy was preferred under certain conditions.

As discussed in Section 3.2, the two methods produced similar results regarding the total area of each strategy (see Table 3). The HFIS suggested a bit more rain tank/green roof uptake instead of business as usual (6%). In regard to the spatial distribution of the strategies, 69.7% of the resort planning generated by the HFIS uses the exact same strategy as that used by MCDM. Moreover, the two methods suggested different strategies (other than business as usual) with respect to 3.26% of the planning area.

Table 3. Comparison of the results from HFIS and MCDM.

Comparison criteria		HFIS	MCDM
	Business as ususal	49.7%	55.6%
Strategy Choice	rain water tanks or green roofs	28.7%	22.7%
	the rain garden or bioretention cells	21.6%	21.7%
Overlay area of choosing same strategy (other than BAU)		69.7%	
User-defined decisions		34	141

4. Conclusions

In this study, a decision-making framework was developed and verified to offer better decision support with fewer user-defined data (to reduce uncertainty). The framework consists of two parts: a fuzzified GIS system and a hierarchical FIS system. The developed framework and the traditional MCDM method were applied on a planning program at Yangchen Lake Resort, Suzhou, Jiangsu, China.

In this study, with the grid size we selected (1 ha), the fuzzified GIS maps maintained a majority of the dominant characteristics of the polygons. The process, by considering the surrounding environment and making reasonable adjustments, also proved to be efficient in revealing non-significant information.

The sponge urban planning map generated by the two methods shares similar strategy choices: BAU: 55.6% (MDCM) or 49.7% (HFIS); rain tanks or green roofs: 22.7% (MDCM) or 28.7% (HFIS); rain gardens or bioretention cells: 21.7% (MDCM) or 21.6% (HFIS). The spatial distribution of strategies (other than BAU) have 69.7% in common.

Regarding the user-defined data, the two methods both require the selection and identification of criteria and their classes' ranges. In this study, the MCDM methods required 25 pairwise comparison matrixes, which consist of 141 manual evaluations to decide the importance between the two criteria. The HFIS required the design of 34 decision rules to determine which strategy was preferred under certain conditions.

Acknowledgments: The research is co-funded by the National Key Technologies R&D Program (No. 2015BAL02B05) and the Priority Academic Program Development of the Jiangsu Higher Education Institution. Thanks to Professor Dafang Fu for the creative leadership and for motivating us to finish the research. Thanks also to Yajun Wang, Shibo Hao, Yangke Li, Hong Zhang, Gongjun Huang, Zhongshuai Jiang, Kun Wei, Jinhui Yang, and Zhongxiang Zhang, who dedicate their time and passion in these two difficult field investigations. Thanks to Chenli Wu, who encouraged us to finish the work.

Author Contributions: Junyu Zhang and Dafang Fu conceived and designed the experiments. Junyu Zhang and Yajun Wang conducted the data collection and performed the experiment; Junyu Zhang and Rajendra Prasad Singh analyzed the data and wrote the paper.

Conflicts of Interest: The authors declare no conflict of interest.

References

1. Xia, J.; Zhang, Y.; Xiong, L.; He, S.; Wang, L.; Yu, Z. Opportunities and challenges of the Sponge City construction related to urban water issues in China. *Sci. China Earth Sci.* **2017**, *60*, 652–658. [CrossRef]
2. Urbonas, B.; Stahre, P. *Stormwater: Best Management Practices and Detention for Water Quality, Drainage, and CSO Management*; Prentice Hall: Upper Saddle River, NJ, USA, 1993.
3. Fairlie, S. *Low Impact Development: Planning and People in a Sustainable Countryside*; Jon Carpenter Chipping Norton: Chipping Norton, UK, 1996.
4. Wong, T.H.; Brown, R.R. The water sensitive city: Principles for practice. *Water Sci. Technol.* **2009**, *60*, 673–682. [CrossRef] [PubMed]
5. Albano, R.; Mancusi, L.; Abbate, A. Improving flood risk analysis for effectively supporting the implementation of flood risk management plans: The case study of "Serio" Valley. *Environ. Sci. Policy* **2017**, *75*, 158–172. [CrossRef]

6.   Inam, A.; Adamowski, J.; Halbe, J.; Malard, J.; Albano, R.; Prasher, S. Coupling of a distributed stakeholder-built system dynamics socio-economic model with SAHYSMOD for sustainable soil salinity management part 1: Model development. *J. Hydrol.* **2017**, *551*. [CrossRef]

7.   Urich, C.; Sitzenfrei, R.; Kleidorfer, M.; Bach, P.M.; McCarthy, D.T.; Deletic, A.; Rauch, W. Evolution of urban drainage networks in DAnCE4Water. In Proceedings of the 9th International Conference on Urban Drainage Modelling, Belgrade, Serbia, 4–6 September 2012.

8.   Ferguson, B.C.; Brown, R.R.; Frantzeskaki, N.; de Haan, F.J.; Deletic, A. The enabling institutional context for integrated water management: Lessons from Melbourne. *Water Res.* **2013**, *47*, 7300–7314. [CrossRef] [PubMed]

9.   Hall, J.W.; Lempert, R.J.; Keller, K.; Hackbarth, A.; Mijere, C.; McInerney, D.J. Robust climate policies under uncertainty: A comparison of robust decision making and info-gap methods. *Risk Anal.* **2012**, *32*, 1657–1672. [CrossRef] [PubMed]

10.   Chang, N.B.; Qi, C.; Yang, Y.J. Optimal expansion of a drinking water infrastructure system with respect to carbon footprint, cost-effectiveness and water demand. *J. Environ. Manag.* **2012**, *110*, 194–206. [CrossRef] [PubMed]

11.   Chen, Y.; Paydar, Z. Evaluation of potential irrigation expansion using a spatial fuzzy multi-criteria decision framework. *Environ. Model. Softw.* **2012**, *38*, 147–157. [CrossRef]

12.   Jeong, J.S.; García-Moruno, L.; Hernández-Blanco, J. A site planning approach for rural buildings into a landscape using a spatial multi-criteria decision analysis methodology. *Land Use Policy* **2013**, *32*, 108–118. [CrossRef]

13.   Rahman, R.; Saha, S.K. Remote sensing, spatial multi criteria evaluation (SMCE) and analytical hierarchy process (AHP) in optimal cropping pattern planning for a flood prone area. *J. Spat. Sci.* **2008**, *53*, 161–177. [CrossRef]

14.   Van Niekerk, A.; du Plessis, D.; Boonzaaier, I.; Spocter, M.; Ferreira, S.; Loots, L.; Donaldson, R. Development of a multi-criteria spatial planning support system for growth potential modelling in the Western Cape, South Africa. *Land Use Policy* **2016**, *50*, 179–193. [CrossRef]

15.   Araya-Munoz, D.; Metzger, M.J.; Stuart, N.; Wilson, A.M.W.; Carvajal, D. A spatial fuzzy logic approach to urban multi-hazard impact assessment in Concepcion, Chile. *Sci. Total Environ.* **2017**, *576*, 508–519. [CrossRef] [PubMed]

16.   Pourghasemi, H.R.; Moradi, H.R.; Fatemi Aghda, S.M.; Gokceoglu, C.; Pradhan, B. GIS-based landslide susceptibility mapping with probabilistic likelihood ratio and spatial multi-criteria evaluation models (North of Tehran, Iran). *Arab. J. Geosci.* **2013**, *7*, 1857–1878. [CrossRef]

17.   Radmehr, A.; Araghinejad, S. Flood vulnerability analysis by fuzzy spatial multi criteria decision making. *Water Resour. Manag.* **2015**, *29*, 4427–4445. [CrossRef]

18.   Rahman, M.R.; Shi, Z.H.; Chongfa, C. Assessing regional environmental quality by integrated use of remote sensing, GIS, and spatial multi-criteria evaluation for prioritization of environmental restoration. *Environ. Monit. Assess.* **2014**, *186*, 6993–7009. [CrossRef] [PubMed]

19.   Saaty, T.L. *The Analytic Hierarchy Process: Planning, Priority Setting, Resource Allocation*; McGraw-Hill: New York, NY, USA, 1980.

20.   Zadeh, L.A. Fuzzy logic = computing with words. *IEEE Trans. Fuzzy Syst.* **1996**, *4*, 103–111. [CrossRef]

21.   Lee, G.; Jun, K.S.; Chung, E.-S. Robust spatial flood vulnerability assessment for Han River using fuzzy TOPSIS with α-cut level set. *Expert Syst. Appl.* **2014**, *41*, 644–654. [CrossRef]

22.   Lee, M.-J.; Kang, J.E.; Kim, G. Application of fuzzy combination operators to flood vulnerability assessments in Seoul, Korea. *Geocarto Int.* **2015**, 1–24. [CrossRef]

23.   Rezaei, F.; Safavi, H.R.; Ahmadi, A. Groundwater vulnerability assessment using fuzzy logic: A case study in the Zayandehrood aquifers, Iran. *Environ Manag.* **2013**, *51*, 267–277. [CrossRef] [PubMed]

24.   Şener, E.; Şener, Ş. Evaluation of groundwater vulnerability to pollution using fuzzy analytic hierarchy process method. *Environ. Earth Sci.* **2015**, *73*, 8405–8424. [CrossRef]

25.   Singh, P.K.; Nair, A. Livelihood vulnerability assessment to climate variability and change using fuzzy cognitive mapping approach. *Clim. Chang.* **2014**, *127*, 475–491. [CrossRef]

26.   Gray, S.R.J.; Gagnon, A.S.; Gray, S.A.; O'Dwyer, B.; O'Mahony, C.; Muir, D.; Devoy, R.J.N.; Falaleeva, M.; Gault, J. Are coastal managers detecting the problem? Assessing stakeholder perception of climate vulnerability using Fuzzy Cognitive Mapping. *Ocean Coast. Manag.* **2014**, *94*, 74–89. [CrossRef]

27. Navas, J.M.; Telfer, T.C.; Ross, L.G. Spatial modeling of environmental vulnerability of marine finfish aquaculture using GIS-based neuro-fuzzy techniques. *Mar. Pollut. Bull.* **2011**, *62*, 1786–1799. [CrossRef] [PubMed]

28. Talebian, A.; Shafahi, Y. The treatment of uncertainty in the dynamic origin–destination estimation problem using a fuzzy approach. *Trans. Plan. Technol.* **2015**, *38*, 795–815. [CrossRef]

29. Teh, L.C.; Teh, L.S. A fuzzy logic approach to marine spatial management. *Environ. Manag.* **2011**, *47*, 536–545. [CrossRef] [PubMed]

30. Zhang, J.; Wang, K.; Chen, X.; Zhu, W. Combining a fuzzy matter-element model with a geographic information system in eco-environmental sensitivity and distribution of land use planning. *Int. J. Environ. Res. Public Health* **2011**, *8*, 1206–1221. [CrossRef] [PubMed]

31. Sivanandam, S.; Sumathi, S.; Deepa, S. *Introduction to Fuzzy Logic Using Matlab*; Springer: Berlin, Germany, 2007; Volume 1.

water

MDPI

Article

Scoping for the Operation of Agile Urban Adaptation for Secondary Cities of the Global South: Possibilities in Pune, India

Mohanasundar Radhakrishnan [1,*], Tejas M. Pathak [1,2], Kenneth Irvine [1] and Assela Pathirana [1]

[1]   Water Science and Engineering Department, IHE Delft Institute for Water Education, 2611 AX Delft, The Netherlands; Tej.urbantech@gmail.com (T.M.P.); k.irvine@un-ihe.org (K.I.); a.pathirana@un-ihe.org (A.P.)

[2]   Tej Urbantech Llp Company, 411038 Pune, India

*   Correspondence: m.radhakrishnan@un-ihe.org; Tel.: +31-647-618-689

Received: 8 November 2017; Accepted: 29 November 2017; Published: 2 December 2017

Abstract: Urban areas, especially in developing countries, are adapting to deficits in infrastructure and basic services (Type I adaptation) and to adaptation gaps in response to current and future climatic, societal and economic change (Type II adaptation). The responses to these adaptations needs can be integrated and implemented using an "agile urban adaptation process", i.e., an adaptive planning process quickly adapting to change in a flexible manner in short planning horizons, where the requirements and responses evolve through evolutionary development, early delivery, continuous improvement and collaboration between self-organizing and cross-functional teams. This paper focuses on how to move from the current conceptual stage to developing practical knowledge for the operation of agile urban adaptation. Scoping methodology comprises (i) understanding and structuring the adaptation context; (ii) exploring the four agile elements—balancing type I & II adaptation needs, flexibility, range of scenarios and involvement of stakeholders—in the adaptation context; (iii) a detailed SWOT analysis (strength, weakness, opportunities and threat) of adaptation responses; (iv) mapping relationships and synergies between the adaptation responses; and (v) preparing agility score cards for adaptation responses. The scoping exercise revealed that the agile adaptation process can move from concept to operation in Pune, India where the city is improving the basic services and adapting to climate change. For example: conventional adaptation responses such as city greening and check-dams across the rivers have agile characteristics; these responses are synergetic with other adaptation responses; and, there is a possibility to compare conventional adaptation responses based on agile characteristics. This scoping exercise also reveals that urban agile adaptation is not about implementing novel adaptation responses but understanding, planning and implementing conventional adaptation responses using an agile perspective. Urban agile adaptation is also about mainstreaming agile ideas using traditional adaptation responses. Hence, it is possible to apply agile the urban adaptation process using conventional adaptation responses in urban areas which address adaptation deficits related to infrastructure development as well as climate and socio-economic adaptation.

Keywords: agile adaptation; cities; climate adaptation; implementation; flooding; urban areas

1. Introduction

Urban areas are adapting to overcome the difficulties and make use of opportunities arising from climate change and other changes, which are both uncertain and complex [1]. Urban areas, especially in developing countries—the secondary cities of global south (SCGS)—are adapting to deficits in infrastructure and basic services, i.e., Type I adaptation; and, to future adaptation needs that are due to climate change, social change and economic change, i.e., Type II adaptation [2]. The type I and type II adaptation responses are not implemented in consideration with each other leading to unnecessary expenditure, inconvenience and waste of opportunities and time [3–5]. However, there are difficulties in integrating Type I and Type II adaptation responses. Type I responses are characterised by urgency as they address the current absence or shortage of urban services, whereas Type II adaptation responses are characterised by uncertainty as they address long term changes [5]. Integration of adaptation responses is gaining prominence after the Paris 2015 accord as governments at various levels all over the world have started planning and implementing adaptation responses [6]. Also, understanding and integrating adaptation responses are important for achieving the sustainable development goals (SDG) [7], as one adaptation response can contribute towards achieving more than one goal [8]. For example, an urban wetland can contribute towards: sustainable cites (Goal 11), as it enhances liveability; climate action (Goal 13), as a buffer in the event of floods; life below water (Goal 14); and life on land (Goal 15). Also integrating responses is relevant in the context of "sponge cities", where various strategies and actions are expected to deal with excess and shortage of water [9,10]. The planning and implementation practises which are prevalent in other cities such as Copenhagen, Rotterdam and Singapore; and approaches advocated by international agencies such as United Nations Development Programme (UNDP), Sustainable Development Solutions Network (SDSN) can be used for (i) guiding the emerging approaches; (ii) overcoming the challenges; and, (iii) making use of the opportunities in the context of sponge cities.

Cities can learn from their experience or from the experience of other cities to address their adaptation deficits and adaptation gaps. For example, cities can learn from the August 2017 Houston floods on what went wrong in terms of urban planning, managing flood risks and why the city was so under prepared [11]. Leapfrogging is the ability of cities to rapidly transition to sustainable development by learning from the mistakes of other cities and adopting more efficient and ecologically friendly practices [8,12–14]. For example, developing wetlands for flood risk management is leapfrogging, as wet lands contribute to ecological improvement and also can lead to indirect economic benefits, unlike a dike or sluice gate which can be economically beneficial but affects ecology. The use of sustainable adaptation responses, such as using wetlands as part of management of urban fabric in the cities of developing countries, is a recent phenomenon, which was introduced upon realising that the conventional responses do not always lead to sustainable development and help in achieving multiple objectives [15]. Hence, leapfrogging potential is a very important criteria for assessing adaptation responses in a SCGS, especially in the context of sponge cities. Although leapfrogging can enable learning of planning and implementation practices from developed country context, it can be effective only if the there is a comprehensive understanding of the local adaptation situation and adaptation responses.

Understanding the local setting to provide appropriate context for action and structuring the local adaptation problems are essential for integrating adaptation responses (e.g., localising the SDG [8], structuring adaptation context and responses [4,16]). There are concepts such as transformative adaptation [17,18] and clumsy solutions [12], and processes based on flexibility such as adaptation pathways [19] and dynamic adaptive policy pathways [20] that enable sequencing and integration of various adaptation responses. However implementation challenges persist, such as proactive analysis of implementation issues during planning stage, contested values, changing goals and objectives and integration of adaptation responses across spatial scales and temporal scales [16,21]. Recent approaches such as agile urban adaptation [13] and flexible adaptation planning processes [22] address some of these challenges in implementing adaptation responses.

Agile adaptation is an adaptive planning process that adapts to change in a flexible manner over short planning horizons, where the requirements and responses evolve through evolutionary development and involves early delivery, continuous improvement and collaboration between self-organizing and cross-functional teams [13,23,24]. Agile adaptation is also seen as an implementation and operational tactic to achieve a set of desired outcomes by quickly adapting to changing needs and making use of opportunities by the stakeholders involved in adaptation planning and implementation [25]. Although application of agility has empirical evidence in software and automobile industry [23,26–28], it is yet to be implemented for urban planning and adaptation management. Pathirana et al. [13] have developed an agile urban adaptation planning process and theoretically demonstrated its application in a complex setting in Can Tho city, Vietnam. The agile urban adaptation process has been presented as a tactic to sustain continuous adaptation and continuous learning in short cycles, using conventional adaptation responses such as dike construction, urban drainage and household measures [13]. The four agile elements identified by Pathirana et al. [13] as essential for the application of agile urban adaptation are: (i) harmonizing type I & II adaptation needs; (ii) flexibility of adaptation responses; (iii) addressing plausible scenarios; and (iv) continuous involvement of stakeholders in the decision making process. However, the way forward to use the agile adaptation using adaptation responses that are already planned following a traditional approach is missing.

Pathirana et al. [13] has demonstrated that it is possible to apply agile adaptation in urban areas. However, the application is still at a programme-tactical level to identify the local adaptation possibilities and pre-requisites to implement an agile adaptation strategy. The pre-requisites for becoming agile at city and household levels is the identification of agile attributes of adaptation responses. At present, there is a lack of knowledge on agility attributes at a project level. Lack of practical knowledge to use agile urban adaptation is a significant gap for application. Hence this paper focuses on developing practical knowledge for using agile urban adaptation. A simple methodology based on tools and concepts that are frequently used by city managers, planners and engineers will be helpful in understanding the concept of agility, select agile adaptation responses and implement them. Hence a simple methodology, straightforward and comprehensive to city engineers, for using agile urban adaptation has been developed. The methodology comprises the following steps: understanding the local adaptation setting inspired by SDGs [8]; SWOT analysis (strength, weakness, opportunities and threat) that is widely used in business, industry and strategic planning [29]; and analysis based on the four agile elements [13]. Also this methodology has been tested in Pune, India—a SCGS—which is improving civic amenities and adapting to climate change.

2. Methodology

Using the agile adaptation process requires the understanding of the adaptation responses, stakeholders, motivations and capacities at different levels from global to household [4,13]. Globally, the sustainable development goals—comprising climate action as a goal—have gained prominence, where the emphasis is on implementing adaptation responses [6,7]. This necessitates understanding the (i) attributes of adaptation response at the local point of application or implementation; (ii) interaction among the responses; (iii) strength and weakness of the responses; and, (iv) opportunities and threats to the adaptation responses in the local point of implementation. The steps in the methodology for using agile urban adaptation process are: (i) understanding the local adaptation needs and adaptation responses (i.e., local setting) inspired by SDGs [8]; (ii) SWOT analysis of adaptation responses; (iii) mapping of synergies and relationship between responses [30,31]; and (iv) preparing agility score cards for adaptation responses [13]. The agility score card comprises the four agile elements, as well as criteria on the leapfrogging potential of the adaptation responses. SWOT analysis is a simple, open and transparent process which can be easily done by the stakeholders in the city who are involved in the planning and implementation of the adaptation responses. Also, developing the qualitative agility score card for adaptation measures is simple, transparent and based on identifiable characteristics such as flexibility, Type II adaptation needs, stakeholder involvement and plausible scenarios. The methodology is presented in Figure 1. Although the operational process appears linear, every individual step involves a number of iterations.

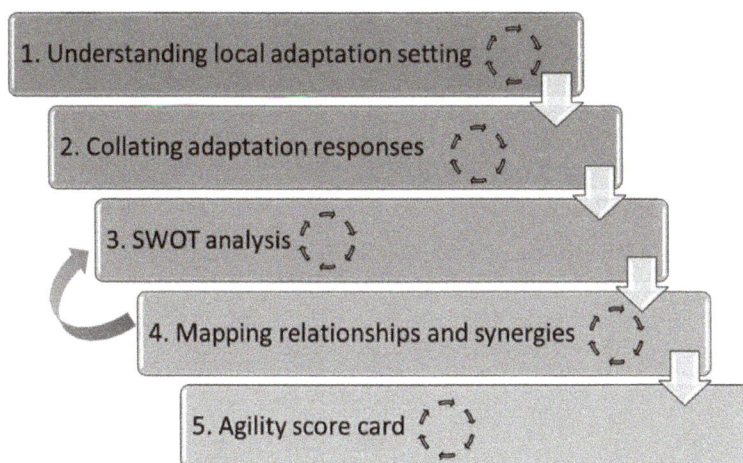

Figure 1. Scoping for using an agile urban adaptation process using a five step operational process.

Understanding the local adaptation setting (Step 1): This process is carried out by reviewing (i) the city's existing urban planning and adaptation planning documents; (ii) relevant literature which assesses the climate, social, economic and political developments; (iii) compliance and guidance documents on infrastructure and basic service levels of cities, such as service level bench mark (SLB) of Indian cities [32]; (iv) partnerships and cooperation agreements on development and adaptation with national and international agencies; and (v) consultation and interviews with stakeholders. This step enables a comprehensive understanding of adaptation and development gaps in the city and helps in structuring the local adaptation problem [4] and localising the SDGs [8]. Such an understanding will also help in identifying major drivers of adaptation, the relationship among adaptation drivers and prioritise the adaptation or development needs.

Collating adaptation responses (Step 2): The various adaptation and development responses planned and implemented in the city are collated using the same set of documents mentioned in Step 1. One useful and ready to access source to start with the collation of basic information about adaptation responses is the annual city budget documents, i.e., the fiscal planning documents and the annual audited statements from the previous years of the city or municipality. Analysing the budget allocated in the current budget plan, the past budget plans and the past annual expenditure will also reveal the trends in expenditure incurred and the preference of adaptation responses. Also, the detailed characteristics of responses—such as need for the responses, nature of responses, purpose of the responses, scale of responses, stakeholders, functional life time of the responses—are collected in order to understand the responses.

SWOT analysis (Step 3): The strengths, weakness, opportunity and threat analysis of the various adaptation responses (collated in Step 2) are carried out. SWOT analysis should be carried out only after understanding and structuring the local adaptation setting (Step 1) to enable a comprehensive analysis. Also, it is strongly recommended to carry out the SWOT analysis together with all the stakeholders as the opinion from a diverse group would reveal the hidden strengths, weakness, opportunities and threats making the result more comprehensive and representative. The diverse group of stakeholders involved in adaptation planning and implementation are urban planners, city engineers, provincial governments, people representatives, service providers such as civic amenity contractors and policy makers. The SWOT analysis will also bring out the multiple benefits of the adaptation response, such as the wetland example which has disaster reduction, liveability enhancement, land improvement and water ecosystems benefits.

Mapping relationships and Synergies between adaptation responses (Step 4): Understanding the adaptation setting helps in identifying the relationships and synergies among the adaptation responses. For example, in case of Can Tho City, Vietnam, it is reported that linking of planned dike enhancement responses with the autonomous household level responses increases the functional life span of dikes [33]. However, linking these responses has an adverse impact on the water quality due to mixing of flood water with sewage [34]. Thus, mapping of relationships between the adaptation responses reveal the hidden threat and opportunities, which further strengthens the SWOT analysis. The 30-year infrastructure development strategy document developed by the Victorian Government, Australia provides a good guidance on mapping the relationships between the adaptation responses [30,35]. Further mapping relationships among the responses helps in identify the opportunities for mainstreaming to implement adaptation responses together with other infrastructure components [36]. Identifying the synergies can be across spatial and temporal scales involving multiple community and sectors [37].

Agility score card for adaptation responses (Step 5): Based on the four agile elements a qualitative score card can be prepared for all the adaptation responses considered so that they can be compared with each other and to understand the overall agility of the city's responses. The score card comprises five criteria. The four criteria representing the four agile elements are: (i) scope for addressing Type I and II needs; (ii) flexibility of adaptation response; (iii) validity across a range of scenarios; and (iv) involvement of stakeholders in all the stages of planning, implementation, operation and monitoring of the adaptation response. In addition to the four agile elements, 'leapfrogging potential' has been included as a fifth criteria due to its crucial importance for developing cities.

This methodology is tested in Pune City in India. The main outcome of the application is how to identify the presence of agile elements and switch to an agile urban adaptation practice in a conventional adaptation context comprising conventional adaptation responses. The focus in this paper is limited to adaptation responses which enhance urban flood resilience.

3. Case Study Application

Pune (Figure 2), the ninth largest Indian city located close to Mumbai, is an important educational, industrial and military hub [38]. Spread over 276 sq.km, the total population of Pune city was about 3.1 million in 2011 [38]. Pune is a hilly city and is located in the confluence of two rivers Mula and Mutha [39]. According to the Government of India's service level bench marking (SLB), Pune ranks high among the Indian cities in terms of basic city service levels such as provision of water and sanitation, electrification, etc., [32]. Disease outbreak, lack of affordable housing, earthquake, water insecurity and pluvial flooding are some of the major threats to Pune [40]. Although Pune Municipal Corporation is responsible for running the city a number of agencies at state, national and international level also contribute to the development of Pune either because of their mandate or through special projects (Table 1).

Understanding the local adaptation context in Pune: The current developmental needs (Type I) of Pune are evident from the gaps in the service level bench marks (Table 2). There is a gap with respect to sewage treatment, improvements to road side drains, ecological improvements and water security. The adaptation needs (Type II) of Pune city is likely to be affected by changing climate, population and economy. The changing climate, represented by four equally likely scenarios [41], is likely to influence precipitation which can become more erratic with heavier extreme rain fall events and longer periods of dry spells [42]. This has a direct impact on the water security and pluvial flooding related problems in Pune [43]. Furthermore, the population in Pune at Year 2027 is likely to be at 5.7 million in a low growth rate scenario and 6.2 million in a high growth rate scenario [44]. There is uncertainty in number of people migrating to Pune and uncertainty in the changes to land use patterns [44]. Hence, the impact of flooding such as number of people displaced or the economic damages due to flooding is influenced jointly by climate change, demographic change and land use change in Pune. This leads to a multiple adaptation needs context with Type II adaptation. However, the magnitude of the impact and responses are likely to vary depending on the scenario.

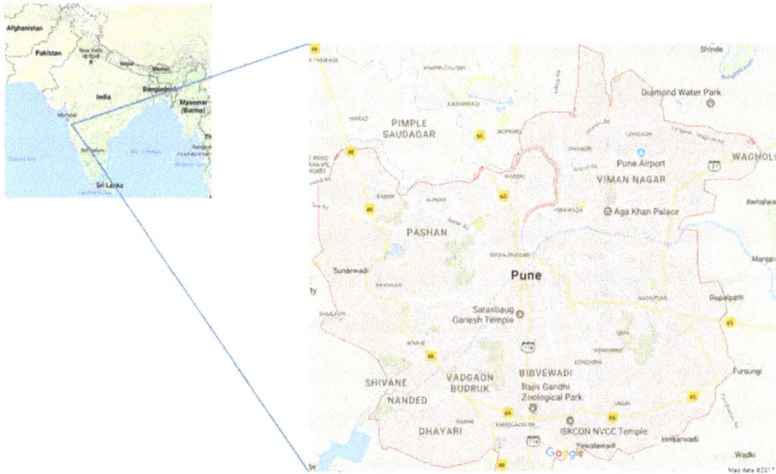

Figure 2. Map showing location of Pune in India (Source: Google maps www.goo.gl/ED9SHA).

Table 1. List of various agencies and Projects involved in planning and implementation of adaptation responses in Pune.

AMRUT	Atal Mission for Rejuvenation and Urban Transformation
Cantt	Cantonment
CPCB	Central Pollution Control Board
CPHEEO	Central Public Health and Environmental Engineering Organisation
CWPRS	Central Water and Power Research Station
GoI	Government of India
GoM	Government of Maharashtra
IPCC	Intergovernmental Panel on Climate Change
JICA	Japan International Cooperation Agency
MIDC	Maharashtra Industrial Development Corporation
MoUD	Ministry of Urban Development
MSRDC	Maharashtra State Road Development Corporation
NGO	Non-Governmental Organisation
NHAI	National Highways Authority of India
NRCD	National River Conservation Directorate
PCMC	Pimpri Chinchwad Municipal Corporation
PMC	Pune Municipal Corporation
PMRDA	Pune Metropolitan Region Development Authority
RFD	River Front Development
ZP	Zilla Parishad (District Council)

Table 2. Gaps in infrastructure and basic service levels in Pune.

Service Level Parameters for Indian Cities	Desired Service Level Benchmark	Current Service Levels of Pune *	Reference
Water Supply			
Coverage of Water Supply connections	100%	90%	ESR [45]
Per Capita Supply of Water	135 L/pers/day	90–120	Smart City Plan [46]
Extent of Non-Revenue Water	15%	30%	Smart City Plan [46]
Extent of Metering	100%	30%	ESR [45]
Continuity of Water supplied	24 h/day	5 h/day	ESR [45]
Efficiency in redressal of customer complaints	80%	100%	ESR [45]

Table 2. *Cont.*

Service Level Parameters for Indian Cities	Desired Service Level Benchmark	Current Service Levels of Pune *	Reference
Water Supply			
Quality of Water Supplied	100%	100%	ESR [45]
Cost Recovery	100%	71%	City development plan [47]
Efficiency in Collection of Water Charges	90%	91%	ESR [45]
Solid Waste Management			
Household Level Coverage	100%	60%	ESR [45]
Efficiency in Collection of Solid Waste	100%	100%	ESR [45]
Extent of Segregation of Solid Waste	100%	44%	ESR [45]
Extent of Solid Waste recovered	80%	80%	ESR [45]
Extent of Scientifically sound Disposal	100%	50%	ESR [45]
Extent of Cost Recovery	100%	80%	ESR [45]
Efficiency in Collection of SWM Charges	90%	80%	ESR [45]
Efficiency in Redressal of Customer Complaints	80%	91%	ESR [45]
Sewerage			
Coverage of Toilets	100%	98%	ESR [45]
Coverage of Sewerage Network	100%	99%	ESR [45]
Collection efficiency of Sewerage Network	100%	80%	ESR [45]
Adequacy of Sewage Treatment Capacity	100%	76%	ESR [45]
Quality of Sewage Treatment	100%	100%	ESR [45]
Extent of Reuse and Recycling of Sewage	20%	8%	ESR [45]
Extent of cost recovery in waste water management	100%	90%	ESR [45]
Efficiency in redressal of customer complaints	80%	80%	ESR [45]
Efficiency in Collection of Sewage Water Charges	90%	80%	ESR [45]
Storm Water Drainage			
Coverage	100%	55%	City development plan [47]
Incidence of water logging	0 numbers	52 Nos.	City development plan [47]

* The current service level bench marks are obtained from the reports that are prepared by Pune Municipal Corporation or by consultants commissioned by Pune Municipal Corporation. Hence there is a chance of over reporting, which has to be verified.

Collating adaptation responses: Although most of the adaptation responses in Pune focus on Type I adaptation needs (i.e., adaptation deficit), they can be expanded or modified to satisfy the Type II needs, which is adaptation gap. Most of the Type I adaptation needs in Pune such as gaps in water supply and drainage are overcome through basic infrastructure service initiatives, which is evident from Pune's budgetary plans [48]. The Smart cities missions, Atal Mission for Renewal and Urban Transformation (AMRUT), National River Conservation Directorate (NRCD)—Japan International Cooperation Agency (JICA) initiative, Swach Bharat Mission/Clean India Mission, Heritage City Development and Augmentation Yojana/plan (HRIDAY) are some of the programmes through which Type I and II gaps are being dealt with in Pune [38]. The adaptation responses relevant to water security and pluvial flooding (Figure 3) under the aforementioned programmes are: (i) additional water intake from Mulshi Dam [48]; (ii) vision 24 × 7 water supply [46]; (iii) rain water harvesting [48]; (iv) storm water drains [48]; (v) increase in sewage treatment capacity [49]; (vi) expansion of sewer trunk mains [49]; (vii) strengthening river embankments [49]; (viii) plantations, gardens and open spaces [49]; and (ix) check dams across rivers [49]. The collated adaptation responses fall into three board categories: (i) ensuring water supply and quality; (ii) preventing flooding and river; and (iii) river front development, which can cater to both Type I and II adaptation needs. The details of adaptation measures such as nature of adaptation and location of the measures are shown in Figure 3. Various stakeholders are involved in every adaptation response (Figure 4). Understanding the attributes of the adaptation responses based on the programme, ownership, funding, engineering, procurement and construction (EPC), operation and maintenance (O&M) enables the mapping of responsibilities of stakeholders (Figure 5). This can facilitate the understanding on the extent of stakeholder involvement. Being aware of the spatial and temporal distribution of the responses, programmes, funding source and stakeholders helps in establishing the relationships between the responses.

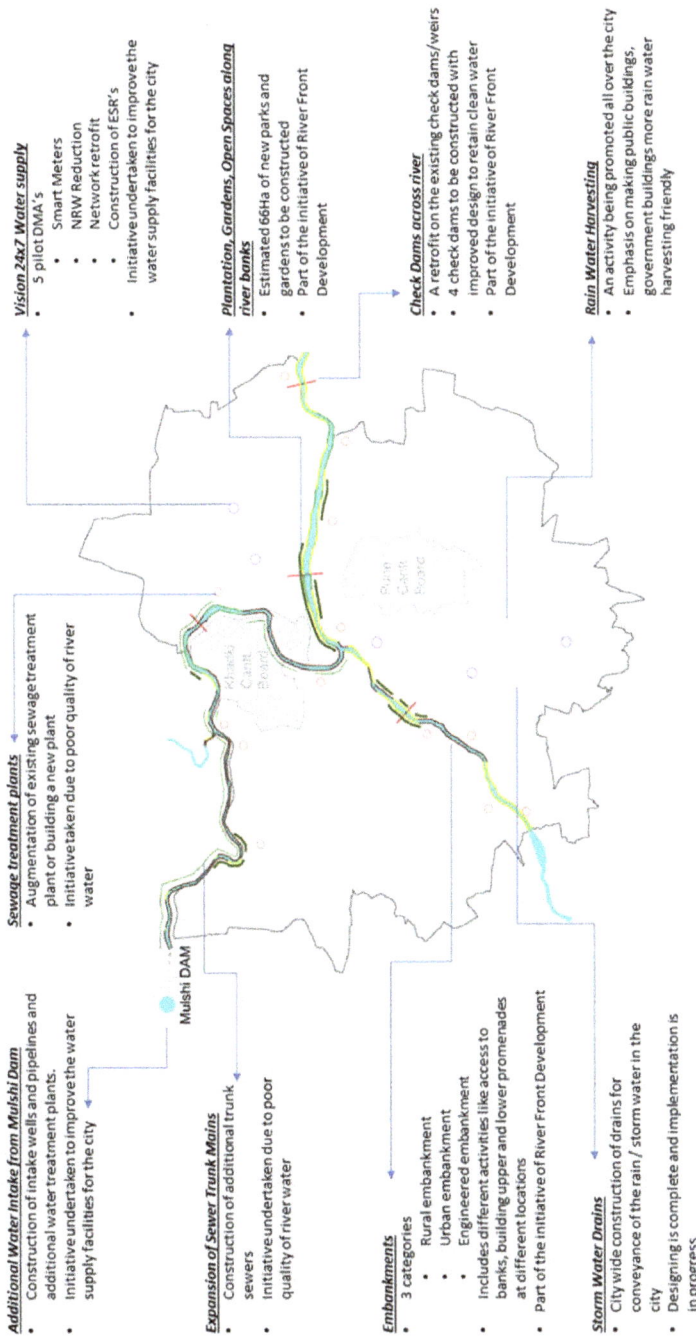

Figure 3. Adaptation responses for Pune, collated from existing planning documents.

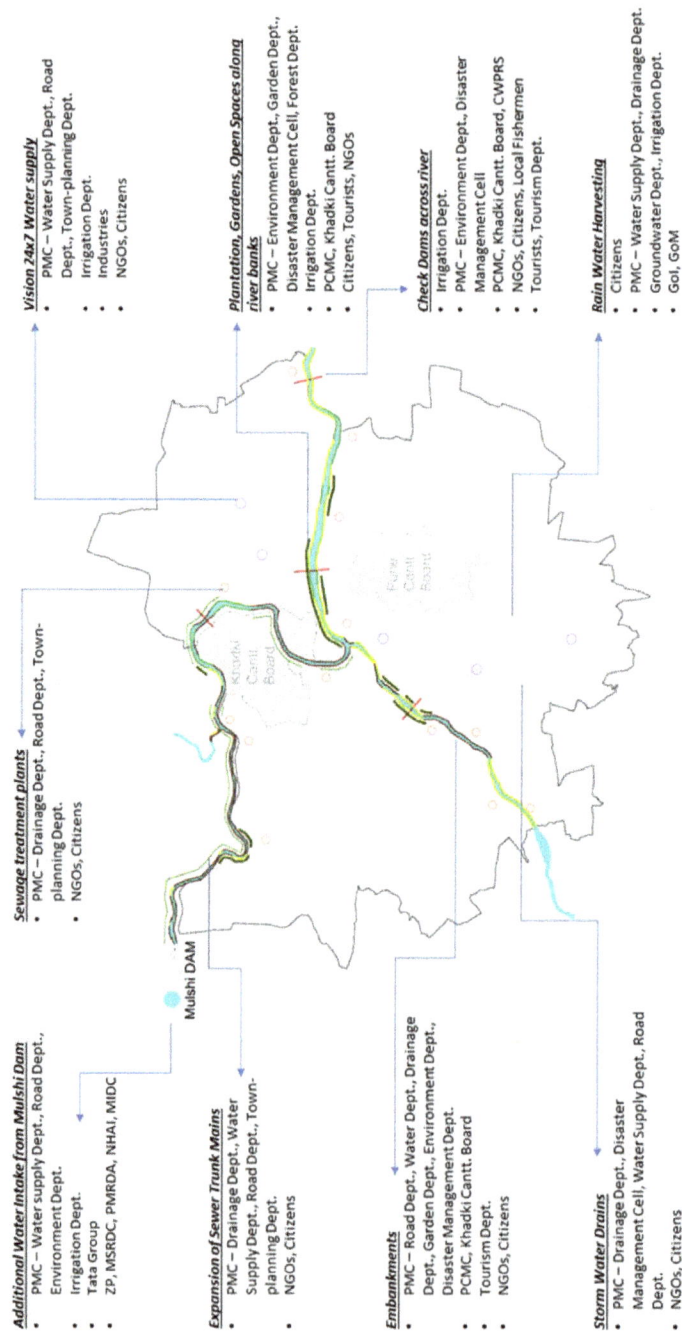

Figure 4. Stakeholders involved in adaptation responses.

Additional Water Intake from Mulshi Dam
- PMC – Water supply Dept., Road Dept.,
  Environment Dept.
- Irrigation Dept.
- Tata Group
- ZP, MSRDC, PMRDA, NHAI, MIDC

Expansion of Sewer Trunk Mains
- PMC – Drainage Dept., Water
  Supply Dept., Road Dept., Town-
  planning Dept.
- NGOs, Citizens

Embankments
- PMC – Road Dept., Water Dept., Drainage
  Dept., Garden Dept., Environment Dept.,
  Disaster Management Dept.
- PCMC, Khadki Cantt. Board
- Tourism Dept.
- NGOs, Citizens

Storm Water Drains
- PMC – Drainage Dept., Disaster
  Management Cell, Water Supply Dept., Road
  Dept.
- NGOs, Citizens

Sewage treatment plants
- PMC – Drainage Dept., Road Dept., Town-
  planning Dept.
- NGOs, Citizens

Vision 24x7 Water supply
- PMC – Water Supply Dept., Road
  Dept., Town-planning Dept.
- Irrigation Dept.
- Industries
- NGOs, Citizens

*Plantation, Gardens, Open Spaces along
river banks*
- PMC – Environment Dept., Garden Dept.,
  Disaster Management Cell, Forest Dept.
- Irrigation Dept.
- PCMC, Khadki Cantt. Board
- Citizens, Tourists, NGOs

Check Dams across river
- Irrigation Dept.
- PMC – Environment Dept., Disaster
  Management Cell
- PCMC, Khadki Cantt. Board, CWPRS
- NGOs, Citizens, Local Fishermen
- Tourists, Tourism Dept.

Rain Water Harvesting
- Citizens
- PMC – Water Supply Dept., Drainage Dept.
- Groundwater Dept., Irrigation Dept.
- GoI, GoM

Mulshi DAM

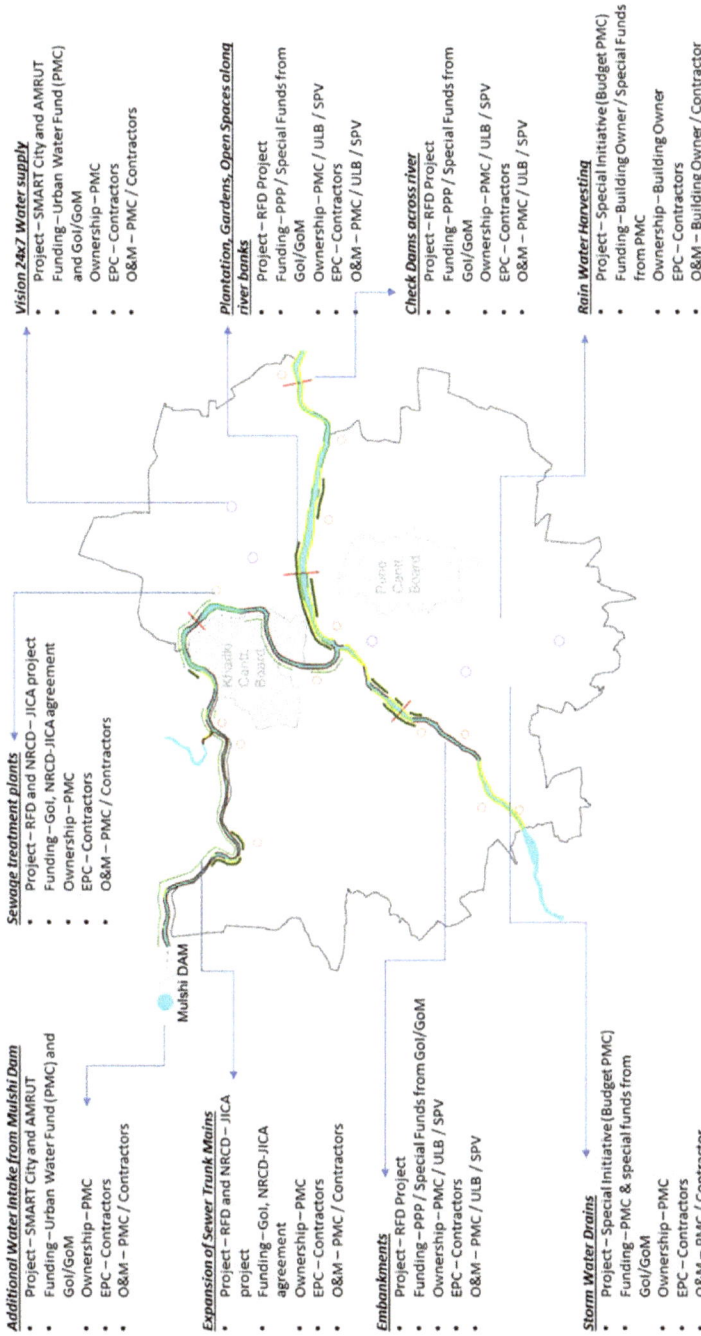

Figure 5. Funding and ownership details of adaptation responses.

Table 3. SWOT analysis of Adaptation responses.

	Additional Water Intake from Mulshi Dam	Sewage Treatment Plants	24 × 7 Water Supply
Strength	Assured water availability and water transport	Improved river water quality	Improved service levels
Weakness	Dependency on single source Too many competing stakeholders	Land acquisition and public acceptance for facility in the vicinity	Lack of coordination between Planning and water department
Opportunity	Possibilities to develop a regional water plan and secondary water sources Possibilities for institutional arrangements between stakeholders	Possibilities to design and implement nature based treatment system using gardens, wet lands and open space	Possibilities for institutional arrangements between stakeholders Possibilities to postpone water intake works
Threat	Failure will have cascading effect Land acquisition and implementation hurdles might lead to delay and cost escalations	High probability for plant to reach peak capacity before the end of design period	Likely changes in population density and land use patterns

	Expansion of Trunk Sewer	Check Dams in River	Gardens and Open spaces along the river
Strength	Better public health and river water quality Guaranteed funds	Increased water availability, flood control and regulated silting	Improved biodiversity, embankment strength and flood protection
Weakness	Construction hurdles in densely populated areas	Land acquisition O&M difficulties	Land acquisition O&M difficulties
Opportunity	Mainstreaming with roads, electric cables, etc., Possibilities for stakeholder institutional arrangements	Construction compatibility with embankments and recreation facilities	Mainstreaming with transportation, recreation, urban farming and open space facilities
Threat	Dumping of solid waste might lead to O&M difficulties High path dependency leading to future complications.	Hindrance to navigation Lack of warning system, preparedness and operational protocol during cloudburst.	Increased public access to riparian areas is a threat to ecosystems.

	Embankments	Strom Water Drains	Rainwater harvesting
Strength	Better access to river side leading to increase in public and tourist activity	Reduced water logging, road accidents and health issues	Improved ground water levels
Weakness	Relocating existing infrastructure such as Dobhi ghats	Removal of encroachments and availability of funds	High dependency on proactive public participation
Opportunity	Mainstreaming with roads, electric cables, gardens, etc., Possibilities for institutional arrangements between stakeholders	Mainstreaming with roads, electric cables, gardens, etc., Possibilities for institutional arrangements between stakeholders	Reduced dependence on city water supply Tax/water tariff incentives
Threat	Land acquisition issues leading to cost overruns and time delays	Land acquisition issues leading to cost overruns and time delays	Lack of adequate maintenance and water quality issues.

SWOT analysis: The strengths, weakness, opportunities and threats of all the adaptation responses in Pune have been analysed (Table 3). From Table 3 it can be seen that the strength of the responses are specific to the objectives of the responses, whereas common weaknesses, opportunities and threats can be seen across the responses. For example, difficulty in land acquisition is a common weakness across responses such as check dams, storm water drains and open area development. Whereas difficulty in land acquisition is a threat to intake works and embankments, engagement with stakeholders is an opportunity across most of the responses. A weak or inadequate stakeholder engagement is a threat to intake well adaptation response. In addition to enhancing the understanding adaptation measures, the SWOT analysis also helps in understanding the implementation bottlenecks (Figure 6) as well the possible collaborations across the adaptation responses (Figure 7).

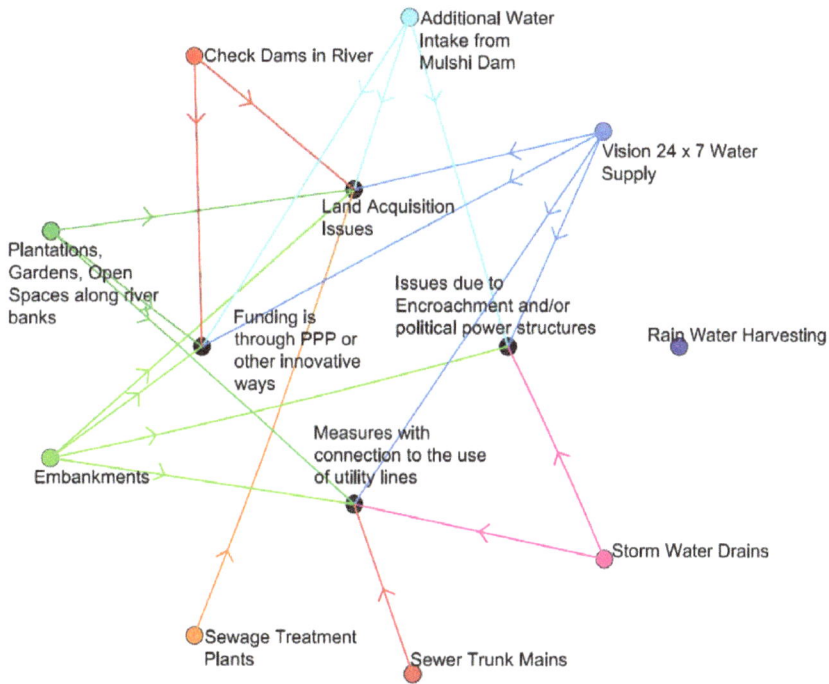

Figure 6. Common implementation bottlenecks in implementing adaptation responses. The arrows originate from the adaptation responses and terminate at the probable bottlenecks which are likely to hinder the response during planning or implementation.

From Figure 6 it can be seen that the common bottlenecks for implementation of the adaptation responses are land acquisition, funding mechanisms, clearing encroachments in project sites and interference with utility services (such as telecommunications, electrical distribution and traffic, as the water pipes and sewers share the same service corridor). However, becoming aware of the bottlenecks across the adaptation responses also gives an opportunity to understand and resolve them. The sewage treatment plants that are planned at multiple locations in Pune are likely to face objections from people living in the vicinity due to foul odour [50]. Also the open space plan and embankment project are likely to face objections from the community residing in the vicinity of river and involved in professional laundry services, as implementing these responses can lead to displacement and livelihood issues [50]. However, from Figure 7 it can be seen that there are possibilities to improve stakeholder participation, improve coordination between utility departments, and adopt comprehensive flood management in Pune, which is based on the commonalities between the adaptation responses.

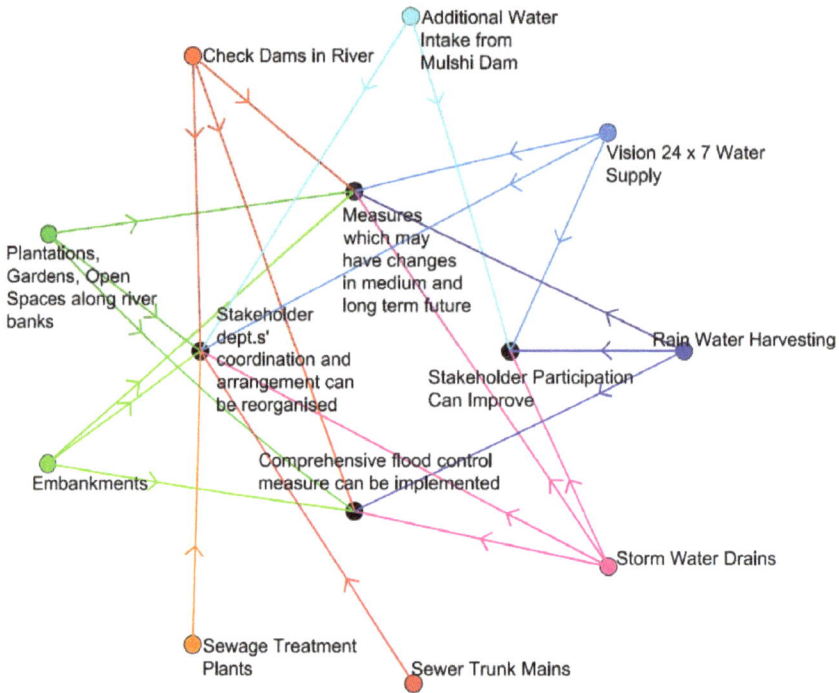

Figure 7. Possible collaboration between adaptation responses (The arrows represent the contribution of response to the common goal, i.e., the arrows originate from the adaptation responses and terminate at the favourable outcome(s) that the adaptation responses are likely to yield during planning or implementation).

Mapping relationships and Synergies between adaptation responses: Understanding the adaptation context and SWOT analysis of adaptation responses in Pune helped in identifying the relationships and synergies among the adaptation responses (Figure 8). For example, the intake works, 24 × 7 water supply and rainwater harvesting responses are related as they all cater to various water demands, storm water drains are related to the embankments and open space management plans as there is a spatial and functional overlap between these responses that can be modified if open areas are designed to detain storm water during intense rainfall. Similarly, the embankments can be related to the functionality of open areas in the city.

Understanding the relationships between adaptation responses (Figure 8) can lead to the identification of synergies between the adaptation responses (Figure 9). For example, supplementing water supply through rainwater harvesting has direct relevance to 24 × 7 water supply, which can lead to change in design of water supply mains and intake well design. This can lead to possible cost savings in terms of reduced water consumption from the mains and savings in energy to deliver water. Amalgamating storm water drains, sewer mains, river embankments and open space management plans can lead to change in design of the measures, coordinated implementation, operation and maintenance. There is good scope for realising the synergy in form of ecosystem benefits if the open space management plan can be coordinated with the sewerage treatment plants and embankments.

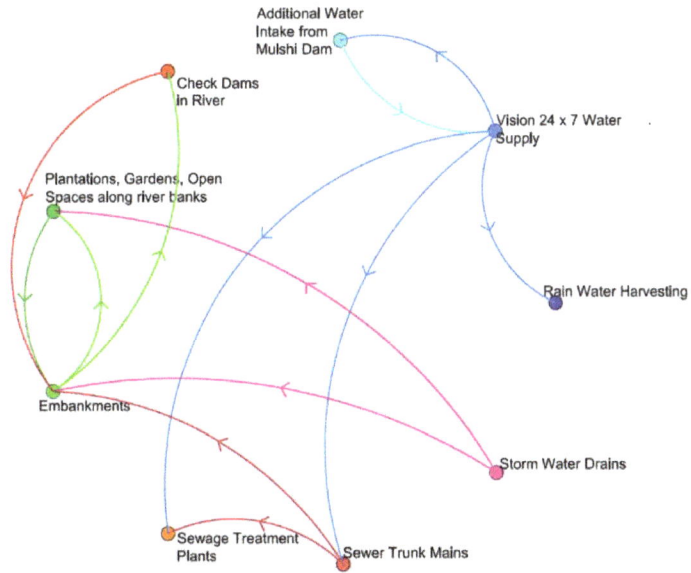

Figure 8. Mapping relationships between adaptation responses The arrows originating from an adaptation responses and terminating at other the adaptation response(s) is the representation of relationship between the responses, i.e., change in one response is most likely to lead change in the other response.

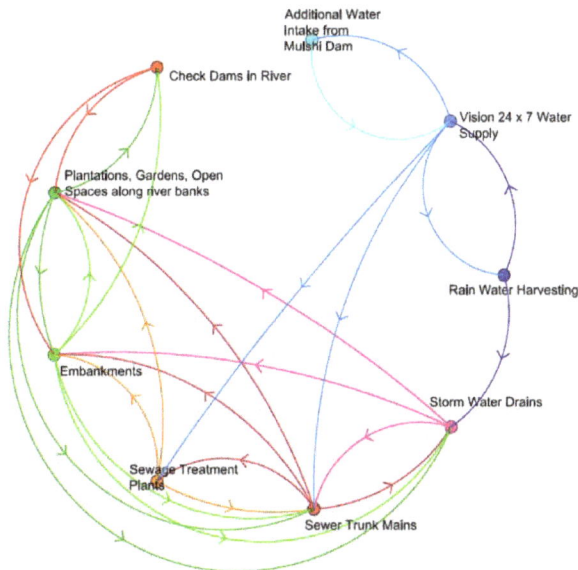

Figure 9. Synergies between adaptation responses. The arrows originating from an adaptation responses and terminating at other the adaptation response(s) is the representation of synergy between the responses, i.e., planning and implementing the responses together is likely to yield a greater benefit than planning and implementing in isolation.

Agility score card for adaptation responses: An agility score card based on four agile elements and leapfrogging potential was prepared for all the adaptation responses in Pune. The agility scores for all the adaptation measures in Pune were allocated based on the expert judgement of the authors. The expert judgement of the authors relied on: (i) the authors' own experience in planning and implementing adaptation responses; and (ii) interviews with various stakeholders in Pune who are involved in the planning and implementation of adaptation responses. Hence the agility scores are highly subjective. The range of qualitative agility scores is 0–5. Zero indicates that it is not possible to incorporate the agile element; one indicates the possibility to find or incorporate agile element is low; two indicate that the chances are medium; three indicate that the chances are good; four indicates that the chances are very good; and five indicates the best opportunity to incorporate the agile element in the adaptation response. The agility sore card for adaptation responses in Pune is presented in Figure 10.

The additional water intake from Mulshi dam has very limited flexibility as it an established practice to incorporate redundancy in the engineering design of pipelines and intake wells based on a future water demand [51]. However, structural flexibility is provided in certain aspects, like the design of bays for pumps which can be easily changed with change in demand. A score of one was assigned to flexibility of water intake works. This response is designed with an objective of satisfying water demand and will become redundant in the event of a bad water quality at the reservoir. A score of three was assigned to the performance of intake works in plausible scenarios. The design of intake facilities does not account for climate change and only considers population increase. A score of one was assigned to Type II adaptation needs consideration. It was evident from discussion with PMC officials that the planning and designing of intake works has been discussed and debated with various departments such as irrigations and dams, roads and with revenue departments due to water allocation, road cutting and land acquisition activities associated with this work [50]. The land owners along the pipe alignment were consulted, but consultations with end users in the city are missing. Hence, the stakeholder consultation component was assigned a score of three. There is no significant contribution to leapfrogging as this response is designed based on a conventional water supply– demand approach, except in terms of use of advanced technology in terms of providing energy efficient pumps and smart operation of pumps. Hence a leapfrogging score of one is given to this response.

Also, from Figure 10 it can be seen that responses such as gardens and open spaces have high agility scores of four across all agility elements. The response can be implemented in a flexible manner spatially and temporally. The open spaces can cater to a variety of plausible scenarios and can address the type II needs. There is also stakeholder involvement in terms of consultation, operation and in maintenance [43,46,50]. This measure helps in leapfrogging as it improves the environment and ecological aspects of the city.

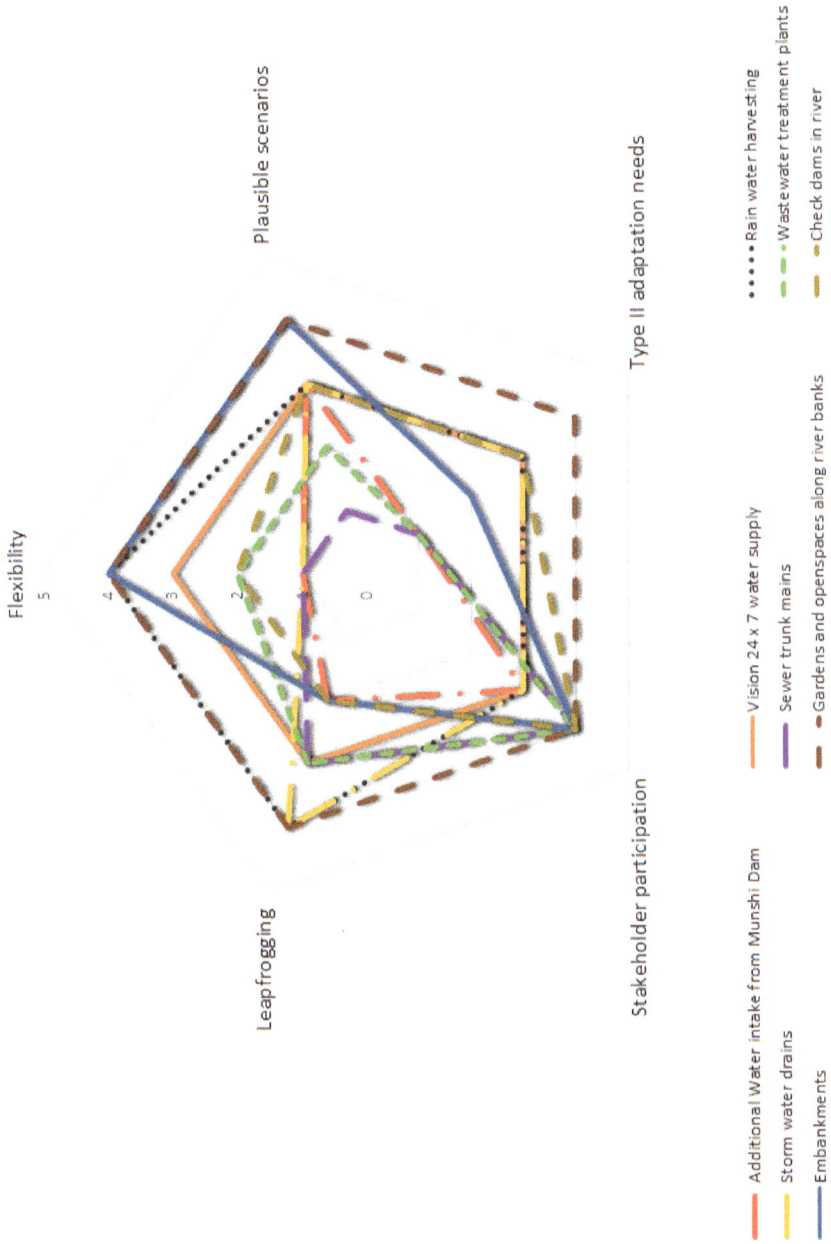

Figure 10. Agility score card for adaptation responses.

4. Discussions

Cities have already connected adaptation goals with development needs at a strategic level (e.g., Surat, Semarang [52]) but are striving hard to strike a balance between development (Type I) and adaptation (Type II) [53]. From the application of the agile urban adaptation process in Pune it is evident that Type I and Type II needs can be integrated at planning and implementation levels. SWOT analysis, mapping of relationships, synergies and the agility score card for responses reveal that agile urban adaptation can be implemented using conventional adaptation responses.

Conventional responses and unconventional process: The scoping exercise in Pune has revealed that conventional adaptation responses such as rain water harvesting, open area development and check dams have agile characteristics. Qualitative analysis of agile elements reveal that the nature of agile characteristics varies across adaptation responses. For example, the check dam has a high score with respect to stakeholder participation but a low score for flexibility. The embankments have high flexibility but very low leapfrogging potential. Overall, there is a high score for stakeholder involvement and there is an equal spread in the scores for other agile elements. This understanding can help in redesigning the responses so that the agility scores can be increased or help in creating additional responses that can make the overall adaptation process more agile. Being flexible and continuously engaging with stakeholders throughout the process also increases the accessibility and responsiveness among stakeholders. This can lead to the evolution of clumsy responses, i.e., adaptation responses which emerge from the local adaptation setting with outcomes that are beneficial and acceptable to all the stakeholders [12].

Simplicity of operational process: From the scoping exercise in Pune, it can be seen that a SWOT analysis and stakeholder mapping of the responses enabled the identification of the relationships and synergies between measures. Although the agility scores are qualitative, there is scope for developing a quantitative scoring method, similar to multiple benefits assessment tools (e.g., BeST [54]). Further identification of the leapfrogging potential is simple and can be guided using the principles of localising the SDGs [8].

Legitimising agile urban adaptation: Lack of motivation and ability among the stakeholders is considered as one the reasons for non-implementation of adaptation responses [21]. SDG initiatives offer motivation and a conducive environment for using agile urban adaptation, which can fast-track the achievement of SDGs. Agility offers the ability to stakeholders to implement adaptation responses in a quick and flexible manner [13]. Although agile adaptation is unlikely to create setbacks, changing frequently and too quickly could be seen as acting without clarity and lack of confidence by some of the stakeholders [55]. This might lead to the lack of political legitimacy and loss of mandate of the political and administrative authority. Policy backing for the agile adaptation process for the purpose of achieving SDGs will enhance its credibility. Agile urban adaptation will be more effective and assertive if it can secure political and institutional legitimacy. A comprehensive insight of resource allocation and timelines for planning, implementation, operation and maintenance of adaptation responses can be obtained through the analysis of the current, historical annual municipal budgets and annual audited statements. Assessing the responses from the aforementioned reports using the agility scorecard will help planning in advance for the next budget cycle, or in case there are special funds or grants which are lapsable.

Although it is theoretically possible to adapt short time steps such as in cycles of few weeks like in software development in an agile manner, we need to consider at least six months to one year lead time for approval and listing in the municipal budget. There are administrative and procedural hurdles in implementing response when they are not mentioned in the municipal budget for the planning year or approved by the municipal council. However, some of the adaptation responses can be implemented through approved actions such as operation and maintenance (O&M) of municipal assets. O&M is a broad budget head with no specific mention about locations or nature of interventions. The approved annual city budgets usually have considerable resources allocated for O&M, including retrofitting and replacement of assets, which can be utilised for agile implementation.

The scoping exercise in Pune reveals that working towards legitimacy might face little or no hindrance as agility can be practised using conventional adaptation responses within the existing urban institutional framework, i.e., the agility scores of agile elements of responses can be increased within the existing urban institutional framework. The legitimacy aspects can be further explored.

Relevance to sponge cities: The adaptation context and adaptation responses in Pune are comparable to that of sponge cities as sponge cites also aim at satisfying the adaptation deficits and adaptation needs, whereas the aspect of harmonising the deficits and need are not obvious. However, harmonising the adaptation deficits and needs is a necessity in sponge cities. Hence, it can be argued that the agile urban adaptation process is also valid and applied in the context of sponge cities. Also in case of sponge cities, planning and implementation of adaptation responses is left to cities although the overall guidance is from three central ministries [9]. The responsibility and ownership at city level for the planning and implementation of responses are ideal for the application of agile adaptation process, localisation and attainment of SDGs and the evolution of clumsy responses in sponge cities. City of Rotterdam in the Netherlands has some good examples in this regard, where there is a good coordination between various departments. For example the water department of Rotterdam proactively engages with other infrastructure departments whenever a major infrastructure change happens and tries to embed a water component as part of the change [56].

5. Conclusions

The paper focused on developing and testing a methodology to ascertain the scope for the operation of agile urban adaptation. The scoping exercise in Pune reveals that urban agile adaptation is not about implementing novel adaptation responses but planning and implementing the conventional adaptation responses in a different manner. For example, conventional adaptation responses such as city greening and check dams across the rivers have agile characteristics, these responses are synergetic with other adaptation responses and there is a possibility to compare conventional adaptation responses based on agile characteristics. Not doing different things but doing things differently, which also resonates well with the localising the SDGs in the cities [8] and can also pave way for legitimising the agile approach. Furthermore, the scoping exercise reveals the widely used SWOT analysis facilities mapping of synergies and relationships between the adaptation responses. Agility can help in realising the synergies between Type I and Type II adaptation needs and also enables leapfrogging towards sustainability. Furthermore, the scoping exercise and testing of agile adaptation in Pune synthesised important practical insights towards the application of scientific knowledge developed by Pathirana et al. [13]. This operational knowledge is crucial as focus on adaptation has started transitioning from planning to implementation of adaptation responses [6]. Hence, it is possible to apply the agile urban adaptation process using conventional adaptation responses in urban areas which address adaptation deficits related to infrastructure development as well as climate and socio-economic adaptation.

Acknowledgments: The authors would like to thank Pune Municipal Corporation, Non-governmental agencies, various agencies of the State Government of Maharashtra and Government of India for sharing the data and for the consultations.

Author Contributions: The paper is based on the results obtained from the research of Tejas M. Pathak, which was done for the partial fulfilment of requirements for the Master of Science degree at the IHE Delft Institute for Water Education under the supervision of Assela Pathirana, Mohanasundar Radhakrishnan and Keneth Irvine. The first draft was written by Mohanasundar Radhakrishnan and Tejas M. Pathak. Assela Pathriana and Keneth Irvine helped to review and improve the paper.

Conflicts of Interest: The authors declare no conflict of interest.

References

1. Revi, A.D.E.; Satterthwaite, F.; Aragón-Durand, J.; Corfee-Morlot, R.B.R.; Kiunsi, M.; Pelling, M.; Roberts, D.C.; Solecki, W. *Urban Areas, in Climate Change 2014: Impacts, Adaptation, and Vulnerability. Part A: Global and Sectoral Aspects. Contribution of Working Group II to the Fifth Assessment Report of the Intergovernmental Panel of Climate Change*; Balbus, J., Cardona, O.D., Eds.; Cambridge University Press: Cambridge, UK; New York, NY, USA, 2014; pp. 535–612.
2. Burton, I. Climate Change and Adaptation Deficit. In Proceedings of the International Conference on Adaptation Science, Management and Policy Options, Lijiang, Yunnan, China, 17–19 May 2004.
3. Pathirana, A.; Radhakrishnan, M.; Quan, N.H.; Zevenbergen, C. Managing urban water systems with significant adaptation deficits—Unified framework for secondary cities: Part I—Conceptual framework. *Clim. Chang.* **2017**, 1–14. [CrossRef]
4. Tessler, Z.D.; Vörösmarty, C.J.; Grossberg, M.; Gladkova, I.; Aizenman, H.; Syvitski, J.P.M.; Foufoula-Georgiou, E. Profiling risk and sustainability in coastal deltas of the world. *Science* **2015**, *349*, 638–643. [CrossRef] [PubMed]
5. UNEP. *The Adaptation Gap Report 2014*; United Nations Environment Programme (UNEP): Nairobi, Kenya, 2014; p. 68.
6. Klein, R.J.T.; Adams, K.M.; Dzebo, A.; Davis, M.; Siebert, C.K. *Advancing Climate Adaptation Practices and Solutions: Emerging Research Priorities*; Working Paper; Davis, M., Ed.; Stockholm Enviroment Institute: Stockholm, Sweden, 2017; p. 25.
7. United Nations. *Transforming Our World: The 2030 Agenda for Sustainable Development*; United Nations: New York, NY, USA, 2015.
8. Kanuri, C.; Revi, A.; Espey, J.; Kuhle, H. *Getting Started with the SDGs in Cities—A Guide for Stakeholder*; Sustainble Development Solutions Network: Paris, France, 2016.
9. Embassy of the Kingdom of The Netherlands. *Factsheet Sponge City Construction in China*; Embassy of the Kingdom of The Netherlands: Beijing, China, 2016.
10. Li, H.; Ding, L.; Ren, M.; Li, C.; Wang, H. Sponge city construction in China: A survey of the challenges and opportunities. *Water* **2017**, *9*, 594. [CrossRef]
11. Bajaj, V.; Jessia, M.; Thompson, S.A. How Houston's Growth Created the Perfect Flood Conditions. *The New York Times*, 5 September 2017.
12. Thompson, M.; Beck, M.B. *Coping with Change: Urban Resilience, Sustainability, Adaptability and Path Dependence*; Future of cities; HM Government Office for Science: London, UK, 2015.
13. Pathirana, A.; Radhakrishnan, M.; Ashley, R.; Quan, N.H.; Zevenbergen, C. Managing urban water systems with significant adaptation deficits—Unified framework for secondary cities: Part II—The pratice. *Clim. Chang.* **2017**, 1–18. [CrossRef]
14. Poustie, M.S.; Frantzeskaki, N.; Brown, R.R. A transition scenario for leapfrogging to a sustainable urban water future in Port Vila, Vanuatu. *Technol. Forecast. Soc. Chang.* **2016**, *105*, 129–139. [CrossRef]
15. Ferguson, B.C.; Frantzeskaki, N.; Brown, R.R. A strategic program for transitioning to a Water Sensitive City. *Landsc. Urb. Plan.* **2013**, *117*, 32–45. [CrossRef]
16. Bosomworth, K.; Leith, P.; Harwood, A.; Wallis, P.J. What's the problem in adaptation pathways planning? The potential of a diagnostic problem-structuring approach. *Environ. Sci. Policy* **2017**, *76*, 23–28.
17. European Environment Agency. *Urban Adaptation to Climate Change in Europe: Transforming Cities in a Changing Climate*; European Environment Agency: Copenhagen, Denmark, 2016; p. 135.
18. Lonsdale, K.; Pringle, P.; Turner, B. *Transformative Adaptation: What It Is, Why It Matters and What Is Needed*; UK Climate Change Impacts Programme (UKCIP): Oxford, UK, 2015; p. 40.
19. Haasnoot, M.; Middelkoop, H.; Offermans, A.; Beek, E.; Deursen, W.P.A.V. Exploring pathways for sustainable water management in river deltas in a changing environment. *Clim. Chang.* **2012**, *115*, 795–819. [CrossRef]
20. Haasnoot, M.; Kwakkel, J.H.; Walker, W.E.; Maat, J.T. Dynamic adaptive policy pathways: A method for crafting robust decisions for a deeply uncertain world. *Glob. Environ. Chang.* **2013**, *23*, 485–498. [CrossRef]
21. Phi, H.L.; Hermans, L.M.; Douven, W.J.A.M.; van Halsema, G.E.; Khan, M.F. A framework to assess plan implementation maturity with an application to flood management in Vietnam. *Water Int.* **2015**, *40*, 984–1003. [CrossRef]

22. Radhakrishnan, M.; Ashley, R.; Gersonius, B.; Pathirana, A.; Zevenbergen, C. *Flexibility in Adaptation Planning: Guidelines for When, Where & How to Embed and Value Flexibility in an Urban Flood Resilience Context;* CRCWSC—Cooperative Research Centre for Water Sensitive Cities: Melbourne, Australia, 2016.

23. Fowler, M.; Highsmith, J. The Agile Manifesto. *Softw. Dev.* **2001**, *9*, 28–35.

24. Leffingwell, D. *Agile Software Requirements: Lean Requirements Practices for Teams, Programs, and the Enterprise;* Cockburn, A., Highsmith, J., Eds.; Pearson education, Inc. rights and contracts department: Boston, MA, USA, 2010.

25. Wendler, R. The Structure of Agility from Different Perspectives. In Proceedings of the 2013 Federated Conference on Computer Science and Information Systems (FedCSIS), Kraków, Poland, 8–11 September 2013.

26. Vinodh, S.; Devadasan, S.; Vimal, K.; Kumar, D. Design of agile supply chain assessment model and its case study in an Indian automotive components manufacturing organization. *J. Manuf. Syst.* **2013**, *32*, 620–631. [CrossRef]

27. Schulz, A.P.; Fricke, E. Incorporating flexibility, agility, robustness, and adaptability within the design of integrated systems-key to success? In Proceedings of the 18th Digital Avionics Systems Conference, St. Louis, MO, USA, 24–29 October 1999.

28. Sánchez, A.M.; Pérez, M.P. Supply chain flexibility and firm performance: A conceptual model and empirical study in the automotive industry. *Int. J. Oper. Prod. Manag.* **2005**, *25*, 681–700. [CrossRef]

29. Jackson, S.E.; Joshi, A.; Erhardt, N.L. Recent research on team and organizational diversity: SWOT analysis and implications. *J. Manag.* **2003**, *29*, 801–830.

30. Infrastructure Victoria. *All Things Considered;* Infrastructure Victoria: Melbourne, Australia, 2016.

31. Infrastructure Victoria. *Victoria's Draft 30-Year Infrastruture Strategy;* Infrastructure Victoria: Melbourne, Australia, 2016.

32. Ministry of Urban Development. *Service Level Bench Mark for Pune;* Ministry of Urban Development, Goverment of India: New Delhi, India, 2017.

33. Radhakrishnan, M.; Quan, N.H.; Gersonius, B.; Pathirana, A.; Vinh, K.Q.; Ashley, M.R.; Zevenbergen, C. Coping capacities for improving adaptation pathways for flood protection in Can Tho, Vietnam. *Clim. Chang.* **2017**, 1–13. [CrossRef]

34. Nguyen, H.Q.; Radhakrishnan, M.; Huynh, T.T.N.; Baino-Salingay, M.L.; Ho, L.P.; Steen, P.V.D.; Pathirana, A. Water Quality Dynamics of Urban Water Bodies during Flooding in Can Tho City, Vietnam. *Water* **2017**, *9*, 260. [CrossRef]

35. Infrastructure Victoria. *Draft Options Book 2016;* Infrastructure Victoria: Melbourne, Australia, 2016.

36. Rijke, J.; Ashley, M.R.; Sakic, R. *Adaptation Mainstreaming for Achieving Flood Resilience in Cities, in Socio-Technical Flood Resilience in Water Sensitive Cities—Adaptation Across Spatial and Temporal Scales;* CRCWSC—Cooperative Research Centre for Water Sensitive Cities: Melbourne, Australia, 2016.

37. Serrao-Neumann, S.; Crick, F.; Harman, B.; Schuch, G.; Choy, D.L. Maximising synergies between disaster risk reduction and climate change adaptation: Potential enablers for improved planning outcomes. *Environ. Sci. Policy* **2015**, *50*, 46–61. [CrossRef]

38. Ministry of Urban Development. *SMART Cities, City Profile—Pune, G.o.I.;* Ministry of Urban Development, Goverment of India: New Delhi, India, 2016. Available online: http://smartcities.gov.in/upload/uploadfiles/files/Maharashtra_Pune.pdf (accessed on 19 March 2017).

39. Deshpande, A.; Karuna, V.; Prabhu, D.; Kiran, U.; Thatte, M.; Garde, M.; Waghmare, R.; Chitale, M. *Socio-Economic Survey of Pune 2008–2009;* Deshpande, A., Ed.; Pune Municipal Corporation: Pune, India, 2009.

40. 100 Resilient Cities. City Strategies. Available online: http://www.100resilientcities.org/strategies#/-_/ (accessed on 27 March 2017).

41. Intergovernmental Panel on Climate Change (IPCC). *Working Group I Contribution to the IPCC Fifth Assessment Report, Climate Change 2013: The Physical Science Basis, Summary for Policymakers;* IPCC: Geneva, Switzerland, 2013.

42. Dupuis, J.; Knoepfel, P. The Adaptation Policy Paradox: the Implementation Deficit of Policies Framed as Climate Change Adaptation. *Ecol. Soc.* **2013**, *18*. [CrossRef]

43. M/s HCP Design, Planning and Management Pvt. Ltd (HPC). *Pune River Development Project: Concept Master Plan;* Pune Muncipal Corporation: Ahmedabad, India, 2016.

44. Pune Municipal Corporation. *Draft Development Plan for Pune City (Old Limit) 2007–2027;* Pune Municipal Corporation: Pune, India, 2007.

45. Pune Municipal Corporation. *Environmental Status Report 2015–16*; Department, E., Ed.; Pune Municipal Corporation: Pune, India, 2016.

46. Smart City Cell. *Reimagining Pune: Mission Smart City*; Pune Muicipal Corporation; Mckinsky and Company: Pune, India, 2016.

47. Voyants Solutions. *City Development Plan*; Pune Municipal Corporation: Pune, India, 2012.

48. Pune Municipal Corporation. *Pune Municipal Corporation Budget 2016–17*; Pune Municipal Corporation: Pune, India, 2016.

49. HCP Design, Planning and Management Pvt. Ltd. *Pune River Development Project Concept Masterplan*; Pune Municipal Corporation: Pune, India, 2016. Available online: https://pmc.gov.in/informpdf/green%20Pune/Riverfront_Website.pdf (accessed on 2 December 2016).

50. Pathak, T. *Consultations with Pune Municipal Corporation Officials*; Pathak, T., Ed.; Pune Municipal Corporation: Pune, India, 2017.

51. Central Public Health and Environmental Engineering and Organisation. *Manual on Water Supply and Treatment*; Central Public Health and Environmental Engineering Organisation: New Delhi, India, 1999; Available online: http://cpheeo.nic.in/Watersupply.htm (accessed on 15 September 2016).

52. Carmin, J.; Dodman, D.; Chu, E. *Urban Climate Adaptation and Leadership*; Organisation for Economic Co-operation and Development (OECD) Publishing: Paris, France, 2013.

53. Chu, E.; Anguelovski, I.; Roberts, D. Climate adaptation as strategic urbanism: Assessing opportunities and uncertainties for equity and inclusive development in cities. *Cities* **2017**, *60*, 378–387. [CrossRef]

54. Horton, B.; Digman, C.J.; Ashley, R.M.; Gill, E. *BeST (Benefits of SuDS Tool) W045c BeST—Technical Guidance. Release Version 3*; Construction Industry Research and Information Association (CIRIA): London, UK, 2016.

55. Buuren, A.; Driessen, P.; Teisman, G.; Rijswick, M. Toward legitimate governance strategies for climate adaptation in the Netherlands: Combining insights from a legal, planning, and network perspective. *Reg. Environ. Chang.* **2013**, *14*, 1021–1033. [CrossRef]

56. Atelier Groenblauw. Urban Blue-Green Grids for Sustainable and Resilient Cities. 2017. Available online: http://www.urbangreenbluegrids.com/about/introduction-to-green-blue-urban-grids/ (accessed on 20 July 2017).

water

MDPI

Article

Spatial Evaluation of Multiple Benefits to Encourage Multi-Functional Design of Sustainable Drainage in Blue-Green Cities

Richard Andrew Fenner

Centre for Sustainable Development, Department of Engineering, Cambridge University,
Cambridge CB2 1PZ, UK; raf37@cam.ac.uk

Received: 4 October 2017; Accepted: 4 December 2017; Published: 7 December 2017

Abstract: Urban drainage systems that incorporate elements of green infrastructure (SuDS/GI) are central features in Blue-Green and Sponge Cities. Such approaches provide effective control of stormwater management whilst generating a range of other benefits. However these benefits often occur coincidentally and are not developed or maximised in the original design. Of all the benefits that may accrue, the *relevant dominant benefits* relating to specific locations and socio-environmental circumstances need to be established, so that flood management functions can be co-designed with these wider benefits to ensure both are achieved during system operation. The paper reviews a number of tools which can evaluate the multiple benefits of SuDS/GI interventions in a variety of ways and introduces new concepts of *benefit intensity* and *benefit profile*. Examples of how these concepts can be applied is provided in a case study of proposed SuDS/GI assets in the central area of Newcastle; UK. Ways in which SuDS/GI features can be actively extended to develop desired relevant dominant benefits are discussed; e.g., by (i) careful consideration of tree and vegetation planting to trap air pollution; (ii) extending linear SuDS systems such as swales to enhance urban connectivity of green space; and (iii) managing green roofs for the effective attenuation of noise or carbon sequestration. The paper concludes that more pro-active development of multiple benefits is possible through careful co-design to achieve the full extent of urban enhancement SuDS/GI schemes can offer.

Keywords: Sustainable Drainage Systems; green infrastructure; Blue-Green cities; Sponge Cities; multiple benefits; benefit intensity

1. Introduction

The purpose of this paper is to examine how the multiple benefits arising from urban drainage systems that incorporate elements of green infrastructure can be pro-actively considered during the design and selection process. This is important if flood mitigation benefits and the potential for wider positive impacts are to be simultaneously achieved.

In urban areas Sustainable Drainage Systems (SuDS) may form part of a network of Green Infrastructure (GI) that provide elements of the natural environment through provision of areas of vegetated open space. Such water management and flood mitigation interventions can deliver a wide range of other benefits which contribute to environmental, economic and social improvements in urban areas. For example, SuDS/GI solutions can contribute to ecosystem services and provide other benefits such as cost effective public health measures [1,2]. For example, it has been suggested the frequency of exposure to natural settings and the extent of vegetation cover may be important in improving human well-being, specifically in the realm of mental health [3]. A recent UK briefing note identifies the following benefits for urban green infrastructure: urban temperature regulation, improving air quality, reducing surface water flooding, reducing pollution of urban water courses, noise reduction, carbon storage, habitat and urban biodiversity, pollination and provision of community food [4]. Other benefits which can be important drivers for developers adopting SuDS/GI infrastructure are uplift in property value adjacent to Blue-Green space and enhanced amenity and recreational opportunities for residents [5]. CIRIA have summarised these multiple benefits as having the potential to provide direct economic value, amenity or aesthetic value, added environmental and ecosystem value, and social value [6]. Some of these benefits (such as pollination) accrue to the immediately adjacent urban areas. Others (such as carbon storage) may contribute less directly, but are exerted over longer time periods at regional, national and international scales.

Trends in urban drainage design have moved towards source control measures using natural drainage processes through vegetated above ground features such as swales, storage ponds and rain gardens. Therefore these wider potential benefits are being more actively discussed in the discourse around Blue-Green cities in Europe [7], Sponge Cities in China [8] and Water Sensitive Urban Design in Australia [9] (as defined in Table 1).

Table 1. Comparison of terms commonly used in different geographic regions.

Term	Short Definition
Blue Green Cities	A Blue-Green city aims to recreate a naturally oriented water cycle while contributing to the amenity of the city by bringing water management and green infrastructure together [10].
Sponge Cities	A Sponge city refers to sustainable urban development including flood control, water conservation, water quality improvement and natural eco-system protection in which a city's water system operates like a sponge to absorb, store, infiltrate and purify rainwater and release it for reuse when needed [8].
Water Sensitive Urban design	Water Sensitive Urban Design encompasses all aspects of integrated urban water management and is significant shift in the way water related environmental resources and water infrastructure are considered in the planning and design of cites, at all scales and densities [11].

The benefits that are most relevant in a given location will depend on the contextual environmental, social and economic characteristics of the immediate area being served. It has been suggested that communities are more likely to support green interventions if they enhance cultural services [12]. However, trade-offs may have to be made between the level of provisions of different benefits [13,14]. It should also be accepted that not all green infrastructure interventions have positive outcomes. Some negative effects potentially may arise, for example from the introduction of pests and diseases and perceived increases in crime in areas with increased vegetation cover [15]. Also inequalities in terms of the section of a community to which benefits accrue have been recognised based on factors such as age, gender and socioeconomic status [16].

In the UK co-ordinated strategies to foster the multi-functional benefits of green infrastructure solutions are being encouraged [4]. However only 48% of local authorities have current green space strategies (reducing from 76% in 2014) and even less have green infrastructure strategies for the creation of new sites. Nevertheless cities such as Birmingham, Manchester and London are producing green infrastructure plans as adaptation measures to manage flooding, climate change and other risks. But maximising the wider range of benefits achievable from green infrastructure requires the right physical interventions in the right place. However, little collaboration has been found between the statutory and non-statutory players in GI planning and the group of organisations managing SuDS. In particular the complex interdependencies between the urban components and stormwater management using SuDS/GI have not yet effectively been translated into governance interactions between the variety of responsible agencies and wider stakeholders [15].

Often urban infrastructure components are analysed on different scales and frequently this is done independently to other components. This is highlighted by the fact that in the UK there are no policies/documents concerning the integration of both SuDS and GI elements. Furthermore there are no planning rules on urban forms which are based on the available evidence for ecosystem service provision [4]. Fundamentally this results from a failure to act at a systems level, partly arising from the administrative arrangements reflecting discrete responsibilities for different types of asset groups. The lack of effective UK legislation, such as a UK SuDS Approval Body (SAB) as intended in the 2010 Flood and Water Management Act, and little regulatory control on SuDS design, construction and maintenance, are also cited as significant barriers that hamper progress [17].

Despite the known benefits of SuDS/GI to urban drainage and flooding problems, widespread implementation of such schemes has been restricted by uncertainties regarding hydrological performance and service delivery, and a lack of confidence that decision makers and communities will accept, support and take ownership of such infrastructure [18]. Barriers to sustainable water management include scientific, technological/technical, institutional, legal, managerial, political, monetary and social, but with social-institutional barriers typically posing the greatest hindrance. Sheer resistance to change represents a particularly relevant socio-institutional barrier for more widespread uptake of SuDS/GI [10]. Surprisingly in the UK these approaches are still regarded as "novel", despite the many successful schemes that are in operation.

A study of barriers to the uptake of SuDS/GI in Newcastle, UK was undertaken through a series of semi-structured interviews with 19 professional stakeholders, with the 5 most prevalent barriers being classified as socio-political [10]. Respondents were asked how these barriers could be overcome. The most prevalent response (63%) was the promotion of multifunctional space and identification of the multiple benefits. Respondents commented that:

> "If [a SuDS scheme] is similar in cost but you can highlight all these other benefits that link with our sustainability target, our air quality improvements, then straight away they would be happy to sign off as a project".

And

> "... you need to think about the multi-functional use of space!".

This raises the challenge of how the multiple benefits of SuDS/GI can be identified and quantified (and monetised), particularly with respect to the benefits that accrue during the everyday non-flood state (e.g., carbon sequestration, habitat and amenity improvements). This state has been defined as the first domain in the 3 Point Approach to urban flood management, reflecting day-to-day performance when there is little or no rain [19]. The optimal approach to creating multifunctional and resilient infrastructure may be through changing how SuDS/GI are planned and delivered towards greater collaborative working and co-funding from organisations and departments with a wide range of remits [10]. Most significantly this multi-functionality should be acknowledged at the institutional level.

The remainder of this paper examines how the potential multifunctionality of SuDS/GI assets can be actively embraced in the design process, first by drawing on a range of tools to identify where the most important benefits lie and then discussing examples of how specific modifications to different kinds of installation can be made to deliver such enhanced performance.

2. Propositions to Address Benefits in SuDS/GI Design

The foregoing provides a background context to three propositions which will be explored in detail in the remainder of this paper. These are listed as follows:

2.1. Proposition 1: Multiple Benefits from SuDS/GI Assets Emerge Coincidentally

Systematic procedures for pro-actively developing drainage infrastructure to deliver a specified range of predetermined desirable multiple benefits are rare. For example in the UK, in developing Sustainable Drainage Systems little if any direct attention is given to the planning of wider benefits that can be expected to be associated with a mature installation. The focus is on achieving the primary function of capturing, storing and infiltrating rainfall and reducing surface runoff flows. Thus multiple benefits may emerge sporadically, coincidentally or even accidentally from SuDS/GI interventions.

GI approaches to achieve general urban greening and improving visual and recreational amenity is at the heart of water sensitive urban design. However consideration of wider potential performance is beyond the remit of most SuDS designers who may wish to see specific evidence of possible benefits, before explicitly addressing how these may be actively delivered. A range of benefit evaluation tools exist to support suitable choices at the design stage, and these are briefly reviewed Section 3, where new concepts of benefit intensity and benefit profile are also introduced. A stronger justification for implementing SuDS/GI approaches can be made if a simultaneous and explicit case can be demonstrated for the wider multiple benefits they can achieve, rather than relying that they may occur fortuitously once an asset has been installed.

2.2. Proposition 2: Agreement Is Needed on Relevant Dominant Benefits

It is clear from many studies that not all benefits occur simultaneously, and some benefits may preclude the establishment of others [14,20]. For example, in a flood plain restoration scheme in the Johnson Creek watershed in Portland Oregon, recreational benefits were found to be in direct conflict with habitat and biodiversity benefits [21]. Therefore not all the multiple benefits from SuDS/GI are realisable all of the time. What is important is to establish a smaller subset of benefits that provide the greatest benefit uplift, uniquely agreed for each site's location specific circumstances. The uplift in each benefit category should be referenced against initial prevailing condition states based on local and contextual environmental performance which already pre-exist in a given location. In this way the **dominant** benefits which provide the greatest discernible change can be established. Just as importantly the principal beneficiaries of this change can be identified. A methodology for achieving this is outlined later in this paper.

A parallel activity would be to engage with local stakeholders to explore priorities and preferences through engagement in Learning and Action Alliances (e.g., [22]) and other forum. It is then a simple step to weight the benefit categories following a systematic survey of the communities and users who will be affected by the flood mitigation proposal. This modification to a neutrally weighted analysis, could help define which are the **relevant** benefits in each location.

2.3. Proposition 3: Systems Should Be Co-Designed to Deliver Both a Flood Control Function and Wider Multiple Benefits

The purpose of refining the benefit evaluation is to create a positive feedback loop into the asset design (Figure 1). The notion of circular or iterative design is not new, and the adjusted or refined design can thus be informed by an understanding of the relevant dominant benefits that might be achieved from a proposed SuDS solution. This will allow the principle drainage function of SuDS/GI to be co-designed with the delivery of the subset of relevant dominant benefits pertaining to site specific contexts and circumstances. In this way the important benefits may be actively enhanced through initial design modifications and the specification of subsequent maintenance and management strategies. Examples of how such refinements could be made are discussed towards the end of this paper in Section 5.

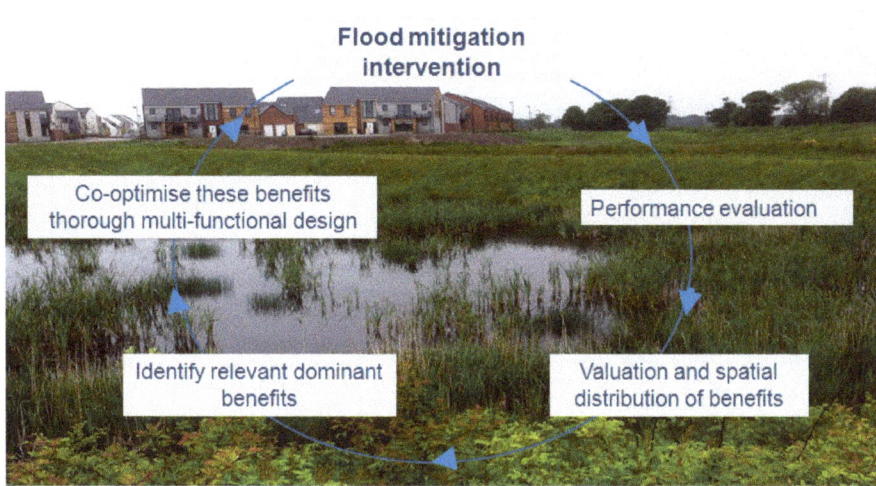

Figure 1. A positive feedback loop to reinforce multi-functional design of relevant dominant benefits in urban drainage systems that incorporate elements of green infrastructure's (SuDS/GI).

3. Benefit Evaluation Tools

A range of methodologies and tools have been proposed to establish and quantify the range of benefits discussed earlier. Such tools are required to evaluate both the primary functions of SuDS/GI and their wider benefits so these kind of assets can be proactively co-ordinated to achieve multi-functionality. Much useful analysis can be found in the literature on ecosystem services to help improve the understanding of multiple benefits which emerge from aspects of urban greening. For example it is now understood that the multi-functional and multi-scale nature of green urban infrastructure can lead to interactions between these benefits at different scales [23].

Methodologies which have been used include life cycle assessment [24,25], scenario planning [26], expert knowledge [27] and modelling using tools such as i-Tree and EnviroAtlas [28,29]. Jayasooriya. and Ng reviewed 20 modelling tools for managing urban flooding and the economics of GI practices. They noted there is a trend for recent tools to include a Geographic Information System (GIS) interface, and called for more tools to incorporate the range of ecosystem services and social benefits which SuDS/GI practices can provide [30]. Recently techniques have begun to emerge which represent the spatial distribution of ecosystem services by normalising each benefit value to a common scale and aggregating these spatially in a GIS platform [31–33].

A novel example of this kind of spatio-temporal approach to Blue-Green infrastructure design and modelling has been produced by the Urban Europe Green/Blue Cities Project. This is based on an integral, multi-scalar and adaptive Blue-Green infrastructure design methodology [34]. By focusing specifically on Kiruna in Sweden and Zwolle in the Netherlands a set of design principles were developed to support the delivery of 'nature' in cities. This was considered at a range of scales to utilise the wider opportunities 'natural systems' can bring to create and sustain better places for people and ecosystems. A decision support system was provided combining both spatial and temporal modelling so the additional value that Blue-Green infrastructure delivers to the quality of urban living could be advanced in terms of identifying the innovation practices required. In both cities a water sensitive urban structure was modelled showing priorities of investment and transformation of the built environment. The range of ecosystem services brought by Blue-Green infrastructure were considered crucial components for the mainstreaming of SuDS/GI solutions.

Some tools provide a cost benefit analysis through a structured assessment to help quantify and evaluate the monetary value of each benefit. However, Spengenberg and Settele have cautioned that monetised results are context and method dependent, and can fail to reflect complex interactions between benefits, and where value transfer is adopted large uncertainties can accrue [35].

A recent tool to help make the business case for SuDS has been developed in the UK by CIRIA (Benefits of SuDS Tool: BeST) (http://www.susdrain.org/resources/best.html) which enables a financial assessment of related Blue-Green solutions to be made [6]. BeST provides a structured approach to evaluating a wide range of benefits and is based on a simple structure, that begins with a screening and qualitative assessment to identify those benefits to evaluate further. On completion of the evaluation, the tool provides a series of graphs and charts based on Ecosystem Services (ESS) and Triple Bottom Line (TBL) (criteria. (N.B. Ecosystem Services criteria relate to supporting, provisioning, regulating, and cultural services); Triple Bottom Line criteria relate to environmental, social and economic domains). BeST doesn't account for every individual circumstance or site specific nuance, relying on the user to contextualise the scheme into the framework of the tool, nor does it provide a detailed distributional analysis of where the benefits will accrue. Application of the tool in the UK has revealed that many of the more substantial benefits from SuDS/GI are not those related to flood, drought or water quality, but those which accrue to amenity and human health, especially for distributed interventions across a catchment [5]. Reinforcing the propositions made earlier in this paper, these authors conclude that opportunities to maximise wider benefits to society are lost where the approach to 'drainage' uses the traditional paradigm of just getting the water away quickly and safely.

To help determine which are the **relevant** benefits to actively pursue in a given location, an Adaptation Support Tool has been developed by van de Ven [36] This can be used to inform stakeholders on the adaptation options they have and where and how effective and costly these would be. The tool contains 72 blue, green and grey adaptation measures to choose from and provides an effectiveness appraisal based on performance metrics detailing storage, peak flow reduction, heat stress reduction, water quality effects and calculates the costs of each adaptation measure and of the total package. Through using a touch-table with a professional facilitator, participants become fully aware of the ecological, recreational and social effects and co-benefits of specific adaptation solutions. The spatial, social, ecological and economic consequences of proposed measures are extensively discussed and decisions on the value of a range of co-benefits of Blue-Green infrastructure can be collectively generated to guide specific designs.

4. Spatial Evaluation Using Concepts of Benefit Intensity and Benefit Profile

Benefit evaluation can be particularly effective when it is used to assess the benefit uplift in a specific location against a chosen initial condition state, or against an alternative (e.g., piped) drainage strategy. This can help make a rational comparison between the magnitude of benefits that can be achieved and hence can identify those benefits that are worth actively pursuing and enhancing through multi-functional design. Furthermore if the spatial distribution of benefits are analysed then it is possible to identify where (and to whom) the overall aggregate benefit occurs and how this is valued by community stakeholders and beneficiaries can be established.

In developing such a methodology Hoang, Fenner and Skandarian emphasise 5 key points [20]:

(i) The general impacts of SuDS and associated Blue-Green infrastructure may include both benefits and dis-benefits and these are context-dependent.

(ii) Trade-offs may occur between different benefit categories for a range of installation types, and these in turn are also influenced by specific local contexts and prevailing background environmental conditions.

(iii) Many of the added benefits are incremental and need to be assessed in relation to the level of similar services which pre-existed in each specific location (i.e., in relation to an initial condition state), and the rate they develop over time.

(iv)   It can be difficult to compare directly across non-commensurate benefit categories to establish the relative contribution that each can deliver in specific of local circumstances, individual site characteristics and against preferences of local communities.

(v)   Benefits can accrue to different stakeholder groups other than the asset owner and these are distributed across spatial scales from local to regional to global.

The benefits or dis-benefits arising from SuDS and Blue-Green Infrastructure can span various categories. Each of these categories can be characterised by a different metric. This makes direct comparison of the relative performance of benefits difficult. Furthermore, the absolute values of each metric offer little insight to the contribution of each benefit, in relation to pre-existing services/benefits in a specific location. Normalising the benefits according to each location's conditions and context helps identify the extent of the benefit uplift relative to an initial condition state and this allows a balanced comparison across the categories to be made.

One approach has been to normalise ecosystem services on a scale of 0 to 10, by representing the ecosystem service relative to the maximum value achieved in any single discrete land parcel across the site of interest [32]. This technique of normalising as a ratio to the maximum value attainable in the site of interest is transferable to representing benefit categories across grid squares in a GIS platform.

Benefit Profile and Benefit Intensity

Building on this approach, Morgan and Fenner [37] have introduced concepts of benefit profile and benefit intensity, described here as follows:

A Benefit Profile can be created by comparing the normalised impacts across multiple benefit categories to reflect the relative extent of the benefit uplift and the total area affected in each category, combined with the effectiveness/potential of each benefit achieved.

The normalised values could be computed as simple ratios, depending on the purpose of the evaluation, such as:

- Benefits from SuDS/GI infrastructure solutions: benefits from alternative (piped?) solutions.
- Benefits after installation of an asset: pre-existing benefits before the installation of an asset.
- Potential benefits at some future time: realised benefit occurring now, etc.

The benefit profile can be presented as a bubble chart which incorporates the concept of how much potential benefit is realised by the scheme (Figure 2). This allows both the *size* of each benefit to be compared together with the *extent* to which it is accruing. Some benefits might be restricted to small local areas only whilst others may be more widely distributed across a wider area of the overall site. Each benefit category can also be represented as a dis-benefit, so for example, habitat creation may also raise concerns over the possibility of harbouring disease vectors.

The total benefit score for each benefit category is the 'headline' figure accumulated across all grid squares and identifies if there has been an improvement or not beyond the initial condition state. This is plotted against the area over which a benefit is generated so the extent over which each benefit has influence can be compared. The size of each bubble then represents the effectiveness of the benefit related to its specific location. A small bubble suggests that in the specific location being considered there is only a low potential for improvement in that benefit category. For example, despite extensive planting of vegetated surfaces and the increase in leaf area which provide the potential for significant pollutant trapping, little uplift in pollution mitigation would be seen if air quality was initially already in a good state, (with no pollutants in the atmosphere to trap as the site is far from high densities of road traffic). A large bubble approaching the full size indicates that all the potential benefit is being realised across the site. From this representation, it becomes clear which are the dominant benefits worthy of actively pursuing through the design process. In the hypothetical example shown in Figure 2 this would be flood mitigation (being the primary function of the drainage installation) and access to green space.

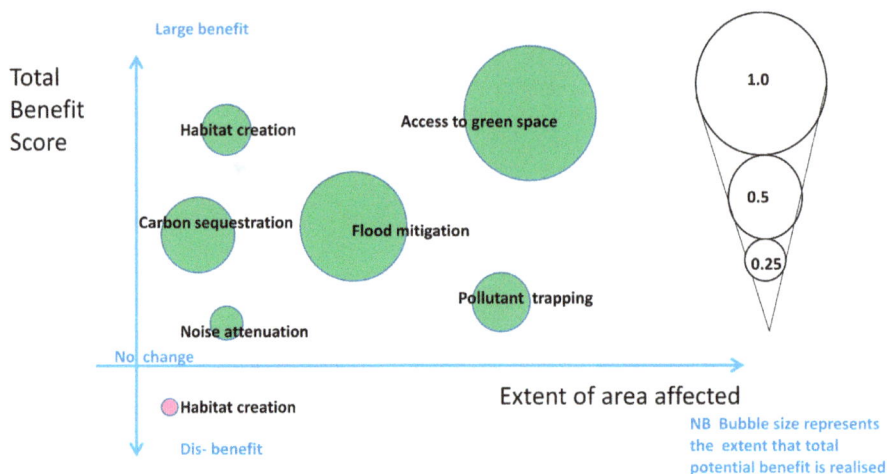

Figure 2. Idealised Benefit profile represented as a bubble chart.

Benefit Intensity is the spatial distribution of the unweighted accumulation of multiple benefits emanating from a SuDS/GI intervention(s) over the adjacent urban area. It is generated by the summation of all the normalised benefit scores in each grid square. As a further refinement, a weighted spatial distribution can also be developed to reflect the way local communities and stakeholders prioritise the importance/significance of each benefit category.

A set of benefit evaluation tools for ArcGIS 10.X has been developed, from which the above parameters can be calculated. The tools are designed to work with Ordnance Survey data commonly available in UK cities and are available at available at: http://www.bluegreencities.ac.uk/bluegreencities/index.aspx. The grid square resolution over which the calculations are performed can vary from 1 m × 1 m to 30 m × 30 m, depending on the purpose and scope of the analysis. These tools have been used to generate benefit intensities and benefit profiles arising from flood mitigation interventions in a number of cities including Portland, USA [20] and Newcastle, UK [37]. In the latter location, six benefit categories were modelled to demonstrate the approach, including access to greenspace, air pollution trapping (PM_{10}), carbon sequestration, flood damage avoided, habitat size and noise attenuation.

Simple computer models were used to identify the extent of each of these benefits in each grid square which were pre-existing under an initial condition state. The calculations were then repeated to compute the level of benefits which could be achieved after the introduction of the SuDS/GI asset(s). Scores in each grid square were then generated for each benefit based on the relative change between these two states.

Raster maps were produced for each benefit category showing both the before and after condition preceding and following the SuDS/GI drainage intervention. For access to green space a raster cost distance calculation was performed twice for all green space greater than 500 m^2 and for all green space between 50–500 m^2. Pollutant trapping was based on the dispersion of PM_{10} taking into account distance from the road network, with different land covers absorbing or blocking PM_{10} particles. Carbon sequestration was also based on twelve land cover types with higher values assigned to woodland and lower rates to grassland, and man-made surfaces assigned a zero score. Flood damage avoided was computed utilising the CityCat Urban Flood Model (developed by Newcastle University, Newcastle, UK) and multiple return periods were considered to find the annualised damage risk. Different depth-damage curves were constructed for each land use and building class data. Habitat size considered the value of having connected green spaces and the GIS tool identified clusters of interconnected green space and estimated the number of species supported by each cluster, with large areas calculated as supporting species density of around $500/m^2$ while smaller areas could be as low as $15/m^2$. Finally noise attenuation was based on traffic noise emanating from the road network and a cost-distance calculation was performed for each of seven noise levels based on attenuation with distance, terrain and surface material and obstacles.

Each of the six individual benefit distributions can then be added together for an overall multiple benefit distribution, to portray the Benefit Intensity across the area where the SuDS/GI infrastructure is currently or intended to be installed.

This approach provides two advantages to existing methods of benefit evaluation. Firstly, the spatial distribution of benefits is considered allowing beneficiaries to be identified in the adjacent communities. Secondly, the benefits are context specific. For example, a small addition of green space in an area with little green space may derive a greater benefit uplift than a large addition of greenspace in an area that is already very green; such as in areas having the presence of pre-existing park land.

The primary output of the tool are maps of benefit intensity identifying where the greatest improvements have been achieved for each benefit category individually and for all the benefits collectively. These maps can be interrogated for information on specific locations or presented as aggregate benefit profiles so that the relevant dominant benefits in a given location can be clearly identified. Once these benefits are established (and confirmed as having merit and meeting the needs of the local community) design and site management practices for the SuDS installation can be adjusted or modified where necessary to optimise the benefits which have been identified as having the most impact.

An example is shown in Figure 3 of how the distribution of individual benefits arsing from a range of proposed interventions (including green roofs, permeable paving, swales, and street trees) in the urban core of Newcastle city centre can be cumulatively depicted as an overall benefit intensity. A further step would be to weight the different benefit categories on the basis of stakeholder preferences following a systematic survey of the communities and users affected by the proposals. This modification to the neutrally weighted analysis presented here could help confirm which dominant benefits should be optimised through initial design and subsequent maintenance and management strategies.

Figure 4 shows the benefit profile for the Newcastle urban core. In this case, a small but effective increase in carbon sequestration is achieved from the increase in natural surfaces such as green roofs. There is a moderate increase in access to greenspace, as green roofs were assumed to be publically inaccessible and so did not contribute to the access benefit score. Finally, a moderate reduction in flood damage is generated, mostly attributable to the performance of the proposed swale along St James' Boulevard. (N.B. The minor flood dis-benefit shown in the benefit profile is caused by a slight misalignment between the flood modelling and the land use map).

The approach is flexible such that other benefit layers could be easily added provided they can be represented by a spatial component. The method can also be adapted to incorporate more advanced models of the benefit categories considered. The method can identify beneficiaries and which parts of a community receives the benefits and can be used to contrast multiple design options rapidly. In doing so it would be suitable for use in the early stages of designing a SuDS/GI scheme. The benefit intensity maps produced may also be useful during consultation with stakeholders and for wider urban planning purposes.

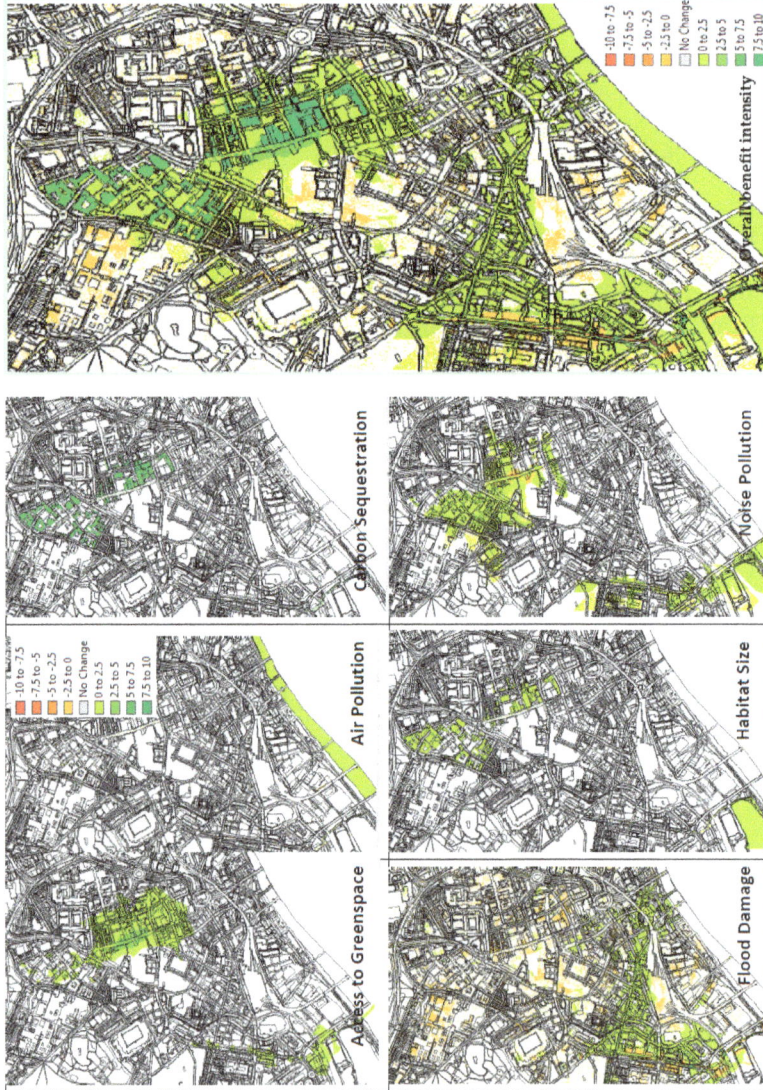

Figure 3. Spatial distribution of individual benefits and accumulation into an overall benefit intensity of SuDS interventions in Newcastle urban core.

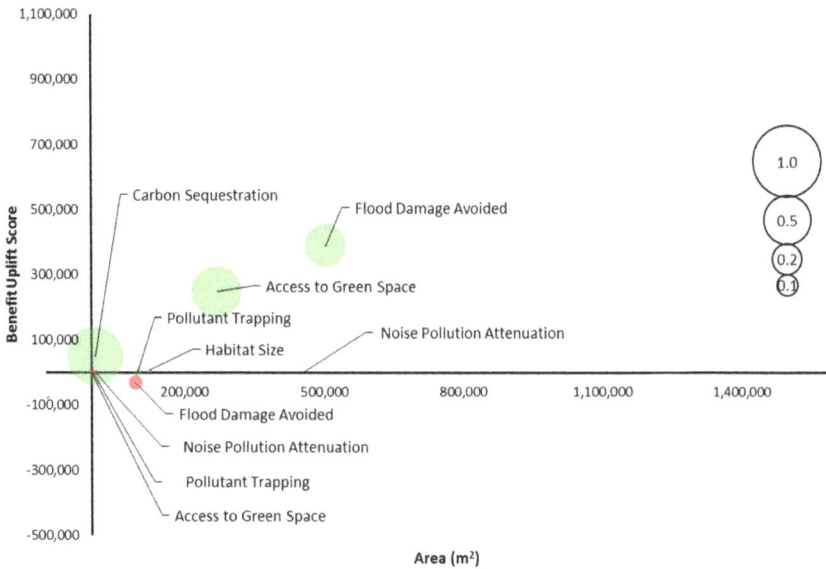

Figure 4. Benefit profile for Newcastle urban core (after Morgan and Fenner [37]).

5. Discussion of Design Modifications for Typical SuDS/GI Assets

Proposition 3 suggested that stormwater management systems should be co-designed to deliver both a flood control function and wider multiple benefits. The first step in achieving this requires the identification of a focussed subset of site specific **relevant dominant benefits**, which can be selected by using the tools described above. The next step is to co-design systems from this broader perspective to achieve these desirable multiple benefits. The nature of the design modifications which may need to be made to achieve this are discussed below for three typical SuDS/GI schemes: green streets, bioswales and green roofs.

5.1. Green Streets: Pollutant Trapping Benefits

Green streets are frequently designed for stormwater management purposes, but they also have the potential to reduce air pollution through trapping on leaf surfaces. In areas that suffer from significant levels of PM_{10} air pollution due to the close proximity to high levels of road traffic, this potential for pollutant trapping may be considered as an important additional benefit. However the installation of street trees must be carefully designed and maintained to enhance, and not worsen, local air quality. Studies have shown that adding vegetation, especially trees, to hot-spots of air pollution (e.g., poorly ventilated areas along streets) can *increase* pollutant concentrations by further restricting air flow and exchange [38]. Conroy has shown the strategic placement and design of green infrastructure is necessary to enhance PM_{10} capture and to prevent the possibility of green infrastructure inadvertently worsening air quality [39]. When trees are planted within street canyons, careful management of the crowns, including pruning, is recommended to promote air flow [40]. Recommendations include: adequate space between the tree crowns and nearby walls, tree height shouldn't exceed the height of nearby buildings, and planting fewer trees is better [38,41]. For green streets within street canyons, designers may want to limit the number of trees in bioretention cells and suspended pavement systems and focus on using stormwater measures based on herbaceous vegetation (e.g., bio-retention cells and bio-swales). The close proximity of the vegetation to the PM_{10} emission sources maximizes the efficiency of interception and deposition; also, the smaller size of vegetation in installations such as street gardens does not significantly hinder air flow within the street [42].

Care must be taken not to limit the upward flow and dispersion of air and pollutants, therefore it is vital for stormwater and air quality researchers to determine the performance, optimal design, and best placement of vegetated stormwater control measures within green streets with regard to reducing PM_{10} concentrations [39].

5.2. Marginal Additional Investments: Leverage Multiple Benefits

Actively seeking a wider set of enhanced benefits can sometimes be achieved by small marginal additional investments [43]. By applying the BeST tool in Roundhay, Ashley showed that the value provided by a SuDS scheme was mostly from benefits other than those directly managing stormwater [5]. 43% of the financial benefits for Roundhay accrued from increased property values. Analysis of the potential for leveraging further benefits for Roundhay by marginal additional investments showed that increasing the numbers of trees from 250 to 1000, increases residents' views over green space for up to 500 people, and health benefits increased proportionally to some 25% compared with the original 6% of total value. Nevertheless, for Roundhay, flood risk reduction provided the largest continuing benefit longer-term, as the amenity benefits no longer were considered part of the overall benefits after the first year, as these benefits accrue immediately following construction [5]. Furthermore the scheme has to be adequately maintained to prevent a decline of amenity values in the future.

5.3. Swales and Linear SuDS/GI Installation: Greenspace Connectivity Benefits

Urban green infrastructure usually consists of a fragmented mosaic of diverse smaller patches of vegetation with different uses within which large areas such as parks are set. Connecting such areas together in the form of urban corridors can have substantial benefits both in amenity terms and for the linear movement of wildlife. With some additional thought at the design stage some linear SuDS/GI schemes such as swales might be marginally extended to provide the mechanism for linking these patches together. The new development in Camborne, Cambridgeshire, UK linked three villages, Upper Camborne, Greater Camborne and Lower Camborne, and was built on 40 ha of former arable land. The master plan for the development protected the remaining areas of semi-natural habitat and linked them together with a variety of new green infrastructure, such as woodland planting, meadows, lakes, amenity grassland, playing fields, allotments and formal areas [4]. The scheme includes 10 miles of new hedgerows and green infrastructure which accounts for 240 ha of the development such that the levels of biodiversity on the site are now higher than when it was arable farmland.

A useful tool to quantify this ability to join up areas of green space is the Integral Index of Connectivity (*IIC*) defined by Pascual-Hortal L. and Sauroa S., and is expressed as [44]:

$$IIC = \frac{\sum_{i=1}^{n} \sum_{j=1}^{n} \frac{a_i \times a_j}{1+nl_{ij}}}{A_L^2}$$

where: a_i = Area of each habitat patch
nl_{ij} = Topological distance between patches *I* and *j*
A_L = Area of Study

Within the Camborne development is Lamb Drove, a small area devoted to the introduction of a range of SuDS measures, which have been systematically monitored over a long period of time. Applying this index to the Lamb Drove installations showed a benefit uplift in the connectivity of 1.34, from the SuDS interventions in the immediate area [45]. Co-designing SuDS/GI schemes to provide both a flood mitigation benefit and improved green space connectivity is a clear example of where multi-functionality can be achieved.

5.4. Green Roofs: Noise Attenuation and Carbon Sequestration Benefits

Hoang and Fenner showed the benefits that can accrue from green roofs will be dependent on how the installations are managed [13]. These benefits can include hydrological performance regarding runoff retention and storage, carbon sequestration, noise attenuation, building and urban cooling, pollutant trapping, food production, biodiversity and social and amenity benefits. However different functions of a green roof can prevail under different conditions. The main determining factors of these conditions are the heat and water budget of the roof. The functions have strong dependencies on soil moisture and characteristics of the vegetation and the soil cover. Also different green roof functions prevail under different physical conditions and that green roof functions may occur in phases or continuously. In particular, high soil moisture can diminish the functioning of noise attenuation, and limits the capacity for stormwater uptake but could enhance photosynthesis, carbon sequestration and heat exchange of the roof. The nature and density of planting are other important variables which can achieve different outcomes. So again an element of co-design is required to achieve the relevant dominant benefits required.

6. Conclusions

Ashley R. et al. have made the following observation [5]:

"If the added benefits to society that can be provided by using SuDS/GI are to be realised then the 'drainage' perspective needs to be supplanted by one in which Blue-Green infrastructure is seen as a starting point for the planning of land use and property development or renovation; i.e., it is no longer 'drainage' that we need; rather what can be achieved is too valuable to be so pigeonholed".

This paper has argued that more needs to be done to pro-actively consider the wider outcomes and benefits that can provide multiple positive gains from the implementation of SuDS/GI solutions for urban drainage and urban flood mitigation. This will involve a systematic appraisal of which benefit opportunities offer the most potential gain in a given location, and meet the expectations and desires of local stakeholders. Such an approach will avoid such benefits being left to occur coincidentally or by chance and so maximise the case for the adoption of SuDS/GI assets by addressing the aims and goals across a wider urban planning framework. To make this case sufficiently convincing the tools that are being developed to evaluate the multiple benefits from SuDS/GI need to be regularly applied so that these **relevant dominant benefits** can be identified for each location and context specific circumstances. This can drive more effective co-design so that areas of truly multi-functional green space can be realised in so many urban locations.

Acknowledgments: This research was performed as part an interdisciplinary project programme undertaken by the Blue-Green Cities (BGC) Research Consortium [46]. The BGC is funded by the UK Engineering and Physical Sciences Research Council under grant EP/K01366 1/1, with additional contributions from the Environment Agency and Rivers Agency (Northern Ireland) and National Science Foundation. Associated research data can be downloaded from http://dx.doi.org/10.17639/nott.59. The views expressed are those of the author and so do not necessarily represent those of the wider consortium. The contributions of Lan Hoang, Malcolm Morgan and Emily O'Donnell are acknowledged in developing the ideas reported here.

Conflicts of Interest: The author declares no conflict of interest.

References

1. Van den Bosch, M.; Nieuwenhuijse, M. No time to lose; green ten cities now. *Environ. Int.* **2017**, *99*, 343–350. [CrossRef] [PubMed]
2. Shanahan, D. Toward improved public health outcomes from urban nature. *Am. J. Public Health* **2015**, *105*, 470–474. [CrossRef] [PubMed]

3. Cox, D.T.C.; Shanahan, D.F.; Hudson, H.L.; Plummer, K.E.; Siriwardena, G.M.; Fuller, R.A.; Anderson, K.; Hancock, S.; Gaston, K.J. Doses of Neighbourhood Nature: The benefits of mental health of living with nature. *Biosciences* **2017**, *67*, 147–155.
4. Wentworth, J. *Urban Green Infrastructure Ecosystem Services*; POSTbrief from UK Parliamentary Office of Science and Technology: London, UK, 26 July 2017.
5. Ashley, R.; Horton, B.; Digman, C.; Gersonius, B.; Shaffer, P.; Bayliss, A.; Bacchin, T. It's not drainage any more: It's too valuable. In Proceedings of the 14th International Conference on Urban Drainage, Prague, Czech Republic, 10–15 September 2017.
6. Ciria. *RP 993: Demonstrating the Multiple, Benefits of SudS—A Business Case*; Ciria: London, UK, 2015. Available online: http://www.susdrain.org/resources/best.html (accessed on 25 May 2016).
7. Lawson, E.; Thorne, C.; Ahilan, S.; Allen, D.; Arthur, S.; Everett, G.; Fenner, R.; Glenis, V.; Guan, D.; Hoang, L.; et al. Delivering and evaluating the multiple flood risk benefits in Blue-Green cities: An interdisciplinary approach. In *Flood Recovery, Innovation and Response IV*; Proverbs, D., Brebbia, C.A., Eds.; WIT Press: Poznan, Poland, 2014; Volume 184, pp. 113–124.
8. Li, H.; Ding, L.; Ren, M.; Li, C.; Wang, H. Sponge City Construction in China: A Survey of the Challenges and Opportunities. *Water* **2017**, *9*, 594. [CrossRef]
9. Wong, T.H.F. (Ed.) *Stormwater Management in a Water Sensitive City*; Melbourne Australia Centre for Water Sensitive Cities: Clayton, Australia, 2012; ISBN 978-1-921912-01-6.
10. O'Donnell, E.; Lamond, J.E.; Thorne, C.T. Recognising barriers to implementation of Blue-Green Infrastructure: A Newcastle case study. *Urban Water J.* **2017**, *14*, 964–971. [CrossRef]
11. Fletcher, T.; Shuster, W.; Hunt, W.; Ashley, R.; Butler, D.; Arthur, S.; Trowsdale, S.; Barraud, S.; Semadeni-Davis, A.; Bertrand-Krajewski, J.-L.; et al. SuDS, LID, BMPOs, WSUD and more—The evolution and application of terminology surrounding urban drainage. *Urban Water J.* **2015**. [CrossRef]
12. Andersson, E.; Tengo, M.; McPhearson, T.; Kremer, P. Cultural ecosystem services as a gateway for improving urban sustainability. *Ecosyst. Serv.* **2015**, *12*, 165–168. [CrossRef]
13. Hoang, L.; Fenner, R.A. Systems interactions of green roofs in Blue-Green cities. In Proceedings of the 13th International Conference on Urban Drainage, Sarawak, Malaysia, 8–12 September 2014.
14. Langemeyer, J.; Gómez-Baggethun, E.; Haase, D.; Scheuer, S.; Elmqvist, T. Bridging the gap between ecosystem service assessments and land use planning through Multi-Criteria Decision Analysis (MCDA). *Environ. Sci. Policy* **2016**, *62*, 45–56. [CrossRef]
15. Hoang, L.; Fenner, R.A. System interactions of flood risk strategies using Sustainable Urban Drainage Systems and Green Infrastructure. *Urban Water J.* **2015**. [CrossRef]
16. Von Dohren, P.; Haase, D. Ecosystem disservices research: A review of the state of the art with a focus of cities. *Ecol. Indic.* **2015**, *52*, 490–497. [CrossRef]
17. Ashley, R.M.; Walker, A.L.; D'Arcy, B.; Wilson, S.; Illman, S.; Shaffer, P.; Woods-Ballard, P.; Chatfield, C. UK sustainable drainage systems past present and future. In *Proceedings of the Civil Engineering*; ICE Publishing: London, UK, 2015; Volume 168, pp. 125–130.
18. Thorne, C.R.; Lawson, E.C.; Ozawa, C.; Hamlin, S.L.; Smith, S.L. Overcoming uncertainty and barriers to adoption of Blue-Green infrastructure for urban flood risk management. *J. Flood Risk Manag.* **2015**, 1–13. [CrossRef]
19. Fratini, C.F.; Geldof, G.D.; Kluck, J.; Mikkelsen, P.S. Three Points Approach (3PA) for urban flood risk management: A tool to support climate change adaptation through transdisciplinarity and multifunctionality. *Urban Water J.* **2012**, *9*, 317–331. [CrossRef]
20. Hoang, L.; Fenner, R.A.; Skandarian, M. Towards A New Approach for Evaluating the Multiple Benefits of Urban Flooding Management Practice. *J. Flood Risk Manag.* **2016**. [CrossRef]
21. Fenner, R.A.; Hoang, L. Institutional perspectives on impacts and benefits of an urban flood management project, Portland Oregon. In Proceedings of the 14th International Conference on Urban Drainage, Prague, Czech Republic, 10–15 September 2017.
22. Lawson, E. Learning and Action Alliances: Defining and Establishing (Fact Sheet). Available online: http://www.bluegreencities.ac.uk/documents/laas-defining-and-establishing.pdf (accessed on 6 December 2017).
23. Demuzere, M.; Orru, K.; Heidrich, O.; Olazabal, E.; Geneletti, D.; Orr, H.; Bhave, A.G.; Mittal, N.; Felie, E.; Faehnle, M. Mitigating and adapting to climate change: Multi-functional and multi-scale assessment of green infrastructure. *J. Environ. Manag.* **2014**, *146*, 107–115. [CrossRef] [PubMed]

24. Casal-Campos, A.; Fu, G.; Butler, D. *The Whole Life Carbon Footprint of Green Infrastructure: A Call for Integration*; NOVA-TECH: Lyon, France, 2013.
25. Flynn, K.M.; Traver, R.G. Green infrastructure life cycle assessment: A bio-infiltration case study. *Ecol. Eng.* **2013**, *55*, 9–22. [CrossRef]
26. Hilde, T.; Paterson, R. Integrating ecosystem services analysis into scenario planning practice: Accounting for street tree benefits with i-Tree valuation in Central Texas. *J. Environ. Manag.* **2014**, *146*, 524–534. [CrossRef] [PubMed]
27. Kopperoinen, L.; Itkonen, P.; Niemelä, J. Using expert knowledge in combining green infrastructure and ecosystem services in land use planning: An insight into a new place-based methodology. *Landsc. Ecol.* **2014**, *29*, 1361–1375. [CrossRef]
28. Kim, G.; Miller, P.A.; Nowak, D.J. Assessing urban vacant land ecosystem services: Urban vacant land as green infrastructure in the City of Roanoke, Virginia. *Urban For. Urban Green.* **2015**, *14*, 519–526. [CrossRef]
29. Pickard, B.R.; Daniel, J.; Mehaffey, M.; Jackson, L.E.; Neale, A. EnviroAtlas: A new geospatial tool to foster ecosystem services science and resource management. *Ecosyst. Serv.* **2015**, *14*, 45–55. [CrossRef]
30. Jayasooriya, V.M.; Ng, A.W.M. Tools for modelling of urban flood management and economics of green infrastructure practices: A review. *Water Air Soil Pollut.* **2014**, *225*, 2055. [CrossRef]
31. Dobbs, C.; Kendal, D.; Nitschke, C.R. Multiple ecosystem services and disservices of the urban forest establishing their connections with landscape structure and socio-demographics. *Ecol. Indic.* **2014**, *43*, 44–55. [CrossRef]
32. Lauf, S.; Haase, D.; Kleinshmit, B. Linkages between ecosystem services provisioning, urban growth and shrinkage—A modelling approach assessing ecosystem service trade-offs. *Ecol. Indic.* **2014**, *42*, 73–74. [CrossRef]
33. Turner, K.G.; Odgaard, M.V.; Bocher, P.K.; Dalgaar, T.; Svenning, J.-C. Bundling ecosystem services in Denmark: Trade-offs and synergies is a cultural landscape. *Landsc. Urban Plan.* **2014**, *125*, 89–104. [CrossRef]
34. Kuzniecow, B.T.; Ashley, R.; van Timmeren, A.; La Fleur, F.; Blecken, G.; Viklander, M. Green blue infrastructure design: An advanced spatio temporal model. In Proceedings of the 14th International Conference on Urban Drainage September (Paper ICUD-0587), Prague, Czech Republic, 10–15 September 2017.
35. Spengenberg, J.; Settele, J. Precisely incorrect? Monetising the value of ecosystem services. *Ecol. Complex.* **2010**, *7*, 327–337. [CrossRef]
36. Van de Ven, F.; Brolsma, R.; McEvoy, S. An adaptation support tool for climate resilient urban planning: Lessons learned from applications. In Proceedings of the 14th International Conference on Urban Drainage, Prague, Czech Republic, 10–15 September 2017.
37. Morgan, M.; Fenner, R.A. Spatial evaluation of the multiple benefits of sustainable drainage systems. *Inst. Civ. Eng. J. Water Manag.* **2017**. [CrossRef]
38. Gromke, C.; Ruck, B. Pollutant concentrations in street canyons of different aspect ratio with avenues of trees for various wind directions. *Bound. Layer Meteorol.* **2012**, *144*, 41–64. [CrossRef]
39. Conroy, K.; Hunt, W.; Kumar, P.; Anderson, A. Air quality considerations for stormwater green street design. In Proceedings of the 14th International Conference on Urban Drainage, Prague, Czech Republic, 10–15 September 2017.
40. Wania, A.; Bruse, M.; Blond, N.; Weber, C. Analysing the influence of different street vegetation on traffic-induced particle dispersion using microscale simulations. *J. Environ. Manag.* **2012**, *94*, 91–101. [CrossRef] [PubMed]
41. Vos, P.E.; Maiheu, B.; Vankerkom, J.; Janssen, S. Improving local air quality in cities: To tree or not to tree? *Environ. Pollut.* **2013**, *183*, 113–122. [CrossRef] [PubMed]
42. Litschke, T.; Kuttler, W. On the reduction of urban particle concentration by vegetation—A review. *Meteorol. Z.* **2008**, *17*, 229–240. [CrossRef]
43. Horton, B.; Digman, C.; Ashley, R.M. Benefit of SuDS Tool (BeST) User Manual (Release 3). 2016. Available online: http://www.susdrain.org/resources/best.html (accessed on 6 December 2017).
44. Pascual-Hortal, L.; Sauroa, S. Comparison and development of new graph-based landscape connectivity indices: Towards the prioritization of habitat patches and corridors for conservation. *Landsc. Ecol.* **2006**, *21*, 959–967. [CrossRef]

45. Navaro, D. Evaluation of Sustainable Urban Drainage Systems Using a Novel Benefit Intensity Methodology. Master's Thesis, Cambridge University, Cambridge, UK, 2014. Unpublished.
46. BGC (Blue–Green Cities). 2017. Available online: http://www.bluegreencities.ac.uk/index.aspx (accessed on 6 December 2017).

water

MDPI

Article

Factors Influencing Stormwater Mitigation in Permeable Pavement

Chun Yan Liu and Ting Fong May Chui *

Department of Civil Engineering, The University of Hong Kong, Hong Kong, China;
liuchunyanpeter@yahoo.com.hk
* Correspondence: maychui@hku.hk; Tel.: +852-2219-4687

Received: 6 November 2017; Accepted: 11 December 2017; Published: 18 December 2017

Abstract: Permeable pavement (PP) is used worldwide to mitigate surface runoff in urban areas. Various studies have examined the factors governing the hydrologic performance of PP. However, relatively little is known about the relative importance of these governing factors and the long-term hydrologic performance of PP. This study applied numerical models—calibrated and validated using existing experimental results—to simulate hundreds of event-based and two long-term rainfall scenarios for two designs of PP. Based on the event-based simulation results, rainfall intensity, rainfall volume, thickness of the storage layer and the hydraulic conductivity of the subgrade were identified as the most influential factors in PP runoff reduction. Over the long term, PP performed significantly better in a relatively drier climate (e.g., New York), reducing nearly 90% of runoff volume compared to 70% in a relatively wetter climate (e.g., Hong Kong). The two designs of PP examined performed differently, and the difference was more apparent in the relatively wetter climate. This study generated insights that will help the design and implementation of PP to mitigate stormwater worldwide.

Keywords: sponge city; sustainable drainage system; porous pavement; stormwater management; low impact development; SWMM

1. Introduction

Natural ground cover allows infiltration [1,2], which stabilises groundwater supply and purifies storm water. Surface runoff is reduced to a relatively low level, because most rainfall events are not large enough to fully saturate the soil. However, urbanisation worldwide has replaced the natural land cover with infrastructure, parking lots, streets and sidewalks, which has greatly disrupted the natural hydrological cycle [2,3]. Excessive use of impervious cover has created problems, such as flooding, water quality degradation and riverbank erosion [1–5]. To mitigate these problems, permeable pavement (PP) has been developed and used worldwide for many years as a form of stormwater control. It aims to mimic natural landscape features, allowing runoff to infiltrate and percolate through its pores, where it is filtered by different layers of aggregates before entering the soil and the environment. PP has been proven to reduce the volume and peak flow of runoff, and delay the timing of peak flow [6–8]. In rainfall events of less than 5 mm, the peak runoff was reduced by 95% and the total rainfall volume decreased by 90% [7]. Studies in North Carolina showed that the curve number—an index that indicates the amount of runoff generated from a given land use for a given storm—of PP, ranged from 45 to 89, while that of normal impervious pavement topped the list at 98 [6]. Some other notable benefits of PP include mitigation of the heat island effect and aesthetic enhancement [2].

A typical structure of PP consists of a surface layer, such as permeable concrete (PC), permeable asphalt (PA), permeable interlocking concrete pavers (PICP), or concrete grid pavers (CGP). Underneath the surface layer lies the storage layer, which usually consists of open-graded aggregates, such as ASTM No. 57 [9], or some other size of gravel or sand. The bottommost layer is the subgrade, which is basically the in-situ soil and can thus be of a wide range of conditions. Perforated drainage pipes can be installed within the storage layer or the subgrade. They are used when the subgrade cannot effectively drain away the infiltrated water [10] and are recommended when the infiltration rate of the in-situ soil is less than 15 mm/h [4].

PP can be effective in reducing runoff, but its performance is highly dependent on rainfall conditions. With certain PP designs, both rainfall intensity and rainfall depth are positively related to runoff [11]. Rainfall and runoff, however, do not have simple and linear relationships. Rainfall intensity has an influence when the infiltration rate of the surface layer is less than the rainfall intensity. In such conditions, PP cannot absorb more stormwater, even though the deeper layer of PP has not 'filled up'. However, runoff is sometimes governed by rainfall depth, as the entire PP can be fully saturated with water flooding up to the surface, producing runoff [11]. Several studies concluded that the amount of runoff is more dependent on rainfall intensity than rainfall depth [8,11–13]. In other words, a longer rainfall event, of lower intensity, is less likely to trigger runoff than a shorter event, of high intensity. However, which factor (rainfall intensity or depth) has most influence depends on the actual rainfall pattern and the PP design. An experiment on a PP-refined parking lot showed that when rainfall intensity was greater than 50 mm/h, PP performed, more or less, like impervious asphalt [14]. When the rainfall intensity was reduced to 25 mm/h, PP resulted in a 40–50% reduction in runoff volume and about a 90% reduction in peak runoff [12].

PP, in wetter antecedent conditions, gives a significantly larger amount of runoff and a higher peak runoff [10–12]. In the absence of preceding rainfall, PP initiated runoff 5–10 min later and produced a flood peak at least 5% smaller than the values achieved, with a preceding rainfall of 119 mm/h [9]. It was also demonstrated in a laboratory study, that an initially air-dried PP could retain 55% of a 1 h, 15 mm storm, while merely 30% of the storm was retained by an initially wet PP [15]. Furthermore, experiments on PP showed that, even when a second rainfall event was less intense than a first, runoff produced by the later event was greater [12].

The design of PP is also important. As expected, the thicker the storage layer, the more effective the PP is in reducing runoff [16]. In terms of surface layers, PC and PICP were found to better reduce runoff and to have higher infiltration rates than CGP and PA [6,12,17]. However, the difference was only observable during extreme rainfall (e.g., >130 mm/h) [4]. Furthermore, the steeper the PP surface, the more runoff it generates [18]. A slope of around 1% is ideal, and it should not exceed 5% for structural stability [4]. Furthermore, PP relies on subgrade soil or an underdrain to drain away the water within the storage layer. High conductivity of the subgrade soil [19] or the presence of an underdrain prevents the PP from being 'filled up', resulting in less runoff [18].

Some factors influential to PP runoff have been thoroughly investigated (e.g., antecedent level, surface infiltration rate), and the general relationships between runoff and certain design parameters (e.g., storage thickness, subgrade conductivity and surface slope) have been experimentally confirmed. However, their relative importance has seldom been examined. This study therefore used numerical modelling to simulate the hydrologic performance of PP under different rainfall conditions and PP designs, with the aim of identifying the relative importance of each factor to runoff reduction. In addition to short-term design storm based simulations, long-term simulations that made use of tens of years of actual rainfall data were performed, to analyse the long-term runoff reduction of PP.

2. Materials and Methods

2.1. Model Development and Calibration

This study applied the Storm Water Management Model version 5.0 (SWMM 5, United States Environmental Protection Agency, Washington, DC, USA), which generates dynamic rainfall–runoff simulation models for single and continuous rainfall events. SWMM 5 can model the hydrologic performance of several types of low impact development controls, including PP. The numerical models developed in this study were based on rainfall-runoff experiments carried out at the Mentougou experimental station of the Beijing Hydraulic Research Institute [11]. Four treatments were used in these experiments (A, B, C and D). Numerical models were only developed for Treatments A, C and D, as Treatments B and C were very similar in structure. Both Treatments A and C were permeable pavements with permeable concrete blocks at the surface. Treatment A used only sand as the storage layer, while treatment C used concrete without sand plus aggregate as the storage layer. Treatment D was impervious, and acted as the control. The cross-sections of the selected permeable pavements and the control are listed in Table 1.

The experimental parameters from [11] were used, or common values were assumed if the data were missing, as listed in Table 2. However, there were two exceptions. First, the area of each treatment in [11] was only 6 m^2, while the area simulated in the models was 6000 m^2, to mimic the size of a real-life PP system (e.g., a large parking lot). Higher numerical precision was achieved with the SWMM, because the outputs are always rounded to 2 decimal points, irrespective of the PP size. The simulated PP was assumed not to receive runoff from the surrounding area. Second, the layer of concrete sub-base without sand in Treatment C was regarded as an aggregate, with a thickness scaled according to its relative porosity, and was modelled with the actual aggregate sub-base underneath as a single layer. This simplification caused minimal computational error, because the concrete layer was thin.

For calibration, the models were subjected to a relatively large rainfall event (a rainfall intensity of 59.36 mm/h for one hour). After fine-tuning some of the parameters, the calibrated models were subjected to the same rainfall intensity, but for two hours, for validation. The accuracy of the model was evaluated by computing the correlation coefficient (R-squared) and the Nash–Sutcliffe model efficiency coefficient (NSE), using the experimental and simulated runoff of the PP.

Table 1. Cross-sections of the selected permeable pavements and the control (Treatments A, C and D) used for modelling. The numbers in brackets indicate the thickness in mm. Information extracted from [11].

Treatment A	Treatment C	Treatment D
Porous concrete block paving (60)	Porous concrete block paving (60)	Impervious surfaces (60)
Sand sub-base (200)	Concrete sub-base without sands (50)	Concrete (150)
Subgrade (640)	Aggregate sub-base (200)	Subgrade (740)
	Subgrade (740)	

Table 2. Input parameters of the pavement types for calibration and validation.

	Treatment A			Treatment C			Treatment D	
Surface and pavement blocks	Manning's roughness	0.1 [1]	Surface & pavement blocks	Manning's roughness	0.06 [1]	Surface	Manning's roughness	0.1 [1]
	Slope (%)	1.5 [1]		Slope (%)	1.35 [1]		Slope (%)	1.5 [1]
	Thickness (mm)	60		Thickness (mm)	60			
	Conductivity (mm/h)	2840		Conductivity (mm/h)	2840		Conductivity (mm/h)	0
Sand	Thickness (mm)	200	Aggregate	Effective thickness (mm) [2]	240		Not applicable	
	Porosity	0.3 [1]		Void ratio	0.2 [1]			
Subgrade soil	Conductivity (mm/h)	0.821	Subgrade soil	Conductivity (mm/h)	0.821			

Note: [1] Obtained from calibration. Others extracted directly from [11]. [2] Effective thickness combined the thickness of the actual aggregate layer and the scaled thickness of the concrete layer.

2.2. Event-Based Simulations

Based on the calibrated and validated models, eight parameters were varied, to understand their influence on runoff generation, as summarised in Table 3. At least two representative values for each of the eight parameters were examined, producing a total of 384 runoff hydrographs for Treatments A and C. The rainfall parameters were selected to represent large and extreme events, with reference to rainfall conditions in different continental regions. For example, in Hong Kong, China, rainfall of 60 mm/h that lasts for 1 and 2 h, represents events with return periods of less than 2 and 5 years, respectively, while rainfall of 120 mm/h that lasts for 1 and 2 h, represents events of return periods of 20 and more than 100 years, respectively [20]. However, in Auckland, New Zealand, and New York City, USA, rainfall of 60 mm/h that lasts for 1 h already has a return period of more than 10 years, while the other simulated events have return periods that are generally more than 100 years [21,22]. A consistent comparison of performance could be performed across scenarios, because only the parameters of interest were varied and this variation was systematic. For each hydrograph, three performance indicators—peak runoff, time to runoff initiation and runoff duration—were extracted. The correlation of each of the eight parameters with each of the performance indicators was analysed.

Table 3. Range of parameter values for Treatments A and C during the event-based simulations.

Parameters	Symbol	Input Values
Rainfall intensity (mm/h)	i	60, 120
Rainfall volume (mm)	V	60, 120, 240
Rainfall duration (h)	t	1, 2
Slope of pavement surface (%)	s	1, 3
Manning's roughness	n	0.05, 0.1
Thickness of storage layer (mm)	d	75, 200, 400
Porosity of storage layer	ϕ	0.35, 0.50 (Treatment A) 0.26, 0.46 (Treatment C)
Hydraulic conductivity of subgrade soil (mm/h)	k	2, 25

2.3. Long-Term Simulations

To understand the average performance of PP over a prolonged period, long-term simulations (i.e., 10 years or longer) were also performed, using the models of Treatments A, C and D. Long-term continuous simulations, unlike event-based simulations, can factor in antecedent conditions. Real rainfall data was used as the model inputs to better capture the actual rainfall characteristics (e.g., temporal pattern, including consecutive rainfall events). As rainfall conditions influence PP performance, two locations of different rainfall characteristics, namely, Hong Kong, China and New York, US, were selected. Twelve years of data (1 January 2003–31 December 2014) from Lok Ma Chau Station (LMC) in Hong Kong and around 45 years of data (4 April 1969–31 December 2013) from John. F. Kennedy International Airport (JFK) in New York were used. The LMC station represents a sub-tropical climate, whereas the JFK station represents a warm temperate climate, with annual average rainfall depths of 1780 mm and 1074 mm, respectively. Apart from having a higher annual average rainfall, the year-to-year fluctuation is also larger at LMC than at JFK. The two stations have similar fractions (i.e., 39.9% and 38.9%) of small to medium rainfall events, with a depth of less than 5 mm. However, 7.2% and 3.0% of rainfall events at LMC are between 100 and 200 mm and above 200 mm, respectively, while the corresponding figures for JFK are only 0.6% and 0.1%. Therefore, there are larger event-to-event variations at LMC than at JFK. The implementation of PP is at an early stage in Hong Kong and there is no existing design guideline. The PP models of Treatments A and C are rather generic, and the parameters are generally within the recommended values for New York [23,24].

3. Results and Discussion

3.1. Calibration and Validation

The simulated results generally matched the experimental results. Figure 1 shows the calibration and validation of Treatment A as an example, demonstrating a good match in the time to initiation, peak, rising and falling limbs. The R-squared and the NSE for all three treatments are summarised in Table 4, and were, in general, very high. The R-squared and the NSE of Treatment C were lower during calibration, but were still very high during validation. Overall, the models were found to be well calibrated and validated, and thus appropriate for use in the event-based and long-term simulations.

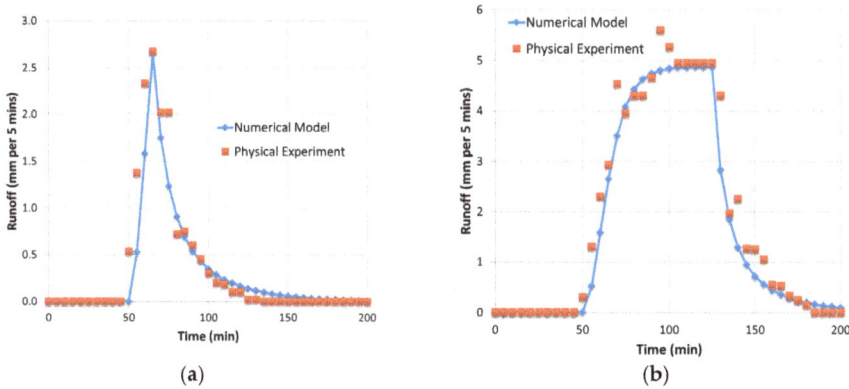

Figure 1. (a) Calibration and (b) validation of Treatment A.

Table 4. Accuracy of calibration and validation.

Process	Treatment	R-Squared	Nash–Sutcliffe Model Efficiency Coefficient
	A	0.91	0.88
Calibration	C	0.62	0.51
	D	0.96	0.95
	A	0.97	0.97
Validation	C	0.96	0.95
	D	0.97	0.97

3.2. Event-Based Simulations

The correlations between each of the eight parameters and each of the three performance indicators are given in Table 5. The parameters for Treatments A and C correlated in a similar pattern, with all three performance indicators.

In terms of rainfall characteristics, rainfall intensity, rainfall volume and rainfall duration were all correlated. Rainfall intensity and rainfall volume were correlated with all three performance indicators, while rainfall duration governed runoff duration. As suggested in the introduction, rainfall intensity generally played a more dominant role in runoff generation than rainfall volume, which was confirmed by the slightly higher correlations between peak runoff and time to runoff initiation with rainfall intensity compared to their correlations with rainfall volume. Rainfall duration, however, did not correlate with these two performance indicators. The general conclusion was, therefore, that runoff from PP is mostly governed by rainfall intensity, followed by rainfall volume and then by rainfall duration.

Among the five design parameters, storage depth was the most influential factor, followed by the conductivity of the subgrade soil. Storage depth strongly governed the time to runoff initiation, and also influenced runoff duration. Its influence on peak runoff was similar to that of subgrade conductivity, which also correlated strongly with runoff duration. Based on these results, engineers should focus more on the selection and optimisation of the storage depth and subgrade conductivity in PP design, as the other factors do not matter as much. This is particularly true if the design rainfall is of a high volume (e.g., continuous rain), because both storage depth and subgrade conductivity enhances the subsurface storage capacity and the dissipation of water to restore that capacity. In contrast, surface parameters (e.g., surface slope and Manning's roughness) are more related to high-intensity events (e.g., short intense storms) but are not as influential.

Table 5. Correlation matrix for Treatments A and C.

Treatment	Parameters	Peak Runoff	Time to Runoff Initiation	Runoff Duration
A	i (mm/h)	0.73	−0.43	0.36
	V (mm)	0.71	−0.41	0.57
	t (h)	0.28	−0.16	0.46
	s	0.03	−0.01	−0.09
	n	−0.04	0.00	0.12
	d (mm)	−0.37	0.68	−0.50
	ϕ	−0.09	0.24	−0.16
	k (mm/h)	−0.25	0.16	−0.46
C	i (mm/h)	0.76	−0.46	0.34
	V (mm)	0.69	−0.44	0.57
	t (h)	0.22	−0.20	0.46
	s	0.03	−0.00	−0.10
	n	−0.03	−0.00	0.12
	d (mm)	−0.33	0.66	−0.50
	ϕ	−0.13	0.22	−0.18
	k (mm/h)	−0.27	0.13	−0.45

Note: Degree of association: strongly negative, fairly strongly negative, fairly strongly positive, strongly positive.

3.3. Long-Term Simulations

In the long-term simulation with the LMC data, Treatment A generally absorbed entire rainfall events of less than 10 mm, and partially reduced runoff in events larger than 10 mm. However, the reduction was not significant for events larger than 20 mm. Treatment C in general reduced runoff more significantly than Treatment A, effectively draining away events up to 20 mm, and the reduction was not significant, only in events larger than 40 mm. As in the event-based simulations, the two most influential design parameters were storage depth and subgrade conductivity. Treatments A and C had the same subgrade conductivity, and therefore the differences in their performance can be mostly attributed to the different storage depths of 200 and 240 mm. Both treatments generally reduced less runoff in wet antecedent conditions (i.e., preceding rainfall), as discussed in the introduction. However, it is challenging to determine the exact impact of the antecedent conditions on the two treatments from the simulations with long-term real rainfall data, because each rainfall event is unique, in terms of its size and temporal distribution, and also, the two treatments would have had different antecedent conditions for each event. Overall, similar observations were made in the JFK simulation.

For ease of comparison, the total volumetric reduction percentages were used to compare the foregoing results with the simulation results for Treatment D (the control). For LMC, throughout the 12 years of simulation, Treatment A and Treatment C, on average, gave volumetric reductions of 63 and 74%, respectively. For JFK, the two treatments reduced appreciably more runoff than at LMC, with volumetric reductions of 90% and 96% in the relatively drier climate of New York. The two treatments performed similarly in the relatively drier climate. It is therefore suggested that it is more important to optimise the design of PP in a location with a wetter climate like that of LMC. This recommendation is consistent with the conclusion drawn from the event-based simulations, that the more influential design parameters (storage depth and subgrade conductivity) should be more carefully selected if the design rainfall is of high volume.

In addition to analysing the aggregate performance across years, the performance during individual rainfall events was also analysed. Two storm events were considered to be independent if they were separated by a dry spell of 24 h [25]. Figure 2 shows the negative relationships between the volumetric runoff reduction percentages and rainfall depths for Treatments A and C. There was a 100% reduction at low rainfall depths, and an exponential decrease in reduction as the rainfall depth increased. The performance at smaller rainfall depths was similar for both LMC and JFK, while there was a lower volumetric runoff reduction at LMC during extreme events with high rainfall depths. Curves have been added to Figure 2a,b to indicate the general lower bounds of the volumetric runoff reduction percentages for rainfall depths greater than 28 and 51 mm, respectively, for Treatments A and C.

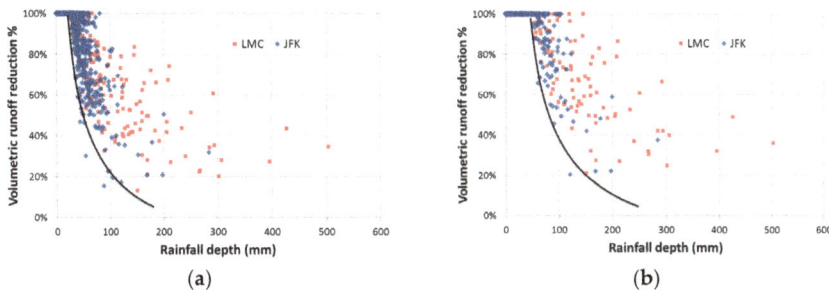

Figure 2. Relationships between percentage of volumetric runoff reduction and rainfall depth for Treatments A (**a**) and C (**b**).

Table 6 presents the percentage of rainfall events at each level of volumetric runoff reduction. Most rainfall events were fully absorbed by the PP (i.e., 76.7% and 91.4% of the events at LMC and JFK for Treatment A, and 93.2% and 97.6% of the events at LMC and JFK for Treatment C). Only a small percentage of events had volumetric runoff reductions of less than 40% (i.e., 3.2% and 0.4% of events at LMC and JFK for Treatment A, and 0.3% and 0.1% of events at LMC and JFK for Treatment C).

Table 6. Percentage of rainfall events at each level of volumetric runoff reduction.

Volumetric Runoff Reduction %	Lok Ma Chau Station (LMC)		John F Kennedy International Airport (JFK)	
	Treatment A	Treatment C	Treatment A	Treatment C
100%	76.7%	93.2%	91.4%	97.6%
85–100%	4.2%	2.1%	3.1%	1.0%
60–85%	9.2%	2.7%	3.8%	0.8%
40–60%	6.8%	1.6%	1.3%	0.4%
0–40%	3.2%	0.3%	0.4%	0.1%

4. Conclusions

This study applied numerical modelling to identify the relative importance of each rainfall and design parameter to runoff reduction in PP. Based on a correlation analysis, the runoff from PP was found to be mostly governed by rainfall intensity, followed by rainfall volume and rainfall duration. In terms of PP design, the depth of the storage layer and the hydraulic conductivity of the subgrade soil had the most influence on runoff reduction, while other parameters had a minimal influence.

The results of the long-term simulations indicated that PP performed significantly better in a relatively drier climate (New York), reducing nearly 90% of the runoff volume compared to 70% in a relatively wetter climate (Hong Kong). The two designs of PP examined (Treatments A and C) performed differently, and the difference was more apparent in the wetter climate, suggesting that detailed design and optimisation of PP are more worthwhile in this environment. Analysis of individual events, within the long-term simulations, showed that most events (over 75% for Treatment A in Hong Kong, and over 90% in the other cases) were fully absorbed by the PP. Only a small percentage of events (i.e., less than 5% for Treatment A in Hong Kong and less than 1% in the other cases) had volumetric reductions of less than 40%.

Overall, this study generated insights on the relative importance of various factors governing runoff reduction and the long-term hydrologic performance of PP, which will help the design and implementation of PP to mitigate stormwater worldwide. It covered several rainfall conditions and PP design parameters, and both event-based and long-term hydrologic performance of PP. The scenarios considered were representative, but by no means exhaustive. Future work should consider more parameters and wider ranges of parameter values. Further questions that should be examined include the impact of antecedent moisture on the performance of different designs of PP and the optimal PP design for different climatic conditions. Finally, this study could only examine hundreds of scenarios, because it used numerical modelling. Although the models were well calibrated and validated, the results are still based on the assumptions built into the SWMM.

Acknowledgments: The work described in this paper was funded by a grant from the Research Grants Council of the Hong Kong Special Administrative Region, China (Project No. HKU17255516).

Author Contributions: Both authors conceived and designed the numerical experiments; C. Y. Liu performed the numerical experiments and analyzed the data; both authors wrote the paper.

Conflicts of Interest: The authors declare no conflict of interest.

References

1. San Mateo Countywide Water Pollution Prevention Program. *Stormwater Technical Guidance*; Version 3.2.; San Mateo Countywide Water Pollution Prevention Program: San Mateo, CA, USA, 2013.
2. Interpave. *Guide to the Design, Construction and Maintenance of Concrete Block Permeable Pavements*, 6th ed.; Interpave, The Precast Concrete Paving & Kerb Association, Product Association of the British Precast Concrete Federation Ltd.: Leicester, UK, 2010.
3. Moglen, G. Hydrology and impervious areas. *J. Hydrol. Eng.* **2009**, *14*, 303–304. [CrossRef]
4. Toronto and Region Conservation Authority (TRCA); Credit Valley Conservation Authority (CVC). *Low Impact Development Stormwater Management Planning and Design Guide*; Version 1.0; TRCA: Toronto, ON, Canada; CVC: Mississauga, ON, USA, 2010.
5. Todeschini, S. Hydrologic and environmental impacts of imperviousness in an industrial catchment of Northern Italy. *J. Hydrol. Eng.* **2016**, *21*. [CrossRef]
6. Bean, E.Z.; Hunt, W.F.; Bidelspach, D.A. Evaluation of four permeable pavement sites in eastern North Carolina for runoff reduction and water quality impacts. *J. Irrig. Drain. Eng.* **2007**, *133*, 583–592. [CrossRef]
7. Brattebo, B.O.; Booth, D.B. Long-term stormwater quantity and quality performance of permeable pavement systems. *Water Res.* **2003**, *37*, 4369–4376. [CrossRef]
8. Hunt, B.; Stevens, S.; Mayes, D. Permeable pavement use and research at two sites in Eastern North Carolina. In *Global Solutions for Urban Drainage, Proceedings of the 9th International Conference on Urban Drainage, Portland, OR, USA, 8–13 September 2002*; ASCE: Reston, VA, USA, 2002.
9. Ferguson, B.K. *Porous Pavements*; CRC Press: Boca Raton, FL, USA, 2005.
10. Dreelin, E.A.; Fowler, L.; Carroll, C.R. A test of porous pavement effectiveness on clay soil during natural storm. *Water Res.* **2006**, *40*, 799–805. [CrossRef] [PubMed]
11. Hou, L.; Feng, S.; Hou, Z.; Ding, Y.; Zhang, S. Experimental study on rainfall-runoff relation for porous pavements. *Hydrol. Res.* **2008**, *39*, 181–190. [CrossRef]
12. Collins, K.A.; Hunt, W.F.; Hathaway, J.M. Hydrologic comparison of four types of permeable pavement and standard asphalt in eastern North Carolina. *J. Hydrol. Eng.* **2008**, *13*, 1146–1157. [CrossRef]
13. Valavala, S.; Montes, F.; Haselbach, L.M. Area-rated rational coefficients for Portland cement pervious concrete pavement. *J. Hydrol. Eng.* **2006**, *11*, 257–260. [CrossRef]

14. Rushton, B.T. Low impact parking lot reduces runoff and pollutant loads. *J. Water Resour. Plan. Manag.* **2001**, *127*, 172–179. [CrossRef]

15. Andersen, C.T.; Foster, I.D.L.; Pratt, C.J. Role of urban surfaces permeable pavements in regulating drainage and evaporation: Development of a laboratory simulation experiment. *Hydrol. Process.* **1999**, *13*, 597–609. [CrossRef]

16. Yoo, C.; Ku, J.M.; Jun, C.; Zhu, J.H. Simulation of infiltration facilities using the SEEP/W model and quantification of flood runoff reduction effect by the decrease in CN. *Water Sci. Technol.* **2016**, *74*, 118–129. [CrossRef] [PubMed]

17. Sañudo-Fontaneda, L.A.; Charlesworth, S.M.; Castro-Fresno, D.; Andres-Valeri, V.C.A.; Rodriguez-Hernandez, J. Water quality and quantity assessment of pervious pavements performance in experimental car park areas. *Water Sci. Technol.* **2014**, *69*, 1526–1533. [CrossRef] [PubMed]

18. Straet, F.; Beckers, E.; Degre, A. Hydraulic behavior of greened porous pavements: A physical study. In Proceedings of the 11th International Conference on Urban Drainage, Edinburgh, UK, 31 August–5 September 2008.

19. Chui, T.F.M.; Liu, X.; Zhan, W. Assessing cost-effectiveness of specific LID practice designs in response to large storm events. *J. Hydrol.* **2016**, *533*, 353–364. [CrossRef]

20. Geotechnical Engineering Office (GEO); Civil Engineering and Development Department; The Government of the Hong Kong Special Administrative Region. *New Intensity-Duration-Frequency Curves for Slope Drainage Design*; GEO Technical Guidance Note No. 30; GEO: Hong Kong, China, 2011.

21. Auckland Regional Council (ARC). *Stormwater Management Devices: Design Guidelines Manual*; ARC Technical Publication 10; ARC: Auckland, New Zealand, 2003.

22. Cornell University. Extreme Precipitation in New York and New England. An Interactive Web Tool for Extreame Precipitation Analysis. Available online: http://precip.eas.cornell.edu (accessed on 12 October 2017).

23. New York City Department of Transportation. *Street Design Manual*, 2nd ed.; New York City Department of Transportation: New York, NY, USA, 2015.

24. Basch, E.; Brana, R.; Briggs, E.; Chang, C.; Iyalla, A.; Logsdon, D.; Meinke, R.; Moomjy, M.; Price, O.D.; Sinckler, S. Roadmap for Pervious Pavement in New York City. In *A Strategic Plan for the New York City Department of Transportation*; The Earth Institute, Columbia University: New York, NY, USA, 2012.

25. Vandenberghe, S.; Verhoest, N.; Buyse, E.; Baets, B.D. A stochastic design rainfall generator based on copulas and mass curves. *Hydrol. Earth Syst. Sci.* **2010**, *14*, 2429–2442. [CrossRef]

water

MDPI

Article

Cross-Analysis of Land and Runoff Variations in Response to Urbanization on Basin, Watershed, and City Scales with/without Green Infrastructures

Jin-Cheng Fu [1], Jiun-Huei Jang [2,*], Chun-Mao Huang [3], Wen-Yen Lin [3] and Chia-Cheng Yeh [1]

[1] National Science and Technology Center for Disaster Reduction, Taipei 23143, Taiwan; jcfu@ncdr.nat.gov.tw (J.-C.F.); andrew@ncdr.nat.gov.tw (C.-C.Y.)
[2] Department of Hydraulic and Ocean Engineering, National Cheng Kung University, Tainan 70101, Taiwan
[3] Department of Urban Planning and Disaster Management, Ming Chuan University, Taoyuan 33348, Taiwan; usskiddddg-996@hotmail.com (C.-M.H.); wylin01@mail.mcu.edu.tw (W.-Y.L.)
*   Correspondence: jamesjang@mail.ncku.edu.tw; Tel.: +886-6-2757575 (ext. 63212)

Received: 13 December 2017; Accepted: 22 January 2018; Published: 26 January 2018

Abstract: Evaluating land and runoff variations caused by urbanization is crucial to ensure the safety of people living in highly developed areas. Based on spatial scales, runoff analysis involves different methods associated with the interpretation of land cover and land use, the application of hydrological models, and the consideration of flood mitigation measures. Most studies have focused on analyzing the phenomenon on a certain scale by using a single data source and a specific model without discussing mutual influences. In this study, the runoff changes caused by urbanization are assessed and cross-analyzed on three sizes of study areas in the Zhuoshui River Basin in Taiwan, including basin (large), watershed (medium), and city (small) scales. The results demonstrate that, on the basin scale, land-cover changes interpreted from satellite images are very helpful for identifying the watersheds with urbanization hotspots that might have larger runoff outputs. However, on the watershed scale, the resolution of the land-cover data is too low, and land-cover data should be replaced by investigated land-use data for sophisticated hydrological modeling. The mixed usage of land-cover and land-use data is not recommended because large discrepancies occur when determining hydrological parameters for runoff simulation. According to present and future land-use scenarios, the influence of urbanization on runoff is simulated by HEC-1 and SWMM on watershed and city scales, respectively. The results of both models are in agreement and show that runoff peaks will obviously increase as a result of urbanization from 2008 to 2030. For low return periods, the increase in runoff as a result of urbanization is more significant and the city's contribution to runoff is much larger than its area. Through statistical regression, the watershed runoff simulated by HEC-1 can be perfectly predicted by the city runoff simulated by SWMM in combination with other land/rainfall parameters. On the city scale, the installation of LID satisfactorily reduces the runoff peaks to pre-urbanization levels for low return periods, but the effects of LID are not as positive and are debatable for higher return periods. These findings can be used to realize the applicability and limitations of different approaches for analyzing and mitigating urbanization-induced runoff in the process of constructing a sponge city.

Keywords: urbanization; land-cover; land-use; surface runoff; low impact development; sponge city

---

1. Introduction

Along with urbanization, disaster mitigation becomes more and more important with the growth of population, industry, and economic activities. The increase of impervious area in cities reduces water infiltration and soil conservation, thereby increasing the likelihood of flooding and shortening the time to runoff peak [1,2]. Therefore, analyzing the surface runoff variation from land-cover and land-use changes pre- and post-urbanization is crucial when developing flood mitigation measures for urban areas. Research has determined that urbanization greatly influences surface runoff. Liu et al. [3] adopted 4-y rainfall data to calculate surface runoff generated by different land-use classes in the Alzette basin of the Steinsel River, Luxembourg, showing that the urban area occupied only 20% of the total area but generated runoff accounting for 29.3% of the total runoff, whereas other land-use classes (such as farmland and woodland) generated less proportions of runoff even though they cover a larger area. For the Buji River Basin in China, Shi et al. [4] indicated that the runoff peak increased by 13.4% when the urbanized area increased from 2.02 km^2 (3.5%) in 1980 to 33.58 km^2 (58.72%) in 2000.

Land-cover and land-use data are different in acquisition and usage. Land cover refers to the area ratio of a region covered by forests, agricultural land, water bodies, or other types of landscapes, whereas land use refers to how an area is used by people (e.g., building, conservation, park). Land-cover changes are usually determined by interpreting multi-period satellite images or remote-sensed data assisted by spatial analysis tools such as GIS (Geographical Information System) for map overlaying [5,6]; this is cost-effective and thus suitable for periodically updating hydrological parameters on a large scale [7–9]. Land-use changes cannot be determined from image processing but are determined from a filed survey or the investigation of environmental, economic, and social activities [10–12]. Thus, land-use data are often localized and related to policies regulating human activities on a small scale [13,14].

The runoff changes caused by land-use and land-cover variations can be estimated by applying hydrological models that incorporate geographical characteristics. Rainfall-runoff models, such as HEC-HMS (Hydrologic Engineering Center-Hydrologic Modeling System, Army Corps of Engineers, Washington, DC, USA) developed by Hydrologic Engineering Center of the US Army Corps of Engineers [15], are usually coupled with soil infiltration models in conjunction with the GIS to analyze runoff distributions on basin scales [16,17]. The precision of this approach is restricted by the resolution of basin size and can be subject to empirical judgement in translating soil conditions into hydrological parameters. To realize the runoff variation in areas much smaller than a river basin, such as a highly urbanized community or village, the hydrodynamic simulation of rainfall-runoff processes is necessary. A variety of hydrological models, such as the SWMM (Storm Water Management Model, Environmental Protection Agency, Washington, DC, USA), MIKE (Institute for Water and Environment, Horsholm, Denmark), and INFOWORKS (Innovyze, Conroe, TX, USA), are available for rainfall-runoff analysis in urban areas [18,19]. Among them, models such as SWMM [20] are prevalent for simulating runoff generated by single or continuous rainfall to address water quality and quantity problems related to an urban drainage system. For example, Guan et al. [21] used the SWMM to simulate hydrological changes when an area develops from rural to residential; Xu and Zhao [18] used SWMM to demonstrate that surface runoff increased by 3.5 times in the process of urbanization for a small catchment in Beijing (Liangshui River Basin, 131 km^2).

In the last two decades, the idea of "sponge city" has been put into practice in the United Sates [22], the United Kingdom [23], Canada [24], Australia [25], and other countries by designing low-impact, green, and sustainable urban drainage systems in highly urbanized areas. The purpose of sponge city is to reduce stormwater runoff, restore the ecological environment, and preserve natural resources through land-use changes that increase absorption, storage, and purification of rainwater [26–28]. To construct a sponge city, many researchers have suggested that the green infrastructure, also called LID (Low Impact Development) approach, is more efficient than traditional systems in reducing initial runoff [28–31]. The SWMM is recommended as the most suitable tool for evaluating the performance of LID facilities [32–35].

From the aforementioned literature review, runoff variations and mitigation in response to land-use and land-cover changes caused by urbanization have been widely studied. However, depending on the research objective, most studies have analyzed the phenomenon either on a large basin scale or on a small city scale by using specific land data and hydrological models. An overall study of the influences of urbanization on runoff across data, models, and scales has not been conducted. Therefore, certain questions require to be answered, for example, (a) is it appropriate to mix land-cover and land-use data for hydrological analysis? (b) how do hydrological models affect runoff results when the same land data are input? and (c) how do LID approaches benefits flood mitigation on different spatial scales? The purpose of this study is to assess and cross-analyze land and runoff changes on basin (large), watershed (medium), and city (small) scales in the Zhuoshui River Basin in Taiwan. On the basin scale, the historical land-cover and runoff changes are analyzed through the classification of satellite images and the application of the HEC-1 model. On the watershed scale, sub-basins with a higher degree of urbanization are pinpointed, and the results from the other two scales are accumulated and cross-analyzed. On the city scale, runoff changes under present and future land-use scenarios with and without LID measures are quantified through the application of SWMM. The procedure enables a comprehensive test of the applicability and limitation of relevant methodologies and provide valuable information for future studies.

2. Study Areas

On the basin scale, the Zhuoshui River Basin is selected as the study area. It is located in central Taiwan and has the longest mainstream of 187 km and the second largest catchment of 3157 km^2 in Taiwan. The Zhuoshui River originates from the Hehuan Mountain at an attitude of 3220 m and flows westward into the Taiwan Strait while converging with many tributaries that divide the entire river basin into 21 watersheds, as revealed by geographical processing. Among these watersheds, the watershed located at the intersection of the Zhoushui River and its tributary the Qingshui River, is selected as the medium-scale study area where an aggregation of artificial structures has been observed. Lying in the middle of the selected watershed, the Jushan urban planning district is further selected as the smallest study area for city-scale land and runoff analyses. Urban planning for Jushan was first formulated under the Japanese colonial rule in 1917, although the initial content was nothing but a simple road construction project covering 53.8 hectares. In 1979, the urban planning area was expanded to 418.1 hectares and has been subsequently revised a total of 13 times since then [36]. Figure 1 illustrates the study areas on basin scale, watershed, and city scales.

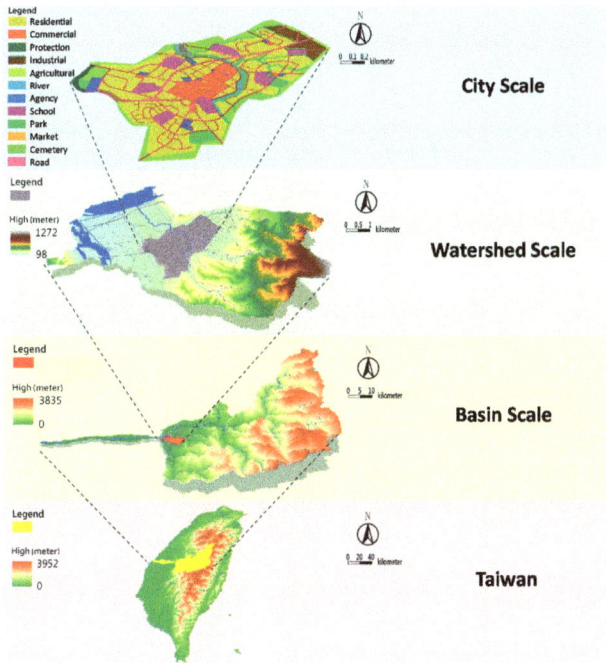

Figure 1. Study areas on different scales.

3. Methodology

In this study, based on the scales of the study areas, different data, models, and time periods are employed for analysis of land and runoff variations, which are listed in Table 1 and described in the following paragraphs. In the table, the runoff models are chosen according to their applicability. The SWMM is developed for simulating water quantity and quality in urban areas with drainage systems, in which the rainfall-runoff process is determined by methods suitable for small catchments. The HEC-1 (Hydrologic Engineering Center-1, Army Corps of Engineers, Washington, DC, USA) model is designed for surface runoff simulation on a river basin scale without considering drainage systems. Compared with SWMM, the HEC-1 has more options on the methods describing the processes of evaporation, infiltration, rainfall accumulation, and runoff routing.

Table 1. Study subjects on different scales.

Subject	Basin	Watershed	City
Land Data	Land-cover	Land-cover/Land-use	Land-use
Runoff Model	HEC-1	HEC-1	SWMM
Time Period	History	History/Present/Future	Present/Future/Future with LID

3.1. Basin Scale

To determine the land-cover change for the Zhoushui River Basin, historical ortho-images taken by the satellite FORMOSA II are adopted for analysis. The FORMOSA II is a sun-synchronous satellite located at an attitude of 891 km above the ground surface, and it provides high-resolution images (2 m for black/white images and 8 m for multi-spectral color images) twice a day for Taiwan area. The satellite images from 1998 to 2013 are collected and divided into smaller segments for interpreting and extracting land-cover characteristics. Using an object-oriented classification method (i.e., SVM,

Support Vector Machine [37]), 15 image indexes including NDVI (Normalized Difference Vegetation Index), SAVI (Soil Adjusted Vegetation Index), mean Brightness, mean Blue, mean Red, mean Green, mean NIR (mean Near Infrared), std. Blue, std. Red, std. Green, std. NIR (standard deviation Near Infrared), pix-base Blue, pix-base Red, pix-base Green, and R to RGB (Red, Green, Blue) are identified for classifying the land-cover into four types: water, built-up land, bare land, and vegetation. The classification results are then displayed by the area transfer matrix to investigate land-cover changes between each period [38].

The urbanization level is measured by determining whether an aggregation of artificial structures is present in the Zhoushui River Basin through Spatial Autocorrelation Analysis (SAA) [39] and Spatial Hotspot Analysis (SHA) [40]. The SAA uses the Moran's I and Z-score [41] as indexes to estimate the significance of homogeneity in structure distributions. If an area has a value of Moran's I > 0 with a Z-score > 1.96, the hypothesis of structure aggregation will be accepted under 5% of significance level. Once the SAA confirms the pattern of structure aggregation, the SHA employs the G_i index defined by Getis and Ord [42] to determine the hotspots of these artificial structures. The more the G_i exceeds 0, the higher the structure aggregation level in an area.

After completing the interpretation of land-cover change, the HEC-1 model, developed by the US Army Corps of Engineers [43], is employed for runoff simulation on the basin scale for different periods. Being a lumped model, the HEC-1 divided the whole basin into various watersheds connected to each other by channel networks, in which the runoff is transported in the forms of overland flow and channel flow depending on topographical characteristics. The watershed divisions are determined by processing DEM (Digital Elevation Model) data on a GIS platform. The CN (Curve Number) method proposed by US Soil Conservation Service (SCS) [44] is employed to estimate infiltration losses for determining the effective rainfall substituted into the HEC-1 for runoff simulation. The CN is an empirical value that increases with impermeability; thus, it is a suitable parameter that reflects urbanization level. The CN values for the four types of land-cover determined earlier are referred to in the table summarized in the SCS technical book [44].

3.2. Watershed Scale

In addition to historical land-cover changes discovered from the satellite images, present and future land-use scenarios are investigated on the watershed scale. The national land-use survey data obtained by the National Land Surveying and Mapping Center of Taiwan in 2008 [45] are selected as the present land-use scenario. The designed land uses in 2030, which are illustrated in the urban planning project proposed by the Nantou County government in Taiwan [36], are adopted as future scenario. Because these scenarios are based on official materials, they are in some ways more realistic than model predictions when considering social and economic influences. The investigated land-use scenarios are then put into the HEC-1 model for calculating present and future runoff variations. Thus, on the watershed scale, the land-runoff relationships are determined for the past, present, and future.

3.3. City Scale

On the city scale, runoff changes induced by land-use development, relocation, and renewal are more localized and rapidly changing; therefore, they require more sophisticated hydrodynamic models. The SWMM (ver. 5.1 with LID module) developed by the US Environmental Protection Agency is adopted for runoff simulation on the city scale. In response to a possible increase in runoff due to urban development, flood mitigation measures based on the concept of LID are introduced. Common LID techniques include permeable pavements, green roofs, rain gardens, bio-retention cells, tree box filters, grass swales, infiltration gutters, and rain barrels. Permeable pavements are porous layers that increase the infiltration and evaporation of surface water; green roofs use vegetation planted on roofs to reduce surface runoff and delay the occurrence of runoff peak; rain gardens and bio-retention cells use landscaping and plants to create retention pools that accumulate stormwater for small areas; tree box filters collect runoff from curbside entrances to store and filter water through plants before discharging it into sewer systems; grass swales or infiltration gutters are ditches with grass or permeable covers that increase infiltration while transporting runoff; and rain barrels are small rainwater harvesting systems that collect and retain rainfall water from roofs. In this study, various LID measures are installed based on their applicability.

4. Results and Discussion

4.1. Land-Cover and Runoff Changes on the Basin Scale

Table 2 summarizes the areas of water, built-up land, bare land, and vegetation in 1998, 2003, 2008, and 2013. Area changes for vegetation and water are not significant in terms of ratio because approximately 91.5% of the total land is covered by vegetation, 4% by water, 3.5% by bare land, and only 1% by built-up land. The built-up area showed a constant increase from 1990 hectares in 1998 to 3023 hectares in 2013, with a total increase of 51.91%. The area of bare land slightly increased from 8551 hectares in 1998 to 9650 hectares in 2003, and substantially increased to 14,561 hectares in 2008, followed by a decrease to 10,347 hectares in 2013, resulting in a total increase of 21% from 1998 to 2013. In contrast, the area covered by vegetation had the opposite pattern by decreasing constantly from 277,445 hectares in 1998 to 270,717 hectares in 2008 before subsequently increasing to 274,707 hectares in 2013. This finding indicates that there was a switch between the bare land and vegetation land from 1998 to 2013.

Tables 3–6 display the land-cover transfer matrixes between 1998–2003, 2003–2008, and 2008–2013, respectively. From Table 3, the increase in built-up land, with a peak increase of 40.28% from 1998 to 2003, can be attributed to construction on vegetation land and bare land. Moreover, the increase in bare land displayed in Tables 3 and 4 is a result of the deterioration of vegetation land, particularly from 2003 to 2008 when 10,170 hectares of vegetation land were converted into bare land. However, in Table 5, a rapid recover of vegetation land from bare land between 2008 and 2013 is seen, with 6008 hectares of bare land being transformed into vegetation land. These phenomena can be explained by the occurrence of a devastating earthquake in 1999, with its epicenter located in the Chi-Chi area in the study river basin. During the Chi-Chi earthquake, slope lands became weaker and unstable, causing collapses and landslides that washed away vegetation covers in the typhoon seasons of the following 10 years. The recovery of vegetation land between 2008 and 2013 indicates that the slope land has regained stability. The significant increase in the built-up area between 1998 and 2003 can also be attributed to the Chi-Chi earthquake owing to the relocation and reconstruction of the buildings destroyed in the disaster. Thus, in the study area, the occurrence of a major earthquake speeds up urbanization by forcing more undeveloped land to become built-up areas.

To determine the spatial aggregation of artificial structures in the Zhuoshui River Basin, SAA is conducted in grid units with a resolution of 500 m. Table 6 lists the Moran's indexes, Z-score, and p-value for 1998, 2003, 2008, and 2013. With all Moran's I > 0, Z-scores > 2.58, and p-values ~0, it demonstrates that the artificial structures in Zhoushui river basin are not distributed randomly in space, but are highly accumulated. Figure 2 shows the hotspots of the artificial structures in 1998, 2003, 2008, and 2013, in which the grey lines are the boundaries of the 21 watersheds and the red color denotes the grids with higher G_i (Getis) values. The hotspots of artificial structures were concentrated in the watersheds around the bottleneck between the upstream catchment and the downstream river channel, particularly in watersheds numbered W19 and W20, which are thus chosen as the subjects for the medium-scale studies.

Table 2. Changes in land-cover areas against time on basin scale.

Land-Cover Types	1998 (ha)	2003 (ha)	2008 (ha)	2013 (ha)	1998–2013 Change Ratio
Water	12,264	12,284	12,170	12,173	−0.74%
Built-up Land	1990	2792	2802	3023	51.91%
Bare Land	8551	9650	14,561	10,347	21.00%
Vegetation	277,445	275,524	270,717	274,707	−0.99%

Table 3. Land-cover transfer matrix from 1998 to 2003.

2003 / 1998	Water (ha)	Built-Up Land (ha)	Bare Land (ha)	Vegetation (ha)	Total (ha)	1998–2003 Change Ratio
Water	11,209	86	376	614	12,284	0.16%
Built-up Land	80	1637	248	827	2792	40.28%
Bare land	237	66	3074	6273	9650	12.84%
Vegetation	738	201	4854	269,731	275,524	−0.69%
Total	12,264	1990	8552	277,445	300,250	

Table 4. Land-cover transfer matrix from 2003 to 2008.

2008 / 2003	Water (ha)	Built-Up Land (ha)	Bare Land (ha)	Vegetation (ha)	Total (ha)	2003–2008 Change Ratio
Water	11,935	77	85	73	12,170	−0.93%
Built-up Land	137	2249	30	386	2802	0.37%
Bare Land	98	82	4211	10,170	14,561	50.89%
Vegetation	114	384	5324	264,895	270,717	−1.74%
Total	12,284	2792	9650	275,524	300,250	

Table 5. Land-cover transfer matrix from 2008 to 2013.

2013 / 2008	Water (ha)	Built-Up Land (ha)	Bare Land (ha)	Vegetation (ha)	Total (ha)	2008–2013 Change Ratio
Water	12,024	22	25	102	12,173	0.03%
Built-up Land	16	2208	275	524	3023	7.87%
Bare Land	92	330	8253	1673	10,347	−28.94%
Vegetation	38	242	6008	268,419	274,707	1.47%
Total	12,170	2802	14,561	270,717	300,250	

Table 6. Spatial autocorrelation of artificial structures from 1998 to 2013.

Index	1998	2003	2008	2013
Moran's I	0.67	0.76	0.60	0.63
Z-score	94.90	98.20	103.92	118.60
p-value	0.00	0.00	0.00	0.00

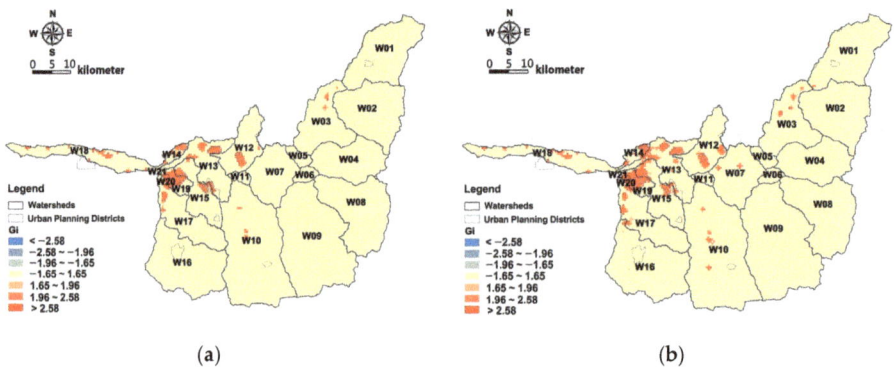

(a)

(b)

Figure 2. *Cont.*

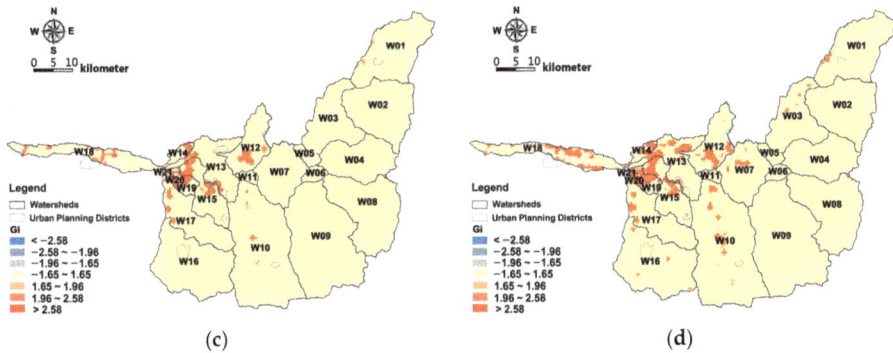

Figure 2. Hotspots of artificial structures on basin scalein (**a**) 1998; (**b**) 2003; (**c**) 2008; (**d**) 2013.

To initiate the HEC-1 model for runoff simulation, the 24-h rainfall amounts under different return periods are obtained from frequency analysis based on the historical records of the Chi-Chi rain station. The 24-h rainfall amounts under 2-, 5-, 25-, 50-, and 100-y return periods are 204.61, 308.29, 473.97, 543.38, and 612.50 mm, respectively. The rainfall hyetographs served for runoff simulation are determined by the intensity-duration-frequency relationships suggested by Horner and Flynt [46] according to the design handbook published by the Water Resource Agency, Taiwan [47]. Table 7 lists the parameters for the HEC-1 model and Table 8 summarizes the CN, the area ratio of impermeability (IMP), and 100-y peak runoff discharge (Q_p^{100}) on different scales. Overall, no significant changes are observed in CN, IMP, and Q_p^{100} on the basin scale. This may be due to the fact that the land-cover are classified into four groups only, with vegetation land accounting for 91.5% of the total land, which limits the variation of CN and consequent runoff. Table 8 also indicates that, because the W20 watershed has higher values of CN and IMP, it generates runoff of 0.24 cms/ha, which is much higher than the basin average 0.16 cms/ha.

Table 7. Parameters for HEC-1 model.

Watershed	Area (km²)	Length-W (m)	Slope-W	Length-C (m)	Slope-C	Roughness-W	n-C
W01	222.84	2328	0.63	47,855	0.05	0.20	0.02
W02	223.20	2619	0.61	42,612	0.05	0.20	0.02
W03	159.35	1665	0.65	47,856	0.05	0.20	0.02
W04	167.32	1836	0.66	45,557	0.06	0.20	0.02
W05	21.16	306	0.71	34,579	0.02	0.20	0.02
W06	19.20	1101	0.73	8721	0.19	0.20	0.02
W07	144.80	5708	0.66	12,684	0.08	0.20	0.02
W08	255.09	2924	0.65	43,617	0.06	0.20	0.02
W09	416.28	23,867	0.72	8721	0.19	0.20	0.02
W10	434.60	4840	0.65	44,893	0.08	0.20	0.02
W11	13.70	1089	0.45	6312	0.15	0.20	0.02
W12	111.10	1808	0.39	30,739	0.03	0.20	0.02
W13	125.60	2044	0.27	30,739	0.03	0.20	0.02
W14	19.40	562	0.09	17,291	0.02	0.20	0.02
W15	102.10	1982	0.38	25,752	0.07	0.20	0.02
W16	269.90	3546	0.57	38,056	0.06	0.20	0.02
W17	141.00	8741	0.46	8065	0.02	0.20	0.02
W18	117.60	1282	0.02	45,867	0.01	0.20	0.02
W19	26.70	1657	0.19	8065	0.02	0.20	0.02
W20	7.60	468	0.06	8065	0.02	0.20	0.02
W21	2.40	282	0.11	4318	0.05	0.20	0.02
W23	0.15	276	0.07	270	0.01	0.20	0.02

Note: Length-W: representative length of watershed; Slope-W: representative slope of watershed; Length-C: representative length of channel; Slope-C: representative slope of channel; Roughness-W: overland-flow roughness coefficient; n-C: Manning's n for channel.

Table 8. Variations of CN, IMP, and Q_p^{100} related to land-cover and land-use changes on different study scales.

Index	Land-Cover				Land-Use	
	1998	2003	2008	2013	2008	2030
Basin scale						
Zhuoshui Basin (316,800 ha)						
CN	43.42	44.17	44.64	44.09		
IMP (%)	0.53	0.62	0.69	0.79		
Q_p^{100} (cms)	48,835	49,350	49,379	49,446		
Watershed scale						
W19 (2673 ha)						
CN	51.45	53.18	51.34	52.20	60.55	62.31
IMP (%)	12.91	13.13	14.05	15.00	14.12	18.76
Q_p^{100} (cms)	416	422	417	424	467	486
W20 (755 ha)						
CN	65.38	64.14	63.63	64.92	69.75	76.48
IMP (%)	7.29	8.66	8.79	9.64	13.50	14.73
Q_p^{100} (cms)	182	181	182	183	188	198
City scale						
Jushan Dist. (422 ha)						
IMP (%)					45.78	61.80
Q_p^{100} (cms)					79.62	84.26

4.2. Land-Cover, Land-Use, and Runoff Changes on the Watershed Scale

Present (2008) and future (2030) land-use maps on the watershed scales for W19 and W20 are illustrated in Figure 3. Current land uses are divided into 11 categories, and the future land uses are further refined into 32 categories. By 2030, large areas of forests are set to be transformed into agricultural land and many residential regions will be converted into commercial areas. According to land-use changes from 2008 to 2030, for W19, the CN values increase from 60.55 to 62.31, and the IMP increase from 14.12% to 18.76%; for W20, the CN values increase from 69.75 to 76.48, and the IMP increase from 13.50% to 14.73%, as listed in Table 8. However, in the overlay year of 2008, the CN values interpreted from land-use data are much larger than those interpreted from land-cover data (equal to 51.34 for W19 and 63.63 for W20), which can lead to different runoff estimations. These discrepancies are not surprising and can be attributed to the difference in land classification, as 11 types of land-uses exists but only four types of land-covers are classified. This finding indicates that uncertainty analysis should be performed if different sources of land data are incorporated for hydrological analysis; otherwise, their results should be regarded separately. In this paper, runoff variations resulted from land-cover and land-use changes are separately discussed.

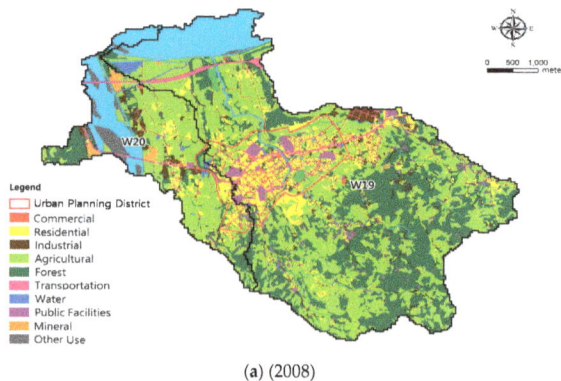

(a) (2008)

Figure 3. *Cont.*

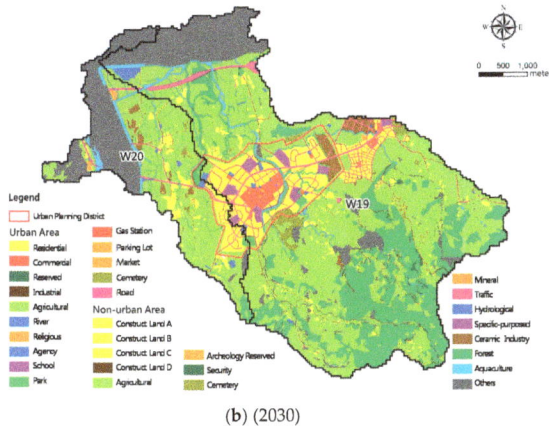

(**b**) (2030)

Figure 3. Present and future land-use maps on watershed scale: (**a**) present; (**b**) future.

Figure 4 shows a comparison of present and future runoff peaks under different return periods for W19 and W20, respectively. Peak runoff are demonstrated to increase in the future under all return periods for both W19 and W20. Although the increments of peak runoff are smaller at low return periods, their change ratios are instead larger. For both watersheds, the peak runoff under 2-y return periods increase by more than 10% and then drop to a level of 5% under return periods larger than 25 years. The runoff hydrographs under 2-y return period for W19 and W20 are displayed in Figure 5, showing that future runoff peaks not only increase but also shift forward; the integration of runoff hydrographs against time shows that the total increases in runoff volume are 212,400 m^3 and 140,400 m^3 for W19 and W20, equivalent to 85 and 56 standard swimming pools, respectively. Such an increase presents a non-negligible extra risk of flooding if additional flood control measures are not implemented.

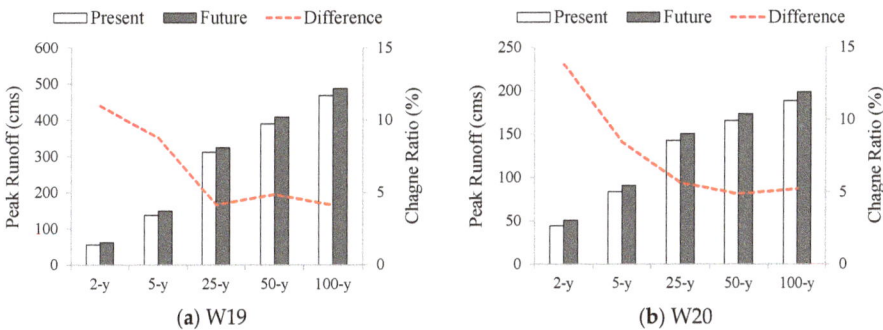

Figure 4. Present and future runoff peaks on watershed scale for (**a**) W19 and (**b**) W20.

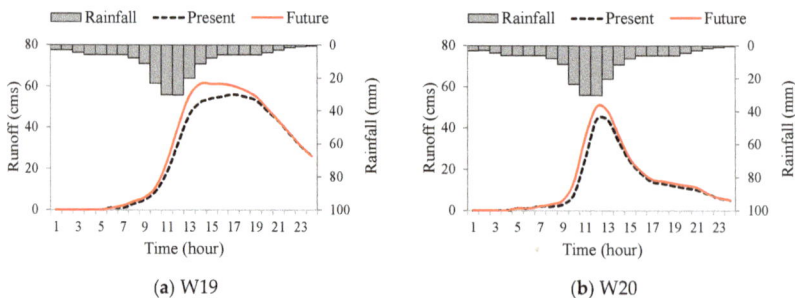

Figure 5. Runoff hydrographs under 2-y return period for watersheds (**a**) W19 and (**b**) W20.

4.3. Land-Use and Runoff Changes on the City Scale

According to the sewer system report for Jushan in 2008 [48], the Jushan District is divided into seven drainage sections as shown in Figure 6 and the land-use maps for these sections at present and in the future are compared in Figure 7. The figure indicates that, by 2030, the commercial area will greatly expand and will be mostly concentrated in S4; a large industrial block will be developed at the borders of S1 and S2; and the residential areas will be increased in almost every section. Figure 8 indicates that the IMP for all the seven sections increase simultaneously, from 23–69% at present to 36–88% in the future, in which the S1 has the highest increase of 58%. The variations of average IMP and Q_p^{100} on the city scale are also listed in Table 8, which shows that this increased IMP will raise Q_p^{100} from 79.62cms to 84.26cms by 2030. Table 9 summarizes the parameters for SWMM.

Table 9. Parameters for SWMM.

Drainage Section	Area (ha)	Slope (%)	n-Imp	n-Per	S-Imp (mm)	S-Per (mm)
S1	48.84	0.7	0.01	0.1	0.05	0.05
S2	57.7	0.7	0.01	0.1	0.05	0.05
S3	69.45	0.7	0.01	0.1	0.05	0.05
S4	58.96	1.45	0.01	0.1	0.05	0.05
S5	92.76	1.45	0.01	0.1	0.05	0.05
S6	54.44	1.45	0.01	0.1	0.05	0.05

Note: n-Imp: Manning's n for impervious land; n-Per: Manning's n for pervious land; S-Imp: depression storage for impervious land; S-Per: depression storage for pervious land.

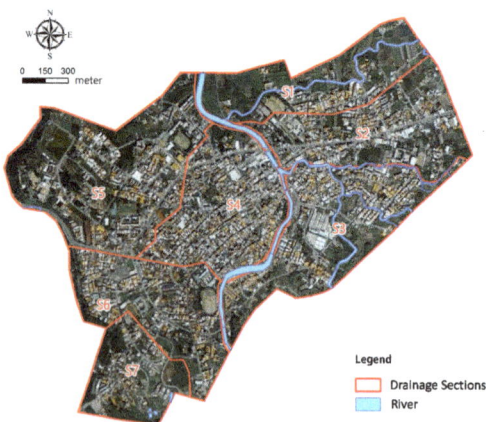

Figure 6. Drainage divisions on city scale.

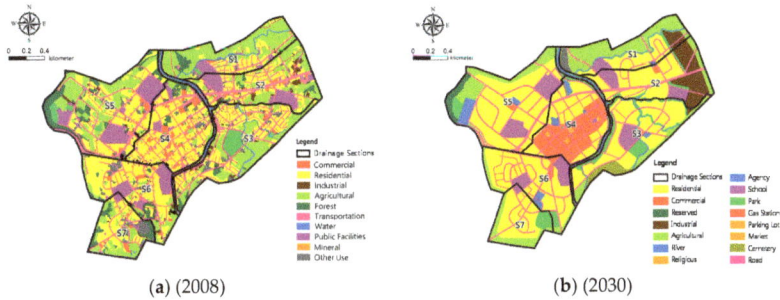

(**a**) (2008)   (**b**) (2030)

Figure 7. Present and future land-uses on city scale: (**a**) present; (**b**) future.

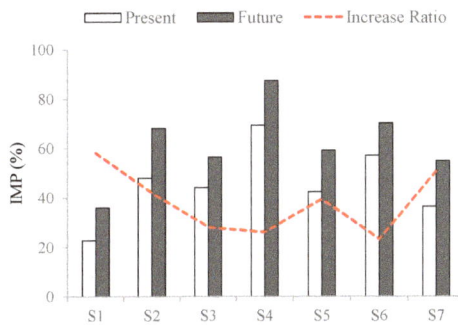

Figure 8. Comparison between present and future IMP on city scale.

In response to the increasing of flood risk, various LID measures are designed and incorporated into the SWMM on the city scale to evaluate their effectiveness of runoff reduction. To reduce public resistance, the LID measures are installed on government-owned lands or facilities. Some lands or facilities such as cemeteries, sewage treatment plants, and gas stations are discounted for LID because of their particular usages. Table 10 summarizes the areas of different land-uses and the area ratios in which different LID facilities are installed. Rain gardens and bio-retention cells are installed in less built-up areas such as parks, parking lots, and plazas, while green roofs are mainly set up on the tops of markets, agencies, and schools. Using the LID control editor built in the SWMM, parameters of the LID facilities are given by different ground layers including surface, soil, and storage, as summarized in Table 11. The surface and soil layers are porous, allowing water to pass through; the storage layer contains water within the soil void and when the storage is full, excessive water will overflow. In this study, no underdrains are considered. The IMP for each drainage section is adjusted for runoff simulation according to the area occupied by each LID facility.

Table 10. Area ratios of different LID facilities with respect to land-uses.

Land-Use Types	Area (ha)	Bio-Retention Cell (%)	Rain Garden (%)	Green Roof (%)	Infiltration Gutter (%)	Permeable Pavement (%)
School	29.03	-	10	30	15	5
Park	11.82	90	-	-	-	-
Agency	7.17	35	-	50	-	-
Social	7.17	-	-	30	15	5
Market	0.95	-	10	80	-	-
Tourist	0.66	35	-	50	-	-
Parking Lot	0.58	-	35	-	-	-
Plaza	0.29	-	35	-	-	-

Table 11. Parameters for LID facilities.

Item		Bio-Retention Cell	Rain Garden	Green Roof	Infiltration Gutter	Permeable Pavement
Surface	Berm Height (m)	0.15	0.15	0.04	-	-
Soil	Thickness (m)	0.45	0.45	0.10	-	0.25
	Porosity	0.25	-	0.30	-	0.25
Storage	Thickness (m)	0.15	-	-	0.10	0.45
	Void Ratio	0.25	-	-	0.25	0.25

Figure 9 shows the simulated runoff peaks and runoff volumes for different return periods under three land-use scenarios: present, future, and future with LID. Without LID, the runoff peaks and runoff volumes are found to increase simultaneously for all return periods due to the increase in IMP in future urban planning. The increased amounts at high return periods are slightly larger than those at low return periods, but not obvious. After introducing the LID facilities, Figure 9a shows that the runoff peaks for 2-y and 5-y return periods will be effectively reduced to a level even less than the present condition. However, for the 25-, 50-, and 100-y return periods, the introduction of LID instead leads to slight increases in runoff peaks. In fact, this phenomenon has been discovered in previous papers (e.g., MaChtcheon, et al., 2012; Tao, et al., 2017), however, currently, no satisfactory explanation has been provided. Fortunately, Figure 9b shows that LID does help to reduce runoff volumes no matter at high or low discharges. From the aforementioned case study, it is certain to say that the introduction of LID benefits flood mitigation at low discharges, but as discharge increases, the effectiveness of LID becomes trivial or unfavorable; consequently, LID should be implemented in conjunction with other flood mitigation measures.

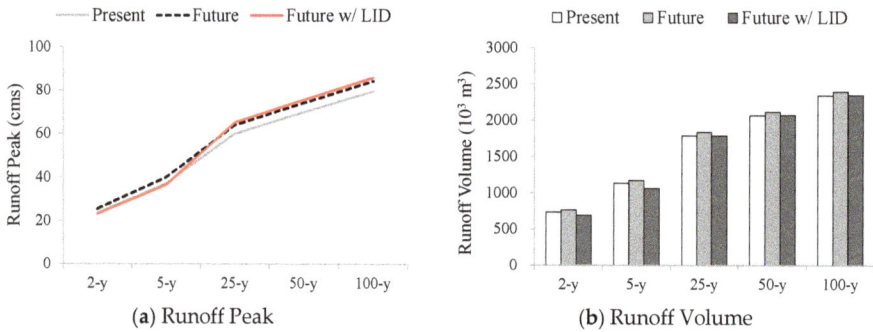

Figure 9. (a) Runoff peak and (b) runoff volume for different return periods under land-use scenarios of present, future, and future with LID.

Table 12 summarizes the runoff on watershed scale simulated by HEC-1 and those on the city scale simulated by SWMM, in which V_{W19}, V_{W20}, and V_{city} are the runoff volumes generated in W19, W20, and Jushan District, respectively. At low return periods, the V_{city} has a larger share in the total watershed runoff $V_{W19} + V_{W20}$ than at high return periods. Considering that Jushan District only occupies 12% of the total watershed area, the city runoff volume at 2-y return period almost doubles no matter at present or in the future. This finding is quite reasonable because, compared with the city areas, rural areas have higher infiltration rates so that lower portions of rainfall will be transformed into effective runoff at initial stages. Through cross-analysis of the runoff results produced by HEC-1 and SWMM, the concomitant usage of different hydrological models on different scales is practical if the same land-use data are applied. Overall, at 2-y return period, LID can reduce approximately 10%

of the city runoff volume, equivalent to 2% of the watershed runoff volume and 0.02% of the basin runoff volume.

By conducting statistical analysis on the results in Table 12, the relationship between city runoff and watershed runoff can be expressed by the following equation:

$$\frac{V_{watershed}}{V_{city}} = \left(\frac{A_{watershed}}{A_{city}}\right)^a \times \left(\frac{IMP_{city}}{IMP_{watershed}}\right)^b \times \left(\frac{R_{city}}{R^{100}_{watershed}}\right)^c \tag{1}$$

in which $V_{watershed}$ and V_{city} are runoff volumes on watershed and city scales, respectively; $IMP_{watershed}$ and IMP_{city} are the area ratios of impermeability on watershed and city scales, respectively; R_{city} is the observed rainfall on city scale; $R^{100}_{watershed}$ is the rainfall on watershed scale under 100-y return period. The coefficients a, b, c are determined by multiple regression analysis which yields $a = 3.01$, $b = 0.71$, $c = 0.37$. With correlation coefficient $R = 0.99$ and Root-Mean-Square-Error $RMSE = 0.27$, the $V_{watershed}$ values simulated by hydrological model are in good agreement with those predicted by the equation, as shown in Figure 10. Being a general expression linking city runoff to watershed runoff, Equation (1) can be applied to any watershed, not only for the specific region in this study. When the ratio of $A_{watershed}/A_{city}$ increases, the watershed will receive more rainfall water and generate more runoff compared with the city. A larger value of $IMP_{city}/IMP_{watershed}$ represents that impermeable ratio at city increases compared with that at watershed; thus, the city will make a larger contribution to watershed runoff. The ratio of $R_{city}/R^{100}_{watershed}$ allows us to predict watershed runoff from the rainfall observation at city; meanwhile, it reflects the phenomenon that the contribution of city runoff decreases at high discharges. The influence of land use will be implicitly included in the coefficients of a, b, c which vary with watershed conditions.

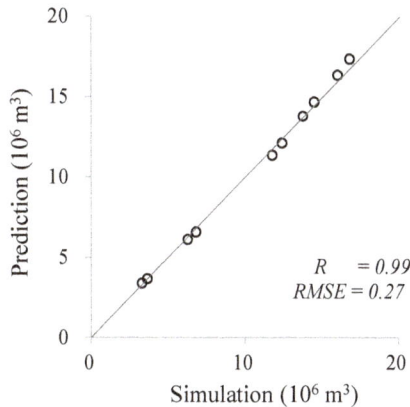

Figure 10. Comparison between the simulated and predicted runoff volumes on watershed scale.

Table 12. Comparison of runoff simulated under different land-use scenarios on watershed and city scales.

Title	2-y	5-y	25-y	50-y	100-y
			Present		
V_{W19} (10³ m³)	2250	4428	8622	10,105	11,840
V_{W20} (10³ m³)	1022	1811	3136	3654	4190
V_{city} (10³ m³)	731	1137	1793	2069	2345
$V_{city}/(V_{W19} + V_{W20})$ (%)	22	18	15	15	15

Table 12. *Cont.*

Title	2-y	5-y	25-y	50-y	100-y
	Future				
V_{W19} (10^3 m^3)	2462	4788	9050	10,606	12,366
V_{W20} (10^3 m^3)	1163	1994	3348	3884	4446
V_{city} (10^3 m^3)	762	1175	1839	2118	2396
$V_{city}/(V_{W19} + V_{W20})$ (%)	21	17	15	15	14
	Future with LID				
V_{city} (10^3 m^3)	687	1061	1790	2072	2347

5. Conclusions

Analyzing runoff variations from land-cover and land-use changes caused by urbanization is critical for flood mitigation. However, analysis should not be conducted only on a city scale but also on larger scales because a city is also a part of a wider watershed or river basin. In this study, land and runoff variations resulting from urbanization are analyzed on the basin scale (large), watershed scale (medium), and city scale (small) for the Zhuoshui River Basin, Taiwan. Based on the scale sizes, different data and models are employed for land interpretation and runoff simulation in the past, at present, and in the future (Table 1). Overall, patterns of land-runoff changes due to urbanization are discovered on the basin scale; discrepancies raised from the usage of different land data and hydrological models are evaluated on the watershed scale; and the impacts of LID facilities on flood reduction in response to urbanization are quantified on the city scale. The research findings for the different scales are summarized as follows:

1. On the basin scale, the land-cover transfer matrixes are established from the interpretation of multi-period satellite images. The results show that a large area of vegetation land (10,170 hectares) became bare land from 1998 to 2008, accompanied by a significant increase (40%) in built-up area from 1998 to 2003. This suggests that the major Chi-Chi earthquake, occurred in 1999, increased the pace of urbanization in the process of reconstruction and relocation at a speed faster than the recovery of natural land covers. Through hotspot analysis, built-up areas were discovered to be mainly concentrated in the Jushan and Chi-Chi districts in the study river basin, which was the earthquake epicenter. However, this increase in built-up area is less than 1% of the entire river basin area; therefore, it has no influence on the variation of simulated runoff on the basin scale.

2. On the watershed scale, the sub-basins containing Jushan District are selected to simulate the runoff variations with respect to land-cover changes from 1998 to 2013 and land-use changes from 2008 to 2030. Again, no significant land-cover and runoff variations were found from 1998 to 2013, but from 2008 to 2030, the increases in CN and IMP caused by land-use changes lead to 5–10% increases in runoff peak under different return periods. However, in the overlay year of 2008 when the land-cover and land-use data are both available, discrepancies in runoff estimations are found because the values of CN and IMP interpreted from land-use data are much larger than those interpreted from land-cover data. Thus, mixed usage of satellite-interpreted land-cover data and manual-investigated land-use data is not recommended for hydrological analysis unless their uncertainties can be quantified on the same basis. In contrast, the concomitant usage of HEC-1 and SWMM shows no contradiction in runoff results if the same land-use data are applied. According to the results by HEC-1 and SWMM, a regressive relationship is derived for predicting watershed runoff from city runoff multiplied by the parameters of area, IMP, and rainfall raised to certain powers. This regression provides a general expression of runoff scaling between different spatial scales. The good agreement between the simulation and prediction demonstrates that the cross-analysis of land/runoff by multiple models is valuable.

3. On the city scale, experimental LID facilities are installed on governmental lands in an attempt to reduce the runoff peaks for Jushan District. The results show that the LID facilities effectively reduce the runoff volumes for all return periods, and the runoff peaks at low return periods will be reduced to a level even less than that at present. However, for high return periods, the introduction of LID instead slightly increases the runoff peaks. Although the "side effects" of LID on increasing runoff at high discharges have been previously discovered, they have not been systematically discussed and require follow-up studies for clarification. Statistically, under 2-y return period, the introduction of LID reduces up to 10% of city runoff volumes, which equals to 2% and 0.02% of watershed and basin runoff volumes, respectively. According to these findings, LID approaches are better regarded as local flood mitigation measures in low-discharge conditions.

Acknowledgments: The authors express sincere gratitude to the National Land Survey and Mapping Center and National Science and Technology Center for Disaster Reduction of Taiwan for providing satellite images and land-use data.

Author Contributions: J.-C.F., W.-Y.L. and J.-H.J. designed the experiments; J.-C.F., W.-Y.L., and C.-M.H. conducted the experiments; C.-C.Y., J.-C.F., C.-M.H. and J.-H.J. analyzed the data; J.-H.J. and J.-C.F. examined and concluded the results; J.-C.F., C.-M.H. and J.-H.J. wrote the paper.

Conflicts of Interest: The authors declare no conflict of interest.

References

1. Nirupama, N.; Simonovic, S.P. Increase of flood risk due to urbanization: A Canadian example. *Nat. Hazards* **2007**, *40*, 25–41. [CrossRef]
2. Saghafian, B.; Farazjoo, H.; Bozorgy, B.; Yazdandoost, F. Flood intensification due to changes in land use. *Water Resour. Manag.* **2008**, *22*, 1051–1067. [CrossRef]
3. Liu, Y.B.; Gebremeskel, S.; De Smedt, F.; Hoffmann, L.; Pfister, L. Predicting storm runoff from different land-use classes using a geographical information system-based distributed model. *Hydrol. Process.* **2006**, *20*, 533–548. [CrossRef]
4. Shi, P.J.; Yuana, Y.; Zhenga, J.; Wanga, J.A.; Gea, Y.; Qiua, G.Y. The Effect of Land Use/Cover Change on Surface Runoff in Shenzhen Region, China. *CATENA* **2007**, *69*, 31–35. [CrossRef]
5. Shalaby, A.; Tateishi, R. Remote sensing and GIS for mapping and monitoring land cover and land-use changes in the Northwestern coastal zone of Egypt. *Appl. Geogr.* **2007**, *27*, 28–41. [CrossRef]
6. Lambin, E.F.; Baulies, X.; Bockstael, N.E. *Land-Use and Land-Cover Change Implementation Strategy*; IGBP Report No. 48 and IHDP Report No. 10; International Geosphere-Biosphere Programme: Stockholm, Sweden, 2002.
7. Engman, E.T.; Gurney, R.J. *Remote Sensing in Hydrology*; Chapman and Hall: London, UK, 1991.
8. Drayton, R.S.; Wilde, B.M.; Harris, J.H.K. Geographic information system approach to distributed modeling. *Hydrol. Process.* **1992**, *6*, 361–368. [CrossRef]
9. Mattikalli, N.M.; Devereux, B.J.; Richards, K.S. Prediction of river discharge and surface water quality using an integrated geographic information system approach. *Int. J. Remote Sens.* **1996**, *17*, 683–701. [CrossRef]
10. Chomitz, K.M.; Gray, D.A. Roads, land use, and deforestation: A spatial model applied to Belize. *World Bank Econ. Rev.* **1996**, *10*, 487–512. [CrossRef]
11. Martínez, J.M.; Suárez-Seoane, S.; De Luis Calabuig, E. Modelling the risk of land cover change from environmental and socio-economic drivers in heterogeneous and changing landscapes: The role of uncertainty. *Landsc. Urban Plan.* **2011**, *101*, 108–119. [CrossRef]
12. Xian, G.; Crane, M.; Su, J. An analysis of urban development and its environmental impact on the Tampa Bay Watershed. *J. Environ. Manag.* **2007**, *85*, 965–976. [CrossRef] [PubMed]
13. Morita, H.; Hoshino, S.; Kagatsume, M.; Mizuno, K. *An Application of the Land-Use Change Model for the Japan Case Study Area*; IIASA Interim Report IR-97-065; International Institute for Applied Systems Analysis: Laxenburg, Austria, 1997.
14. Serra, P.; Pons, X.; Saurí, D. Land-cover and land-use change in a Mediterranean landscape: A spatial analysis of driving forces integrating biophysical and human factors. *Appl. Geogr.* **2008**, *28*, 189–209. [CrossRef]

15. United States Army Corps of Engineers (USACE). *Hydrologic Modeling System: HEC-HMS Technical Reference Manual*; Hydrologic Engineering Center: Davis, CA, USA, 2000.
16. Nageshwar, R.B.; Wesley, P.J.; Ravikumar, S.D. Hydrologic parameter estimation using geographic information systems. *J. Water Res. Plan. Manag.* **1992**, *11*, 492–512.
17. Schumann, A.H. Development of conceptual semi-distributed hydrological models and estimation of their parameters with the aid of GIS. *Hydrol. Sci. J.* **1993**, *38*, 519–528. [CrossRef]
18. Xu, Z.; Zhao, G. Impact of urbanization on rainfall-runoff processes: Case study in the Liangshui River Basin in Beijing, China. *Proc. Int. Assoc. Hydrol. Sci.* **2016**, *373*, 7–12. [CrossRef]
19. Guo, F.; Hanfei, Q.U.; Zeng, H.; Cong, P.; Geng, X. Flood hazard forecast of Pajiang River flood storage and detention basin based on MIKE21. *J. Nat. Disasters* **2013**, *22*, 144–152.
20. U.S. Environmental Protection Agency (USEPA). *Storm Water Management Model User's Manual Version 5.1*; National Risk Management Laboratory Office of Research and Development: Cincinnati, OH, USA, 2015.
21. Guan, M.; Sillanpää, N.; Koivusalo, H. Modelling and assessment of hydrological changes in a developing urban catchment. *Hydrol. Process.* **2015**, *29*, 2880–2894. [CrossRef]
22. U.S. Environmental Protection Agency (US EPA). *Low-Impact Development Design Strategies: An Integrated Design Approach*; EPA 841-B-00003; US EPA: Washington, DC, USA, 1999.
23. British Columbia Ministry of Environment (BCME). *Stormwater Planning: A Guidebook for British Columbia*; Government of British Columbia: Victoria, UK. Available online: http://www.toolkit.bc.ca/resource/stormwater-planning-guidebook-british-columbia (accessed on 24 January 2018).
24. Olewiler, N. *The Value of Natural Capital in Settled Areas of Canada*; Ducks Unlimited Canada and The Nature Conservancy of Canada: Toronto, ON, Canada, 2004.
25. Sharma, A.K.; Pezzaniti, D.; Myers, B.; Cook, S.; Tjandraatmadja, G.; Chacko, P.; Chavoshi, S.; Kemp, D.; Leonard, R.; Koth, B. Water sensitive urban design: An investigation of current systems, implementation drivers, community perceptions and potential to supplement urban water services. *Water* **2016**, *8*, 272. [CrossRef]
26. U.S. Environmental Protection Agency (USEPA). *Combined Sewer Overflow (CSO) Control Policy*; Federal Register; United States Environmental Protection Agency: Washington, DC, USA, 1994; Volume 75, pp. 18688–18698.
27. Ministry of Housing and Urban-Rural Development (MHURD). *Technical Guide for Sponge Cities—Water System Construction of Low Impact Development*; Ministry of Housing and Urban-Rural Development: Beijing, China, 2016.
28. Li, H.; Ding, L.; Ren, M.; Li, C.; Wang, H. Sponge city construction in China: A survey of the challenges and opportunities. *Water* **2017**, *9*, 594. [CrossRef]
29. McCutcheon, M.; Wride, D.; Reinicke, J. An evaluation of modeling green infrastructure using LID controls. *J. Water Manag. Model.* **2012**, 193–205. [CrossRef]
30. Chen, B.; Liu, J.; She, N.; Xu, K. Optimization of low impact development facilities in Beijing CITIC complex. In Proceedings of the International Low Impact Development Conference, Houston, TX, USA, 19–21 January 2015; pp. 342–351.
31. Lucas, W.C.; Sample, D.J. Reducing combined sewer overflows by using outlet controls for Green Stormwater Infrastructure: Case study in Richmond, Virginia. *J. Hydrol.* **2015**, *520*, 473–488. [CrossRef]
32. Guo, C.Y. Detention basin sizing for small urban catchments. *J. Water Resour. Plan. Manag.* **2012**, *125*, 1–5.
33. Jayasooriya, V.M.; Ng, A.W.M. Tools for modeling of stormwater management and economics of green infrastructure practices: A review. *Water Air Soil Pollut.* **2014**, *225*, 2055–2061. [CrossRef]
34. Peng, H.Q.; Liu, Y.; Wang, H.W.; Ma, L.M. Assessment of the service performance of drainage system and transformation of pipeline network based on urban combined sewer system model. *Environ. Sci. Pollut. Res.* **2015**, *22*, 15712–15721. [CrossRef] [PubMed]
35. Tao, J.; Li, Z.; Peng, X.; Ying, G. Quantitative analysis of impact of green stormwater infrastructures on combined sewer overflow control and urban flooding control. *Front. Environ. Sci. Eng.* **2017**, *11*, 11. [CrossRef]
36. Nantou County. *Revision of Urban Planning Project for Jushan Township*; Nantou County: Nantou City, Taiwan, 2013. (In Chinese)
37. Vapnik, V.N. *The Nature of Statistical Learning Theory*; Springer: New York, NY, USA, 1995.

38. Selçuk, R. Analyzing land use/land cover changes using remote sensing and GIS in Rize, North-East Turkey. *Sensors* **2008**, *8*, 6188–6202.

39. Cliff, A.C.; Ord, J.K. *Spatial Autocorrelation*; Pion Limited: London, UK, 1973.

40. Ord, J.K.; Getis, A. Local spatial autocorrelation statistics: Distributional issues and an application. *Geogr. Anal.* **1995**, *27*, 286–306. [CrossRef]

41. Moran, P.A.P. Notes on continuous stochastic phenomena. *Biometrika* **1950**, *37*, 17–23. [CrossRef] [PubMed]

42. Getis, A.; Ord, J.K. The analysis of spatial association by use of distance statistics. *Geogr. Anal.* **1992**, *24*, 189–206. [CrossRef]

43. Singh, V.P. *Applied Modeling in Catchment Hydrology*; Water Resources Publications: Littleton, CO, USA, 1982.

44. U.S. Department of Agriculture (USDA). *Urban Hydrology for Small Watersheds*; Technical Release 55 (TR-55); Natural Resources Conservation Service, Conservation Engineering Division: Washington, DC, USA, 1986.

45. National Land Survey and Mapping Center (NLSC). *The Second National Land-Use Survey*; Ministry of Interior: Taipei, Taiwan, 2008.

46. Horner, W.W.; Flynt, F.L. Relation between rainfall and run-off from small urban areas. *Proc. Am. Soc. Civ. Eng.* **1936**, *101*, 140–183.

47. Water Resource Agency (WRA). *Handbook for Hydrological Design*; Water Resource Agency: Taichung, Taiwan, 2001. (In Chinese)

48. Nantou County. *Review and Planning for Jushan Township Sewer System*; Nantou County: Nantou City, Taiwan, 2008. (In Chinese)

![water logo] *water*

MDPI

Case Report

Evaluating the Water Quality Benefits of a Bioswale in Brunswick County, North Carolina (NC), USA

Rebecca A. Purvis [1,*], Ryan J. Winston [2], William F. Hunt [1], Brian Lipscomb [3], Karthik Narayanaswamy [4], Andrew McDaniel [3], Matthew S. Lauffer [3] and Susan Libes [5]

[1] Department of Biological and Agricultural Engineering, North Carolina State University, Raleigh, NC 27695, USA; wfhunt@ncsu.edu
[2] Department of Food, Agricultural, and Biological Engineering, Ohio State University, Columbus, OH 43210, USA; Winston.201@osu.edu
[3] North Carolina Department of Transportation, Raleigh, NC 27610, USA; blipscomb@ncdot.gov (B.L.); ahmcdaniel@ncdot.gov (A.M.); mslauffer@ncdot.gov (M.S.L.)
[4] AECOM, Morrisville, NC 27560, USA; karthik.narayanaswamy@aecom.com
[5] Department of Coastal and Marine Systems Science, Coastal Carolina University, Conway, SC 29528, USA; susan@coastal.edu
* Correspondence: rpurvis@ncsu.edu; Tel.: +1-832-350-2406

Received: 28 November 2017; Accepted: 26 January 2018; Published: 31 January 2018

Abstract: Standard roadside vegetated swales often do not provide consistent pollutant removal. To increase infiltration and pollutant removal, bioswales are designed with an underlying soil media and an underdrain. However, there are little data on the ability of these stormwater control measures (SCMs) to reduce pollutant concentrations. A bioswale treating road runoff was monitored, with volume-proportional, composite stormwater runoff samples taken for the inlet, overflow, and underdrain outflow. Samples were tested for total suspended solids (TSS), total volatile suspended solids (VSS), enterococcus, *E. coli*, and turbidity. Underdrain flow was significantly cleaner than untreated road runoff for all monitored pollutants. As expected, the water quality of overflow was not significantly improved, since little to no interaction with soils occurred for this portion of the water balance. However, overflow bacteria concentrations were similar to those from the underdrain perhaps due to a first flush of bacteria which was treated by the soil media. For all sampling locations, enterococci concentrations were always higher than the USEPA geometric mean recommendation of 35 Most Probable Number (MPN)/100 mL, but there were events where the fecal coliform concentrations was below the USEPA's 200 MPN/100 mL limit. A reduction in TSS concentration was seen for both overflow and underdrain flow, and only the underdrain effluent concentrations were below the North Carolina's high quality water limit of 20 mg/L. Comparing results herein to standard swales, the bioswale has the potential to provide greater treatment and become a popular tool.

Keywords: bacteria; bioinfiltration; infiltration; pathogens; runoff; sediment; urbanization

1. Introduction

Urbanization is a global trend, with 54% of the total population living in urban areas in 2014 and expected to reach 66% by 2050 [1]. Urbanization negatively impacts the environment, notably water quality, due to an increase in impervious cover [2–6]. Urban runoff contains pollutants including suspended solids, heavy metals, nutrients, and pathogens [7–9]. Pathogens have been reported as one of the leading causes for impaired surface waters placed by the United States Environmental Protection Agency (USEPA) [10]. Elevated bacteria levels can lead to economic losses in recreation waters, increased drinking water treatment costs, and potential health concerns [11].

There are many external factors impacting the fate of bacteria in a watershed including: a variety of sources (domestic pets, wild birds and animals, and human waste) [12] and various environmental factors (temperature, light intensity, and predation) [13–16], and treatment within stormwater control measures (SCMs). Removal mechanisms for bacteria include filtration, adsorption to a soil, desiccation, and predation [14]. Biofilm development may enhance adsorption to a soil [17,18]. However, bacteria can be difficult to permanently sequester, due to the potential to reproduce in a soil [13,19].

Little research is available regarding whether bioretention media promotes bacterial sequestration or provides an environment for growth [20]; through growth and resuspension, media could act as a source of bacteria to stormwater runoff. While 'true' pathogens are the biggest concern, fecal indicator bacteria (FIB) are the regulatory metric used to monitor water quality and public health decision making [21–24]. Fecal indicator bacteria (FIB) include *Escherichia coli* (*E. coli*), enterococci, and total and fecal coliforms [16]. While not pathogenic, FIB are associated with fecal matter, thus signaling the potential presence of human pathogens [25]. In addition, FIB are usually found in higher numbers, have a higher survival rate, and are easier and more economical to detect in laboratory testing than true pathogens [26,27]. Understanding how unit processes (for FIB and pathogens) can be employed in SCMs is integral to reducing the impacts of bacteria in stormwater on receiving waters.

Stormwater runoff can be managed using low-impact development (LID) techniques, which targets treatment of a water quality volume at or near the source of runoff [28]. LID techniques attempt to mimic the hydrologic and water quality characteristics of the pre-development watershed [29,30].

One commonly installed LID SCM is a bioretention cell (BRC). Pollutant removal is primarily reliant on the engineered bioretention media, which is generally sand-based with small amounts of silt, clay, and organic matter [31]. The goal of a BRC is to reduce stormwater runoff volume, control peak flows, and improve water quality through filtration, infiltration, and nutrient transformation [31].

Dry swales are shallow, vegetated channels that are generally designed and constructed with a triangular or trapezoidal cross-section and are typically for stormwater conveyance [32,33]. Dry swales have reported mean volume reduction from 23 to 47% [29,34–38], which translates into pollutant load reduction for receiving waters [29,31,38]. Pollutant removal mechanisms employed by standard swales include: sedimentation, filtration, infiltration, and modest amounts of biological and chemical reactions at the soil surface [32,39,40]. Although there is growing literature on the capabilities of swales to reduce runoff [32,38–41], a lack of consistent water quality treatment has been observed in dry swales, in particular for bacteria removal [25,42].

Bioswales are a category of SCM which combine the conveyance function of a traditional grass swale with the filtration and biological treatment mechanisms of bioretention [43]. While similar in appearance to a grassed swale, a bioswale employs an engineered soil media, similar to bioretention media, below the vegetation; the media is underlain by a gravel drainage layer surrounding a perforated underdrain. A bioswale promotes infiltration and filtration through the largely-sand media and underdrain while maintaining stormwater conveyance on the surface during large rainfall events. Natural organic material (NOM) is included in the media mixture to promote chemical transformations and sorption of phosphorus and heavy metals [44,45]. Only a few studies of FIB removal through soil media have been conducted, most of which show up to 1-log reduction in FIB concentrations. Rusciano and Obropta [18] found a 91.5% removal of fecal coliform bacteria through bio-media columns, Garbrecht et al. [46] found *E. coli* reduction coefficients between 32–91% based on the soil type in the column, and Hunt et al. [47] found an average of 69% and 71% removal of fecal coliform and *E. coli*, respectively, from stormwater runoff treated by a bioretention cell. However, virtually no data exist on the performance of bioswales for runoff conveyance, water quality treatment, or bacteria removal capabilities [25,48–50].

While preliminary research on bioswales does show the potential for stormwater runoff volume reduction [51–53], the exact extent of this reduction is not well known. Research is needed to determine how incorporating soil media and an underdrain affect volume reduction and how their pollutant removal mechanisms affect bacteria sequestration and subsequent removal.

2. Materials and Methods

2.1. Lumber River Basin and Lockwoods Folly River Description

The study site was located in Bolivia, North Carolina (NC) in Brunswick County (34°0′16.2972″ N, 78°15′38.7792″ W) and drains into the Lockwoods Folly River, which is located in the Lumber River Basin. Pathogens, nutrients, and sediment loads were all problems in this watershed [54]. High levels of fecal coliform bacteria have caused the Lockwoods Folly River to be included in the USEPA's 303(d) list of impaired waters and have resulted in its closure to shellfishing [55]. The stressors of urbanization are expected to exacerbate these problems as Brunswick County, NC, is the 31st fastest growing county in the United States during 2010–2016 [56].

2.2. Brunswick County Bioswale

2.2.1. Watershed Characteristics

To treat road runoff, a bioswale was installed in the right-of-way on NC 211, approximately 1.6 km east of its intersection with US17. NC 211 is a two-lane state highway with an asphalt wearing course which was in good condition during the study period. The drainage area was 0.74 hectares, 44% of which was directly connected impervious area. Pervious areas were the existing grassed shoulders, in good condition and on a 4:1 horizontal distance:vertical distance (H:V) slope with highly transmissive, sandy soils. Diffuse stormwater runoff from the northern lane of the two-lane road discharged directly onto the grass shoulder, which acts as a vegetated filter strip, allowing for initial settling of sediment and particulate-borne contaminants and for some infiltration. A portion of this channel was removed to install the forebay and bioswale. The concrete-lined channel first drained into a forebay (Figure 1), which served to dissipate energy and prevent erosion in the bioswale. The bioswale commenced immediately downslope of the forebay.

Figure 1. Watershed, grassed shoulder, and concrete-lined channel leading to the forebay of the bioswale.

2.2.2. Bioswale Design

The forebay is 2.7 m wide by 10.7 m long. The initial 6.1 m was a triangular channel on a 3% slope. The plunge pool consisted of the latter 4.6 m of the forebay and has a depth of approximately 0.15 m. The slope into and out of the pool was 6:1 (H:V). The entire forebay was lined with class A rip-rap (50 to 150 mm diameter stone [57]) to a depth of 0.2 m. The high flow media (Table 1), approximately 0.9 m deep, began 1.8 m past the start of the rip-rap lined channel and continued under the forebay to ensure that the system completely drained inter-event.

Table 1. High flow media characteristics.

Characteristic	Value
Hydraulic conductivity (K_{sat})	2540 mm h^{-1}
Peat Moss	15% by volume
Total Carbon	>85%
Carbon to Nitrogen Ratio	15:1 to 23:1
Lignin Content	49–52%
Humic Acid	>18%
pH	6.0–7.0
Moisture Content	30–50%
Passing 2.0 mm sieve	95–100%
Passing 1.0 mm sieve	>80%
Sand-Fine	<5%
Sand-Medium	10–15%
Sand-Coarse	15–25%
Sand-Very Coarse	40–45%
Gravel	10–20%
Clay/Silts	<2%

To create the bioswale, a trench with a width of 1.2 m, depth of 0.8 m, and length of 30.5 m was excavated, starting at the end of the plunge pool. Once excavated, the entire trench, including under the rip-rap channel and plunge pool, was lined with a high flow fabric prior to being backfilled. This fabric ensured the high flow media remained within the system, but allowed water that passed through the media to infiltrate into the underlying soil. The first 11 m, starting at the end of the plunge pool, was completely (all 0.9 m) filled with only the high flow media.

After the first 11 m, a perforated underdrain was installed. The trapezoidal base of the ditch was filled with a 5 cm layer of ASTM standard #57 stone (2.36 to 37.5 mm stone size [58]), serving as internal water storage (IWS), which has been shown to substantially improve runoff reduction within bioretention cells by promoting inter-event exfiltration [59,60]. Then, 18 m of perforated high-density polyethylene pipe (HDPE) pipe (0.2 m diameter) (Advanced Drainage Systems, Inc., Raleigh, NC, USA) was placed over the stone layer, creating an IWS zone of 5 cm. The pipe was covered with 5 cm ASTM #57 stone. Next, a pea gravel layer was placed on top of the ASTM #57 stone. Finally, a fiberglass mesh screen was placed around the pea gravel (Figure 2). These 'choking' layers of gravel and media limited soil media movement to the underdrain. The remaining trench volume was filled with the high flow media and covered with a thin-cut warm-season sod. The resulting bioswale consisted of a triangular channel with 4:1 H:V side slopes and a total length of 42 m. A longitudinal cross-section of the full bioswale system can be seen in Figure 3, with design characteristics in Table 2.

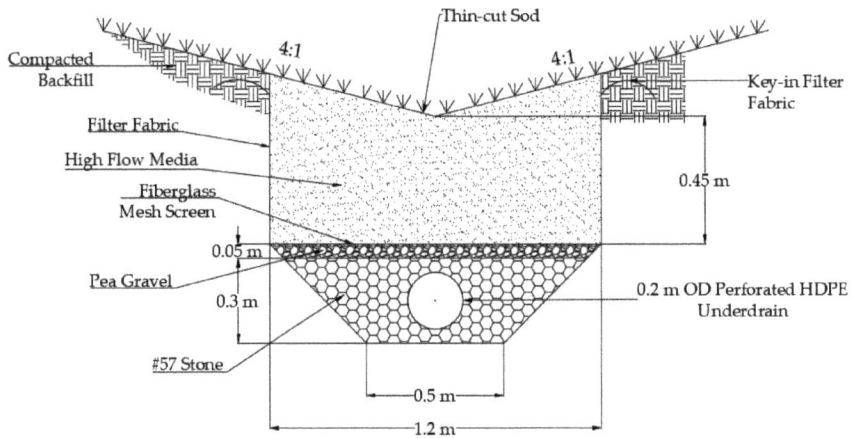

Figure 2. Horizontal cross-section of the underdrained portion (flow going into page). Note-not to scale.

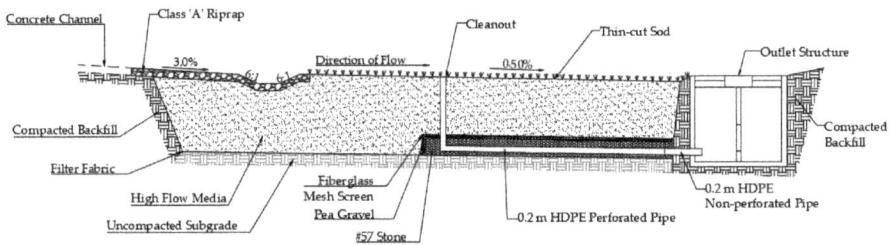

Figure 3. Full structure longitudinal cross-section. Note: not to scale.

Table 2. Bioswale design characteristics.

Characteristic	Value
Rip-rap channel length	6 m
Rip-rap channel slope	3%
Plunge pool length	4.6 m
Plunge pool depth	0.15 m
Underdrain length	18.3 m
Underdrain diameter	0.2 m
Media depth	0.45–0.9 m
Total length	42 m
Surface geometry	Triangular
Surface side slopes	4:1
Media void storage	22.7 m^3
Surface storage	14.2 m^3

The bioswale underdrain and surface flow discharged into the existing outlet structure (Figure 4). The first chamber housed the monitoring equipment for the underdrain; surface flow was prevented from mixing with underdrainage. Bioswale overflow was monitored in the downstream chamber. The outlet structure was elevated 15 cm above the swale, resulting in up to 67% of the bioswale surface area being inundated during a storm. The maximum cumulative storage within the bioswale at the brink of overflow was 36.9 m^3, 14.2 m^3 of surface storage and 22.7 m^3 of soil void space.

Figure 4. Bioswale outlet structure, with the upstream grate housing the underdrain monitoring and the downstream grate housing the overflow monitoring. The completed bioswale is present in the background.

2.2.3. Climatic and Water Quality Data Collection

As stormwater entered the bioswale (rip-rap), a wooden board was used to pool water for sample aliquot collection. Inlet aliquots were collected using a Teledyne ISCO 6712 (Lincoln, NE, USA) automated sampler. Sampling was triggered by two nearby rain gauges, the first enabled the sampler after 2.54 mm of rainfall had occurred, while the second triggered the sampler to obtain 200 mL aliquots after each additional 1 mm of rainfall. Thus, the maximum rainfall depth which could be sampled was 52 mm. However, the 1-mm trigger was increased in anticipation of several larger storm events, and the actual maximum rainfall depth for a sampled storm event was 92 mm Since rainfall depth is considered a good predictor of runoff volume in urbanized watersheds, these samples were considered flow-proportional [61].

At the underdrain and overflow monitoring points in the outlet structure, purpose-built weirs were installed to measure discharge with time. Each weir had an ISCO 6712 automated sampler with a 730 bubbler module, which measured flow depth over the weir plate (Figure 5). The sampler converted the flow depth to a corresponding flow rate using the following equations.

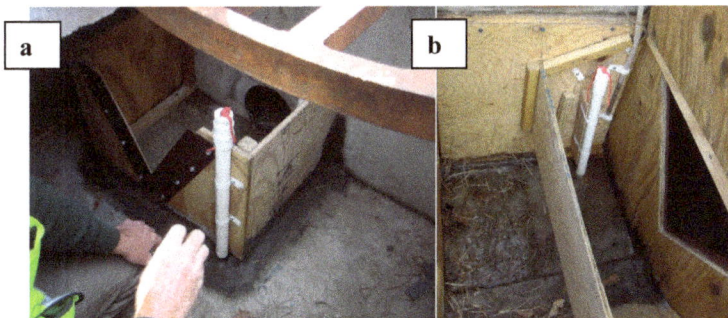

Figure 5. (**a**) Sixty-degree v-notch weir installed to measure drainage from the underdrain and pond water for sampling (flow direction from background to foreground); (**b**) Ninety-degree v-notch weir installed to measure overflow/bypass with a baffle (center) to still flow for sampling (flow direction from left to right).

The equation for the 60° underdrain weir is as follows:

$$Q = 796.7 \times H^{2.5} \tag{1}$$

where Q is the flow rate in L s^{-1} and H is the flow depth in m.

The equation for the 90° overflow weir is as follows:

$$Q = 1380 \times H^{2.5} \tag{2}$$

where Q is the flow rate in L s^{-1} and H is the flow depth in m.

Flow rate data at these two monitoring points were integrated with time to determine runoff volume passing each weir on 2-min intervals Equation (3). Automated samplers then obtained volume-proportional 200 mL aliquots on pre-programmed intervals (e.g., every 500 L) throughout the hydrograph. The sample tubing was installed behind the weir plate to ensure samples were collected before overflowing the weir.

$$V = Q \times t \tag{3}$$

where V is the corresponding volume in L, Q is the corresponding flow rate in L s^{-1}, and t is the time step of 120 s (2 min) sampling interval.

Samples were triggered across the hydrograph based on volume passing over the weir and represented, at minimum, 80% of the total flow volume, characterizing (essentially) the entire pollutograph. Composite samples were analyzed for water quality only if paired inlet and outlet samples were obtained.

The rainfall depth in the on-site manual rain gauge was checked during each sampling mission to compare against the tipping bucket rain gauge data. Rainfall and water quality data were collected over a 1-year period (25 February 2014 through 26 February 2015).

2.3. Water Quality Analysis

Stormwater runoff samples were obtained from the ISCO samplers within 24 h of the cessation of rainfall. The 10 L composite sample bottles were shaken vigorously to re-suspend sediment and sub-sampled into laboratory containers for transit. The remaining sample volume in the composite sample jar was discarded and the bottle washed with deionized water and replaced within the ISCO sampler for the next storm event. Samples were placed on ice immediately after sub-sampling and chilled to less than 4 °C for transit to the Environmental Quality Laboratory at Coastal Carolina University. All samples were measured for conductivity, turbidity, total suspended solids (TSS), volatile suspended solids (VSS), enterococci (Ent), and fecal coliform (FC) using Standard Methods [62–66] except enterococci [67] Lab duplicates and field duplicates were analyzed for all water quality parameters (TSS, turbidity, VSS, fecal coliform, and enterococci) for the inlet sample, because this was the location with the largest collected sample volume. All duplicates were within 20% relative percent difference.

2.4. Statistical Analysis

The water quality data were statistically analyzed to compare paired influent and effluent water quality for five parameters: TSS, turbidity, VSS, fecal coliform, and enterococci. Each data set was tested for normality using the Anderson-Darling procedure using $\alpha = 0.05$. For all water quality parameters, at least one data set (either inlet, underdrain, or overflow) was not normally distributed and was unable to be transformed using log or squared transformations. Kendall's tau non-parametric rank correlation, therefore, determined statistically significant correlations between pollutants. Tests were also run to determine any correlations between runoff flow concentrations (inflow, underdrain, and overflow) and rainfall characteristics (rainfall depth and antecedent dry period). To assess the effects of treatment, or lack thereof, in the filter media, statistical comparisons, using Wilcoxon Signed

Rank Test, were made between the inlet and underdrain and inlet and overflow data sets. A criterion of 95% confidence ($\alpha = 0.05$) was used for all tests. Statistical analyses were performed using the R software (v. 3.4.3) (R Core Team, Vienna, Austria) [68].

Concentration reductions (CR) were calculated Equation (4) for each pollutant and outlet monitoring point using USEPA's efficiency ratio [69]:

$$CR = \left(1 - \frac{mean\ outlet\ concentration}{mean\ inlet\ concentration}\right) * 100\% \tag{4}$$

The geometric mean was used for enterococci and fecal coliform; the arithmetic mean was used for TSS, VSS, and turbidity.

Probability plots for enterococci and fecal coliform were created to evaluate the bioswale across all influent and outflow concentrations. The probability was calculated using Equation (5):

$$P = \frac{i - 0.5}{n} \tag{5}$$

where P is the probability of an observation, i is the rank of the observation, and n is the number of observations in the data set [70].

3. Results and Discussion

3.1. Storm Event Characteristics

A total of 15 storm events were sampled for water quality. These storms ranged in rainfall depth from 13.2 to 91.7 mm (mean 38.1 mm), with an antecedent dry period (ADP) from 0.35 to 7.8 days (mean 4.92 days) (Table 3). Sampled storm events were collected throughout the year, with 3–4 events captured in each season. However, not all events had enough runoff volume, overflow in particular, to be analyzed for all contaminants (Table 3).

Table 3. Rainfall characteristics for each storm sampling event.

Storm Sampling Event	Date	Rainfall Depth (mm)	Antecedent Dry Period (days)	Sampled for Inlet Flow?	Sampled for Underdrain Flow?	Sampled for Overflow?
1	4/16/2014	18.0	0.35	B, S, T	B, S, T	-
2	5/16/2014	90.9	MD	B, S, T	B, S, T	-
3	6/21/2014	34.0	7.07	B, S, T	B, S, t	-
4	6/24/2014	18.8	1.39	B, S, T	B, S, T	-
5	7/4/2014	69.3	4.56	B, S, T	B, S, T	B, S, T
6	7/25/2014	40.1	MD	B, S, T	B, S, T	B, S, T
7	9/6/2014	25.4	6.79	B, S, T	B, S, T	-
8	9/30/2014	13.2	3.34	B, T	B, T	B, T
9	11/1/2014	25.4	MD	B, S, T	B, S, T	B, T
10	11/23/2014	54.6	6.44	B, S, T	B, S, T	B, S, T
11	1/12/2015	22.9	7.80	B, S, T	B, S, T	B, T
12	1/24/2015	91.7	5.19	B, S, T	B, S, T	B, S, T
13	2/17/2015	17.3	6.23	B, S, T	B, S, T	-
14	2/23/2015	16.5	MD	B, S, T	B, S, T	B, T
15	2/26/2015	33.0	MD	B, S, T	B, S, T	B, S, T

MD: missing datasets, unable to calculate antecedent dry period; B: bacteria (fecal coliform, enterococcus); S: sediment (TSS, VSS); T: turbidity.

3.2. Impact on Pathogen Indicator Species and Sediment Removal

Non-parametric statistical tests between the runoff concentrations at all monitoring points and the storm characteristics of rainfall depth and antecedent dry period found only two significant correlations (a = 0.05). The positive correlations were between rainfall depth and underdrain VSS concentration ($p = 0.043$) and rainfall depth and overflow fecal coliform concentration ($p = 0.012$). These results highlight the complexity of basing the bioswale's performance on the storm characteristics of rainfall depth and antecedent dry period.

Outflow concentrations for all five pollutants examined were lower, but statistically significant reductions were only observed when comparing underdrain to influent concentrations (Table 4). A principal pollutant removal mechanism of a bioswale is filtration [71,72], which explains why TSS and VSS concentrations from underdrains were very low (4.2 and 1.6 mg/L, respectively). There was less impact observed in overflow since it does not undergo filtration through the media. However, the bioswale's forebay and vegetation aids in reducing the runoff velocity, allowing for sedimentation [35,73], thus (non-significantly) reducing TSS concentrations in overflow to 32 mg/L compared to 35 mg/L influent. Overflow results herein can be compared to those of other 'standard' swales and were similar, which ranged from 8–70 mg/L TSS [32,35,73,74].

Table 4. Means for pathogen indicator species and sediments and water quality standards.

Sampling Location	Enterococci [1] (MPN/100 mL)	Fecal Coliform [1] (MPN/100 mL)	TSS [2] (mg/L)	VSS [2] (mg/L)	Turbidity [2] (NTU)
Inflow	3451 {903}	320 {126}	35.1 {1.9}	12.4 {0.6}	23.3 {0.5}
Underdrain	**1411** (0.004) {290}	*111* (0.021) {61}	**4.2** (0.000) {0.2}	**1.6** (0.000) {0.04}	**14.8** (0.000) {0.3}
Overflow	1549 (0.455) {549}	*79* (0.180) {30}	31.6 (0.313) {4.7}	9.7 (0.313) {1.3}	22.4 (0.326) {1.3}
North Carolina Limits	35 [a]	200 [b]	20.0 [b]	-	50.0 [b]

[1] Geometric mean, [2] Arithmetic mean, [a] [11], [b] [75], Bolded values were significant reductions with respect to inflow concentrations, italicized values were below the U.S. EPA limits, (*p*-value compared to inflow concentration), {standard error of mean}.

The bioswale reduced fecal coliform concentrations to values less than the U.S. EPA water quality limit of 200 Most Probable Number (MPN)/100 mL [75], but was unable to meet the enterococci swimming limit of 35 MPN/100 mL [11] (Table 5). However, influent concentrations were 100-fold higher than the federal standards for enterococci and only 1.5 times higher for fecal coliform. Both inflow and outflow also met state standards for turbidity (50 Nephelometric Turbidity Units (NTU)). Underdrain TSS concentrations (mean 4 mg/L) were less than typical water quality standards (20 mg/L), but those of overflow (32 mg/L) were not.

Table 5. Percent reduction of mean from inlet for each pollutant from surface and underdrain samples.

Sampling Location	Enterococci [1]	Fecal Coliform [1]	TSS [2]	VSS [2]	Turbidity [2]
Underdrain	59%	65%	88%	87%	36%
Overflow	55%	75%	10%	21%	4%

[1] Geometric mean, [2] Arithmetic mean, Bolded values were significant reductions.

The largest concentration reductions were observed for underdrain TSS (88%) and VSS (87%) because these pollutants were presumably filtered by the fill media (Table 5). Sediment trapping efficiencies were similar to those for bioretention [47] and Austin sand filters [76]. Reductions from inflow to (non-filtered) overflow for these two pollutants were much lower (10% for TSS and 21% for VSS), suggesting water exceeding a bioswale's capacity for filtration will receive little treatment. Results associated with turbidity were similar albeit influent turbidity was already quite low (23 NTUs); a more noticeable reduction in turbidity was associated with filtered underdrain discharges. Both fecal coliform and enterococci reductions were substantial (>50%) for both underdrain and overflow monitoring points. However, only underdrain effluent was significantly improved compared to influent, perhaps due to the small sample size (*n* = 5) for the overflow monitoring point.

Figure 6 illustrates that for all storms sampled and at all sampling locations, concentrations of enterococci exceeded the water quality standard of 35 MPN/100 mL. However, approximately 33% of inflow concentrations, 67% of overflow concentrations, and 73% of underdrain concentrations were less than the 200 MPN/100 mL water quality standard for fecal coliform (Figure 7). Up to 1-log difference in fecal coliform at the inlet compared to the underdrain and overflow was observed, supporting the significant treatment of fecal coliform within the bioswale. One interesting finding is that for both enterococci and fecal coliform, the distributions of underdrain and overflow concentrations were quite similar (Figures 6 and 7). One plausible explanation is that bacteria in this study were mobilized primarily during the first flush [77]. Overflow occurs only after the rainfall exceeds the bioswale's soil and surface storage capacity. When overflow begins, a substantial portion of the bacteria may thus have already been settled and/or filtered out by the soil media (i.e., cleaner inflow at the time overflow begins). While overflow is not treated to the same extent as underdrain flow, lower bacterial concentrations in stormwater runoff after the 'dirtiest' water has been mobilized from the watershed may have resulted in similar distributions of underdrain and overflow FIB concentrations.

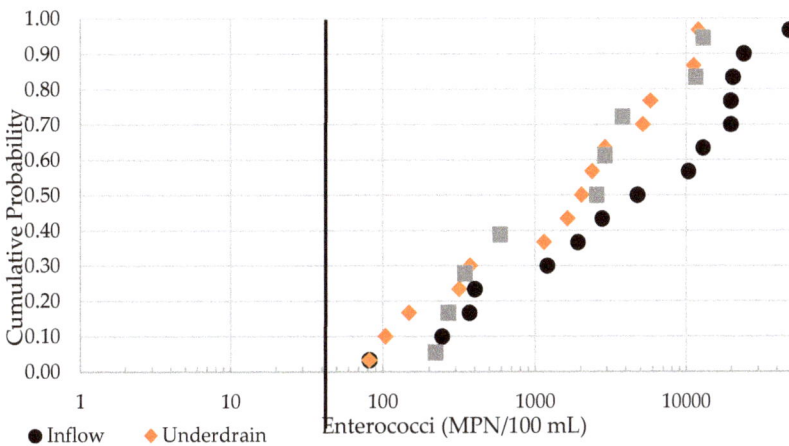

Figure 6. Enterococci probability plot with water quality standard of 35 MPN/100 mL.

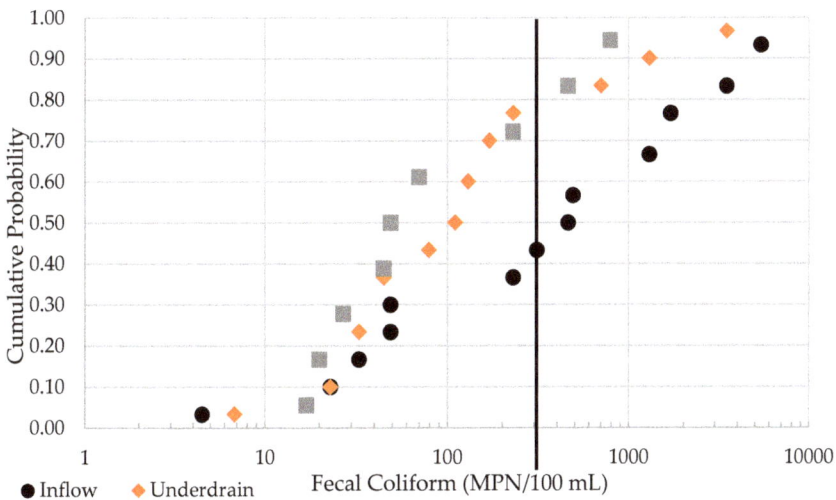

Figure 7. Fecal coliform probability plot with water quality standard of 200 MPN/100 mL.

Results demonstrate the complexities associated with (1) understanding bioswale treatment and (2) determining whether an SCM, including a bioswale, is a 'good' practice for FIB treatment. While no previously peer-reviewed studies have reported how a bioswale impacts pollutants, many studies have evaluated the effectiveness of other filtration-based SCMs, in particular bioretention. Taken cumulatively, studies illustrate a wide range of concentration "reductions" for enterococci, fecal coliform, and TSS for various SCMs (Table 6). The bioswale concentrations herein (particularly those of the filtered effluent from the underdrain) were in the range of or modestly higher than those of other SCMs. Thus, bioswales appear capable of achieving similar results to that of other more commonly-employed SCMs, such as bioretention.

Table 6. Summary of studies reporting bacteria and sediment concentration reductions from SCMs.

Author	Location	SCM Type	Ent. CR (%)	FC CR (%)	TSS CR (%)
Herein	Brunswick County, NC	Bioswale Overflow	55	75	10
		Bioswale Underdrain	59	65	88
Hathaway and Hunt [78]	Wilmington, NC	Bioretention cell	89	-	-
		Bioretention cell	−1	-	-
Passeport et al. [79]	Alamance County, NC	Bioretention cell	-	95	-
		Bioretention cell	-	85	-
Davis [80]	College Park, MD	Bioretention cell	-	-	47 [a]
		Bioretention cell	-	-	62 [a]
Hunt et al. [47]	Charlotte, NC	Bioretention cell	-	69 [a]	60 [a]
Hathaway and Hunt [78]	Wilmington, NC	Wet pond	90	-	-
		Wet pond	87	-	-
Hathaway and Hunt [78]	Wilmington, NC	Wetland	69	-	-
		Wetland	41	-	-
Davies and Bavor [81]	Sydney, Australia	Wetland	85	-	-
Krometis et al. [82]	Central NC	Wet Retention Pond	−108	−41	-
		Wet Retention Pond	36	31	-
Mallin et al. [83]	New Hanover County, NC	Wet Detention Pond	-	86 [a]	65 [a]
		Wet Detention Pond	-	56 [a]	−37 [a]
		Wet Detention Pond	-	−15 [a]	−22 [a]

Ent. CR: Enterococci concentration reduction; FC CR: Fecal coliform concentration reduction; TSS CR: Total suspended sediment concentration reduction; [a] Concentration reduction (CR) manually calculated based on concentrations provided by corresponding author(s).

3.3. Statistically Significant Correlations

Several significant correlations were found to impact the bioswale's performance (Figure 8) (p-value < 0.05 shows significance). The inflow concentration of enterococci and fecal coliform were positively correlated with both TSS and VSS. Thus, if sediment concentrations were high, one would likely observe higher enterococci and fecal coliform concentrations, which is logical since bacteria are often sediment-bound [84,85]. This potentially provides a framework for choosing locations to retrofit SCMs when targeting bacteria removal in a watershed: by simply measuring TSS or VSS concentrations, which is inherently less expensive and time consuming. Drainage areas that produce relatively higher sediment concentrations would appear to be good candidates for fecal coliform- and enterococcus-reducing SCMs, such as bioswales.

	Inlet ENT	Inlet FC	Inlet TSS	Inlet VSS	Inlet Turb	Under ENT	Under FC	Under TSS	Under VSS	Under Turb	Over ENT	Over FC	Over TSS	Over VSS	Over Turb
Inlet ENT	0	0	0	0	0.15	0	0.06	0.16	0.48	0.81	0	0.18	0.59	0.59	0.91
Inlet FC		0	0	0	0.06	0.01	0	0.26	0.57	0.5	0.01	0.12	0.24	0.24	0.46
Inlet TSS			0	0	0.01	0.03	0.04	0.08	0.29	0.46	0.03	0.03	0.41	0.41	0.73
Inlet VSS				0	0	0.04	0.04	0.08	0.31	0.29	0.02	0.05	0.41	0.41	0.64
Inlet Turb					0	0.2	0.02	0.05	0.1	0.08	0.38	0.17	0.24	0.24	0
Under ENT						0	0.03	0.04	0.2	0.79	0	0.09	0.6	0.6	0.88
Under FC							0	0.1	0.22	0.33	0.15	0.42	0.12	0.12	0.1
Under TSS								0	0	0.12	0.05	0.02	0.41	0.41	0.45
Under VSS									0	0.02	0.2	0.03	0.12	0.12	0.2
Under Turb										0	0.69	0.38	0.01	0.01	0.04
Over ENT											0	0.06	0.76	0.76	0.87
Over FC												0	0.59	0.59	0.62
Over TSS													0	0.01	0.04
Over VSS														0	0.04
Over Turb															0

Figure 8. Kendall's Tau *p*-value correlation matrix for all pollutants (a = 0.05).

4. Summary and Conclusions

Concentrations for the five pollutants examined (turbidity, TSS, VSS, enterococci and fecal coliform) were uniformly lower in underdrain flow than in inflow. Greater than 55% removal was observed for FIB and greater than 85% removal for TSS, suggesting filtration is an effective removal mechanism for these pollutants. Overflow concentrations were similar to those from the underdrain; overflow was not treated to the same extent as drainage, but that overflowed occurred after the first flush, and therefore may have had lower bacterial concentrations. Fecal coliform outflow concentrations often met targets established by the USEPA [11] for recreational waters, but this was never the case for enterococci. A cause might be the markedly higher inflow concentrations measured for the latter indicator species. Turbidity targets were also met for both underdrain flow and overflow, but only underdrain outflow achieved TSS thresholds [75].

Synthesizing the results of this bioswale and comparing them to other SCMs indicates that the practice might be a popular tool. Bioswales fits within existing rights-of-way and water that infiltrates the media was measured to be cleaner than that of surface flow for common pollutants. This case study suggests that bioswales function should be more closely examined as a function of the ratio of watershed size to bioswale length, the impact of slope, drainage area properties, soil media type, etc. This will allow for design standards for bioswales to be crafted.

Acknowledgments: This material is based on work supported by the North Carolina Department of Transportation (NCDOT) Grant 2016-18 and the National Science Foundation (NSF) Grant DGE-1252376. Field sample collection and water quality analysis was performed by Jeff Barley (CCU's Environmental Quality Lab).

Author Contributions: Matthew S. Lauffer, Andrew McDaniel, Ryan J. Winston, Brian Lipscomb and William F. Hunt conceived and designed the experiments; Ryan J. Winston and Susan Libes performed the experiments; Rebecca A. Purvis, Ryan J. Winston, and William F. Hunt analyzed the data; Coastal Carolina University contributed reagents/materials/analysis tools; Rebecca A. Purvis, Ryan J. Winston, William F. Hunt, and Karthik Narayanaswamy wrote the paper.

Conflicts of Interest: The authors declare no conflict of interest. The sponsors of the project paid for the design and construction of the bioswale. The sponsors also reviewed the manuscript prior to submission, but otherwise had no function in the design of the study; in the collection, analyses or interpretation of data; in the writing of the manuscript, or in the decision to publish the results.

References

1. United Nations. *World Urbanization Prospects: The 2014 Revision, Highlights*; United Nations Department of Economic and Social Affairs, Population Division: New York, NY, USA, 2014. Available online: https://esa.un.org/unpd/wup/publications/files/wup2014-highlights.pdf (accessed on 1 October 2017).
2. Morisawa, M.; LaFlure, E. Hydraulic geometry, stream equilibrium and urbanization. In *Adjustments of the Fluvial Systems, Proceedings of the 10th annual Geomorphology Symposium Series, Binghampton, New York, NY, USA, 21–22 September 1979*; Rhodes, D.D., Williams, G.P., Eds.; Kendall/Hunt Publishing Co., Inc.: Dubuque, IA, USA, 1979.
3. Arnold, C.L.; Boison, P.J.; Patton, P.C. Sawmill Brook: An example of rapid geomorphic change related to urbanization. *J. Geol.* **1982**, *90*, 155–166. [CrossRef]
4. Bannerman, R.T.; Owens, D.W.; Dodds, R.B.; Hornewer, N.J. Sources of pollutants in Wisconsin stormwater. *Water Sci. Technol.* **1993**, *28*, 241–259.
5. Brabec, E.; Schulte, S.; Richards, P.L. Impervious surfaces and water quality: A review of current literature and its implications for watershed planning. *J. Plan. Lit.* **2002**, *16*, 499–514. [CrossRef]
6. Todeschinie, S. Hydrologic and Environmental Impacts of Imperviousness in an Industrial Catchment of Northern Italy. *J. Hydrol. Eng.* **2016**, *21*, 05016013. [CrossRef]
7. Thomson, N.R.; McBean, E.A.; Snodgrass, W.; Monstrenko, I.B. Highway stormwater runoff quality: Development of surrogate parameter relationships. *Water Air Soil Pollut.* **1997**, *94*, 307–347. [CrossRef]
8. Opher, T.; Friedler, E. Factors affection highway runoff quality. *Urban Water J.* **2010**, *7*, 155–172. [CrossRef]
9. Ingvertsen, S.T.; Cederkvist, K.; Régent, Y.; Sommer, H.; Magid, J.; Jensen, M.B. Assessment of existing roadside swales and engineered filter soil: I. Characterization and lifetime expectancy. *J. Environ. Qual.* **2012**, *41*, 1960–1969. [CrossRef] [PubMed]
10. U.S. Environmental Protection Agency (USEPA). *National Summary of State Information*; U.S. Environmental Protection Agency (USEPA): Washington, DC, USA, 2016. Available online: https://www.iaspub.epa.gov/waters10/attains_nation_cy.control (accessed on 5 June 2017).
11. U.S. Environmental Protection Agency (USEPA). *2012 Recreational Water Quality Criteria*; EPA-820-F-12-061; USEPA Office of Water: Washington, DC, USA, 2012.
12. Mallin, M.A.; Williams, K.E.; Esham, E.C.; Lowe, R.P. Effect of human development on bacteriological water quality in coastal watersheds. *Ecol. Appl.* **2000**, *10*, 1047–1056. [CrossRef]
13. U.S. Environmental Protection Agency (USEPA). *Protocol for Developing Pathogen TMDLs*; EPA-841-R-00-002; USEPA Office of Water: Washington, DC, USA, 2001.
14. Stevik, T.K.; Aa, K.; Ausland, G.; Hanssen, J.F. Retention and removal of pathogenic bacteria in wastewater percolating through porous media: A review. *Water Res.* **2004**, *38*, 1355–1367. [CrossRef] [PubMed]
15. Arnone, R.D.; Walling, J.P. Waterborne pathogens in urban watersheds. *J. Water Health* **2007**, *5*, 149–162. [CrossRef] [PubMed]
16. Struck, S.D.; Selvakumar, A.; Borst, M. Prediction of Effluent Quality from Retention Ponds and Constructed Wetlands for Managing Bacterial Stressors in Storm-Water Runoff. *J. Irrig. Drain. Eng.* **2008**, *134*, 567–578. [CrossRef]
17. Weber-Shirk, M.L.; Dick, R.I. Physical-Chemical Mechanisms in Slow Sand Filters. *J. Am. Water Works Assoc.* **1997**, *89*, 87–100.

18. Rusciano, G.M.; Obropta, C.C. Bioretention column study: Fecal coliform and total suspended solids reductions. *Trans. ASABE* **2007**, *50*, 1261–1269. [CrossRef]
19. Sherer, B.M.; Miner, R.; Moore, J.A.; Buckhouse, J.C. Indicator bacterial survival in stream sediments. *J. Environ. Qual.* **1992**, *21*, 591–595. [CrossRef]
20. Mohanty, S.K.; Torkelson, A.A.; Dodd, H.; Nelson, K.L.; Boehm, A.B. Engineering Solutions to Improve the Removal of Fecal Indicator Bacteria by Bioinfiltration Systems during Intermittent Flow of Stormwater. *Environ. Sci. Technol.* **2013**, *47*, 10791–10798. [CrossRef] [PubMed]
21. Leclerc, H.; Mossel, D.A.A.; Edberg, S.C.; Struijk, C.B. Advances in the bacteriology of the coliform group: Their suitability as markers of microbial water safety. *Annu. Rev. Microbiol.* **2001**, *55*, 201–234. [CrossRef] [PubMed]
22. Savichtcheva, O.; Okabe, S. Alternative indicators of fecal pollution: Relations with pathogens and conventional indicators, current methodologies for direct pathogen monitoring and future application perspectives. *Water Res.* **2006**, *40*, 2463–2476. [CrossRef] [PubMed]
23. Pan, X.; Jones, K.D. Seasonal variation of fecal indicator bacteria in storm events within the US stormwater database. *Water Sci. Technol.* **2012**, *65*, 1076–1080. [CrossRef] [PubMed]
24. O'Neill, S.; Adhikari, A.R.; Gautam, M.R.; Acharya, K. Bacterial contamination due to point and nonpoint source pollution in a rapidly growing urban center in an arid region. *Urban Water J.* **2013**, *10*, 411–421. [CrossRef]
25. Hathaway, J.M.; Hunt, W.F.; Graves, A.K.; Wright, J.D. Field evaluation of bioretention indicator bacteria sequestration in Wilmington, NC. *J. Environ. Eng.* **2011**, *137*, 1103–1113. [CrossRef]
26. Li, Y.L.; Deletic, A.; Alcazar, L.; Bratieres, K.; Fletcher, T.D.; McCarthy, D.T. Removal of *Clostridium perfringens*, *Escherichia coli*, and F-RNA coliphanges by stormwater biofilters. *Ecol. Eng.* **2012**, *49*, 137–145. [CrossRef]
27. Rowny, J.G.; Stewart, J.R. Characterization of nonpoint source microbial contamination in an urbanizing watershed serving as a municipal water supply. *Water Res.* **2012**, *46*, 6143–6153. [CrossRef] [PubMed]
28. Fletcher, T.D.; Shuster, W.; Hunt, W.F.; Ashley, R.; Butler, D.; Arthur, S.; Trowsdale, S.; Barraud, S.; Semadeni-Davies, A.; Bertrand-Krajewski, J.L.; et al. SUDS, LID, BMPs, WSUD and more—The evolution and application of terminology surrounding urban drainage. *Urban Water J.* **2015**, *12*, 525–542. [CrossRef]
29. Rushton, B.T. Low-impact parking lot design reduces runoff and pollutants loads. *J. Water Resour. Plan. Manag.* **2001**, *127*, 172–179. [CrossRef]
30. Holman-Dodds, J.K.; Bradley, A.A.; Potter, K.W. Evaluation of hydrologic benefits of infiltration based urban storm water management. *J. Am. Water Resour. Assoc.* **2003**, *39*, 205–215. [CrossRef]
31. Hunt, W.F.; Davis, A.P.; Traver, R.G. Meeting hydrologic and water quality goals through targeted bioretention design. *J. Environ. Eng.* **2012**, *138*, 698–707. [CrossRef]
32. Barrett, M.E.; Wals, P.M.; Malina, J.F., Jr.; Charbeneau, R.J. Performance of vegetative controls for treating highway runoff. *J. Environ. Eng.* **1998**, *124*, 1121–1128. [CrossRef]
33. Davis, A.P.; Stagge, J.H.; Jamil, E.; Kim, H. Hydraulic performance of grass swales for managing highway runoff. *Water Res.* **2011**, *46*, 6775–6786. [CrossRef] [PubMed]
34. Deletic, A. Modelling of water and sediment transport over grassed areas. *J. Hydrol.* **2001**, *248*, 168–182. [CrossRef]
35. Bäckström, M. Sediment transport in grassed swales during simulated runoff events. *Water Sci. Technol.* **2002**, *45*, 41–49. [PubMed]
36. Barrett, M.E. Performance comparison of structural stormwater best management practices. *Water Environ. Res.* **2005**, *77*, 78–86. [CrossRef] [PubMed]
37. Ackerman, D.; Stein, E.D. Evaluating the effectiveness of best management practices using dynamic modeling. *J. Environ. Eng.* **2008**, *134*, 628–639. [CrossRef]
38. Knight, E.M.P.; Hunt, W.F., III; Winston, R.J. Side-by-side evaluation of four level spreader-vegetated filter strips and a swale in eastern North Carolina. *J. Soil Water Conserv.* **2013**, *68*, 60–72. [CrossRef]
39. Stagge, J.H.; Davis, A.P.; Jamil, E.; Kim, H. Performance of grass swales for improving water quality from highway runoff. *Water Res.* **2012**, *46*, 6731–6742. [CrossRef] [PubMed]
40. Yu, S.L.; Kuo, J.T.; Fassman, E.A.; Pan, H. Field test of grassed-swale performance in removing runoff pollution. *J. Water Resour. Plan. Manag.* **2001**, *127*, 168–171. [CrossRef]
41. Lucke, T.; Mohamed, M.A.K.; Tindale, N. Pollutant removal and hydraulic retention performance of field grassed swales during runoff simulation experiments. *Water* **2014**, *6*, 1887–1904. [CrossRef]

42. Davis, A.P.; Hunt, W.F.; Traver, R.G.; Clar, M. Bioretention technology: Overview of current practice and future needs. *J. Environ. Eng.* **2009**, *135*, 109–117. [CrossRef]

43. Christianson, R.D.; Barfield, B.J.; Hayes, J.C.; Gasem, K.; Brown, G.O. Modeling effectiveness of bioretention cells for control of stormwater quantity and quality. In *Critical Transitions in Water and Environmental Resources Management, Proceedings of the 2004 World Water and Environmental Resources Congress, Salt Lake City, UT, USA, 27 June–1 July 2004*; American Society of Civil Engineers: Reston, VA, USA, 2004.

44. Davis, A.P.; Shokouhian, M.; Sharma, H.; Minami, C. Laboratory study of biological retention for urban stormwater management. *Water Environ. Res.* **2001**, *73*, 5–14. [CrossRef] [PubMed]

45. Hunt, W.F.; Jarrett, A.R.; Smith, J.T.; Sharkey, L.J. Evaluating bioretention hydrology and nutrient removal at three field sites in North Carolina. *J. Irrig. Drain. Eng.* **2006**, *132*, 600–608. [CrossRef]

46. Garbrecht, K.; Fox, G.A.; Guzman, J.A.; Alexander, D. *E. coli* transport through soil columns: Implications for bioretention cell removal efficiency. *Trans. ASABE* **2009**, *52*, 481–486. [CrossRef]

47. Hunt, W.F.; Smith, J.T.; Jadlocki, S.J.; Hathaway, J.M.; Eubanks, P.R. Pollutant removal and peak flow mitigation by a bioretention cell in urban Charlotte, N.C. *J. Environ. Eng.* **2008**, *134*, 403–408. [CrossRef]

48. Hathaway, J.M.; Hunt, W.F.; Jadlocki, S. Indicator bacteria removal in stormwater best management practices in Charlotte, North Carolina. *J. Environ. Eng.* **2009**, *135*, 1275–1285. [CrossRef]

49. Zinger, Y.; Deletic, A.; Fletcher, T.D.; Breen, P.; Wong, T. A Dual-Mode Biofilter System: Case Study in Kfar Sava, Israel. In Proceedings of the 12th International Conference on Urban Drainage, Porto Alegre, Brazil, 11–16 September 2011.

50. Zhang, L.; Seagren, E.A.; Davis, A.P.; Karns, J.S. Effects of temperature on bacterial transport and destruction in bioretention media: Field and laboratory evaluations. *Water Environ. Res.* **2012**, *84*, 485–496. [CrossRef] [PubMed]

51. McLaughlin, J. NYC bioswales pilot project improves stormwater management. *Clear Waters* **2012**, *2012*, 20–23.

52. Xiao, Q.; McPherson, E.G. Testing a Bioswale to Treat and Reduce Parking Lot Runoff. Center for Urban Forest Research, University of California-Davis, 2009. Available online: https://www.fs.fed.us/psw/topics/urban_forestry/products/psw_cufr761_P47ReportLRes_AC.pdf (accessed on 14 April 2017).

53. Anderson, B.S.; Phillips, B.M.; Voorhees, J.P.; Siegler, K.; Tjeerdema, R. Bioswales reduce contaminants associated with toxicity in urban storm water. *Environ. Toxicol. Chem.* **2016**, *35*, 3124–3134. [CrossRef] [PubMed]

54. Stantec. *Lockwoods Folly River Local Watershed Plan Preliminary Findings Report for North Carolina Ecosystem Enhancement Program*; Stantec Consulting Services Inc.: Raleigh, NC, USA, 2006. Available online: https://www.ncdeq.gov (accessed on 3 May 2017).

55. U.S. Environmental Protection Agency (USEPA). North Carolina Water Quality Assessment Report. 2014. Available online: https://iaspub.epa.gov/waters10/attains_state.control?p_state=NC (accessed on 3 May 2017).

56. U.S. Census Bureau. Resident Population Estimates for the 100 Fastest Growing U.S. Counties with 10,000 or More Population in 2010: 1 April 2010 to 1 July 2016. U.S. Census Bureau Population Division: Washington, DC, USA, 2017. Available online: https://factfinder.census.gov (accessed on 1 October 2017).

57. North Carolina Department of Transportation (NCDOT). Standard Specifications—16 Erosion Control and Roadside Development. North Carolina Department of Transportation: Raleigh, NC, USA, 2012. Available online: https://connect.ncdot.gov/resources/Specifications/Pages/2012StandSpecsMan.aspx?Order=SM-16-1610 (accessed on 15 November 2017).

58. North Carolina Department of Transportation. Division 10 Materials. North Carolina Department of Transportation: Raleigh, NC, USA, 2016. Available online: https://connect.ncdot.gov/resources/specifications/2006%20specifications%20books/10.%20materials.pdf (accessed on 15 November 2017).

59. Brown, R.A.; Hunt, W.F. Underdrain configuration to enhance bioretention exfiltration to reduce pollutant loads. *J. Environ. Eng.* **2011**, *137*, 1082–1091. [CrossRef]

60. Winston, R.J.; Dorsey, J.D.; Hunt, W.F. Quantifying volume reduction and peak flow mitigation for three bioretention cells in clay soils in northeast Ohio. *Sci. Total Environ.* **2016**, *553*, 83–95. [CrossRef] [PubMed]

61. U.S. Environmental Protection Agency (USEPA). *NPDES Storm Water Sampling Guidance Document*; EPA 833-B-92-002; USEPA Office of Water: Washington, DC, USA, 1992.

62. American Public Health Association (APHA). *Conductivity*; SM 2510b-97; APHA: Washington, DC, USA, 1997.
63. American Public Health Association (APHA). *Turbidity*; SM 2130b-01; APHA: Washington, DC, USA, 2001.
64. American Public Health Association (APHA). *Total Suspended Solids*; SM 2540d-97; APHA: Washington, DC, USA, 1997.
65. American Public Health Association (APHA). *Fixed and Volatile Solids*; SM 2540e-97; APHA: Washington, DC, USA, 1997.
66. American Public Health Association (APHA). *Fecal Coliform Procedure*; SM 9221e-99; APHA: Washington, DC, USA, 1999.
67. ASTM. *Standard Test Method for Enterococci in Water Using Enterolert*; ASTM D6503-14; ASTM: West Conshohocken, PA, USA, 2014.
68. R Core Team. *R: A Language and Environment for Statistical Computing*; R Foundation for Statistical Computing: Vienna, Austria, 2013; ISBN 3-900051-07-0.
69. U.S. Environmental Protection Agency (USEPA). *Urban Stormwater BMP Performance Monitoring*; EPA-821-B-02-001; USEPA Office of Water: Washington, DC, USA, 2002.
70. Burton, G.A.; Pitt, R.E. *Stormwater Effects Handbook: A Toolbox for Watershed Managers, Scientists, and Engineers*; CRC: Boca Raton, FL, USA, 2002.
71. Davis, A.P.; Shokouhian, M.; Sharma, H.; Minami, C.; Winogradoff, D. Water quality improvement through bioretention: Lead, copper, and zinc removal. *Water Environ. Res.* **2003**, *75*, 73–82. [CrossRef] [PubMed]
72. Davis, A.P.; Shokouhian, M.; Sharma, H.; Minami, C. Water quality improvement through bioretention media: Nitrogen and phosphorus removal. *Water Environ. Res.* **2006**, *78*, 284–293. [CrossRef] [PubMed]
73. Deletic, A.; Fletcher, T.D. Performance of grass filters used for stormwater treatment—A field and modelling study. *J. Hydrol.* **2006**, *317*, 261–275. [CrossRef]
74. Winston, R.J.; Hunt, W.F.; Kennedy, S.G.; Wright, J.D.; Lauffer, M.S. Field evaluation of storm-water control measures for highway runoff treatment. *J. Environ. Eng.* **2012**, *138*, 101–111. [CrossRef]
75. North Carolina Department of Environmental Quality (NCDEQ). *Surface Waters and Wetland Standards*; 15A NCAC 2B; NCDEQ Division of Water Quality: Raleigh, NC, USA, 2007.
76. Barrett, M.E. Performance, cost, and maintenance requirements of Austin sand filters. *J. Water Resour. Plan. Manag.* **2003**, *129*, 234–242. [CrossRef]
77. Hathaway, J.M.; Hunt, W.F. Evaluation of first flush for indicator bacteria for total suspended solids in urban stormwater runoff. *Water Air Soil Pollut.* **2011**, *217*, 135–147. [CrossRef]
78. Hathaway, J.M.; Hunt, W.F. Indicator Bacteria Performance of Stormwater Control Measures in Wilmington, NC. *J. Irrig. Drain. Eng.* **2012**, *138*, 185–197. [CrossRef]
79. Passeport, E.; Hunt, W.F.; Line, D.E.; Smith, R.A.; Brown, R.A. Field study of the ability of two grassed bioretention cells to reduce storm-water runoff pollution. *J. Irrig. Drain. Eng.* **2009**, *135*, 505–510. [CrossRef]
80. Davis, A.P. Field Performance of Bioretention: Water Quality. *Environ. Eng. Sci.* **2007**, *24*, 1048–1064. [CrossRef]
81. Davies, C.M.; Bavor, H.J. The fate of stormwater-associated bacteria in constructed wetland and water pollution control pond systems. *J. Appl. Microbiol.* **2000**, *89*, 349–360. [CrossRef] [PubMed]
82. Krometis, L.H.; Drummey, P.N.; Characklis, G.W.; Sobsey, M.D. Impact of microbial partitioning on wet retention pond effectiveness. *J. Environ. Eng.* **2009**, *135*, 758–767. [CrossRef]
83. Mallin, M.A.; Ensign, S.H.; Wheeler, T.L.; Mayes, D.B. Pollutant removal efficacy of three wet detention ponds. *J. Environ. Qual.* **2002**, *31*, 654–660. [CrossRef] [PubMed]
84. Characklis, G.W.; Dilts, M.J.; Simmons, O.D., III; Likirdopulos, C.A.; Krometis, L.-A.H.; Sobsey, M.D. Microbial partitioning to settleable particles in stormwater. *Water Res.* **2005**, *39*, 1773–1782. [CrossRef] [PubMed]
85. Krometis, L.-A.H.; Characklis, G.W.; Simmons, O.D., III; Dilts, M.J.; Likirdopulos, C.A.; Sobsey, M.D. Intra-storm variability in microbial partitioning and microbial loading rates. *Water Res.* **2007**, *41*, 506–516. [CrossRef] [PubMed]

water

MDPI

Article

Effect of Saturated Zone on Nitrogen Removal Processes in Stormwater Bioretention Systems

Chuansheng Wang [1], Fan Wang [1], Huapeng Qin [1,*], Xiangfei Zeng [1], Xueran Li [1] and Shaw-Lei Yu [2]

1   Key Laboratory for Urban Habitat Environmental Science and Technology, School of Environment and Energy, Peking University Shenzhen Graduate School, Shenzhen 518055, China; wangcs@sz.pku.edu.cn (C.W.); wangfan@pkusz.edu.cn (F.W.); zengxiangfaye@pku.edu.cn (X.Z.); snowbubu@163.com (X.L.)
2   Department of Civil and Environmental Engineering, School of Engineering and Applied Science, University of Virginia, Charlottesville, VA 22904, USA; sly@virginia.edu
*   Correspondence: qinhp@pkusz.edu.cn

Received: 9 December 2017; Accepted: 2 February 2018; Published: 7 February 2018

Abstract: The introduction of a saturated zone (SZ) has been recommended to address the issue of nitrogen removal fluctuation in the bioretention system, which is one of the most versatile low-impact development facilities for urban stormwater management. Nine experimental columns were used to characterize the nitrogen concentration variations over the outflow during wetting periods and in SZ during the antecedent drying periods (ADPs), as well as compare removal efficiencies of various nitrogen species in systems with different SZ depths under alternate drying and wetting conditions. Results indicated that NO_3^--N concentrations in the outflow showed quasi-logistic curve-shaped variations over time: being low (<0.5 mg/L) in the early process, sharply increasing thereafter, and finally flattening around 3.0 mg/L with NO_3^- leaching; NH_4^+-N and organic nitrogen (ON) concentrations were consistently low around 0.5 mg/L and 1.8 mg/L, respectively during the wetting periods. NH_4^+ removal efficiency in bioretention systems was consistently high around 80%, not varying with the increasing SZ depth; ON removal efficiency had a slight rise from 57% to 84% and NO_3^- removal efficiency was significantly enhanced from −23% to 62% with the SZ depth increasing from 0 to 600 mm. Deeper SZ could store more runoff and promote more denitrification of NO_3^- and mineralization of ON during the ADPs, providing more "old" water with low NO_3^- and ON concentrations for water exchange with "new" inflow of higher NO_3^- and ON concentrations during the wetting periods. The total nitrogen (TN) removal, a combined result of the instantaneous removal through adsorption and retention in the upper soil layer during the wetting periods and the gradual removal via denitrification and mineralization in SZ during the ADPs, was also improved by increasing the SZ depth; TN removal efficiency was elevated from 35% to 73% when the SZ depth increased from zero to 600 mm.

Keywords: bioretention; saturated zone; nitrogen removal; leaching; drying and wetting

1. Introduction

With the rapid development of urbanization, nitrogen pollution in storm runoff has aroused widespread public concerns [1]. It is well known that excessive nitrogen, some from urban or agricultural runoff, is one of the main contaminants contributing to eutrophication in many water bodies [1–4]. Managing stormwater runoff has therefore become an important task in water quality protection. In recent years, low-impact development (LID) has been proposed as an ecologically and economically sustainable approach to stormwater management around the world [5]. Bioretention (also referred to as rain garden or biofiltration), consisting essentially of vegetation, mulch, soil media, sand layer and gravel sump, is a widespread LID technology that has proved to be effective in removing phosphorus, suspended solids, chemical oxygen demand and heavy metals [6–12]. However, previous research indicated that the removal efficiencies of nitrogenous pollutants varied dramatically in bioretention systems [13–16]. While ammonium (NH_4^+) and organic nitrogen (ON) removal were normally effective due to the retention and adsorption processes in media layer of bioretention systems, total nitrogen (TN) removal fluctuated because of nitrate (NO_3^-) leaching in bioretention systems [6,14,17,18].

In recent times, the introduction of a saturated zone (SZ) into bioretention systems has been widely recommended to promote nitrogen removal and address the issue of nitrate removal fluctuation [10,19–23]. This is because SZs can create an anaerobic environment to promote permanent NO_3^- removal via denitrification. Despite the growing interests in bioretention systems with SZs, reported nitrogen removal efficiencies were still not consistent in different studies. For example, some researchers found SZ combined with carbon source could remove NO_3^- from roof runoff by up to 67% [24], others reported that SZ could effectively remove more than 90% of NO_3^- from stormwater runoff [25,26]. However, a few studies revealed that nitrogen removal was even less than 20% and NO_3^- leaching even occurred occasionally in bioretention systems with SZ design [13,18,27]. Therefore, in view of the controversy over the effectiveness of SZs, there is a need to understand the nitrogen removal processes that in essence account for the variabilities of various nitrogen removal efficiencies in different bioretention systems with SZs.

Given that depth is one of the most critical parameters for SZ design, various nitrogen removal efficiencies of bioretention systems with SZ depths of 0, 150, 450, 600 mm were compared by Zinger et al [19]. Their results showed that NO_3^- and TN removal efficiencies increased with the increase of the SZ depth and were up to 99% in systems with 450 and 600 mm deep SZs. However, they mainly focused on the nitrogen removal efficiencies during storm events (wetting periods) and did not investigate the removal processes under alternate drying and wetting conditions that could help probe into the influence mechanism of SZ on nitrogen removal.

This study is a further in-depth investigation of the effect of saturated zone on the nitrogen removal with the objectives to depict the major nitrogen removal processes in bioretention systems and provide implications into SZ design for bioretention systems under alternate drying and wetting conditions. In detail, we have characterized the nitrogen concentration variations over the outflow course during the wetting periods and in SZ during the antecedent drying periods (ADPs); the removal performance of NH_4^+, ON, NO_3^- and TN between systems with different SZ depths were also compared; finally, different removal pathways for different nitrogen species were proposed in bioretention systems with SZs.

2. Methods

2.1. Experimental Set-up

Nine bioretention columns were built at the campus of Peking University Shenzhen Graduate School (PKUSZ), China in 2016. Each of them consisted of five layers from top to bottom: experimental vegetation, mulch layer, mixed soil media, sand bed, and gravel sump, which were all placed in the Polymeric Methyl Methacrylate (PMMA, with good rigidity and durability) cylinder containers (Figure 1a and Table 1). The first set of bioretention systems included three mesocosm bioretention columns 500 mm in diameter (two with 300 mm SZ and one without SZ). The second set included six columns 250 mm in diameter (referred to as "small-scale bioretention systems" thereinafter) with 0–600 mm SZ (Table 1 and Figure 1). Bulrushes (*Phragmites australis*, forty per mesocosm column, ten in each smaller one) were planted as experimental vegetation in the clear top section of each column, while the next four layers were wrapped up in tinfoil to be isolated from outside heat (Figure 1b). The first 50 mm deep mulch layer was comprised of wood chips (from local pine trees), and the next 450 mm deep mixed soil layer was composed of 50% (by weight) sandy loam soil from the campus of PKUSZ, 40% sand (d_{50} of approximately 0.5 mm in size), 10% peat moss (*Pindstrup Sphagnum*). An additional amount of lime (calcium carbonate) was added into this mixed soil layer to achieve a pH of 6.5–7.5 as recommended by several construction manuals [28–30]. Below it, the 120-mm deep transition layer was composed of river sand (d_{50} of 1–2 mm in size). At the bottom, 300 mm high gravel sump combined with carbon source (newspaper, 5% by volume) was installed in the three mesocosm bioretention columns. In six small-scale columns, embedded elbow pipes (diameter 25 mm) at the draining ports were raised to the heights of 0, 200–600 mm to create anaerobic saturated conditions in the gravel sumps.

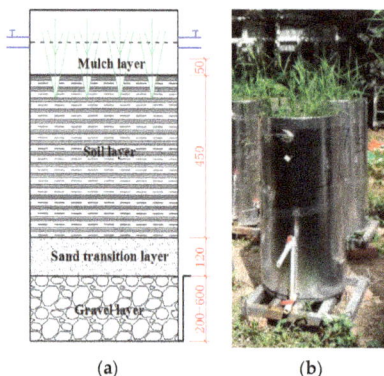

(a)   (b)

Figure 1. Bioretention Columns: Structure details (**a**) and experiment site pictures (**b**). Unit: mm.

Table 1. The media in the bioretention columns.

Media Layer	Depth (mm)	Material
Mulch	50	Wood chips
Soil layer	450	Sandy loam
Transition layer	120	River sand (1–2 mm)
Saturated zone	0, 200, 300, 400, 500, 600	Gravel and Carbon source

The experimental bulrushes were carefully cultivated from root to seedlings under controlled laboratory conditions for two months from December 2015 to February 2016. Then all bioretention columns were placed into a transparent canopy, which ensured enough natural sunlight but avoided rainfall entering the columns.

Based on typical subtropical coastal climate conditions in Shenzhen, a twice-weekly dosing scheme with synthetic runoff (intensity: ~20 mm/h, duration: one hour) was adopted. Each bioretention system was sized at 5% of the catchment area. This dosing method was similar to that used in previous studies [18,28,31]. Synthetic runoff in this study was prepared to mimic local highly polluted runoff characteristics of chemical oxygen demand (COD), organic nitrogen (ON), ammonium-nitrogen (NH_4^+-N), nitrate-nitrogen (NO_3^--N) and pH 6.5–7.5 (Table 2), according to Huang et al [32]. Each column was watered with experimental synthetic runoff (Table 2) to allow plant growth for four weeks until March 2016 to achieve stability.

Table 2. The mean inflow concentrations of pollutants in the synthetic runoff.

Pollutant	Mean Inflow Concentration	Source
Chemical Oxygen Demand	200 mg/L	Glucose ($C_6H_{12}O_6$)
NO_3^--N	2.5 mg/L	Potassium Nitrate (KNO_3)
NH_4^+-N	2.5 mg/L	Ammonium Chloride (NH_4Cl)
ON	5.0 mg/L	3-Aminopropanoic ($C_3H_7NO_2$)

2.2. Experimental Procedure

2.2.1. Mesocosm Bioretention Systems

The three mesocosm bioretention systems were dosed with 70 L synthetic runoff for one hour at three-day intervals for five storm events (also called five wetting periods) during March to April, 2016 (at 23 °C~30 °C). The outflow was monitored over the 1-h wetting period with nine sub-samples taken after draining about 0, 1%, 10%, 16.7%, 33.3%, 50%, 66.7%, 83.3% and 100% of integrated outflow volume. In addition, the entire outflow was thoroughly mixed to form one final composite sample for each column.

After one wetting period, the stored water in SZ was monitored every hour within the first ten-hour drying period and then every 12 h in the remaining ADPs. Various nitrogen removal efficiencies of NH_4^+, NO_3^-, ON and TN between systems with and without SZs were compared. DO in SZ was also detected by sensor online monitor meter (NUL-205, Neuron Logger Sensors, USA) every two hours during the ADPs.

2.2.2. Small-Scale Bioretention Systems

The six small-scale systems with 0–600 mm deep SZ were dosed with 20 L per hour at three-day intervals during April to May, 2016 (at 25 °C~31 °C), and five wetting periods were monitored as well. The outflow was monitored over the 1-h wetting period with five sub-samples taken after draining about 0, 25%, 50%, 75% and 100% of integrated outflow volume. In addition, the entire outflow was mixed thoroughly to form one final composite sample for each column.

2.3. Sample Analysis

The inflow volume and flow rate were controlled by the metering pump to maintain consistence during the wetting periods. According to Standard Methods for the Examination of Water and Wastewater (2012), after collection, the samples were immediately filtered through 0.22-μm membrane filter and then kept frozen at −20 °C to prevent microbial activities before further analysis. NH_4^+, NO_3^- and NO_2^- concentrations were measured using the CleverChem 200+ automatic discontinuous analyzer based on the monitoring methods of Nessler's reagent spectrophotometry, the Hydrazine Sulfate-NEDD spectrophotometry, sulfanilamide and N-(1-naphthalene) ethylenediamine hydrochloride spectrophotometry, respectively. TN was completely converted to NO_3^- by the alkaline potassium persulfate digestion method for measurements, while ON was determined by the formula: ON = TN − NO_3^- − NO_2^- − NH_4^+. In addition, the pH of each sample was monitored by laboratory pH meter (IE438, Mettler, Greifensee, Switzerland).

In this study, the event mean concentration (*EMC*) removal method, recommended by the American Society of Civil Engineers and the Environmental Protection Agency (ASCE-EPA) [33], was used to calculate the removal efficiency:

$$\text{Removal efficiency} = \left[\frac{EMC_i - EMC_0}{EMC_i} \right] \times 100\% \tag{1}$$

where EMC_i and EMC_0 were the *EMC*s of various nitrogen species (NH_4^+, ON, NO_3^- and TN) in the inflow and outflow during the wetting period, respectively.

All figures, including scatter plots and the box-and-whisker plots, were made using OriginPro 2015 (OriginLab Corp., Northampton, MA, USA). Statistical significant difference analyses were performed by the statistical software package PASW Statistics 19.0 (SPSS Inc., Chicago, IL, USA). Two-way ANOVA was used to test the significant difference in the removal efficiencies of various nitrogen species between systems with different SZs, the existence of SZ (i.e., with SZ and without SZ) was selected as the fixed factor and the nitrogen removal efficiencies as the dependents. The $p < 0.05$ was accepted as the threshold of significance.

3. Results

The retention rate about 12% was nearly the same between replicate systems: In the three mesocosm bioretention systems, when the inflow of each event was exactly 70 L in volume, the outflow volume was consistently similar around 61.7 L; In the six small-scale bioretention systems, when the inflow of each event was exactly 20 L in volume, the outflow volume of each column ranged from 17.5 to 17.7 L. The pH of each sample was stable at 7.0 ± 0.5. NO_2^- concentrations in all samples were below detection limit (<0.1 mg/L^{-1}), and this study would focus on the removal efficiencies and time-based concentrations of NH_4^+, NO_3^-, ON and TN.

3.1. Variations in Nitrogen Concentrations over the Outflow Process

Variations in the average NH_4^+, NO_3^-, ON and TN concentrations (for five wetting periods) over the outflow course in the mesocosm bioretention columns in the presence and absence of SZ were presented in Figure 2. NO_3^- concentrations varied over time: being low in the early outflow process, rapidly increasing thereafter, and finally flattening, displaying quasi-logistic curve-shaped variations. NO_3^- leaching with the ultimate concentrations exceeding the inflow NO_3^- concentration of 2.5 mg/L often occurred in the final outflow process during wetting periods. Meanwhile, in the absence of SZ, NO_3^- leaching began earlier over the outflow course. Instead, during each wetting period, NH_4^+ and ON concentrations were consistently low, around 0.5 mg/L and 1.8 mg/L, respectively in bioretention systems with or without the SZ. TN concentrations showed the similar characteristic variations of NO_3^- over the whole outflow process (Figure 2).

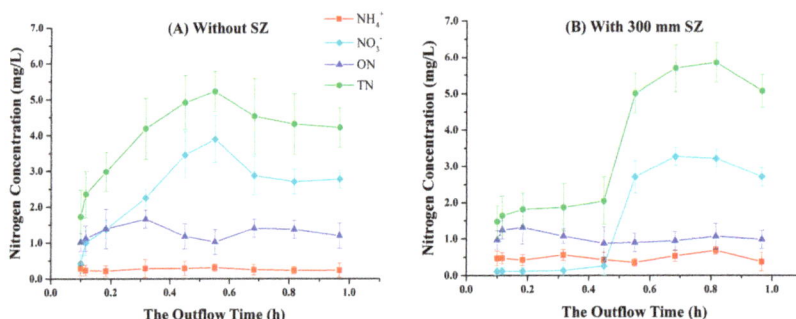

Figure 2. The variations in the average outflow ammonium (NH_4^+), nitrate (NO_3^-), organic nitrogen (ON) and total nitrogen (TN) concentrations for five simulated wetting periods with time. The bars indicate the standard deviations of nitrogen concentration. (**A**) variations of nitrogen concentrations in bioretention columns with 300 mm SZ; (**B**) variations of nitrogen concentrations in bioretention columns without SZ.

3.2. Variations in Nitrogen Concentrations in SZ during Drying Periods

During the 3-day antecedent drying periods (ADPs), DO was consistently low, generally less than 0.5 mg/L in SZ. Variations in the average NH_4^+, NO_3^-, ON and TN concentrations in SZ during ADPs were displayed in Figure 3. Significant removal was observed in SZ with NO_3^- concentrations decreasing from around the inflow concentration of 2.5 mg/L to almost zero within 12 h, while ON removal also mainly occurred within 12 h but the final ON concentrations were around 1.0 mg/L. NH_4^+ concentrations were relatively consistent in SZ without change. TN concentrations revealed similar removal patterns of NO_3^- during the ADPs (Figure 3).

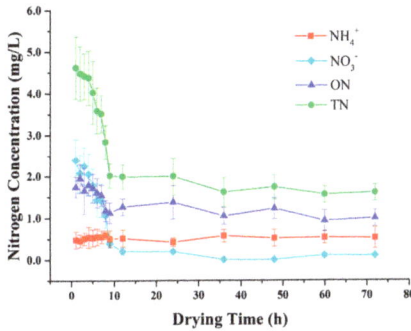

Figure 3. The variations in the average ammonium (NH$_4^+$), nitrate (NO$_3^-$), organic nitrogen (ON) and total nitrogen (TN) concentrations in the saturated zone of bioretention systems during antecedent drying periods (ADPs). The bars indicate the standard deviations of nitrogen concentration.

3.3. Effect of the Presence of SZ on Nitrogen Removal

Nitrogen removal efficiencies in bioretention systems with and without SZs were presented in Figure 4. The deviations from averages of various nitrogen species removal efficiencies in bioretention with 300 mm SZ were obviously less than the bioretention without SZ. The ANOVA analysis revealed that there were significant differences in the nitrogen removal efficiencies between the systems with and without the SZ ($p < 0.01$ for NH$_4^+$, $p < 0.05$ for NO$_3^-$, $p < 0.05$ for ON). In the presence of SZ in bioretention systems, NO$_3^-$ removal was significantly promoted, ON removal was slightly enhanced, while NH$_4^+$ removal efficiency was somewhat lowered. TN removal efficiency reflected the combination of NH$_4^+$, NO$_3^-$ and ON.

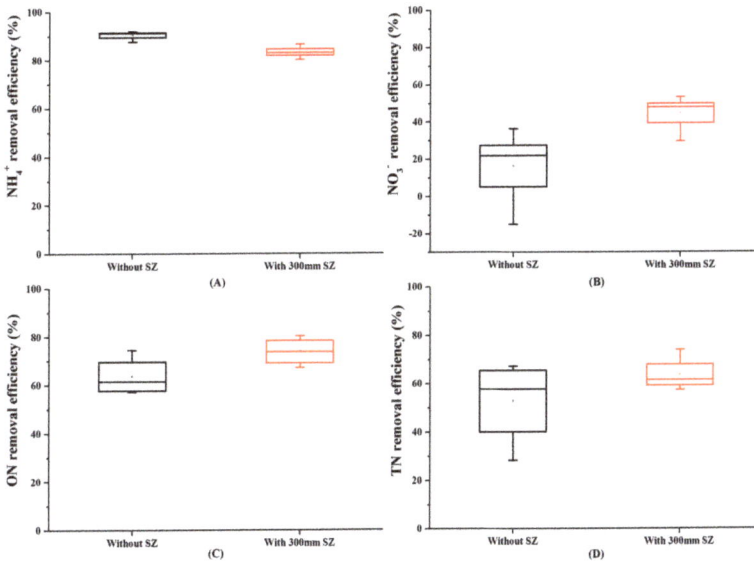

Figure 4. The removal efficiency of ammonium (**A**), nitrate (**B**), organic nitrogen (**C**) and total nitrogen (**D**) between the system with SZ and non-SZ system. The description of the above box- and whisker-plots: the bottom and top of the box are the first and third quartiles, and the band inside the box is the median. The ends of the whiskers represent the minimum and maximum of all of the data.

3.4. Effect of SZ Depths on Nitrogen Removal

In six small-scale bioretention columns with different SZ depths, the average NO_3^- concentrations for five wetting periods showed similar quasi-logistic curve-shaped variations over time (Figures 4 and 5). Likewise, NO_3^- leaching occurred in the latter outflow process without regard to the SZ depth. Furthermore, in bioretention systems with deeper SZ, the early outflow process with low NO_3^- concentrations was extended and NO_3^- leaching was delayed to occur in the final outflow process with lower peak concentrations (Figure 5B). Instead, with the increase of SZ depth, NH_4^+ and ON concentrations did not significantly fluctuate over the outflow course and remained consistent between systems with different SZ depths (Figure 5A,C). TN concentrations showed quasi-logistic curve-shaped variations over the whole outflow process. With the increase of SZ depth, the outflow process with low nitrogen concentrations was extended and the peak of nitrogen concentrations was lowered (Figure 5D).

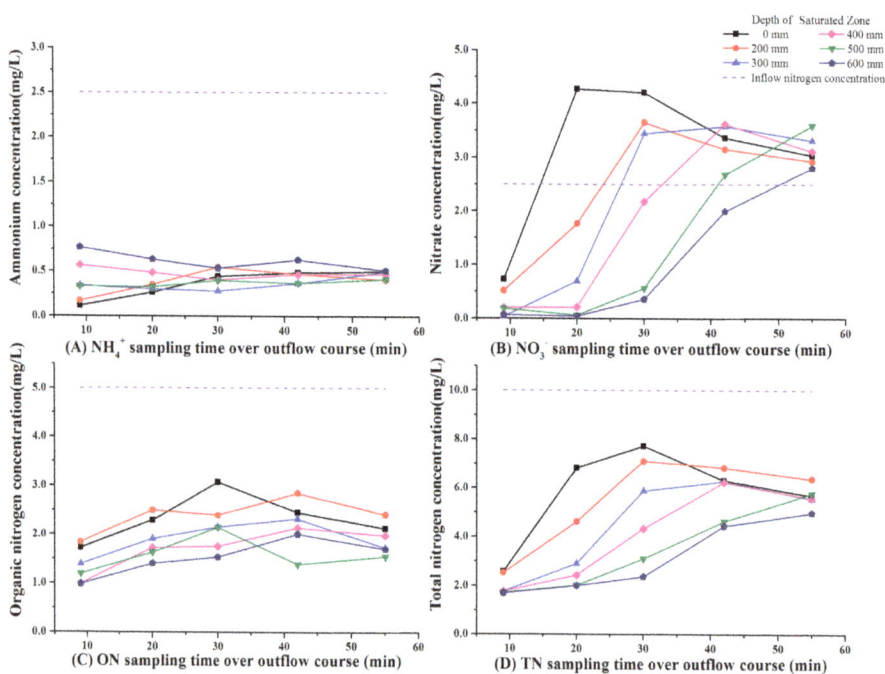

Figure 5. The variations in the average ammonium (**A**), nitrate (**B**), organic nitrogen (**C**) and total nitrogen (**D**) concentrations in the outflow over time during simulated wetting periods with different saturated zone depths.

Removal efficiencies of NH_4^+ were hardly affected by SZ depth, whereas NO_3^-, ON and TN removal was significantly affected by the SZ depth. Particularly, significant difference of nitrogen removal efficiencies existed between the systems with large-depth difference ($p < 0.05$). All bioretention systems were highly effective in NH_4^+ removal with the EMC removal efficiencies ranging from 79% $\pm$ 5% to 87% $\pm$ 5% and their performance was relatively stable in systems with different SZ depths (Figure 6). However, NO_3^- removal efficiency increased significantly with the SZ depth, which varied from −23% $\pm$ 50% (net leaching and high variations) without SZ to 62% $\pm$ 15% with 600 mm deep SZ (Figure 6). Also, ON removal efficiency rose from 57% $\pm$ 10% to 84% $\pm$ 8% when the SZ depth increased from zero to 600 mm (Figure 6). Accordingly, the removal efficiency of TN (the sum of NH_4^+,

NO_3^- and ON) increased from 35% $\pm$ 18% without SZ to 73% $\pm$ 5% with the SZ depth of 600 mm (Figure 6).

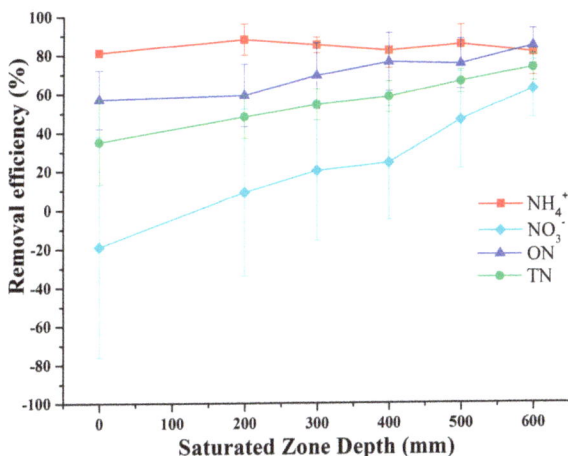

Figure 6. Variations in the average ammonium (NH_4^+), nitrate (NO_3^-), organic nitrogen (ON) and total nitrogen (TN) removal efficiencies for five simulated wetting periods with SZ depths. The bars indicate the standard deviations of nitrogen concentration.

4. Discussion

4.1. Ammonium Removal

NH_4^+ could be removed by volatilization, media adsorption, plant assimilation and microbial nitrification in bioretention systems [34]. In this study, NH_4^+ volatilization should be generally minimal since the pH of mixed soil or sand was around 7.0 $\pm$ 0.5. Adsorption could occur instantaneously after the inflow of synthetic runoff, but mainly in the upper section of bioretention systems with its capacity depending on the physical properties of soil media such as cation exchange capacity [34]. Plant assimilation could also be quickly initiated by the nutrient-laden inflow [35], which mainly occurs in the rhizosphere (approximately the upper 0–30 cm of the soil layer for *Phragmites australis*). Microbial nitrification was not significant during the wetting periods due to the short duration of one hour [17,34], but may mainly account for the NH_4^+ losses from the soil during antecedent drying periods (ADPs) by converting NH_4^+ to NO_3^- (Figure 7). Microbial nitrification is typically an aerobic process that should also be limited to occur in the upper aerobic section. Therefore, NH_4^+ removal should mainly occur in the upper section of bioretention systems, and the introduction of an anaerobic SZ in the bottom had little effect on NH_4^+ removal so that different SZ depths in each bioretention exerted a negligible influence on NH_4^+ removal. This is in line with similarly high NH_4^+ removal efficiencies in bioretention systems with different SZ depths (Figure 6), as reported by the previous studies [6–9,11]. The significant difference ($p < 0.01$ for NH_4^+) between the system with 300 mm SZ and the non-SZ system revealed that the slight decrease of NH_4^+ removal may be due to the effective mineralization of ON in SZ, this is in line with the minor decrease of ON during the ADPs (Figure 3).

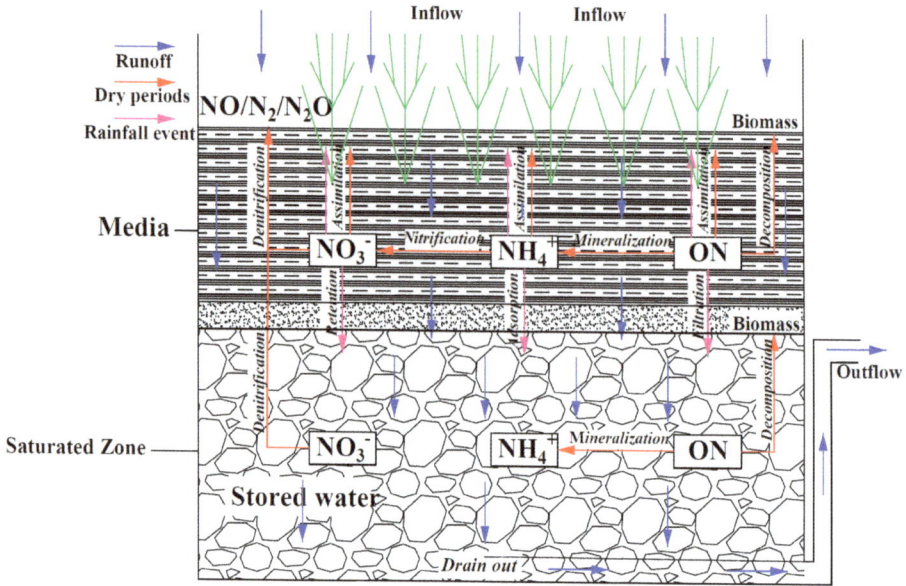

Figure 7. The diagram showing nitrogen processes that mainly occur in SZ and media layer of a bioretention system under alternate drying and wetting periods.

4.2. Nitrate Removal

NO_3^- could possibly be removed by several physical and microbial processes [17,34]. During wetting periods, due to its negative charge, NO_3^- is unlikely to be retained effectively by soil media in that most of the soil particles are negatively charged [34]. Besides, NO_3^- is soluble, and hydrological downward washing of NO_3^- from the soil media could lead to significant leaching of NO_3^-. Microbial denitrification is a permanent nitrogen removal pathway by transforming NO_3^- to gaseous nitrogen forms (NO, N_2O and N_2) via denitrifiers [34] and promoting denitrification becomes a major solution to improving nitrogen removal performance for bioretention systems. Meanwhile, NO_3^- can be produced by nitrifiers that convert NH_4^+ to NO_3^- via nitrification. However, denitrification or nitrification is not likely to occur significantly within the short-time (around 1 h) wetting period [36], but should mainly occur during the ADPs. The introduction of an SZ with stored runoff could pose little effect on nitrification (mostly aerobic process) but significantly affect denitrification by facilitating the growth of anaerobic denitrifiers to promote the removal of NO_3^- during the ADPs [20,34,36–38]. This can explain the growing NO_3^- removal efficiencies with the increasing SZ depths (Figure 6), which can be strongly supported by the significant difference of ANOVA analysis ($p < 0.05$ for NO_3^-).

The occurrence of denitrification in SZ during the ADPs was evident by the quick decrease of NO_3 in Section 3.2 (Figure 3). This could be best illustrated with the additional results in this study with low DO (less than 0.5 mg/L), during the three-day interval, NO_3^- in SZ can be fully removed by improved denitrification [7,8,10]. The low NO_3^- concentrations in the early outflow process (Figure 5B) could reflect the NO_3^- level in the "old" water stored in SZ after ADPs, and during wetting periods, the gradual increase in the outflow of NO_3^- concentrations in the latter process could be a result of the mixing of the "old" water stored in SZ with low NO_3^- concentrations and the "new" inflow transported downward from the upper soil media layer with relatively high NO_3^- concentrations (Figure 5B). With the diminishing effect of "old" stored runoff, the outflow NO_3^- concentrations increased and even exceeded the inflow concentrations with NO_3^- leaching, and the net leaching of

NO_3^- in the final outflow process likely originated from the nitrification with significant amounts of NO_3^- retained in the upper soil media layer during the ADPs.

The deeper SZ in this study was able to store more water to enable more denitrification during ADPs and provides more "old" water of low NO_3^- concentrations for water exchange with "new" inflow of high NO_3^- concentrations. Therefore, during wetting periods, systems with deeper SZ had more outflow of low NO_3^- concentrations and smaller amounts of outflow NO_3^- integrated over the whole outflow course, thus corresponding to higher NO_3^- removal performance (Figure 5B). The effectiveness of elevating NO_3^- removal by elevating SZ depths, is consistent with the results of Zinger et al [19] that showed a steady increase in NO_3^- removal efficiency ranging from 66% to 99% with the SZ depth from 450 mm to 600 mm. However, their up to 99% NO_3^- removal efficiencies in bioretention systems with SZ depths of 450 to 600 mm were consistently higher, probably owing to their low pollutant concentrations with TN concentration of 2.13 mg/L (NH_4^+ conc. of 0.29 mg/L, NO_3^- conc. of 0.74 mg/L, ON conc. of 1.1 mg/L, Melbourne, Australia) compared to that of 10 mg/L (NH_4^+ conc. of 2.5, NO_3^- conc. of 2.5 mg/L, ON conc. of 5.0 mg/L, Shenzhen, China) in this study (Table 2). Also, the SZ in this study was the gravel sump layer and the increase of SZ depth implied increasing the gravel sump thickness and thus, increasing the quantity of stored runoff in this study, which could be used to indicate the functionality of SZ; whereas Zinger et al. [19] simply elevated the elbow pipes of the bioretention systems of fixed structure (the system consists of 400 mm sandy loam with vegetation, 400 mm fine sand, 30 mm transition river sand and 70 mm gravel layers) to different heights to create SZs of different depths and their SZs might encompass the soil and sand layers as well as the same 70 mm deep gravel sumps. Therefore, the advanced SZ design in our study can verify the effectiveness of SZ, and better optimize the design ratio of upper soil layer and bottom SZ for removing nitrogen.

However, though we have acknowledged the importance of a deep SZ, the increase of SZ gravel sump depth may incur additional excavation costs. Therefore, we suggest optimizing the nitrate removal by setting an appropriate depth-ratio of the upper media layer and SZ to promote the complete nitrogen cycle from NH_4^+ to NO_3^- through nitrification and the permanent nitrogen removal of NO_3^- by converting NO_3^- to gaseous nitrogen ($NO/N_2O/N_2$) through denitrification (Figure 7), which also should be compromised with the cost.

4.3. Organic Nitrogen Removal

Likewise, ON removal could be attributed to many complex physical, chemical and microbial processes [7]. During wetting periods, ON could be removed substantially by retention and filtration in media layer and water exchange of stored runoff in SZ (Figure 7). Apart from that, during the 3-day ADPs, the upper soil media layer was mostly aerobic, mineralization (ammonification) can be promoted [34] to convert ON to NH_4^+. The bottom SZ added into bioretention systems formed an anaerobic condition under which mineralization might still occur, which was evident by the decrease of ON within 12-h drying periods (Figure 3). Therefore, ON removal could occur in both these layers and the introduction of SZ could be helpful to remove ON. The effectiveness of SZ on ON removal can be reflected by the significant difference of removal efficiencies between systems ($p < 0.05$ for ON).

The consistently low ON concentrations of each column during a wetting period indicated effective ON removal of bioretention systems (Figure 5C). Moreover, Figure 5C showed that the deeper the SZ was, the lower the ON concentration in the early outflow was. The reason might be that a deeper SZ provided more space for microbial mineralization to transform organic matter, leading to higher removal efficiency during the three-day drying periods (Figure 3). The results attested that ON removal could be slightly improved with the increase of SZ depths of bioretention systems.

4.4. Total Nitrogen Removal

TN removal in bioretention systems is a combined result of NH_4^+, NO_3^-, and ON removal which can be influenced by the hydrological and biotic processes in both the upper media layer and SZ under alternate drying and wetting conditions. As shown in Figure 7, NH_4^+ and ON could be effectively removed instantaneously due to retention and adsorption during wetting periods, while NH_4^+, ON and NO_3^- could mainly be removed via gradual nitrification, mineralization and denitrification processes during the ADPs (Figure 7). Therefore, TN removal in bioretention systems was determined by coupling the instantaneous hydrological water-exchange and chemical absorption during wetting periods with the gradual biotic removal, the former happened very quickly within one-hour wetting period, while the latter mainly occurred in the antecedent drying periods (Figure 3). The quasi-logistic curve-shaped variations of outflow TN concentrations similar to those of outflow NO_3^- concentrations showed that the fluctuations of TN removal were mainly controlled by NO_3^- (Figure 5B,D). Instead, higher NH_4^+ and ON removal with deeper SZ mainly accounted for significantly increased TN removal ($p < 0.05$) owing to their high removal efficiencies (Figure 6).

The deeper SZ contained more "old" stored runoff, which could improve plants' growth with more moisture during the ADPs. Especially, it could form an anaerobic area for biotic removal pathways during the ADPs, and store a lager quantity of water for hydrological water-exchange removal pathways during wetting periods, which are important to NO_3^- and ON removal. As a result, SZ could help solve the fluctuations of nitrogen removal, the extent of which was dependent on the stored runoff quantity during the ADPs. However, in reality, the selection of suitable SZ depths of bioretention systems for nitrogen removal should be compromised with the construction costs according to local pollution and meteorological condition.

5. Conclusions

NH_4^+ removal efficiency in bioretention systems was consistently high through effective adsorption in soil layer, not varying with the increasing SZ depth, whereas ON and NO_3^- removal efficiency significantly increased with the SZ depth increasing from 0 to 600 mm due to increased mineralization and denitrification. NO_3^- concentrations over the outflow process presented quasi-logistic curve-shaped variations over time: being low in the early outflow process, sharply increasing thereafter, and in the final process flattening with the ultimate concentrations exceeding the inflow concentrations. NO_3^- leaching often occurred in the final outflow process without regard to the SZ depth. With the increase of SZ depth, quasi-logistic curve-shaped variations were observed with the longer duration of low NO_3^- concentrations in the early outflow process and less NO_3^- leaching in the latter outflow process. The incorporation of a deeper SZ proved beneficial to nitrogen removal and maintenance of bioretention functionality. The effectiveness of increasing SZ depths to promote TN removal was attributed to larger amounts of stored runoff where denitrification of NO_3^- and mineralization of ON occurred during the ADPs.

One possible direction for future nitrogen removal improvement is to transport the products of nitrification—NO_3^-—from the upper soil media to the SZ occasionally during ADPs to further increase the NO_3^- removal, which, however, requires additional care in the field. Also, more detailed studies and in-depth analyses are needed to quantify the effect of SZ in this "black box"—a bioretention system under alternate drying and wetting conditions.

Acknowledgments: This research was supported by The National Natural Science Foundation of China (41603073), Shenzhen Science and Technology Development Fund Project (JCYJ20150518092928547), National Water Pollution Control and Management Technology Major Projects (No. 2013ZX07501005), and the project of Shenzhen Municipal Development and Reform Commission (Discipline construction of watershed ecological engineering). The authors are deeply grateful to Ming Cheng, Kangmao He, Yanyan Zheng for helping with laboratory work and analysis. Finally, the authors would like to thank, in particular, the invaluable support received from Meiyue Ding over the years.

Author Contributions: Chuansheng Wang, Huapeng Qin and Fan Wang conceived and designed the experiments; Chuansheng Wang, Xueran Li and Xiangfei Zeng performed the experiments; Chuansheng Wang and Fan Wang analyzed the data; Fan Wang and Shaw-Lei Yu contributed experimental analysis tools; Chuansheng Wang and Fan Wang wrote the paper; Huapeng Qin and Shaw-Lei Yu helped review and edit the paper.

Conflicts of Interest: The authors declare no conflict of interest.

References

1. United States Environmental Protection Agency (US EPA). *Renewed Call to Action to Reduce Nutrient Pollution and Support for Incremental Actions to Protect Water Quality and Public Health*; United States Environmental Protection Agency: Washington, DC, USA, 2016. Available online: https://www.epa.gov/sites/production/files/2016-09/documents/renewed-call-nutrient-memo-2016.pdf (accessed on 22 September 2016).

2. Driscoll, C.T.; Whitall, D.R.; Aber, J.D.; Boyer, E.W.; Castro, M.S.; Cronan, C.S.; Groffman, P.; Hopkinson, C.; Lambert, K.; Lambert, K.; et al. Nitrogen Pollution in the Northeastern United States: Sources, Effects, and Management Options. *BioScience* **2003**, *53*, 357–374. [CrossRef]

3. Taylor, G.D.; Fletcher, T.D.; Wong, T.H.F.; Breen, P.F.; Duncan, H.P. Nitrogen composition in urban runoff: Implications for stormwater management. *Water Res.* **2005**, *39*, 1982–1989. [CrossRef] [PubMed]

4. Liu, J.; Sample, D.J.; Bell, C.; Guan, Y. Review and Research Needs of Bioretention Used for the Treatment of Urban Stormwater. *Water* **2014**, *6*, 1069–1099. [CrossRef]

5. Jia, H.; Yao, H.; Shaw, L.Y. Advances in LID BMPs research and practice for urban runoff control in China. *Front. Environ. Sci. Eng.* **2013**, *7*, 709–720. [CrossRef]

6. Davis, A.P.; Shokouhian, M.; Sharma, H.; Minami, C.; Winogradoff, D.A. Water quality improvement through bioretention: Lead, copper, and zinc removal. *Water Environ. Res.* **2003**, *75*, 73–82. [CrossRef] [PubMed]

7. Davis, A.P.; Shokouhian, M.; Sharma, H.; Minami, C. Water quality improvement through bioretention media: Nitrogen and phosphorus removal. *Water Environ. Res.* **2006**, *78*, 284–293. [CrossRef] [PubMed]

8. Hatt, B.E.; Fletcher, T.; Deletic, A. Hydraulic and Pollutant Removal Performance of Fine Media Stormwater Filtration Systems. *Environ. Sci. Technol.* **2008**, *42*, 2535–2541. [CrossRef] [PubMed]

9. Hunt, W.F.; Davis, A.P.; Traver, R.G. Meeting Hydrologic and Water Quality Goals through Targeted Bioretention Design. *J. Environ. Eng.* **2012**, *138*, 698–707. [CrossRef]

10. Lynn, T.J.; Yeh, D.H.; Ergas, S.J. Performance and longevity of denitrifying wood-chip biofilters for stormwater treatment: A microcosm study. *Environ. Eng. Sci.* **2015**, *32*, 321–330. [CrossRef]

11. Palmer, E.T.; Poor, C.J.; Hinman, C.; Stark, J.D. Nitrate and phosphate removal through enhanced bioretention media: Mesocosm study. *Water Environ. Res.* **2013**, *85*, 823–832. [CrossRef] [PubMed]

12. Rycewicz-Borecki, M.; Mclean, J.E.; Dupont, R.R. Nitrogen and phosphorus mass balance, retention and uptake in six plant species grown in stormwater bioretention microcosms. *Ecol. Eng.* **2017**, *99*, 409–416. [CrossRef]

13. Hunt, W.; Jarrett, A.; Smith, J.; Sharkey, L. Evaluating Bioretention Hydrology and Nutrient Removal at Three Field Sites in North Carolina. *J. Irrig. Drain. Eng.* **2006**, *132*, 600–608. [CrossRef]

14. Bratieres, K.; Fletcher, T.D.; Deletic, A.; Zinger, Y.A. Nutrient and sediment removal by stormwater biofilters: A large-scale design optimisation study. *Water Res.* **2008**, *42*, 3930–3940. [CrossRef] [PubMed]

15. Davis, A.P.; Hunt, W.F.; Traver, R.G.; Clar, M. Bioretention technology: Overview of current practice and future needs. *J. Environ. Eng.* **2009**, *135*, 109–117. [CrossRef]

16. Collins, K.A.; Lawrence, T.J.; Stander, E.K.; Jontos, R.J.; Kaushale, S.S.; Newcomer, T.A.; Grimmg, N.B.; Ekberg, M.C. Opportunities and challenges for managing nitrogen in urban stormwater: A review and synthesis. *Ecol. Eng.* **2010**, *36*, 1507–1519. [CrossRef]

17. Sharkey, L.J.; Hunt, W.F. Hydrologic and water quality performance of four bioretention cells in central North Carolina. In Proceedings of the Watershed Management Conference—Managing Watersheds for Human and Natural Impacts: Engineering, Ecological, and Economic Challenges, Williamsburg, VA, USA, 19–22 July 2005; pp. 833–842.

18. Blecken, G.; Zinger, Y.; Deletic, A.; Fletcher, T.; Hedstrom, A.; Viklander, M. Laboratory study on stormwater biofiltration: Nutrient and sediment removal in cold temperatures. *J. Hydrol.* **2010**, *394*, 507–514. [CrossRef]

19. Zinger, T.; Fletcher, T.D.; Deletic, A.; Blecken, G.T.; Viklander, M. Optimisation of the nitrogen retention capacity of stormwater biofiltration systems. Presented at the 6th International Conference on Sustainable Techniques and Strategies in Urban Water Management, Lyon, France, 24–28 June 2007.

20. Zinger, Y.; Deletic, A.; Fletcher, T.D. The effect of various intermittent wet-dry cycles on nitrogen removal capacity in biofilters systems. Presented at the 13th International Rainwater Catchment Systems Conference and 5th International Water Sensitive Urban Design Conference, Sydney, Australia, 21–23 August 2007.

21. Zinger, Y.; Blecken, G.; Fletcher, T.D.; Viklander, M.; Deletic, A. Optimising nitrogen removal in existing stormwater biofilters: Benefits and tradeoffs of a retrofitted saturated zone. *Ecol. Eng.* **2013**, *51*, 75–82. [CrossRef]

22. Soberg, L.C.; Viklander, M.; Blecken, G. Do salt and low temperature impair metal treatment in stormwater bioretention cells with or without a submerged zone. *Sci. Total Environ.* **2016**, *579*, 1588–1599. [CrossRef] [PubMed]

23. Amir, A.; Asher, B. Use of Cotton as a Carbon Source for Denitrification in Biofilters for Groundwater Remediation. *Water* **2017**, *9*, 714. [CrossRef]

24. Dietz, M.E.; Clausen, J.C. Saturation to Improve Pollutant Retention in a Rain Garden. *Environ. Sci. Technol.* **2006**, *40*, 1335–1340. [CrossRef] [PubMed]

25. Zhang, Z.; Rengel, Z.; Liaghati, T.; Antoniette, T.; Meney, K. Influence of plant species and submerged zone with carbon addition on nutrient removal in stormwater biofilter. *Ecol. Eng.* **2011**, *37*, 1833–1841. [CrossRef]

26. Zhang, Z.; Rengel, Z.; Meney, K. Interactive effects of nitrogen and phosphorus loadings on nutrient removal from simulated wastewater using Schoenoplectus validus in wetland microcosms. *Chemosphere* **2008**, *72*, 1823–1828. [CrossRef] [PubMed]

27. Passeport, E.; Hunt, W.F.; Line, D.E. Field study of the ability of two grassed bioretention cells to reduce storm-water runoff pollution. *J. Irrig. Drain. Eng.* **2009**, *135*, 505–510. [CrossRef]

28. Monash University. *Adoption Guidelines for Stormwater Biofiltration Systems: Facility for Advancing Water Biofiltration*; Monash University: Melbourne, Australia, 2009.

29. Melbourne Water. *WSUD Engineering Procedures: Stromwater*; CSIRO Publishing: Melbourne, Australia, 2005.

30. Department of Environmental Resources. *Bioretention Manual. Environmental Services Division*; Department of Environmental Resources: The Prince George's County, MD, USA, 2007.

31. Blecken, G.; Zinger, Y.; Deletic, A.; Fletcher, T.; Viklander, M. Impact of a submerged zone and a carbon source on heavy metal removal in stormwater biofilters. *Ecol. Eng.* **2009**, *35*, 769–778. [CrossRef]

32. Huang, J.; Du, P.; Ao, C.T.; Lei, M.H.; Zhao, D.Q.; Ho, M.; Wang, Z. Characterization of surface runoff from a subtropics urban catchment. *J. Environ. Sci.* **2007**, *19*, 148–152. [CrossRef]

33. American Public Health Association; American Water Works Association; Water Environment Federation. *Standard Methods for the Examination of Water and Wastewater*, 22nd ed.; American Public Health Association, American Water Works Association, Water Environment Federation: Washington, DC, USA, 2012.

34. Payne, E.G.; Fletcher, T.D.; Cook, P.L.; Deletic, A.; Hatt, B.E. Processes and drivers of nitrogen removal in stormwater biofiltration. *Crit. Rev. Environ. Sci. Technol.* **2014**, *44*, 796–846. [CrossRef]

35. Geronimo, F.K.F.; Maniquiz-Redillas, M.C.; Kim, L.H. Fate and removal of nutrients in bioretention systems. *Desalin. Water Treat.* **2015**, *53*, 3072–3079. [CrossRef]

36. Peterson, I.J.; Igielski, S.; Davis, A.P. Enhanced denitrification in bioretention using woodchips as an organic carbon source. *J. Sustain. Water Built Environ.* **2015**, *1*. [CrossRef]

37. Brown, R.A.; Hunt, W.F. Underdrain configuration to enhance bioretention exfiltration to reduce pollutant loads. *J. Environ. Eng.* **2011**, *137*, 1082–1091. [CrossRef]

38. Subramaniam, D.; Mather, P.B.; Russell, S.; Rajapakse, J. Dynamics of Nitrate-Nitrogen Removal in Experimental Stormwater Biofilters under Intermittent Wetting and Drying. *J. Environ. Eng.* **2016**, *142*. [CrossRef]

water

MDPI

Article

Phosphorus Solubilizing and Releasing Bacteria Screening from the Rhizosphere in a Natural Wetland

Ying Cao [1,2], Dafang Fu [1,*], Tingfeng Liu [2], Guang Guo [2] and Zhixin Hu [2]

[1] Department of Municipal Engineering, School of Civil Engineering, Southeast University, Nanjing 210096, China; hjcaoying@njit.edu.cn

[2] Department of Environmental Engineering, Nanjing Institute of Technology, Nanjing 211167, China; hjliutingfeng@njit.edu.cn (T.L.); guoguang007007@163.com (G.G.); zxhu@njit.edu.cn (Z.H.)

* Correspondence: fdf@seu.edu.cn

Received: 24 December 2017; Accepted: 30 January 2018; Published: 12 February 2018

Abstract: Inorganic phosphorus (P)-solubilizing bacteria (IPSB) and organic P-mineralizing bacteria (OPMB) were isolated from bacteria that were first extracted from the rhizosphere soil of a natural wetland and then grown on either tricalcium phosphate or lecithin medium. The solubilizing of inorganic P was the major contribution to P availability, since the isolated bacteria released much more available P from inorganic tricalcium phosphate than lecithin. IPSB No. 5 had the highest P release rate, that is, 0.53 $mg \cdot L^{-1} \cdot h^{-1}$ in 96 h, and R10's release rate was 0.52 $mg \cdot L^{-1} \cdot h^{-1}$ in 10 days. The bacteria were identified as *Pseudomonas* sp. and *Pseudomonas knackmussii*, respectively. R10 released as much as 125.88 $mg \cdot L^{-1}$ dissolved P from tricalcium phosphate medium, while R4 released the most dissolved P from organic P medium among the isolates, with a concentration of 1.88 $mg \cdot L^{-1}$ and a releasing rate of 0.0078 $mg \cdot L^{-1} \cdot h^{-1}$ in ten days. P releasing increased with a pH decrease only when it was from inorganic P, not organic lecithin, and there was no significant correlation between the culture pH and P solubilizing. High-throughput sequencing analysis revealed that the dominant phylum in the studied wetland rhizosphere consisted of *Acidobacteria*, *Proteobacteria*, *Bacteroidetes* and *Chloroflexi*, accounting for 34.9%, 34.2%, 8.8% and 4.8%, respectively.

Keywords: wetland; rhizosphere bacteria; available phosphorus; pH

1. Introduction

Wetlands are useful for controlling pollution related to phosphorus (P) [1–4], as it may be taken up by vegetation before being washed into lakes or rivers. Much research is concerned with the effects of P removal by plants in wetlands [1,5,6], especially with the control of non-point source pollution [7]. However, plants can only absorb P in dissolved forms such as the free orthophosphate ions, $H_2PO_4^-$ and HPO_4^{2-} [8], while the available P in natural soil can be easily converted into insoluble complexes such as iron and aluminum hydrous oxides, crystalline and amorphous aluminum silicate, and calcium carbonate [9]. P can be dissolved by several mechanisms including Fe reduction [10], Ca removal and ligand exchange. Under anaerobic conditions, Fe^{3+} is microbially reduced to soluble Fe^{2+}, which releases into solution the PO_4^{3-} bound to Fe^{2+} oxides [11]. Ca^{2+} or PO_4^{3-} removal, such as that absorbed into the cell, may promote the dissociation of tricalcium phosphate. Organic acids can affect phosphate dissolution by competition for adsorption sites, the dissolution of adsorbents, a change in the surface charge of adsorbents, the creation of new adsorption sites through the adsorption of metal ions, and the retardation of the crystal growth of poorly ordered Al and Fe oxides [12].

There are many studies focusing on P-solubilizing and -releasing bacteria in agricultural fields. Inorganic P-solubilizing bacteria (IPSB), such as *Bacillus thuringiensis*, *Enterobacter intermedium*, *B. megaterium*, *Burkholderia caryophylli*, *Pseudomonas cichorii*, *Pseudomonas syringae*. *Bacillus* sp. and *Burkholderia* sp. [13–16], have been screened and identified for solubilizing insoluble forms of P, in order to make them available for the growth of crops [17,18]. As organic P accounts for 30–80% of the total P in most agricultural soils [19], organic P-mineralizing bacteria (OPMB) have been considered as predominant for P availability. OPMBs like *Bacillus cereus* and *Bacillus megaterium* that were isolated from agricultural soils have also been reported [14].

In natural wetlands, the rhizosphere bacteria's ability to transform P, such as through the release of available P, has not been reported. Bacteria with better P-solubilizing and -releasing abilities for extraction from natural wetlands have not been studied so far. It is not known which bacteria may helpful and how they promote the uptake of P by wetland plants, although it would be helpful to remove extra P from the environment in order to reduce the eutrophication problems. Unlike agricultural soil, wetlands have a much higher level of inorganic P and moisture content, with the inorganic P accounting for 60–97% of the total P in wetland areas; total P was also lower than in fertilized fields [20–24]. Due to the different physicochemical properties and vegetation, the microbial community of wetlands is different from that of farmland soil [10,25,26]. Rhizosphere microorganisms are able to convert insoluble forms of P to an accessible soluble form, where different strains show different efficiencies in P solubilization [16]. Reasons for the differences in the duration of P solubilization and release of bacteria strains may include: (1) different growth rate of the bacteria; (2) the production and secretion of different P solubilizing materials such as H^+ in the form of different acids, reducing substances to convert Fe^{3+} to Fe^{2+}- and Ca^{2+}-bound proteins out of the cell to solubilize tricalcium phosphate; (3) adsorption and absorption of Ca^{2+} or $PO_4{}^{3-}$ into the cell to promote the dissociation of tricalcium phosphate; and (4) production and secretion of related hydrolytic enzymes for releasing P from organic P, such as lecithin.

Certain bacteria species like IPSB and OPMB could be isolated, screened and applied in the rhizosphere of wetlands in order to promote the P absorption of plants, to enhance the P removal efficiency of wetlands. However, phosphorus solubilizing and releasing bacteria screening studies of a natural wetland rhizosphere have not been reported to a great extent so far. In this study, the composition of the bacteria community in the rhizosphere from the sampling wetland was examined using a high-throughput sequencing analysis. IPSB and OPMB were isolated using tricalcium phosphate and lecithin medium from the rhizosphere of *Acorus calamus* L., reed and *Iris tectorum* Maxim. in the studied wetland. The pH values were also recorded during the incubation, because the rate of phosphate uptake decreases as the pH value of the external solution increases [27]. In other words, P availability decreased with increase in the pH value of aquaponic nutrient solutions [28], and high pH values resulted in the formation of insoluble calcium phosphate species.

The objectives of this work are: (i) extract bacteria strains with better P solubilizing and releasing abilities from the natural wetland rhizosphere; (ii) test their ability to solubilize and release P from different compounds (organic P and inorganic P); (iii) identify the bacteria that solubilize and release the most P; and (iv) test the composition of the bacteria community in the studied wetland rhizosphere to show the proportion of the identified P solubilizing and releasing bacteria and to predict the potential P solubilizing and releasing microbes to be cultured and screened. The novelty of this work is the extraction of P solubilizing and releasing bacteria from the rhizosphere of natural wetland plants. These screened bacteria may more adaptable in the wetland environment when applied back to the field because they were wetland rhizosphere species. The work will also provide strains and a research basis for the future use of bacteria to promote the removal of phosphorus pollutants by wetland plants. This work used a high-throughput sequencing analysis to illustrate the composition of the bacteria community in the rhizosphere of the wetland, to understand the species and their proportion of wetland rhizosphere bacteria to the greatest extent and not be restricted to the bacteria that can be cultured in a lab.

2. Material and Methods

2.1. Material

Acorus calamus L., reed and *Iris tectorum* Maxim. were chosen because they are common wetland plants in East China. The experimental vegetation and their rhizosphere material were sampled from the wetland area of Tinyin Lake near the egret natural reserve in Nanjing, China, of which the water quality was monitored every year, showing that the average concentration of total phosphorus was lower than the level III baseline in the Chinese National Surface Water Environmental Quality Standard (GB3838-2002, 2002). Rhizosphere samples were taken from a mixture of soil on the roots (in 5 mm) of the plants that were gathered around the wetland area of the lake.

2.2. Methods

2.2.1. Extraction, Isolation and Purification of Rhizosphere Bacteria

The root surface soil and epidermis were first scraped off the plants, and then treated with a shaking technique to extract liquids with rhizosphere bacteria. The shaking technique here was to add 1 g of root surface soil and epidermis into a sterile conical flask and fill sterile water to 100 mL. The liquid was then placed on a constant temperature oscillator (HZ-81B, Shanghai Yuejin medical instruments factory, Shanghai, China) at 30 °C for 30 min. There was a dilution rate of 10^{-2} extract. The extracts were then prepared under sterile conditions with dilution rates of 10^{-3}, 10^{-4}, 10^{-5} and 10^{-6}, respectively. The separation and purification of microorganisms in the extracts were carried out using plate culture and plate streak methods in the aseptic environment [29] (Superclean bench, CJ-1FD, Shanghai Boxun medical bioinstrument Co., Ltd., Shanghai, China; Biochemistry cultivation cabinet, GZP250S, Shanghai Xinnuo instrument and equipment Co., Ltd., Shanghai, China). The bacterial collective medium contained: beef paste (3 g), NaCl (5 g), peptone (10 g), agar (15 g), distilled water (1000 mL), with a pH value of 7.0–7.2. Different bacteria colonies grown on the plates were selected based on their morphological characteristics such as size, shape, surface condition, color, edge and cross-section shape.

2.2.2. Preparation of P Medium

(1) Inorganic P ($Ca_3(PO_4)_2$) liquid medium (IPLM)

The isolation and purification of IPSB strains were carried out using IPLM based on tricalcium phosphate medium (TPM) [30], which contained: glucose (10 g), $(NH_4)_2SO_4$ (0.5 g), NaCl (0.3 g), KCl (0.3 g), $MgSO_4 \cdot 7H_2O$ (0.3 g), $FeSO_4 \cdot 7H_2O$ (0.03 g), $MnSO_4 \cdot H_2O$ (0.03 g), $Ca_3(PO_4)_2$ (0.7 g) (140 mg·L^{-1} total P) and distilled water (1000 mL) with a pH value of 7.0–7.2.

(2) Organic P (lecithin) liquid medium (OPLM)

A liquid lecithin medium (OPLM) for the OPMB strains was prepared based on the Menkina medium [31], which was composed of: Glucose (10 g), $(NH_4)_2SO_4$ (0.5 g), NaCl (0.3 g), KCl (0.3 g), $MgSO_4 \cdot 7H_2O$ (0.3 g), $FeSO_4 \cdot 7H_2O$ (0.03 g), $MnSO_4 \cdot H_2O$ (0.03 g), $CaCO_3$ (5 g), lecithin (2 g) (85.10 mg·L^{-1} total P), and distilled water 1000 mL, with a pH value of 7.0–7.2.

Inorganic P ($Ca_3(PO_4)_2$) solid medium (IPSM) and organic P (lecithin) solid medium (OPSM) were made by adding 18 g agar to the liquid medium described above. All media were adjusted to proper pH values (7.0–7.2) with 1 mol·L^{-1} NaOH or HCl, and sterilized using an autoclave (LDZX-40KBS, Shanghai shenan medical equipment factory, Shanghai, China) for 20 min at 121 °C.

2.2.3. Measurement of P-Solubilizing and -Releasing Capability

(1) Measurement of P-solubilizing capability of bacteria in solid medium

The preliminarily-screened P-solubilizing bacteria strains were first cultured on IPSM and OPSM respectively, and then incubated at 30 °C for 96 h, then the P solubilizing ring was observed and its size was measured by 2b/a, where a is the total diameter of the ring and colony, and 2b is the difference between the outer and inner rings. The P-solubilizing ring is a clarified area around the bacterial colony on IPSM/OPSM, which indicates that the bacteria may have stronger P-solubilizing ability than those without the ring. By observing whether there is a P solubilizing ring and its size, it is more convenient to screen out the bacteria with a stronger P solubilizing ability from a large number of bacteria.

(2) Measurement of P-solubilizing and -releasing capability of bacteria in liquid medium and pH

To determine the short-term P-solubilizing and -releasing rate of the bacteria, each strain was first cultured in 100 mL liquid medium, and then incubated in a constant-temperature oscillator at 30 °C for 96 h (130 r·min^{-1}), when the pH and soluble P was examined from the supernatant of centrifugal medium ((7000 r·min^{-1}) for 20 min) every 48 h. To determine the long-term P solubilizing and releasing rate, the bacteria strain was first cultured in the liquid medium and then incubated at 30 °C for 10 days, when the concentration of soluble P from the supernatant of centrifugal medium (Centrifuge, TGL-16G-XTA10, Shanghai Anting scientific instrument factory , Shanghai, China) was tested every 24 h.

(3) Analysis of the samples

Total P and soluble phosphates were determined using the ammonium molybdate spectrophotometric method (Spectrophotometer, 752N, Shanghai Youke instrument Co., Ltd., Shanghai, China) (Determination of total phosphorus of water quality ammonium molybdate spectrophotometric method Chinese GB 11893-1989) and represented as P concentration (mg·L^{-1}). Samples of total P were pretreated by potassium persulfate digestion. Three replicates of the control treatment were included in the experiment for each isolate and the results were expressed as mean values.

2.2.4. P-Solubilizing and -Releasing Rate Calculation

The bacterial P-solubilizing and -releasing rate η (mg·L^{-1}·h^{-1}) was calculated using the following formula:

$$\eta = \frac{(C_n - C_0)}{t} \tag{1}$$

where C_0 is the initial soluble P concentration in the liquid medium, C_n is the soluble P concentration in the liquid medium after n hours, and t is the incubation time.

2.2.5. Microbial Community Analysis

A total of 15 g of rhizosphere soil sample was sent to Shanghai Majorbio Bio-pharm Technology Co., Ltd. (Shanghai, China) and analyzed using a high-throughput sequencing analysis. The sequencing type, primer name and sequencing platform were bacterial 16S rRNA, 338F_806R and PE300, respectively.

3. Results and Discussion

3.1. P-Solubilizing Capacity of Inorganic Insoluble P ($Ca_3(PO_4)_2$)

(1) P-solubilizing and -releasing capacity of insoluble $Ca_3(PO_4)_2$ in the solid medium

It could be seen in the experiment with isolated bacteria strains growing on IPSM that only some strains showed the visible P solubilizing ring. Among the 13 strains selected and cultured on IPSM for 96 h, colonies of strain No. 1 to No. 11 formed the P solubilizing ring in the first 24 h, while No. 12 and 13 formed the P solubilizing ring later, after around 96 h and 72 h, respectively, illustrating the differences in the duration of the P-solubilizing and -releasing of bacteria stains. Among these isolates, 15.4%, 30.1%, 30.1% and 23.1% reached the maximum size in 24 h, 48 h, 72 h and 96 h, respectively. In terms of the average ring size grown at 96 h, strains No. 4, 6, 7 and 10 were better in insoluble P-solubilizing and -releasing in the solid medium (Figure 1a).

The insoluble tricalcium phosphate ($Ca_3(PO_4)_2$) in the solid medium appeared turbid when the ring area appeared clarified, which means the tricalcium phosphate was dissolved/consumed. The biochemical interactions may include: (1) the tricalcium phosphate dissolved in the low pH environment caused by acid production from the bacteria; (2) the adsorption or absorption of Ca^{2+} by bacteria thus releasing PO_4^{3-}; and (3) the absorption of PO_4^{3-} by the bacteria. If the ring was caused by the third reason, and the PO_4^{3-} was absorbed by bacteria, and then the bacteria cannot be judged as P-releasing bacteria. Thus, these bacteria have to be incubated in liquid medium and the soluble P level in the solution must be tested. In IPLM, if the insoluble tricalcium phosphate dissolved during the incubation, the higher soluble P level would indicate that the bacteria have a better P-solubilizing and -releasing ability. Otherwise the lower soluble P level when the insoluble tricalcium phosphate dissolved would mean that the bacteria could solubilize P not release much of the available P, which may be taken up by the bacteria. In this case, this type of bacteria cannot be used to promote P uptake by wetland plants since P is not released.

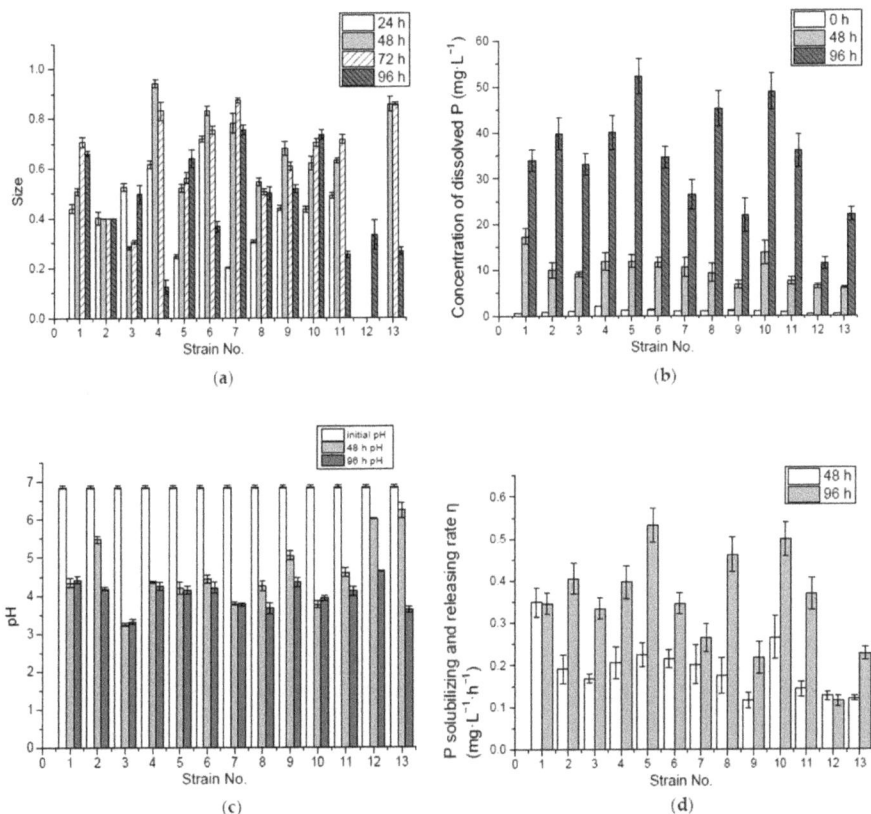

Figure 1. The size (2b/a) of the P solubilizing ring of 13 preliminary screened bacteria from *Acorus calamus* L. and *Iris tectorum* Maxim. (**a**); Concentration of dissolved P (**b**), pH (**c**) and P-solubilizing and -releasing rate (**d**) by 0 hour, 48 h, and 96 h in $Ca_3(PO_4)_2$ liquid medium. Error bars represent the standard errors.

(2)   P-solubilizing and -releasing capacity of insoluble $Ca_3(PO_4)_2$ in liquid medium

In the first 48 h of incubation, No. 1 and 10 released more dissolved P. By 96 h, however, No. 5, 8 and 10 released more in the liquid medium, as shown in Figure 1b. The decrease in pH value helped P release, since orthophosphate is mostly available at pH 3.0–5.5 in the aquaponic nutrient solution [28]. In this experiment, most of the isolates which brought the medium pH value down below 5 in the first 48 h might have a better capability of P solubilization and release. Thus, strains, like No. 12, 13 and 9, that were unable to lower the pH below 5 at the beginning showed poor performance in P release. However, the decrease in pH value was not the only cause of P release, although No. 3 and 7 had lowered the pH value below 4 by the end of the first 48 h, the concentration of the dissolved P in the medium was relatively low, compared with the dissolved P in the media of the other isolates. Take glucose in the medium as substrate, for example, under aerobic conditions the sugar is first metabolized to pyruvic acid through glycolysis, and the intermediate products are fructose 1,6-bisphosphate, glyceraldehyde 3-phosphate, and 2-phosphoglycerate, sequentially, during which two phosphates have been consumed to generate a net yield of two adenosine triphosphate (ATP) molecules. Then the pyruvate from the glycolysis is converted to acetyl-Coenzyme A (acetyl-CoA) by the action of pyruvate dehydrogenase complex, and acetyl-CoA is completely oxidized to CO_2 and water through citric acid cycle. O_2 consumption and pyruvate oxidation were found to be stimulated by some four-carbon dicarboxylic acids (fumarate, succinate, malate and oxaloacetate), five-carbon dicarboxylic acid (a-ketoglutarate), or six-carbon tricarboxylic acids (citrate, isocitrate, cis-aconitate) [29,32]. These citric acid intermediates are important sources of biosynthetic precursors. In aerobic respiration one molecular of glucose may yield 36–38 ATP and P-containing bio-molecules like nucleic acid, phospholipid, polyphosphate and P-containing enzymes are synthesized [29]. Due to this large consumption/biosynthesis of P, the dissolved P in the liquid medium would be low when a large number of bacteria cells or P-containing bio-molecules are produced under aerobic conditions. Under anaerobic conditions, some facultative or anaerobic bacteria species undergo fermentation where glucose is metabolized to pyruvate. As a result of the need to produce high-energy, phosphate-containing, organic compounds (generally in the form of CoA-esters) fermentative bacteria use Nicotinamide adenine dinucleotide (NADH) and other cofactors to produce many different reduced metabolic by-products, often including H_2 and small organic acids and alcohols derived from pyruvate, such as ethanol, acetate, lactate, and butyrate [32].

Many studies have shown that P solubilizing microorganisms can secrete a variety of low-molecular organic acids during metabolism, such as malic acid, propionic acid, lactic acid, acetic acid and citric acid [33]. These organic acid anions can react with calcium ions in the liquid medium to release P from modestly soluble phosphates [34]. Nos. 5, 8 and 10 had higher solubilizing and releasing rates in incubation during the 96 h, which were 0.53 $mg \cdot L^{-1} \cdot h^{-1}$, 0.46 $mg \cdot L^{-1} \cdot h^{-1}$ and 0.49 $mg \cdot L^{-1} \cdot h^{-1}$, respectively. The pH values of super P-releasing isolates (Nos. 5, 8 and 10) in the liquid medium with $Ca_3(PO_4)_2$ eventually reached 4.5–3.8, as shown in the results in Figure 1c, similar to the final pH value of 3.99 found in a culture of *Pseudomonas* K3, which was the best in P solubilizing and screened from the rhizosphere of the calcareous soil [35].

(3)   P-solubilizing and -releasing capacity of isolates from reed in liquid medium with insoluble $Ca_3(PO_4)_2$

Isolate R10 was the best at releasing available P from insoluble $Ca_3(PO_4)_2$ in the liquid medium, releasing 125.88 $mg \cdot L^{-1}$ of solubilized P, accounting for about 90.0% of the total P ten days into the incubation. Other isolates released 47.1–77.0% of the total P in the culture medium. The rest of the total P might be either absorbed by the bacteria and synthesized to organic matter, such as phospholipid, polyphosphate and nucleic acid, or deposited as insoluble tricalcium phosphate, iron (III) phosphate or magnesium phosphate. Some isolates (R1, R2, R3, R7, R8, R9 and R10) released much of the dissolved P at the beginning, possibly due to the fast production of organic acid and carbonic acid from glucose when seeded into the medium. The organic acid might either release hydrogen ions to dissolve

$Ca_3(PO_4)_2$ or combine with calcium ions to release P [36]. Then the concentration of dissolved P kept decreasing for one or two days, which could have been caused by the absorption and assimilation of P in the growth-lag phase of bacterial cells, followed by a period of fast growth, when more acid was produced, leading to the continuously increasing levels of dissolved P (Figure 2).

Figure 2b shows that, although the pH kept decreasing during the incubation, the dissolved P in some cultures started to decline from the middle of the process, implying that it could not have been caused by pH or acid production, but might have resulted from the absorption of P into bacterial cells and the formation of organic P through biosynthesis.

Table 1 shows the maximum concentration of dissolved P and their corresponding time during the incubation process, where R1, R2, R7, R8, R10 and R11 reached their maximum concentration on the tenth day, suggesting that these isolates could have solubilized and released available P for a longer term and at a higher concentration if more total P had existed in the medium. A total of 66.7% of the isolates (R1, R2, R3, R6, R7, R8, R9 and R10) reached the highest rate of dissolution and release P in the first 24 h, with R8 in 48 h, R5 and R11 in 72 h and R12 in 120 h, respectively. R10 had the highest P-releasing rate in the 10 days, with $0.52 \text{ mg·L}^{-1} \cdot \text{h}^{-1}$. The results are summarized in Table 1.

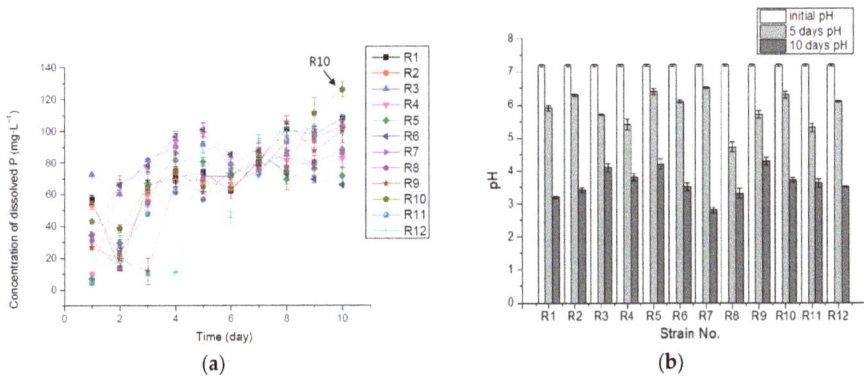

Figure 2. Concentration of dissolved P (mg·L^{-1}) (**a**) and pH (**b**) in 10-day incubation in liquid medium with insoluble $Ca_3(PO_4)_2$.

Table 1. Maximum concentration and rate of P release in 10-day incubation in liquid medium with insoluble $Ca_3(PO_4)_2$.

Strain No.	R1	R2	R3	R4	R5	R6	R7	R8	R9	R10	R11	R12
Max Concentration of dissolved P (mg·L^{-1})	107.77	103.24	91.92	98.71	81.73	100.97	102.11	86.26	105.5	125.88	106.63	100.97
Max Con. Day (day)	10	10	5	5	4	5	10	10	8	10	10	9
Max rate of solubilizing P ($\text{mg·L}^{-1} \cdot \text{h}^{-1}$)	2.37	2.23	3.03	1.06	0.93	1.44	1.46	1.30	1.11	1.80	0.66	0.73
Max rate time (hours)	24	24	24	48	72	24	24	24	24	24	72	120
P releasing rate η in 10 days ($\text{mg·L}^{-1} \cdot \text{h}^{-1}$)	0.45	0.43	0.37	0.35	0.3	0.27	0.43	0.36	0.42	0.52	0.44	0.41

(4) Identification of bacteria species

No. 5 and R10 were selected and sent to Shanghai Majorbio Bio-pharm Technology Co., Ltd. to be sequenced and contrasted, and were then identified as *Pseudomonas* sp. and *Pseudomonas knackmussii*, both belonging to the *Proteobacteria* phylum.

3.2. P-Solubilizing Capacity of Organic P (Lecithin)

(1) P-solubilizing capacity of organic P (lecithin) in solid medium

It was observed that the P-solubilizing rings formed in organic solid medium with lecithin were larger in size and clearer than those in inorganic solid medium. The largest size was formed between 48 and 72 h. No. 2, 6 and 10 were the best in forming a P-solubilizing ring in organic solid medium with lecithin, based on the average size from Figure 3a. However, the solubilized ring around the bacteria colony could only indicate that the "cloudy" lecithin was hydrolyzed. Whether the phosphate in the lecithin molecules was released as soluble P requires further confirmation. The hydrolyzed products can be fatty acids and other lipophilic substances such as arachidonic acid, diacylglycerol, phosphatidic acid, alcohol, inositol 1,4,5-trisphosphate, diacyl glycerol, etc. [37–39] Even when soluble P was released, it could be further absorbed by bacteria. Thus, the soluble P levels in the liquid medium during the bacterial incubation should be tested to analyze the bacteria's P-solubilizing and -releasing ability.

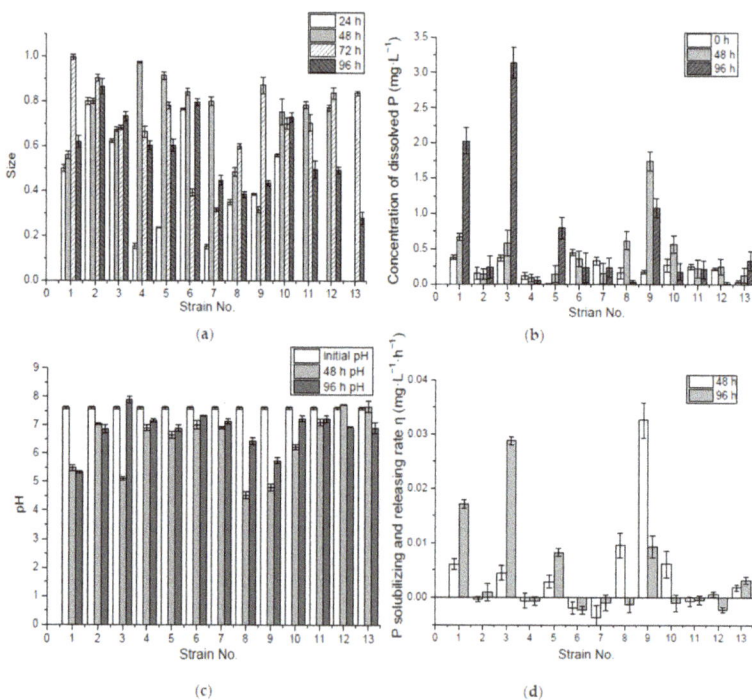

Figure 3. Size of P-solubilizing ring formed by 13 preliminary screened bacteria in the solid medium with lecithin. The microbes were sampled from rhizospheres of *Acorus calamus* L., and *Iris tectorum* Maxim. around Tinyin Lake. (**a**) shows that the ring size varying by 24, 48, 72 and 96 h. Concentration of dissolved P (**b**), pH (**c**) and P solubilizing and releasing rate (**d**) by 0 hour, 48 h and 96 h in liquid medium with organic P (lecithin).

(2) P-solubilizing and -releasing capacity of organic P (lecithin) in liquid medium

The concentration of dissolved P and release rate from lecithin were much lower than those from $Ca_3(PO_4)_2$, where No. 1, 3 and 9 were better at releasing dissolved phosphates. Decreasing the pH or producing acid did not promote the release of dissolved P. The pH value did not decline as much as under incubation in the liquid medium with $Ca_3(PO_4)_2$, and it even increased, which may have resulted from the neutralization of acid by alkaline substances such as choline and ammonia produced by the bacteria. Since lecithin contains not only PO_4^{3-}, but also N in the molecule, during the bacterial metabolism N may be released in the form of choline or ammonia [40]. Many bacteria could produce phospholipase to hydrolyze phospholipids into fatty acids and other lipophilic substances. There are four major classes, termed A, B, C and D, distinguished by the type of reaction that they catalyze. Among these, phospholipase C cleaves before the phosphate, releasing diacylglycerol and a phosphate-containing head group. Phospholipase D cleaves after the phosphate, releasing phosphatidic acid and an alcohol. Bacteria such as *Bacillus cereus* and *Pseudomonas aeruginosa* [41] could produce phospholipase C, and bacteria such as *Stenotrophomonas maltophilia* [42] and *Streptomyces prunicolor* [43] could produce phospholipase D. Figure 3b shows that there was a limitation for most of the isolates to release P since the production volume of dissolved P was small, possibly due to the limitation in production of phospholipase C and D, which can hydrolyze lecithin and release phosphatidic acid, a free head group, such as choline, and further release phosphates [37,44,45] (Figure 3).

(3)   P-solubilizing and -releasing capacity of isolates from reed in liquid medium with organic P (lecithin)

The 10-day incubation pH varied from 5.2 to 7.8, illustrating that even when the P was released, it would combine with Ca^{2+} to form insoluble phosphate, which might thus restrict the concentration of dissolved P in the medium. For R3, R4 and R8 described above, the pH value increased to over 7.5 later in the incubation. These increased alkalinity and dissolved organic compounds (DOC) prevented Ca^{2+} from binding with PO_4^{3-} ions. DOC, in the form of organic matter, has charges related to the pH-dependent characteristics of organic acid functional groups [46]. Under alkaline conditions, the organic matter and bacteria surface, which is made of peptidoglycan, eichoic acid, and capsule polysaccharide, have negatively charged cation exchange sites. Therefore, ions with positive charges, such as Ca^{2+}, were attracted to the organic matter surface due to electrostatic interactions [28], thus helping phosphate dissolution in the medium with a higher pH. The partially alkaline environment was more conducive to releasing P in the organic phosphorus culture (Figure 4).

R2, R3, R4 and R8 kept increasing the dissolved P in the medium during the incubation, as shown in Figure 4. R3, R4 and R8 obtained relatively higher maximum concentrations of P release combined with the corresponding time (10th day) (Table 2), implying that these isolates were better at releasing available P for a longer term and higher concentration in the liquid medium with lecithin. The mineralization of organic P does not increase with the decrease of pH.

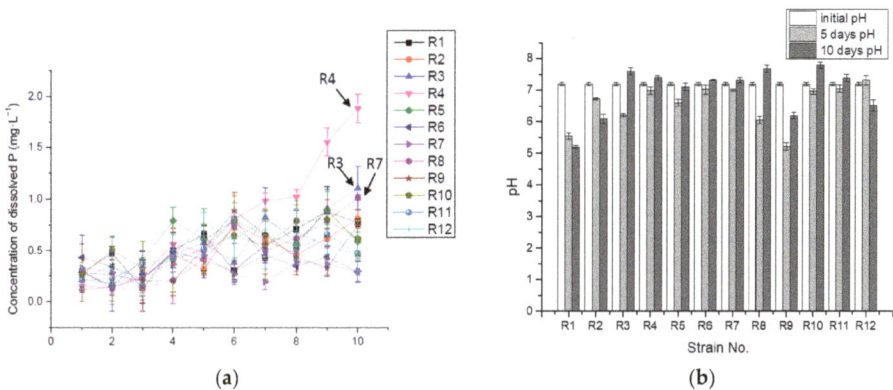

Figure 4. Concentration of dissolved P (mg·L^{-1}) (a) and pH (b) in 10-day incubation in liquid medium with organic P (lecithin).

3.3. Correlation of Solubilized P and pH in Ca₃(PO₄)₂ Liquid Medium

The pure $Ca_3(PO_4)_2$ liquid medium without seeding had a significantly negative linear correlation of P solubilization and pH from about 6.0 and lower. However when pH was higher than this, the liner relationship did not exist. When cultured with the isolated strains, there was no significant correlation between the culture pH and the P-solubilizing amount of the different isolates (Figure 5). Although compared to the P released from the lecithin medium, the pH decreased much more and solubilized more P in $Ca_3(PO_4)_2$ liquid medium, as discussed above. Thus, other factors such as P absorption, secreted organic acid and related enzymes [14], the reaction with organic molecules in the environment and growing patterns of isolated bacteria species are also important in P solubilization in $Ca_3(PO_4)_2$ liquid medium.

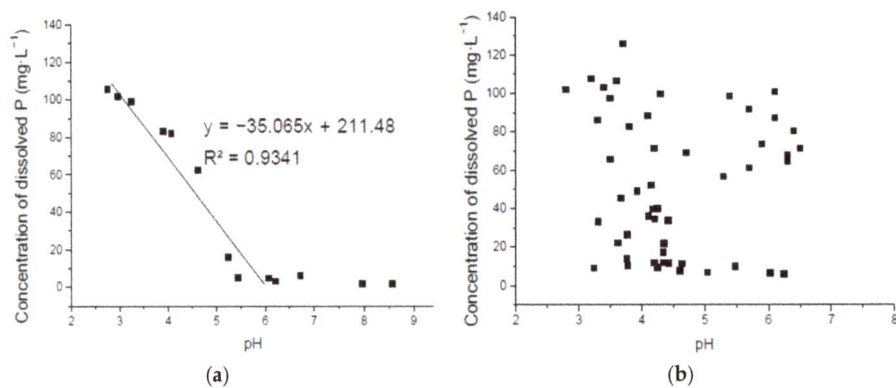

Figure 5. Correlation of solubilized P and pH in $Ca_3(PO_4)_2$ liquid medium without seeding the isolates where the pH was adjusted by hydrochloric acid (**a**) and seeding with the isolated bacteria (**b**).

Since there was only small decrease of pH and no significant correlation between the culture pH and the P mineralized by the OPMB strains grown in lecithin medium [15], the correlation of phosphorus solubilization and pH in lecithin medium was not discussed.

3.4. Microbial Community Analysis

MiSeq analysis of the samples on bacteria 16S rRNA from rhizospheres of the experimental plant yielded 30212 sequences and 445 OTUs. Relatively abundant among the most dominant bacteria genera living in the sampled wetland rhizospheres was *Acidobacteria* (Figure 6). Some were acidophilic, and since they were only recently discovered and the large majority have not been cultured, the ecology and metabolism of these bacteria is not well understood [49]. Their acidophilic ability may account for tolerance at low pH levels and help dissolve mineral P such as $Ca_3(PO_4)_2$. No. 5 and R10, which were identified as *Pseudomonas* sp. and *Pseudomonas knackmussii*, both belonging to the *pseudomonas* genus, accounted for less than 0.1%.

The natural wetland rhizosphere bacteria community was compared with other wetland environments in Table 3. The dominant phylum and its proportion were unique. *Acidobacteria*, *Proteobacteria*, *Bacteroidetes* and *Chloroflexi* were the most abundant in the samples, accounting for 34.9%, 34.2%, 8.8% and 4.8%, respectively. This is different to the Yellow River Delta Wetland and floating treatment wetland where *Proteobacteria* and *Acidobacteria* account for only 6.36% and less than 0.3%, respectively. Thus, in the studied natural wetland, rhizosphere *Acidobacteria* may play an important role in nutrient cycling. Because many of these species are uncultured, new methods other than incubation are required to illustrate their function in P solubilization and release.

Table 2. Maximum concentration and rate of P release for 10-day incubation in liquid medium with organic P (lecithin).

Strain No.	R1	R2	R3	R4	R5	R6	R7	R8	R9	R10	R11	R12
Max Concentration of P (mg·L^{-1})	0.88	0.81	1.10	1.88	0.90	0.49	0.45	1.01	0.89	0.80	0.81	0.79
Max Con. Day (day)	9	10	10	10	9	4	5	10	6	9	6	7
Max rate of releasing P (mg·L^{-1}·h^{-1})	0.0117	0.0053	0.0088	0.0078	0.0118	0.0178	0.0135	0.0126	0.0116	0.0106	0.0087	0.0093
Max rate time (hours)	24	144	24	240	24	24	24	24	24	24	24	24
P releasing rate η in 10 days (mg·L^{-1}·h^{-1})	0.0032	0.0033	0.0046	0.0078	0.0024	0.0012	0.0012	0.0042	0.0031	0.0026	0.0019	0.0019

Table 3. Dominant phylum, high quality sequences, richness and diversity estimators of the rhizosphere sample (CaoR1-B), and comparison with other wetland environments.

Sample	OTU	Sequences	Shannon	Simpson	Ace	Chao	Species	Genus	Dominant Phylum	Proportion of Dominant
CaoR1-B	445	30212	5.521541	0.00652	445	445	274	179	Acidobacteria	34.9%
Yellow river Delta Wetland[47]		30874							Proteobacteria	44.67%
Floating treatment wetland [48]	11477	78242	8.49	0.9931	2105	1603			Proteobacteria	46.4%

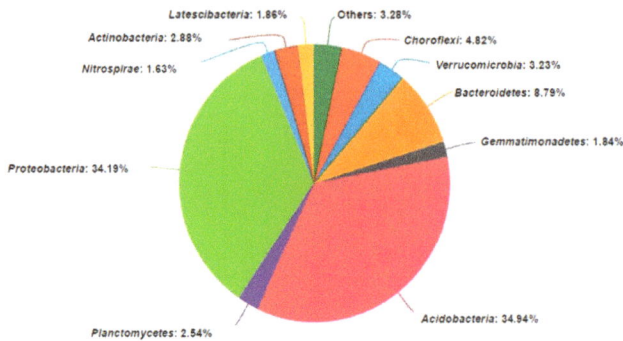

(**a**) Bacteria community analysis pie plot at the Phylum level

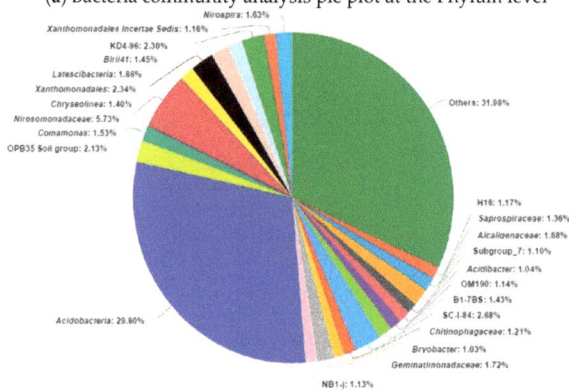

(**b**) Bacteria community analysis pie plot at the Genus level

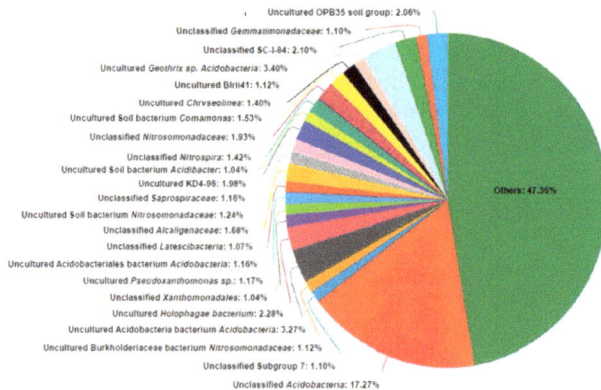

(c) Bacteria community analysis pie plot at the Species level

Figure 6. Bacteria community analysis pie plot at the Phylum (**a**), Genus (**b**) and Species (c) level sampled from rhizosphere of the experimental plant. Genera making up less than 1% of the total composition are defined as "others".

4. Practical Applications and Future Perspectives

Through this study, the effects of various bacteria on the formation of dissolved P in the rhizosphere of wetland plants and the basis of the P cycle in the root–microbial–soil–water system of wetland ecosystems could be further elucidated. The bacteria with better P solubilization and release could be preserved for future applications, for instance, seeding on the rhizosphere of wetland plants or directly in the wetland field to promote plant removal of P in the environment, especially for the removal of plant-unavailable P. This thus provides a new idea and approach to enhance the P pollution control efficiency of wetlands.

The bacteria's P-solubilizing and -releasing effects on other soil forms of P (Fe-P, Al-P, Phytate phosphorus, etc.) could be tested in the future research. The effects of P solubilization and release and plant absorption should be further analyzed when the screened bacteria are applied back in the rhizosphere or in the wetland field. Since the community analysis of the rhizosphere indicated that much of the bacteria species were not cultured, more types of culture technique and screening methods are required to maximize the number of species in the study of P-solubilizing and -releasing bacteria in the rhizosphere of wetlands. Varieties of screening condition should be used, for example, some of the bacteria might have better P-solubilizing and -releasing abilities under anaerobic or facultative conditions. A combination of bacteria species or microbes other than single bacteria strains, fungi or archaea also could be screened and analyzed from the rhizosphere of wetlands and applied to P removal and pollution control in the following studies.

5. Conclusions

The natural wetland rhizospheres in this experiment were dominated by *Acidobacteria*. The isolates' solubilizing and releasing rate of inorganic phosphorus was much higher than that of organic P, demonstrating their different impacts on organic and inorganic P. No. 5, 8 and 10 isolated from *Acorus calamus* L. and *Iris tectorum* Maxim. and R10 from reed released more available P in the $Ca_3(PO_4)_2$ medium, while No. 1, 3 and 9, as well as R3, R4 and R8 were better at releasing available P from lecithin. A decrease in pH would increase P solubilization from tricalcium phosphate, but there was no significant correlation between the culture pH and the P solubilization. The mineralization of organic P did not increase with the decrease in pH.

Acknowledgments: Funding: This work was supported by the National Natural Science Foundation of China [31600091], Nanjing Institute of Technology Innovation Fund Project [CKJB201409], Key State Laboratory of Lakes and Environment Open Fund [2016SKL009] and Teaching Reformation and Construction Fund of Nanjing Institute of Technology [JS201719]. Thanks to Professor Dafang Fu for leading and motivating us to finish the research. We thank Rajendra Prasad Singh for helping the submission and revision of the paper. Thanks also to Feng Yang, Changfei Zhang, Fenghua Gao, Mengyun Zhou, Huan Sun, Qi Mei, Anqi Xu, Caiju Qiu, Derong Yang and Tiansheng Wen who dedicate their time and passion in the difficult field sampling and lab investigation. Thanks to Keqiang Ding and Hongyi Li who encourage us to finish the work.

Author Contributions: Ying Cao and Dafang Fu conceived and designed the experiments; Ying Cao, Tingfeng Liu and Guang Guo performed the experiments; Ying Cao, Dafang Fu and Tingfeng Liu analyzed the data; Ying Cao, Dafang Fu, Tingfeng Liu, Guang Guo, Zhixin Hu contributed reagents/materials/analysis tools; Ying Cao and Dafang Fu wrote the paper.

Conflicts of Interest: The authors declare no conflict of interest.

References

1. Zou, Y.; Zhang, L.; Wang, L.; Zhang, S.; Yu, X. Effects of Aeration, Vegetation, and Iron Input on Total P Removal in a Lacustrine Wetland Receiving Agricultural Drainage. *Water* **2018**, *10*, 61. [CrossRef]
2. Jin, M.; Carlos, J.; McConnell, R.; Hall, G.; Champagne, P. Peat as Substrate for Small-Scale Constructed Wetlands Polishing Secondary Effluents from Municipal Wastewater Treatment Plant. *Water* **2017**, *9*, 928. [CrossRef]
3. Li, W.; Cui, L.; Zhang, Y.; Cai, Z.; Zhang, M.; Xu, W.; Zhao, X.; Lei, Y.; Pan, X.; Li, J.; et al. Using a Backpropagation Artificial Neural Network to Predict Nutrient Removal in Tidal Flow Constructed Wetlands. *Water* **2018**, *10*, 83. [CrossRef]
4. Zhai, J.; Xiao, J.; Rahaman, M.H.; John, Y.; Xiao, J. Seasonal Variation of Nutrient Removal in a Full-Scale Artificial Aerated Hybrid Constructed Wetland. *Water* **2016**, *8*, 551. [CrossRef]

5.   Jia, X.; Otte, M.L.; Liu, Y.; Qin, L.; Tian, X.; Lu, X.; Jiang, M.; Zou, Y. Performance of Iron Plaque of Wetland
      Plants for Regulating Iron, Manganese, and Phosphorus from Agricultural Drainage Water. *Water* **2018**,
      10, 42. [CrossRef]

6.   Ping, Y.; Pan, X.; Cui, L.; Li, W.; Lei, Y.; Zhou, J.; Wei, J. Effects of Plant Growth Form and Water Substrates
      on the Decomposition of Submerged Litter: Evidence of Constructed Wetland Plants in a Greenhouse
      Experiment. *Water* **2017**, *9*, 827. [CrossRef]

7.   Cai, S.; Shi, H.; Pan, X.; Liu, F.; Cui, Y.; Xie, H. Integrating Ecological Restoration of Agricultural Non-Point
      Source Pollution in Poyang Lake Basin in China. *Water* **2017**, *9*, 745. [CrossRef]

8.   Penn, C.; Chagas, I.; Klimeski, A.; Lyngsie, G. A Review of Phosphorus Removal Structures: How to Assess
      and Compare Their Performance. *Water* **2017**, *9*, 583. [CrossRef]

9.   Altomare, C.; Norvell, W.A.; Bjorkman, T.; Harman, G.E. Solubilization of phosphates and
      micronutrients by the plant growth promoting and biocontrol fungus *Trichoderma harzianum* Fifai1295-22.
      Appl. Environ. Microbiol. **1999**, *65*, 2926–2933. [PubMed]

10.   Reddy, K.R.; Delaune, R.D. Biogeochemistry of wetlands: Science and applications. *Soil Sci. Soc. Am. J.* **2008**,
      73, 1779.

11.   Shenker, M.; Seitelbach, S.; Brand, S.; Haim, A.; Litaor, M.I. Redox reactions and phosphorus release in
      re-flooded soils of an altered wetland. *Eur. J. Soil Sci.* **2005**, *56*, 515–525. [CrossRef]

12.   Borggaard, O.K.; Raben-Lange, B.; Gimsing, A.L.; Strobel, B.W. Influence of humic substances on phosphate
      adsorption by aluminum and iron oxides. *Geoderma* **2005**, *127*, 270–279. [CrossRef]

13.   Delfim, J.; Schoebitz, M.; Paulino, L.; Hirzel, J.; Zagal, E. Phosphorus Availability in Wheat, in Volcanic Soils
      Inoculated with Phosphate-Solubilizing *Bacillus thuringiensis*. *Sustainability* **2018**, *10*, 144. [CrossRef]

14.   Kim, C.H.; Han, S.H.; Kim, K.Y.; Cho, B.H.; Kim, Y.H.; Koo, B.S.; Kim, Y.C. Cloning and expression of
      pyrroloquinoline quinone (PQQ) genes from a phosphate-solubilizing bacterium *Enterobacter intermedium*.
      Curr. Microbiol. **2004**, *47*, 6. [CrossRef]

15.   Tao, G.C.; Tian, S.J.; Cai, M.Y.; Guang, H.X. Phosphate-Solubilizing and -Mineralizing Abilities of Bacteria
      Isolated from Soils. *Pedosphere* **2008**, *4*, 515–523. [CrossRef]

16.   Oliveira, C.A.; Alves, V.M.C.; Marriel, I.E.; Gomes, E.A.; Scotti, M.R.; Carneiro, N.P.; Guimarães, C.T.;
      Schaffert, R.E.; Sá, N.M.H. Phosphate solubilizing microorganisms isolated from rhizosphere of maize
      cultivated in an oxisol of the Brazilian Cerrado Biome. *Soil. Biol. Biochem.* **2009**, *41*, 1782–1787. [CrossRef]

17.   Goldstein, A. Recent progress in understanding the moleculargenetics and biochemistry of calcium
      phosphate solubilization by gram negative bacteria. *Biol. Agric. Hortic.* **1995**, *12*, 185–193. [CrossRef]

18.   Illmer, P.; Schinner, F. Solubilization of inorganic phosphateby microorganisms isolated from forest soils.
      Soil. Biol. Biochem. **1992**, *24*, 389–395. [CrossRef]

19.   Li, S.M.; Li, L.; Zhang, F.S.; Tang, C. Acid Phosphatase Role in Chickpea/Maize Intercropping. *Ann. Bot.*
      2004, *94*, 297–303. [CrossRef] [PubMed]

20.   Zhai, J.H.; Zeng, C.S.; Tong, C.; Wang, W.Q.; Liao, J. Organic and Inorganic Phosphorus in Sediments of the
      Min River Estuarine Wetlands: Contents and Profile Distribution. *J. Subtropic. Res. Environ.* **2010**, *5*, 9–14.
      [CrossRef]

21.   Cao, X.Y.; Chong, Y.X.; Yu, G.W.; Zhong, H.T. Difference of P Content in Different Area Substrate of
      Constructed Wetland. *Environ. Sci.* **2012**, *33*, 4033–4039. [CrossRef]

22.   Lu, J.; Gao, B.; Hao, H. Study on the occurrence of phosphorus in the sediments of artificial wetland.
      Spectrosc. Spectr. Anal. **2014**, *11*, 3162–3165. [CrossRef]

23.   Liang, W.; Shao, X.X.; Wu, M.; Li, W.H.; Ye, X.Q.; Jiang, K.Y. Phosphorus fraction in the sediments from
      different vegetation type in hangzhou bay coastal wetlands. *Acta Ecol. Sin.* **2012**, *32*, 5025–5033. [CrossRef]

24.   Kong, L.Z. Study on the Morphological Characteristics and Adsorption Desorption Characteristics of
      Phosphorus in Wetland Soil under Different Tillage Year. Ph.D. Thesis, Anhwei Normal University,
      Wuhu, China, 2014.

25.   Lynch, J.M.; Whipps, J.M. Substrate flow in the rhizosphere. *Plant Soil* **1990**, *129*, 1–10. [CrossRef]

26.   Richardson, A.E.; Simpson, R.J. Soil microorganisms mediating phosphorus availability update on microbial
      phosphorus. *Plant Physiol.* **2011**, *156*, 989. [CrossRef] [PubMed]

27.   White, P.J. Ion Uptake Mechanisms of Individual Cells and Roots: Shortdistance Transport. In *Marschner's
      Mineral Nutrition of Higher Plants*, 3rd ed.; Academic Press: Cambridge, MA, USA, 2012; pp. 7–47.

28. Cerozi, B.S.; Fitzsimmons, K. The effect of pH on phosphorus availability and speciation in an aquaponics nutrient solution. *Bioresour. Technol.* **2016**, *219*, 778–781. [CrossRef] [PubMed]

29. Zhou, Q.Y.; Gao, T.Y. *Environmental Engineering Microbiology*, 3rd ed.; Higher Education Press: Beijing, China, 2010; pp. 380–448. ISBN 978-7-04-022265-4. (In Chinese)

30. Nautiyal, C.S. An efficient microbiological growth medium for screening phosphate solubilizing microorganisms. *FEMS Microbiol. Lett.* **1999**, *170*, 265–270. [CrossRef] [PubMed]

31. Niewolak, S. Occurrence of microorganisms in fertilized lakes. II. Lecithin-mineralizing microorganisms. *Pol. Arch. Hydrobiol.* **1980**, *27*, 53–71.

32. Pepper, L.L.; Gerba, C.P.; Terry, J. *Environmental Microbiology*, 3rd ed.; Academic Press: Cambridge, MA, USA, 2014; ISBN 978-0-12-394626-3.

33. Lin, Q.M.; Wang, H.; Zhao, X.R.; Zhao, Z.J. The solubilizing ability of some bacteria and fungi and its mechanisms. *Microbiol. China* **2001**, *28*, 26–30. [CrossRef]

34. Yu, Q.Y.; Chen, S.Y.; Ma, Z.Y.; Wang, J.F.; Wei, L. Screening of phosphorus bacteria and its effects on maize growth at seedling stage. *Ecol. Environ. Sci.* **2012**, *7*, 1257–1261. [CrossRef]

35. Yu, W.B.; Yang, X.M.; Shen, Q.R.; Xu, Y.C. Mechanism on phosphate solubilization of pseudomonas sp. K_3 and its phosphate solubilization ability under buffering condition. *Plant Nutr. Fertile Sci.* **2010**, *16*, 354–361.

36. Moorberg, C.J.; Vepraskas, M.J.; Niewoehner, C.P. Phosphorus dissolution in the rhizosphere of bald cypress trees in restored wetland soils wetland soils. *Soil Sci. Soc. Am. J.* **2015**, *79*, 343–355. [CrossRef]

37. Kadamur, G.; Ross, E.M. Mammalian phospholipase C. *Annu. Rev. Physiol.* **2013**, *75*, 127–154. [CrossRef] [PubMed]

38. Titball, R.W. Bacterial phospholipase C. *Microbiol. Rev.* **1993**, *57*, 347–366. [PubMed]

39. Essen, L.O.; Perisic, O.; Katan, M.; Wu, Y.; Roberts, M.F.; Williams, R.L. Structural mapping of the catalytic mechanism for a mammalian phosphoinositide-specific phospholipase C. *Biochemistry* **1997**, *36*, 1704. [CrossRef] [PubMed]

40. Kolesnikov, Y.S.; Nokhrina, K.P.; Kretynin, S.V.; Volotovski, I.D.; Martinec, J.; Romanov, G.A.; Kravets, V.S. Molecular structure of phospholipase D and regulatory mechanisms of its activity in plant and animal cells. *Biochem. Biokhimiia* **2012**, *77*, 1–14. [CrossRef] [PubMed]

41. Zhan, Y.S. Screening of Phospholipase Producing Strain C and Study on Its Enzymatic Properties. Ph.D. Thesis, Hunan Agricultural University, Changsha, China, 2010.

42. Dai, S.L.; Zhang, J.; Shang, J. Screening and identification of high efficiency producing bacteria from phospholipase D. *Jiangsu Agric. Sci.* **2013**, *41*, 309–311.

43. Yang, X.L.; Zhang, L.; Wang, Y.D. Screening and identification of high yield phospholipase D strain and optimization of fermentation conditions. *Jiangsu Agric. Sci.* **2016**, *44*, 521–525.

44. Foster, D.A. Targeting Phospholipase D-mediated Survival Signals in Cancer. *Curr. Signal Transduct. Ther.* **2006**, *1*, 295–303. [CrossRef]

45. Balboa, M.A.; Firestein, B.L.; Godson, C.; Bell, K.S.; Insel, P.A. Protein kinase C alpha mediates phospholipase D activation by nucleotides and phorbol ester in Madin-Darby canine kidney cells. Stimulation of phospholipase D is independent of activation of polyphosphoinositide-specific phospholipase C and phospholipase A. *J. Biol. Chem.* **1994**, *269*, 10511–10516. [PubMed]

46. Pierzynski, G.M.; Sims, J.T.; Vance, G.F. *Soils and Environmental Quality*, 3rd ed.; CRC Press: Boca Raton, FL, USA, 2005; pp. 145–146. ISBN 0-8493-1616-2.

47. Li, Q. Study on Soil Microbial Community Structure Based on High Throughput Sequencing Technology. Ph.D. Thesis, Shandong Normal University, Jinan, China, 2015.

48. Gao, L.; Zhou, W.; Huang, J.; He, S.; Yan, Y.; Zhu, W. Nitrogen removal by the enhanced floating treatment wetlands from the secondary effluent. *Bioresour. Technol.* **2017**, *234*, 243–252. [CrossRef] [PubMed]

49. Quaiser, A.; Ochsenreiter, T.; Lanz, C.; Schuster, S.C.; Treusch, A.H.; Eck, J. Acidobacteria form a coherent but highly diverse group within the bacterial domain: Evidence from environmental genomics. *Mol. Microbiol.* **2003**, *50*, 563–575. [CrossRef] [PubMed]

water

MDPI

Article

Sequencing Infrastructure Investments under Deep Uncertainty Using Real Options Analysis

Nishtha Manocha * and Vladan Babovic

Department of Civil and Environmental Engineering, National University of Singapore,
Singapore 117576, Singapore; vladan@nus.edu.sg
* Correspondence: nishtha.m@u.nus.edu

Received: 8 January 2018; Accepted: 12 February 2018; Published: 23 February 2018

Abstract: The adaptation tipping point and adaptation pathway approach developed to make decisions under deep uncertainty do not shed light on which among the multiple available pathways should be chosen as the preferred pathway. This creates the need to extend these approaches by means of suitable tools that can help sequence actions and subsequently enable the outlining of relevant policies. This paper presents two sequencing approaches, namely, the "Build to Target" and "Build Up" approach, to aid in sub-selecting a set of preferred pathways. Both approaches differ in the levels of flexibility they offer. They are exemplified by means of two case studies wherein the Net Present Valuation and the Real Options Analysis are employed as selection criterions. The results demonstrate the benefit of these two approaches when used in conjunction with the adaptation pathways and show how the pathways selected by means of a Build to Target approach generally have a value greater than, or at least the same as, the pathways selected by the Build Up approach. Further, this paper also demonstrates the capacity of Real Options to quantify and capture the economic value of flexibility, which cannot be done by traditional valuation approaches such as Net Present Valuation.

Keywords: deep uncertainty; Real Options Analysis; sequencing approaches

---

1. Introduction

Policymakers today are faced with the difficult task of planning for large-scale infrastructure that can cater to the climatic and socio-economic changes that the future will bring. Large-scale infrastructure is cost intensive, has a life span of at least a few decades, requires long-term planning and is essentially irreversible. Thus, when built as a strategy to cope with a changing climate, infrastructure systems carry an inherent risk. Given the inability to accurately predict the future, infrastructure systems designed and commissioned by employing the "predict then build" approach could fail to meet their objectives if the future turns out to be incompatible with the future initially assumed in the design stage. Therefore, to address this deeply uncertain nature resulting from long-term changes, it is becoming necessary to develop strategies and plans that support flexibility and react more strategically than traditional planning approaches, i.e., identify strategies that perform well under many future scenarios, instead of optimizing a strategy for only a handful [1].

Urban water infrastructure development requires careful long-term planning to reduce the risk of climate change and other uncertainties [2] such as population growth, economic development, urbanization, and a changing social and political environment, etc. In response to this, various approaches have been developed over the years to aid decision makers' in this regard. These are Assumption-based Planning [3,4], Robust Decision Making [5,6], Decision Tree Analysis [7], Adaptive Policy Making [8,9], Adaptation Tipping Point [10–12], Adaptation Pathways [11,13], Dynamic Adaptation Policy Pathways [14,15], Info Gap Robustness Pathway Method [16] and Real Options analysis [17–20]. In addition, there are many other methodologies, tools and techniques that support these approaches. A few examples are: scenario planning [21], assumption-based planning [3], Monte Carlo Analysis [17], Exploratory Modeling Analysis [22] and Info gap decision theory [23]. Among these approaches, the adaptation tipping point and adaptation pathway approach [11,13] are increasingly gaining popularity in view of climate change adaptation.

Adaptation Pathways provide decision makers with an excellent overview of all available actions and possible sequences and combinations that can be made to ensure that the system performs as expected in its lifetime. While this broad overview is very beneficial, it still leaves decision makers with the dilemma of which particular action should be exercised today, which action should be exercised with a delay and which action should be kept as an option. Studies [11] have suggested that the weakness in the adaptation pathway approach is the complexity of the analysis and any simplifications that must be made to make the analysis tractable. Therefore, in order to answer these questions, it becomes necessary to analyse these pathways in greater depth with suitable tools. Further assessment has to be carried out to enable the sub-selection of a set of preferred pathways. This is necessary as decision makers have to outline a plan(s) that will be followed given that the climate changes as expected and to identify policies that have to be implemented both in the short and long term. Pathways that are not selected in the current time frame remain active as options in case the future turns out to be incompatible with selected plans.

The subselection of the pathways can be undertaken in a few different ways depending upon the viewpoints of the management, the data available and the trust or preference in one methodology over the other. The case studies presented in this paper illustrate adaptation pathway maps for flood management. The first case study, the Kent Ridge Case [24] Study, is focused on flood management for a small tropical urban catchment located in Singapore, while the second is a hypothetical case study focused on flood management for the Rhine delta [3]. The first case study presents a basic economic assessment to rank the various adaptation pathways and presents the best performing (economically) adaptation pathways as the preferred subset. Subsequent studies on the Waas Case [15] aim to use multiobjective robust optimization to trim down the solution space. This paper uses these two case studies to explore and provide an alternative approach to improve the manner in which the adaptation pathways are shortlisted. This is done by employing two formal sequencing approaches to outline the preferred set of adaptation pathway(s).

Sequencing approaches have been used for a long time to identify long term plans for infrastructure development. In the context of water resources, management sequencing has been used extensively to identify water supply projects [24]. These include studies that focus on reservoir expansion [25–27], groundwater supply [2] and building a water supply portfolio [28–30]. In these studies, the economic cost calculated by means of traditional Net Present Valuation is used as the primary parameter to guide the sequencing. However, traditional valuation approaches including the Net Present Value (NPV) are inherently flawed in analyses projects with future uncertainties [31]. A real options valuation is a more appropriate tool for decision making under uncertainty. The ability to adapt a solution at a time in the future when more information about the uncertain driver is available allows one to limit the downside of making a wrong decision, and capture the upside of new information and opportunities. This flexibility can be translated into an economic gain which can be captured by Real Options Analysis. Real Options Analysis has previously been used to support decision making under deep uncertainty in a handful of case studies. In the area of infrastructure design and management, applications include analyzing scale versus flexibility [32], architecting a maritime domain protection system [33,34], design of a parking garage [35], design of urban water management systems [20], flood risk management [18,36,37] and design of water supply systems [17].

The objectives of this paper are to:

- Present two formal sequencing approaches to extend the current adaptation tipping point and adaptive pathway approached. These include the 'build up' (BU) and 'build to target' (BTT) approaches [38]. The purpose of the sequencing exercise is to identify a preferred pathway(s) that will be followed given that the climate changes as expected. Given the nature of the adaptive approach, the pathway(s) can be supplemented or updated if the manifestation of reality is other than what was planned for.

- To demonstrate the ability of these approaches by applying them to two case studies in the domain of flood management.
- To assess the impact of choosing different selection criterions (Net Present Value and the Real Options Analysis) on the performance of these approaches and the subsequent sub-selection of the adaptation pathway(s).
- To understand if different perspectives of national development (varied landuse) can have an impact on the economic performance and the shortlisting of the various pathways. This objective is specific to the Kent Ridge Case Study.

The rest of the paper is organized as follows: The overall framework used in this paper is presented in Section 2. The sequencing approaches and the steps involved in using these approaches to extend the adaptation tipping point and the adaptation pathway approach are explained in detail in Section 3. The two case studies are described in Section 4, following which the results are presented and discussed in Section 5. Finally, a summary of the paper is presented in Section 6.

2. Methodology

The sequencing approaches presented in this paper are employed to enable the selection of a subset of preferred pathways once the adaptation pathway maps are available. In the development of adaptation pathway maps, the first step is the description of the case study and the outlining of the current situation in order to better understand the objectives. The second step is the identification and development of a portfolio of adaptation actions that can be implemented to ensure that objectives are met all along the planning horizon. The next step is the definition of scenarios, over which the adaptation actions will be assessed. The fourth step is the setup an assessment model. Thereafter, the tipping points of the adaptation actions in isolation and combination are calculated.

Adaptation Tipping Points are the physical boundary conditions where acceptable technical, environmental, societal or economic standards may be compromised [9]. Once determined, the tipping points are positioned in time using the assessment scenarios and are combined to draw the adaptation pathway maps. This approach is illustrated in Steps 1 through 5 of Figure 1. A combination of the defined Adaptation Tipping Points with assessment scenarios used to develop adaptation pathway maps provides information on the resilience of the infrastructure system to climate change and the potential need for alternative adaptive strategies.

The complete framework employed in this study is presented in Figure 1. Steps 1 through 5 are part of the adaptation tipping point and the adaptation pathway approach presented in previous studies [13,39]; Step 6 is presented in this paper.

The proposed sequencing approaches employed in this paper involve dividing the pathway into a number of phases (P) wherein adaptation actions have to be implemented. A pathway is composed of a set of adaptation actions that meet the study objectives all along the planning horizon (T). The numbers of phases that each pathway has are dependent upon the timing of reaching the adaptation tipping point of the individual adaptation actions involved in the pathway. That is, once an adaptation tipping point is reached, a new phase of investments is required that can add another adaptation action to the current configuration.

The purpose of both of the sequencing process is to identify the combination of adaptation actions at the beginning of each phase so as to maximize the selection criterion while meeting the study objectives.

In order to perform the sequencing based on the two approaches, identify the preferred pathway and compare the two sequencing processes, two economic selection criterions were identified. These are the net present valuation (NPV) and real option assessment (ROA). The sequencing was performed individually for each of these selection criterions to compare the impact of using different selection criterions on the sequencing approaches. A higher value of the selection criterion indicates a better economic performance of the pathway.

Net Present Valuation is the traditional approach for making investment decisions. The NPV approach applies a discount rate calculated either by the opportunity cost, the weighted average cost of capital or other methods to represent all future cash flows in present equivalent economic units. By discounting both the future costs and benefits to the present time, it tries to estimate whether the project's cost can be offset by the expected benefits. The investment rule is to investigate when NPV is greater or equal to zero. For mutually exclusive projects, the rule is to choose the project with the highest NPV. In contrast to this, the Real Options analysis is a relatively new approach that is now gaining traction in the valuation of flexible adaptive investments that must be made under deep uncertainty. Real Options Analysis is able to take into account and value uncertainty and flexibility; as opposed to traditional approaches of economic valuation borrowing concepts originating from the options analysis developed in finance [40], we can apply them to undertake a Real Options Analysis for management of infrastructure systems [33]. The specifics of performing a real options analysis are beyond the scope of this paper. For details on how to perform this assessment, please refer to [31].

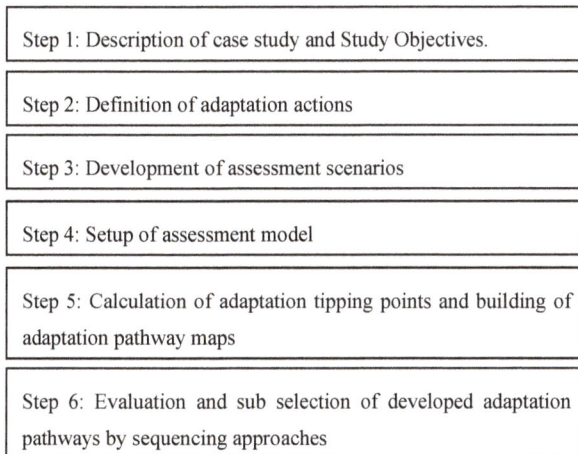

Step 1: Description of case study and Study Objectives.
Step 2: Definition of adaptation actions
Step 3: Development of assessment scenarios
Step 4: Setup of assessment model
Step 5: Calculation of adaptation tipping points and building of adaptation pathway maps
Step 6: Evaluation and sub selection of developed adaptation pathways by sequencing approaches

Figure 1. Overall framework of the study.

3. Sequencing Approaches

In the Build Up (BU) method, the adaptation pathway is identified in a chronological order. Decisions on the adaptation action that has to be implemented are made for one phase at a time. Once a decision about a specific phase is made, the concerned adaptation action is built and remains fixed for the entire planning horizon. By contrast, in the Build to Target (BTT) method, the pathway to be built is selected by maximizing the value of the selection criterion for the final year of the planning horizon. Once the pathway is identified, the individual adaptation actions that make up the pathway are built based on encountering their respective tipping points in reality.

This can be better understood by considering a simple example. Let us say that we have to design a water supply portfolio that can cater to the water demand of a town, by supplying it with "X" liters per day of water in the year 2100. In order to do this, let us assume that the town has a few different adaptation actions it can consider. These include: expanding the water reservoirs to increase storage capacity, building desalinization plants, increasing the domestic rain water harvesting capacity and constructing water recycling facilities. If we approach this problem by the BTT method, we will run simulation models and find the best set of actions, or the best pathway (based on governance perspective, for example, a solution that performs the best economically; technically, the most energy efficient) that can collectively supply the town with "S" liters of water in the year 2100. Let us assume that this best pathway includes increasing the reservoir storage capacity by 30% and having a desalination plant of capacity "X" liters/day. Once this pathway is selected, the timing of bringing these adaptation actions online can be selected by means of assessing the adaptation tipping points (time when a given configuration has a supply lower than the demand) under various user defined scenarios. The timing of implementing the individual adaptation actions can be revisited in the future when more data about the uncertainty driver are available. However, if the same decision is to be made by the BU approach, the decision is taken in phases. The system is first assessed to find the tipping point of the current configuration. Let us assume that the current configuration can supply the town with water until the year 2030. Thus, we now need to identify the particular adaptation action that should be implemented by 2030 to ensure that the town has sufficient water to meet its needs. We do so by running simulations and finding the best suited adaptation action to be implemented in the next phase. Let us assume that the simulations result advocates the increase of domestic rain water harvesting schemes of capacity "Y" liters which have a tipping point in the year 2060. As our planning horizon stretches to the year 2100, we have to identify the adaptation action that has to be implemented in the subsequent phase. In this case, the adaptation action identified is building a desalinization plant of capacity "Z" liters/day which has a tipping point in the year 2105. Thus, the pathway identified by the BU approach is implementing a domestic rain water harvesting scheme of capacity "Y" liters by the year 2030 and bringing online a desalinization plant of capacity "Z" liters/day by the year 2060.

Given the nature of the BTT approach, since the adaptation pathway is identified first, even though the timing of building each phase carries some flexibility, the action to be built at each phase is constricted to align with the selected pathway. In this regard, the Build Up method is more flexible and capable to respond to future changes, such as a changing climate, changing population, changing landuse, etc. However, it also generally results in less favorable selection criterion values compared with those obtained using the Build to Target method for the selected design conditions, which are a disadvantage if the assumed design conditions actually occur [41]. The following sub-sections shed light on the steps involved in performing each sequencing technique.

3.1. Build to Target (BTT) Method

The proposed process of sequencing using the BTT method is shown in Figure 2. The first step is identifying the measures, policies or adaptation actions already in place to meet the study objectives. Subsequently, the pathways developed are ranked based on their selection criterion performance. The pathway that meets the study objectives for the whole planning horizon and also maximizes the selection criterion is chosen as the preferred pathway. Consequently, based on how reality unfolds, the individual actions that make up the pathway phases are built.

3.2. Build Up (BU) Method

The proposed process of using the Build Up (BU) method is shown in Figure 3. The first step is identifying the measures, policies or adaptation actions already in place to meet the study objectives. Following this, the actions to be implemented in the next phases are identified. This identification is based on the average value of the selection criterion of the phase in consideration across the range of assessment scenarios. The value of the selection criterion is maximized at each phase. This is repeated until the identified actions in combination can cater to the entire planning horizon, i.e., have an adaptation tipping point (ATP) greater than or equal to the final year of the planning horizon (T).

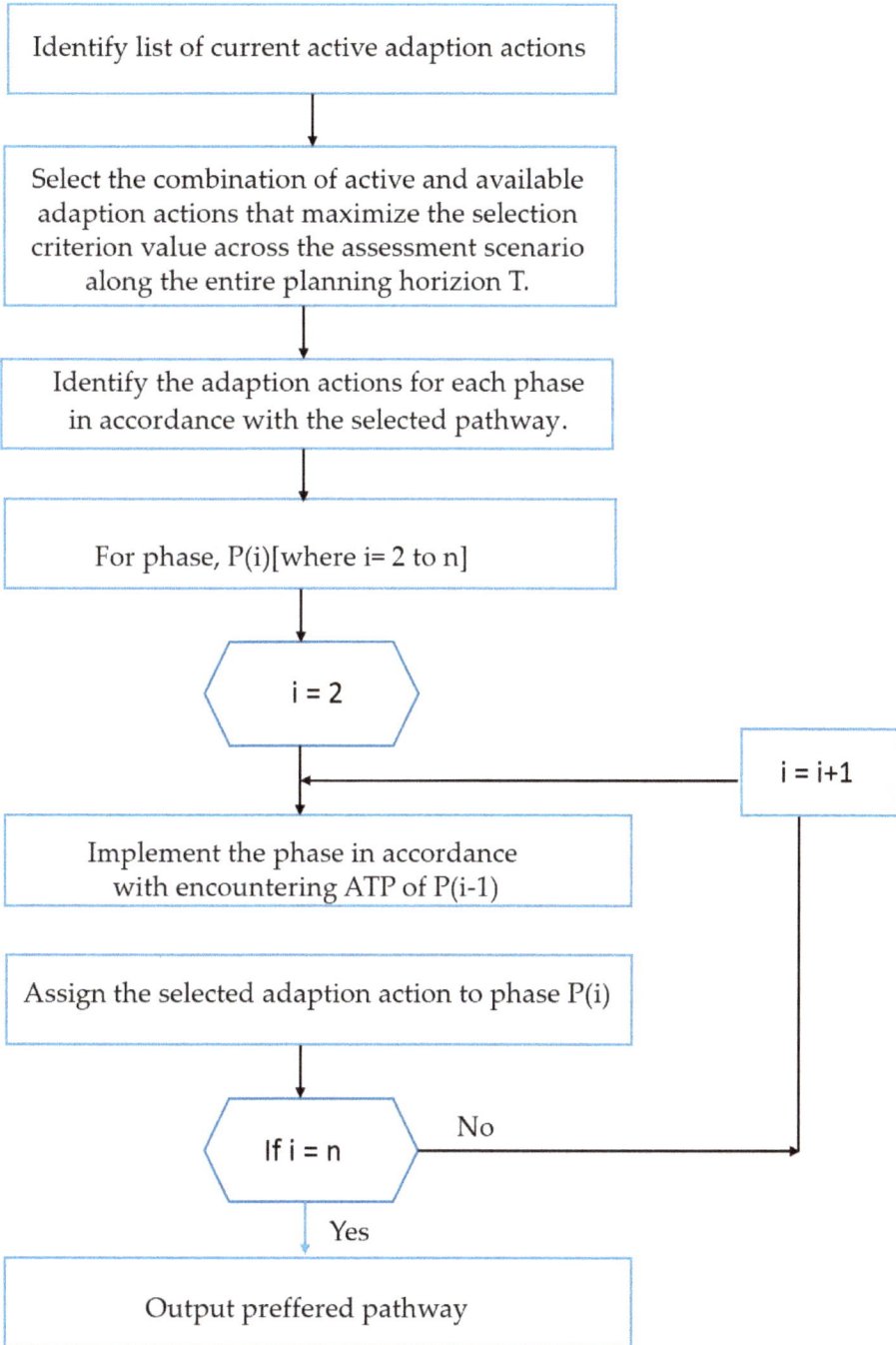

Figure 2. Build to Target (BTT) sequencing approach.

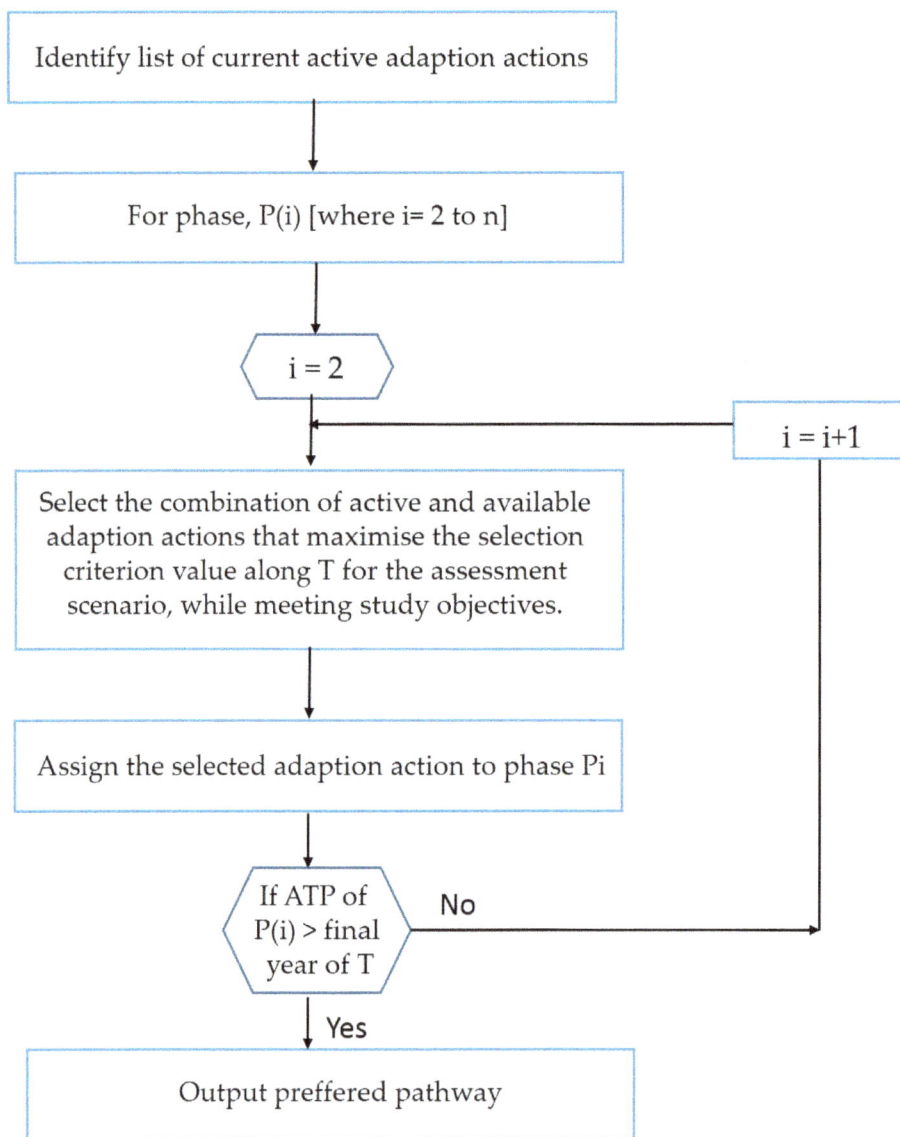

Figure 3. BU Sequencing approach.

4. Description of Case Studies

Two case studies are employed in this paper to present the sequencing approaches. The two case studies are focused on flood management but differ in size and geographical location. This section introduces the two case studies and provides the necessary background information on the same in the context of this paper.

4.1. Kent Ridge Catchment

The Kent Ridge Catchment is a tropical urban catchment located in the south of Singapore. The Kent Ridge Catchment is 85,000 square meters in size and is located within the campus of the National University of Singapore. This catchment contains all the main land use types of Singapore and hence can be considered as reasonably representative from a hydrological point of view.

The adaptation tipping point and the adaptation pathway approach were employed in a previous study [39] to develop adaptive plans for flood management in the Kent Ridge Catchment. The adaptation actions and assessment scenarios explored have been presented here for brevity. For details, please refer to the original study.

The adaptation actions considered in this study are defined in terms of infrastructure that can be built or modified for the purpose of storm water management for dense urban areas. The portfolio of actions includes both traditional grey and innovative green infrastructure solutions, namely: expansion of drainage canals, implementation of green roofs and implementation of porous pavements. The adaptation actions encompass a drainage increase in four configurations (C1–C4), implementation of porous pavements in three configurations (C1–C3) and the implementation of green roofs in three configurations (C1–C3). Further, the actions explored are aligned with the national objectives of Singapore. These actions, in varied configurations, make up the set of adaptation actions assessed in this study. They are described and summarized in Table 1. The notations specified in Table 1 are used to address the actions in the subsequent sections of this paper.

Table 1. Portfolio of Adaptation Actions, Kent Ridge case study (Table adapted from [39]).

Configuration	Notation	Description
Current Drainage Configuration	A	The current configuration is maintained
Drainage Increase C1	B	15% increase from baseline drainage capacity
Drainage Increase C2	C	20% increase from baseline drainage capacity
Drainage Increase C3	D	30% increase from baseline drainage capacity
Drainage Increase C4	E	50% increase from baseline drainage capacity
Porous Pavements C1	F	50% of all available pavements covered
Porous Pavements C2	G	60% of all available pavements covered
Porous Pavements C3	H	80% of all available pavements covered
Green Roofs C1	I	20% of all available roof space covered
Green Roofs C2	J	35% of all available roof space covered
Green Roofs C3	K	50% of all available roof space covered

The scenarios developed for the Kent Ridge Catchment cover a range of possible climatic (Table S1) and land-use futures (Table S2). This is done to study the individual and coupled impact of climatic and anthropogenic influences on the timing of reaching the respective tipping points. The climatic scenarios include the Wet 1 (W1), Wet 2 (W2) and the Dry 1 (D1) climate scenario. In addition to the climate scenarios described, an additional scenario, namely the "All climate scenario" is presented in the results section. This scenario warrants the running of the sequencing process across the three climate scenarios described in Table S1 and maximizing the mean of the selection criterion across these climate scenarios. The current climate is indicated as Business as Usual (BAU).

The landuse scenarios include a Baseline (landuse B), Green (landuse G) and Sustainable Grey landuse SG) landuse scenario. The baseline scenario is representative of the existing practices of land use. The green scenario follows a sustainable approach. The sustainable grey landuse allows for construction of buildings on account of green spaces but mandates retrofitting the developments in a predefined proportion with storm water management solutions such as green roofs and/or porous pavements.

The adaptation tipping points are described in terms of maximum annual rainfall beyond which the given adaptation action ceases to perform as expected (Figure S1) Using the adaptation tipping points, an adaptation pathway map was developed for the Kent Ridge Catchement.

Figure 4 can be read like a map indicating several possible routes (adaptation pathways) to reach a desired point (year 2100) in the future. Each adaptation action (shown on the Y axis) is depicted by a specific color. The X axis represents the ATPs and the corresponding years they are exercised under different climate scenarios. The squares represent interchange points, where one can switch to a different adaptation action. The color of the square indicates the action from which one has to switch. The black bars mean that an ATP is reached and that the current configuration can no longer perform as expected, requiring a switch. Each pathway is defined by the particular sequencing of actions that allows the configuration to meet objectives until the year 2100.

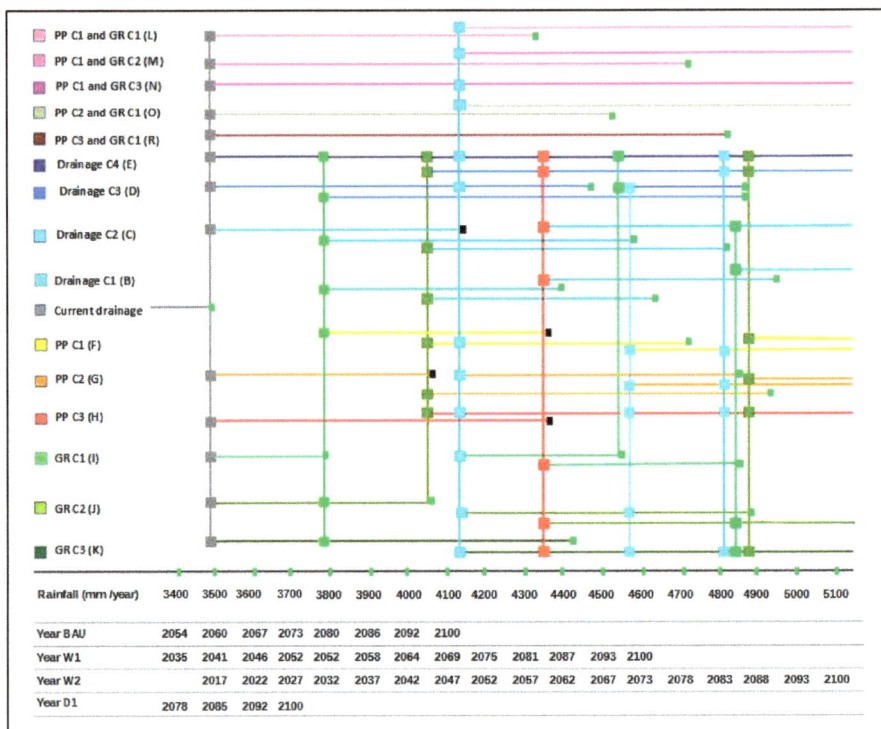

Figure 4. Adaptation Pathways Map, for the current land use and all climate scenarios. Porous Pavements are abbreviated as PP and green roofs are abbreviated as GR. Adapted from [39].

4.2. The Waas Case Study

In response to two flood events of the last 30 years that resulted in a total damage of 2810 billion Euros, a hypothetical Waas case study [11] employed the adaptation tipping point and adaptation pathway approach to develop different flood management solutions in the Rhine delta of the Netherlands. The study generated 30 transient scenarios on the basis of three climate scenarios established by the Royal Dutch Meteorological Institute (KNMI). Namely, the "No Climate Change" scenario, the "G" scenario, and the "Wp" scenario [42]. The G scenario has a temperature rise of 1°C in 2100, the winter-time precipitation increasing by 3.6% and the mean summer precipitation increasing by 2.8%. The Wp scenario has a temperature rise of 2 °C, winter-time precipitation increasing by 14.2%, but the mean summer-time precipitation decreases by 19% [42]. In addition to this, the "All" climate scenario was used to depict a central climate scenario. Adaptation actions were defined Table 2 and their respective adaptation tipping points were calculated for all climate scenarios. The portfolio of adaptation actions is presented here for brevity (Table 2). Pathways over a time frame of 100 years were then generated. Detailed explanation of the methodology of development of the adaptation pathways is available in the study [11].

While the Waas Case study provided the adaptation tipping points (described as a maximum allowable damage of 200 M Euros) for all adaptation actions under all outlined climate scenarios (Figure S2), it provided one adaptation pathway map drawn for the median adaptation tipping point of all scenarios. In order to facilitate this study, using the adaptation tipping points provided, adaptation maps were developed for the Wp, G and No Climate Change scenarios.

Figure 5 can be read like a map indicating several possible routes to get to a desired point (target) in the future. The circles indicate a transfer station to another policy; the blocks indicate a terminal station at which an Adaptation Tipping Point (ATP) is reached. Each adaptation action is indicated by a particular color.

Table 2. Portfolio of Adaptation Actions, Waas case study (Table adapted from [11]).

Configuration	Notation	Description [11]
No Policy		
DH 500	A	Dike height rise to be able to cope with the 1:500 discharge based on measurements
DH 1000	B	Dike height rise to be able to cope with the 1:1000 discharge based on measurements
DH 1.5	C	Dike rise: adapting to 1.5 times the second highest discharge ever measured ('rule of thumb measure')
RfRL	D	'Room for the river'-Large scale: with extra side channels, the river has more space after a threshold discharge is exceeded
RfRS	E	'Room for the river'-Small scale: with extra side channels, the river has more space after a threshold discharge is exceeded
FloatH	F	Floating houses: resulting in damage functions with ten times less damage
FaC	G	Fort cities: extra embankments around the cities
Mound	H	All cities will be raised by 4 m, resulting in houses on an area of elevated ground

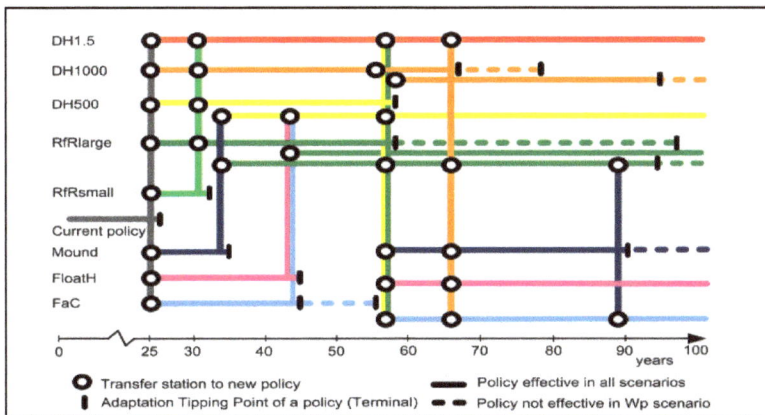

Figure 5. Adaptation pathway map for flood management based on the median value for the sell-by date of policy options for all climate realizations for the Waas Case adapted from [14].

5. Results and Discussion

The adaptation pathway maps in Figures 4 and 5 developed in the case studies described in the previous sections show that in the Kent Ride Case study, assuming a baseline climate, there are 90 possible pathways under the "baseline", 152 possible pathways under the "green" and 245 pathways under the "sustainable grey" landuse scenario. For the Waas Case study we have 32 potential pathways that can be employed to cater to the "No Climate Change", 18 potential pathways to cater to "G" and 36 potential pathways to cater to "Wp" climate scenario. As discussed before, while having an overview provided by the adaptation pathway map is beneficial, having so many solutions puts decision makers in a very difficult situation of selecting one from the multiple available pathways for implementation. To make the existing adaptation tipping point and adaptation pathway approaches more suitable to decision making under uncertainty, we thus need to extend the approaches by appropriate tools and techniques. This paper employed two sequencing approaches to enable this subselection of one or a few pathways that can be built to cater to a changing climate.

The results for the two sequencing approaches (Build Up and Build to Target) for the two selection criterions (Net Present Valuation and Real Options Analysis) for both case studies are presented. The sequencing was performed individually for each selection criterion, the Net Present Value (NPV) and the Real Options Value.

In order to read the results of the Kent Ridge Catchment (Figure 6), please refer to the portfolio of available adaptation actions presented in Table 1 and to read those of the Waas Study Figure 7, please refer to the portfolio of available adaptation actions presented in Table 2. Please note, for ease of readability, only a subset of results has been plotted for discussion; tables showing the complete set of results are available in the supplementary material (Tables S3–S10).

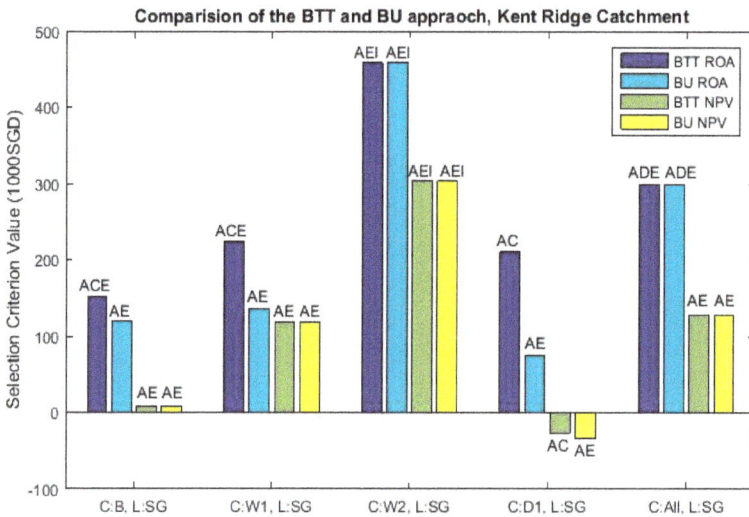

Figure 6. Pathways selected for the Sustainable Grey landuse under different climate (C) scenarios, by means of the BTT and BU approach using the Real Options Value and Net Present Value as selection criterions for the Kent Ridge Catchment.

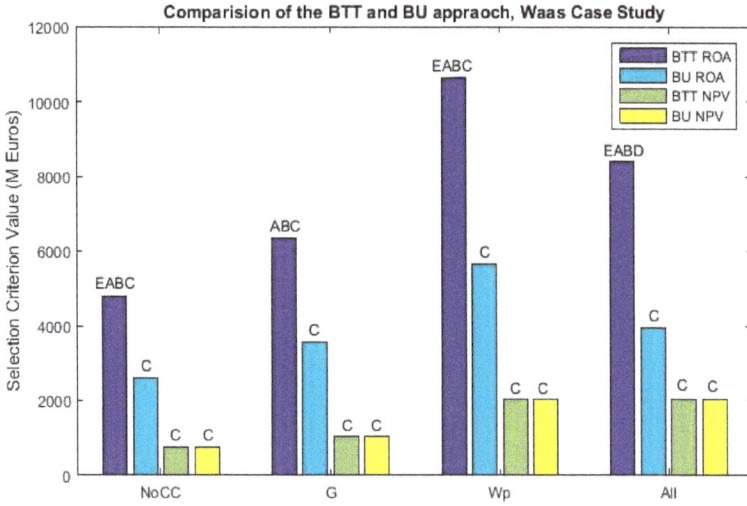

Figure 7. Pathways selected under different climate scenarios, by means of the BTT and BU approach using the Real Options Value and Net Present Value as selection criterions for the Waas Case Study.

Both the Build Up and Build to Target approaches offer different levels of flexibility when used to build and assess adaptation pathways. The Build Up approach lets one start with planning and building for the current or most expected scenario and allows for expanding infrastructure based on the changing climate scenario with time. The selection of the preferred pathway by this approach uses the current state of knowledge as a starting point. It makes current decisions based on what we know today and makes subsequent decisions based on the real future as it unfolds. Although this approach allows easy switching between different climate scenarios, switching to a different climate scenario from this starting point does not lead to building of the pathway that would have been selected if the switched to climate scenario would have been the assessment scenario of choice. The Build to Target approach in contrast warrants the abandoning of a given pathway and revaluating the pathway of choice when dealing with a different assessment scenario at a later time step. The adaptation pathway map developed enables this reassessment to be carried out with relative ease. Thus, while both sequencing approaches help in identification of the preferred pathway, given the characteristic of an adaptation pathway, the identified pathway is not a locked in solution.

Figures 6 and 7 indicate that when comparing the values of the selection criterion of the two approaches, for both case studies it is seen that when Real Options Analysis is the selection criterion of choice, the pathways selected by means of a Build to Target approach generally have a value greater than or the same as that selected by the Build Up approach When NPV is the selection criterion of choice, generally the same pathways are chosen from both approaches. This is due to the nature of the selection criterion.

The traditional NPV assessment misses out on the value of managerial flexibility which is captured and reflected by the Real Options Valuation. This leads to employing only the costs and benefits to maximize the value of the entire pathway (Build to Target) or that of each phase (Build Up) for the entire pathway. Since costs and benefits of each phase are simply added in an NPV assessment, it is expected that an individual action that maximizes the value of the pathway when outlining phases (Build Up) will also be the best candidate when selecting the pathway that maximizes the total value of a given sequence (Build to Target). This once again makes the case for using Real Options Analysis as the selection criterion of choice when making investment decisions under a blanket of uncertainty. Given that the pathway chosen by either approach can be adapted in an unexpected scenario, choosing the Build to Target approach upfront is logical as the pathway of choice has a higher (or same) value of the section criterion when compared to the Build Up approach.

The Real Options Assessment criterion assesses the value of a given action as though it was a financial call option. This assessment inherently accounts for the flexibility that is imparted as a result of using a particular adaptation action. On the other hand, the NPV is a standard capital budgeting exercise that justifies the costs of a given pathway against its expected benefits given no flexibility. The higher the value of each of these criterions, the better the pathway is. The preferred pathways based on each of these criterions are different. The selection criterion value of pathways (Figures 8 and 9) suggests that the maximization of one criterion can only be achieved at the compromise of the second. In other words, maximizing flexibility comes at a cost. This is evident by the fact that a higher Real Options Value is accompanied by a drop in the NPV criterion value.

The Kent Ridge case study has an additional study objective of identifying the preferred landuse in addition to the preferred pathway. The pathways developed for the sustainable grey landuse consistently maximize the value of the selection criterion across the entire set of climate scenarios (Figure 10). This result is very interesting for land scarce nations such as Singapore as it indicates that the sustainable grey landuse scenario economically outperforms the other land use scenarios. Thus, the need for storm water management can be met while simultaneously freeing up land area for other purposes.

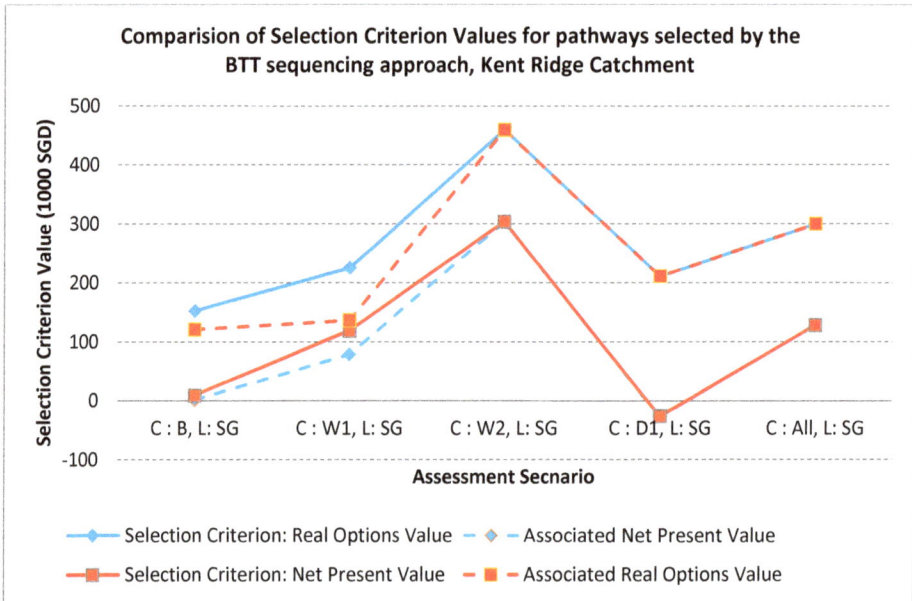

Figure 8. Comparison of Selection Criterion Values for pathways selected by the BTT sequencing approach under different climate (C) and landuse (L) scenarios, Kent Ridge Catchment.

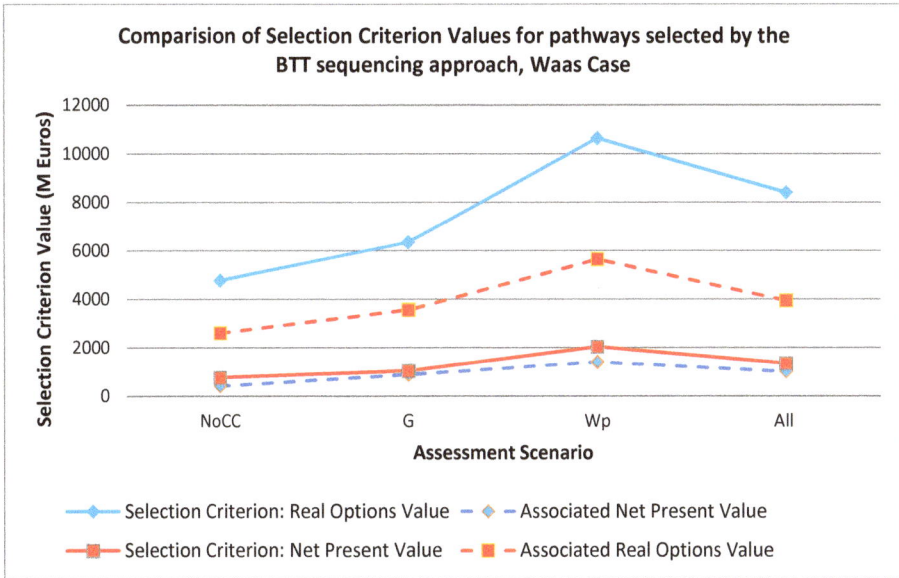

Figure 9. Comparison of Selection Criterion Values for pathways selected by the BTT sequencing approach under different assessment scenarios, Waas Case.

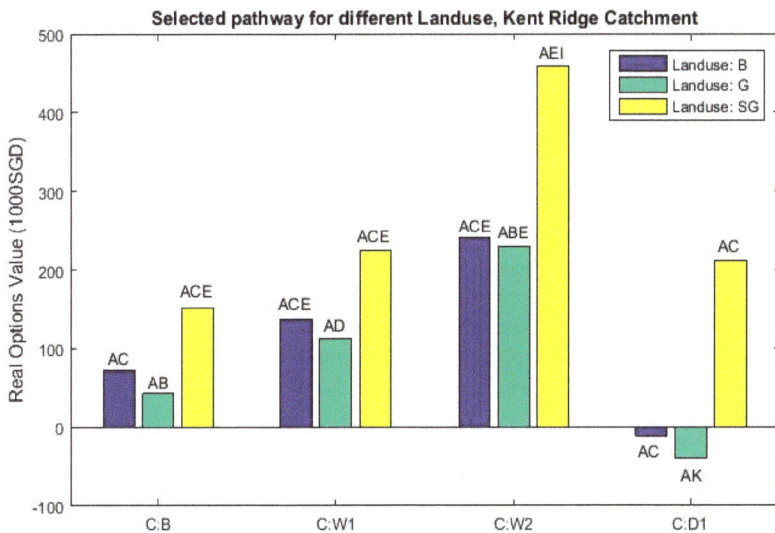

Figure 10. Comparison of the selection criterion (Real Options Analysis) for pathways selected for different climate scenarios (C) and landuse scenarios by the BTT sequencing approach for the Kent Ridge Catchment.

The final question that has to be answered is the selection of the assessment climate scenario for identifying the preferred pathway. For a pathway to be viable for an individual climate assessment scenario, it has to be extendable to adapt to a changing climate if necessary. Thus, if we consider building infrastructure in Kent Ridge to cater to the baseline climate scenario, it is provisioned to

273

timely and economically expand to cater to a W1 or W2 climate scenario when necessary. The same reasoning is used when designing pathways for individual climate scenarios for the Waas case study. Thus, the selected adaptation action for any phase of development can be changed to the other actions available in the adaptation pathways map. The change could be triggered by the need to align actions with the national development policies, the priorities of the stakeholders, new technologies improving the economic performance of a given action, changing objectives or even the introduction of new objectives that have to be met along the planning horizon. In both the case studies, the purpose of running the sequencing over the "All" assessment scenario is to find a pathway that is optimized to perform well across the range of climate scenarios developed. However, the pathway identified for the "All" scenario does not always (for both sequencing approaches) cater to the worst possible climate. This indicates that if this pathway is selected as the preferred pathway, it may have to be supplemented by additional adaptation actions in the future if the worst possible climate materializes, or may incur some losses due to overbuilding if the future reflects the least extreme scenario. The value of the selection criterion for the "All" climate scenario for both case studies lies within the range of values varying between the least to the most extreme climate scenario. While the pathway identified for the "All" climate scenario does offer a trade-off between the value of the selection criterion and the number of scenarios it can cater to, the final choice of which assessment scenario will dictate the selection of the preferred pathway is case dependent. Selecting the pathway that can cater to the most extreme climate scenario is akin to choosing a robust solution, while selecting the pathway for the all scenario is akin to choosing the pathway that offers the best trade-off between flexibility/cost (depending up the selection criterion) and flood robustness. The choice of the climate assessment scenario is thus dependent upon the level of desired robustness that a solution must offer. In cases wherein there is critical infrastructure such as hospitals, airports, military installations, etc., the need for robustness will be greater than in areas which are primarily playgrounds, forests or areas which house low value infrastructure.

6. Conclusions

An adaptation map consists of multiple pathways that provide a very large solution space. Having such a large set of solutions makes it difficult for decision makers to use the adaptation pathway map as is, to guide specific decision on which infrastructure solutions to implement. There is thus a need to further analyze the adaptation pathway maps to identify which action or pathway should be considered for implementation. As the implementation of measures on a large scale require long-term planning, it is important to establish a baseline plan to allow adequate time for implementation of adaptation measures. The identification of this baseline plan is necessary as its understanding can be very beneficial to decision makers, to establish an approximate time scale for adaptation. Given that the further assessment of adaptation pathways to develop a baseline plan requires post processing of the entire adaptation pathway maps developed, currently used informal approaches to aid in the subselection of adaptation pathways are very cumbersome and computationally intensive. This paper aimed to improve the manner in which pathways are shortlisted by presenting two formal sequencing approaches, namely the "Build Up" and the "Build to Target" approaches for pathway subselection. These approaches vary in the degree of flexibility they offer. Each of these approaches used the Net Present Value and the Real Options Analysis as selection criterions. Two case studies, the Kent Ridge and the Waas, were used to exemplify these approaches.

The Build Up approach maximizes the value of the selection criterion for each phase, identifying the adaptation actions for each phase in a sequential manner, while the Build to Target approach identifies the complete pathway that maximizes the value of the selection criterion for the entire planning horizon. Thus, it is intuitive that the Build Up approach is more flexible and responsive to a changing scenario. However, it results in plans with less favorable selection criterion values when compared to plans selected by the Build to Target approach. When comparing the pathways that resulted from the analysis of both of these approaches it was observed that the Build to Target approach shortlisted pathways that carried a selection criterion value that was higher (or at least the same) than the value of pathways outlined by the Build Up approach. Given the characteristic of the Adaptation Pathway map to cater to uncertainties, it is thus recommended that the Built to Target approach be used to outline and shortlist adaptation pathways from a given adaptation pathway map.

The NPV and the Real Options valuation were used as the selection criterion to shortlist pathways by each approach. Given the real option analysis is able to capture the value of managerial flexibility to respond to uncertainty, this paper makes the case of using the real options value as the selection criterion of choice when faced with making investment decision under uncertainty. When comparing the value of the two criterions, we can see that the maximization of one criterion can only be achieved at the compromise of the second. The fact that a higher Real Options Value is accompanied by a drop in the NPV criterion value quantifies the notion that increasing flexibility comes at a cost.

This paper focused on maximizing the value of a single economic criterion, either the NPV or the Real Options value, to enable the sub-selection of the preferred pathway. However, given the nature of the problem, it may not appropriate to use a single criterion to justify the choice of the plan used to combat uncertainty. This notion can be addressed further in future studies by exploring the use of multiple objective genetic optimization in designing and subselecting adaptation pathways.

Supplementary Materials: The following are available online at www.mdpi.com/2073-4441/10/2/229/s1, Figure S1: Adaptation Tipping Points of Adaptation Actions. Please refer to Table 1 in the main paper for the key of action portfolios, Figure S2: Adaptation Tipping Points for the Waas Case, Table S1: Climate Scenarios, Kent Ridge Case Study, Table S2: Land Use Scenarios, Kent Ridge Case Study, Table S3: Results BTT sequencing approach using ROA as selection criterion, Kent Ridge Catchment., Table S4: Results BU sequencing approach using ROA as selection criterion, Kent Ridge Catchment., Table S5: Results BTT sequencing approach using NPV as selection criterion, Kent Ridge Catchment, Table S6: Results BTT sequencing approach using NPV as selection criterion, Kent Ridge Catchment., Table S7: Results BTT sequencing approach using ROA as selection criterion, Waas Case, Table S8: Results BU sequencing approach using ROA as selection criterion, Waas Case, Table S9: Results BTT sequencing approach using NPV as selection criterion, Waas Case, Table S10: Results BU sequencing approach using NPV as selection criterion, Waas Case.

Author Contributions: Nishtha Manocha and Vladan Babovic conceived the idea of applying sequencing approaches to aid pathway sub-selection. Nishtha Manocha performed the sequencing approaches and analysed the results. Nishtha Manocha wrote the paper which was reviewed by Vladan Babovic.

Conflicts of Interest: The authors declare no conflict of interest

References

1. Urich, C.; Rauch, W. Exploring critical pathways for urban water management to identify robust strategies under deep uncertainties. *Water Res.* **2014**, *66*, 374–389. [CrossRef] [PubMed]
2. Chang, N.-B.; Qi, C.; Yang, Y.J. Optimal expansion of a drinking water infrastructure system with respect to carbon footprint, cost-effectiveness and water demand. *J. Environ. Manag.* **2012**, *110*, 194–206. [CrossRef] [PubMed]
3. Dewar, J.A. *Assumption-Based Planning: A Tool for Reducing Avoidable Surprises*; Cambridge University Press: New York, NY, USA, 2002.
4. Dewar, J.A.; Builder, C.H.; Hix, W.M.; Levin, M.H. *Assumption-Based Planning; A Planning Tool for very Uncertain Times*; Rand Corp.: Santa Monica, CA, USA, 1993.
5. Lempert, R.J.; Popper, S.W.; Bankes, S.C. *Shaping the next one Hundred Years: New Methods for Quantitative, Long-Term Policy Analysis and Bibliography*; Rand Corp.: Santa Monica, CA, USA, 2003.
6. Groves, D.G.; Lempert, R.J. A new analytic method for finding policy-relevant scenarios. *Glob. Environ. Chang.* **2007**, *17*, 73–85. [CrossRef]
7. Ranger, N.; Reeder, T.; Lowe, J. Addressing 'deep' uncertainty over long-term climate in major infrastructure projects: four innovations of the Thames Estuary 2100 Project. *EURO J. Decis. Process.* **2013**, *1*, 233–262. [CrossRef]
8. Walker, W.; Haasnoot, M.; Kwakkel, J. Adapt or Perish: A Review of Planning Approaches for Adaptation under Deep Uncertainty. *Sustainability* **2013**, *5*, 955–979. [CrossRef]
9. Walker, W.E.; Rahman, S.A.; Cave, J. Adaptive policies, policy analysis, and policy-making. *Eur. J. Oper. Res.* **2001**, *128*, 282–289. [CrossRef]

10. Reeder, T.; Ranger, N. *How Do You Adapt in an Uncertain World? Lessons from the Thames Estuary 2100 Project*; World Resources Institute: Washington, DC, USA, 2011.

11. Haasnoot, M.; Middelkoop, H.; Offermans, A.; Van Beek, E.; Van Deursen, W.P. Exploring pathways for sustainable water management in river deltas in a changing environment. *Clim. Chang.* **2012**, *115*, 795–819. [CrossRef]

12. Kwadijk, J.C.J.; Haasnoot, M.; Mulder, J.P.M.; Hoogvliet, M.M.C.; Jeuken, A.B.M.; van der Krogt, R.A.A.; van Oostrom, N.G.C.; Schelfhout, H.A.; van Velzen, E.H.; van Waveren, H.; et al. Using adaptation tipping points to prepare for climate change and sea level rise: A case study in the Netherlands. *Wiley Interdiscip. Rev. Clim. Chang.* **2010**, *1*, 729–740. [CrossRef]

13. Haasnoot, M.; Middelkoop, H.; van Beek, E.; van Deursen, W.P.A. A method to develop sustainable water management strategies for an uncertain future. *Sustain. Dev.* **2011**, *19*, 369–381. [CrossRef]

14. Haasnoot, M.; Kwakkel, J.H.; Walker, W.E.; ter Maat, J. Dynamic adaptive policy pathways: A method for crafting robust decisions for a deeply uncertain world. *Glob. Environ. Chang.* **2013**, *23*, 485–498. [CrossRef]

15. Kwakkel, J.H.; Haasnoot, M.; Walker, W.E. Developing dynamic adaptive policy pathways: A computer-assisted approach for developing adaptive strategies for a deeply uncertain world. *Clim. Chang.* **2014**, *132*, 373–386. [CrossRef]

16. Zischg, J.; Goncalves, M.L.; Bacchin, T.K.; Leonhardt, G.; Viklander, M.; van Timmeren, A.; Rauch, W.; Sitzenfrei, R. Info-Gap robustness pathway method for transitioning of urban drainage systems under deep uncertainties. *Water Sci. Technol.* **2017**, *76*, 1272. [CrossRef] [PubMed]

17. Zhang, S.X.; Babovic, V. A real options approach to the design and architecture of water supply systems using innovative water technologies under uncertainty. *J. Hydroinform.* **2012**, *14*, 13. [CrossRef]

18. Gersonius, B.; Ashley, R.; Pathirana, A.; Zevenbergen, C. Climate change uncertainty: Building flexibility into water and flood risk infrastructure. *Clim. Chang.* **2012**, *116*, 411–423. [CrossRef]

19. Hu, J.; Cardin, M.-A. Generating flexibility in the design of engineering systems to enable better sustainability and lifecycle performance. *Res. Eng. Des.* **2015**, *26*, 121–143. [CrossRef]

20. Deng, Y.; Cardin, M.-A.; Babovic, V.; Santhanakrishnan, D.; Schmitter, P.; Meshgi, A. Valuing flexibilities in the design of urban water management systems. *Water Res.* **2013**, *47*, 7162–7174. [CrossRef] [PubMed]

21. Swart, R.J.; Raskin, P.; Robinson, J. The problem of the future: Sustainability science and scenario analysis. *Glob. Environ. Chang.* **2004**, *14*, 137–146. [CrossRef]

22. Bankes, S. Exploratory modeling for policy analysis. *Oper. Res.* **1993**, *41*, 435–449. [CrossRef]

23. Korteling, B.; Dessai, S.; Kapelan, Z. Using information-gap decision theory for water resources planning under severe uncertainty. *Water Resour. Manag.* **2013**, *27*, 1149–1172. [CrossRef]

24. Butcher, W.S.; Haimes, Y.Y.; Hall, W.A. Dynamic programing for the optimal sequencing of water supply projects. *Water Resour. Res.* **1969**, *5*, 1196–1204. [CrossRef]

25. Becker, L.; Yeh, W.W.G. Optimal timing, sequencing, and sizing of multiple reservoir surface water supply facilities. *Water Resour. Res.* **1974**, *10*, 57–62. [CrossRef]

26. Braga, B.P.F.; Conejo, J.G.L.; Becker, L.; Yeh, W.W.-G. Capacity expansion of Sao Paulo water supply. *J. Water Resour. Plan. Manag.* **1985**, *111*, 238–252. [CrossRef]

27. Dandy, G.; Connarty, A.M. Interactions between Water Pricing, Demand Management and the Sequencing of Water Resource Projects. In *Water down under 94: Groundwater/Surface Hydrology Common Interest Papers*; Institution of Engineers: Barton, Australia, 1994.

28. Mulvihill, M.E.; Dracup, J.A. Optimal timing and sizing of a conjunctive urban water supply and waste water system with nonlinear programing. *Water Resour. Res.* **1974**, *10*, 170–175. [CrossRef]

29. Rubinstein, J.; Ortolano, L. Water conservation and capacity expansion. *J. Water Resour. Plan. Manag.* **1984**, *110*, 220–237. [CrossRef]

30. Martin, Q.W. Hierarchical algorithm for water supply expansion. *J. Water Resour. Plan. Manag* **1987**, *113*, 677–695. [CrossRef]

31. Manocha, N.; Babovic, V. Planning Flood Risk Infrastructure Development under Climate Change Uncertainty. *Procedia Eng.* **2016**, *154*, 1406–1413. [CrossRef]

32. Dixit, A.K.; Pindyck, R.S. *The Options Approach to Capital Investment*; Harvard Business School Press: Boston, MA, USA, 1995; p. 105.

33. Buurman, J.; Zhang, S.; Babovic, V. Reducing risk through real options in systems design: The case of architecting a maritime domain protection system. *Risk Anal.* **2009**, *29*, 366–379. [CrossRef] [PubMed]

34. Zhang, S.X.; Babovic, V. An evolutionary real options framework for the design and management of projects and systems with complex real options and exercising conditions. *Decis. Support Syst.* **2011**, *51*, 119–129. [CrossRef]

35. De Neufville, R.; Scholtes, S.; Wang, T. Real Options by Spreadsheet Parking Garage. *J. Infrastruct. Syst.* **2006**, *12*, 107–111. [CrossRef]

36. Woodward, M.; Kapelan, Z.; Gouldby, B. Adaptive Flood Risk Management under Climate Change Uncertainty Using Real Options and Optimization. *Risk Anal.* **2014**, *34*, 75–92. [CrossRef] [PubMed]

37. Buurman, J.; Babovic, V. Adaptation Pathways and Real Options Analysis: An approach to deep uncertainty in climate change adaptation policies. *Policy Soc.* **2016**, *35*, 137–150. [CrossRef]

38. Dandy, G.; Kolokas, L.; Frey, J.; Gransbury, J.; Duncker, A.; Murphy, L. Optimal staging of capital works for large water distribution systems. In Proceedings of the ASCE Conference on Water Resources Planning and Management, Roanoke, VA, USA, 19–22 May 2002.

39. Manocha, N.; Babovic, V. Development and valuation of adaptation pathways for storm water management infrastructure. *Environ. Sci. Policy* **2017**, *77*, 86–97. [CrossRef]

40. Myers, S.C. Finance Theory and Financial Strategy. *Interfaces* **1984**, *14*, 126–137. [CrossRef]

41. Beh, E.H.Y.; Dandy, G.C.; Maier, H.R.; Paton, F.L. Optimal sequencing of water supply options at the regional scale incorporating alternative water supply sources and multiple objectives. *Environ. Model. Softw.* **2014**, *53*, 137–153. [CrossRef]

42. Van den Hurk, B.; Tank, A.K.; Lenderink, G.; van Ulden, A.; van Oldenborgh, G.J.; Katsman, C.; van den Brink, H.; Keller, F.; Bessembinder, J.; Burgers, G.; et al. New climate change scenarios for the Netherlands. *Water Sci. Technol.* **2007**, *56*, 27–33. [CrossRef] [PubMed]

![water logo] *water*

MDPI

Article

Adaptation Tipping Points of a Wetland under a Drying Climate

Amar Nanda [1,2], Leah Beesley [2,3], Luca Locatelli [4], Berry Gersonius [2,5], Matthew R. Hipsey [2,6] and Anas Ghadouani [1,2,*]

[1] Department of Civil, Environmental & Mining Engineering, The University of Western Australia, 35 Stirling Highway, M051, Perth 6009 WA, Australia; 21449794@student.uwa.edu.au

[2] Cooperative Research Centre for Water Sensitive Cities (CRCWSC), Clayton 3800 VIC, Australia; leah.beesley@uwa.edu.au (L.B.); b.gersonius@un-ihe.org (B.G.); matt.hipsey@uwa.edu.au (M.R.H.)

[3] School of Biological Sciences, The University of Western Australia, 35 Stirling Highway, M004, Perth 6009 WA, Australia

[4] Department of Environmental Engineering, Technical University of Denmark, 2800 Kongens Lyngby, Denmark; lulo@env.dtu.dk

[5] UNESCO-IHE, Westvest 7, 2611 AX Delft, The Netherlands

[6] School of Agriculture and Environment, The University of Western Australia, Perth 6009 WA, Australia

* Correspondence: anas.ghadouani@uwa.edu.au; Tel.: +61-8-6488-2687

Received: 12 January 2018; Accepted: 20 February 2018; Published: 24 February 2018

Abstract: Wetlands experience considerable alteration to their hydrology, which typically contributes to a decline in their overall ecological integrity. Wetland management strategies aim to repair wetland hydrology and attenuate wetland loss that is associated with climate change. However, decision makers often lack the data needed to support complex social environmental systems models, making it difficult to assess the effectiveness of current or past practices. Adaptation Tipping Points (ATPs) is a policy-oriented method that can be useful in these situations. Here, a modified ATP framework is presented to assess the suitability of ecosystem management when rigorous ecological data are lacking. We define the effectiveness of the wetland management strategy by its ability to maintain sustainable minimum water levels that are required to support ecological processes. These minimum water requirements are defined in water management and environmental policy of the wetland. Here, we trial the method on Forrestdale Lake, a wetland in a region experiencing a markedly drying climate. ATPs were defined by linking key ecological objectives identified by policy documents to threshold values for water depth. We then used long-term hydrologic data (1978–2012) to assess if and when thresholds were breached. We found that from the mid-1990s, declining wetland water depth breached ATPs for the majority of the wetland objectives. We conclude that the wetland management strategy has been ineffective from the mid-1990s, when the region's climate dried markedly. The extent of legislation, policies, and management authorities across different scales and levels of governance need to be understood to adapt ecosystem management strategies. Empirical verification of the ATP assessment is required to validate the suitability of the method. However, in general we consider ATPs to be a useful desktop method to assess the suitability of management when rigorous ecological data are lacking.

Keywords: ecosystem; wetland; adaptation tipping points; climate change; management strategy

1. Introduction

Ecological systems with high resilience are able to cope with frequent disturbance and remain relatively stable over time, whereas systems with low resilience are likely to transition to altered states, often with reduced function in the wake of disturbance [1]. Systems with low resilience can shift between alternative stable states by an incremental change of conditions that induce a catastrophic (reversible) shift or by perturbations that are large enough to move the system to a lower alternative state with reduced functions [2,3]. Social-ecological systems (SES) have interacting components (e.g., political, social, or ecological) and have many functions that depend on feedback mechanisms between processes that take place at multiple scales [4,5].

Ecosystems are managed to maintain their beneficial ecological functions, but can be vulnerable to altered external processes (e.g., climate change); such processes can shift ecosystems to reduced ecological functions [6]. These complex ecosystems, under the influence of drivers of ecological and social processes, can change and then often display nonlinear behavior with prolonged periods of stability alternating with sudden changes or critical transitions of the socio-ecological system [2,7]). These sudden changes are often not foreseen by management practices due to the nature of changes; these approaches are commonly defined by law-enforced threshold levels along environmental gradients [8]. Interventions to inform policy or management are therefore ineffective or not timely enough to maintain ecosystems with multiple socio-ecological functions in a state of prolonged stability.

Thresholds and tipping points are important focal points for adaptive management [9–12], but often lack data to define exact biophysical thresholds to model the complicated interactions in SES models [13]. However, several ecological indicators [14] and policy-based approaches do exist to determine when the limits of a system are reached, and when future change will become critical for the system. Examples include flood mitigation through adapting infrastructure [15–18], adapting water resources management with decision frameworks [19,20], and institutional adaptation, through the inclusion of capacity building by government agencies [21,22]. Despite the considerable body of the literature, there has been limited focus on: (1) defining thresholds for ecosystem processes, (2) how to inform policies that environmental change has become critical [23], and (3) when interventions are needed to address different key ecosystem processes.

A policy-based approach that defines when and which objectives of a current strategy are being met, is a starting point to adapt existing strategies and formulate new ones, is referred to as the Adaptation Tipping Point (ATP) method [16]. An adaptation tipping point is the moment when the magnitude of change is such that a current management strategy can no longer meet its objectives. As a result, adaptive management is needed to prevent or postpone these ATPs. This method has previously been applied to river restoration and a species re-introduction programme [24–26]; unfortunately, the approach fails to address whether or not current management strategies are sustainable when system behaviour is poorly understood, and when there are time lags that are involved for different subsystems in a larger SES [11]. However, the ATP approach confronts the lack of quantitative and qualitative ecological data sets to infer acceptability of management [10,27,28] by using stakeholder engagement to determine unknown/ill-defined thresholds, and thereby prevents a focus on only existing management strategies [26,27]. To prevent confusion with definitions of tipping points in other fields (e.g., climate sciences, ecology), we will use the term "adaptation tipping point" in this study.

A management strategy needs to be informed about when an ecosystem could shift into an alternate state that will have low resilience when the system is exposed to stressors induced by climate change. Wetlands are ecosystems that are particularly vulnerable to decreased ecological resilience due to factors such as, altered hydrology, invasive species, nutrient loading, and fire regimes, that can cause wetlands to shift from a "clear-water" to "turbid-water" stable state, or from a permanent to a seasonal hydro-regime that inadequately supports ecological processes [2,3]. In light of current management strategies and shortcomings, we are interested in how much hydrological variation an ecosystem can cope with before the durability of a strategy to conserve the ecosystem expires, and when this will occur. The overall aim of this study is to provide a modified ATP framework to identify the effectiveness of ecosystem management strategies; this will be applied by using a case study. The effectiveness of the ecosystem management strategy is defined using three ecosystem functions:

1. hydrological response and variation;
2. temporal scale ecosystem responses; and,
3. recovery rate or alternative stable state of ecological processes.

2. Method

The original five-step ATP methodology includes (Figure 1): (i) the determination of climate change effects on the system; (ii) followed by identifying key objectives and thresholds; (iii) the determination when standards were compromised in the past; (iv) analysing when standards were compromised in the future; and, (v) to repeat step 1–4 for alternative strategies. Further details about the original methodology can be found in [16]. We modified the original methodology to determine ATPs for different socio-ecological objectives and thresholds with the assessment of historical hydrological time series. We expanded step 3 to interpreted ATPs in conjunction with the hydrological response and variation; temporal scale ecosystem responses; and, recovery rate and alternative stable state of ecological processes (Figure 1).

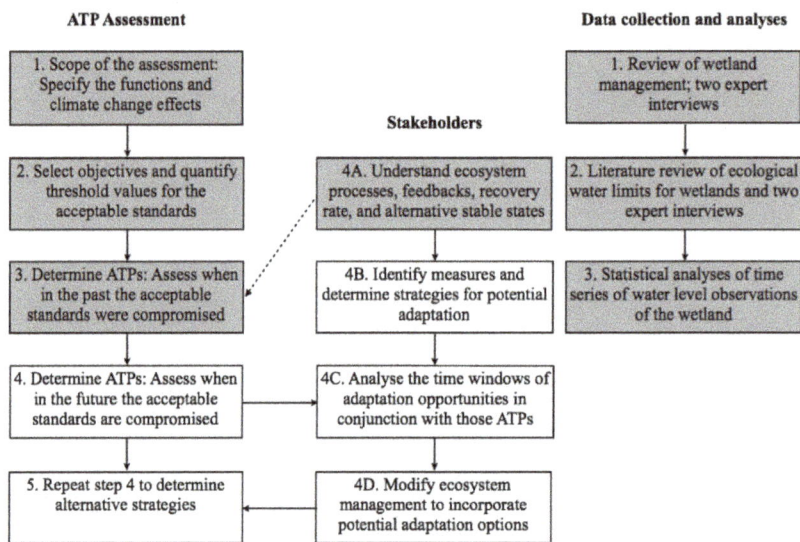

ATP Assessment   **Data collection and analyses**

| 1. Scope of the assessment: Specify the functions and climate change effects | | 1. Review of wetland management; two expert interviews |

Stakeholders

| 2. Select objectives and quantify threshold values for the acceptable standards | 4A. Understand ecosystem processes, feedbacks, recovery rate, and alternative stable states | 2. Literature review of ecological water limits for wetlands and two expert interviews |

| 3. Determine ATPs: Assess when in the past the acceptable standards were compromised | 4B. Identify measures and determine strategies for potential adaptation | 3. Statistical analyses of time series of water level observations of the wetland |

| 4. Determine ATPs: Assess when in the future the acceptable standards are compromised | 4C. Analyse the time windows of adaptation opportunities in conjunction with those ATPs |

| 5. Repeat step 4 to determine alternative strategies | 4D. Modify ecosystem management to incorporate potential adaptation options |

Figure 1. The complete Adaptation Tipping Point (ATP) methodology with an overview of the steps undertaken in this study (indicated with grey boxes), along with the data collection and analyses conducted in this study (Adapted from: [16]).

2.1. Case Study Area

The wetland in our case study area, Forrestdale Lake (Figure 2), is located in the biodiverse region in south-west Western Australia [29], and has been noticeably impacted by anthropogenic factors [30,31]. The wetland supports many waterbirds and its surrounding riparian vegetation supports terrestrial birds, significant reptiles, mammals, and other vertebrate species [32]. The lakes' high biodiversity makes it an important regional conservation area [33]. An estimated 85% of the Swan Coastal Plain (SCP) wetlands have been lost since colonial settlement and are likely to experience increasing hydrological stress due to further decreasing rainfall [32,34,35] and catchment urbanisation [36]. The wetland experiences a Mediterranean climate with a mean annual rainfall of 852 mm in the period 1980–2014. Approximately 80% of the annual precipitation occurs in winter between May and September, with groundwater recharge occurring from June to September [37].

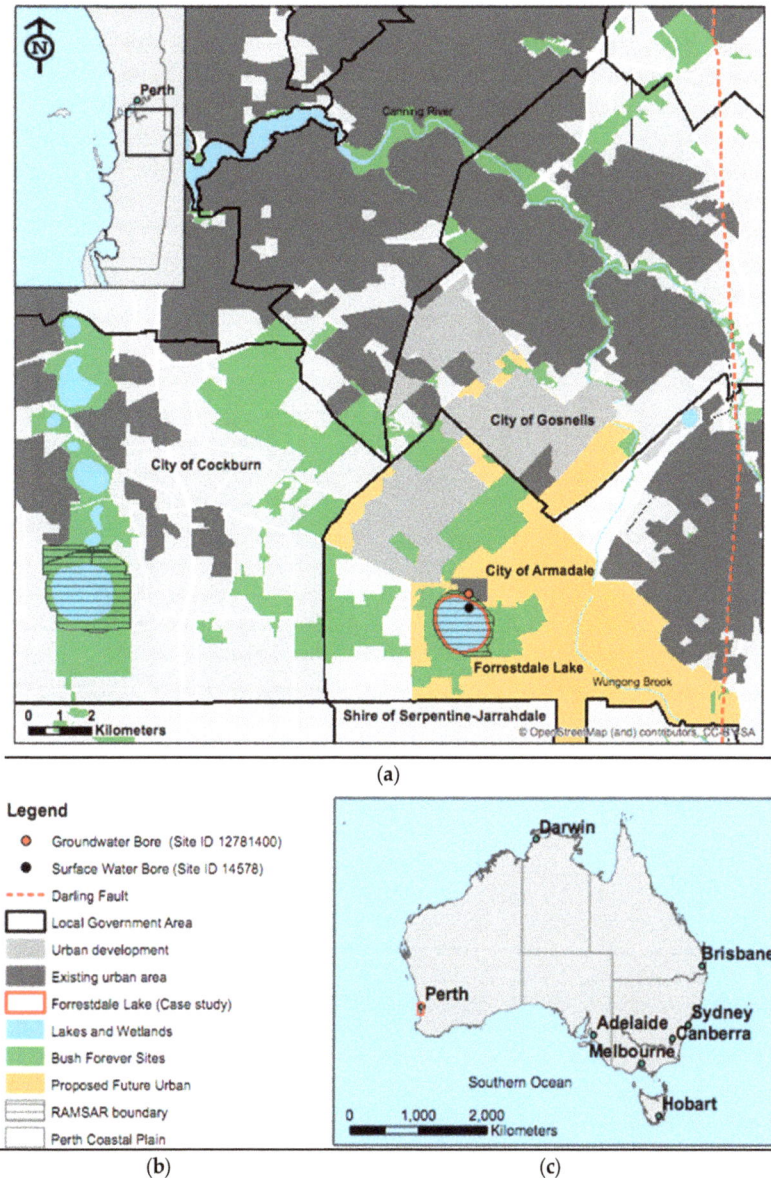

Figure 2. Location of Forrestdale Lake (32° 09′ 30″ S, 115° 56′ 16″ E) within its groundwater catchment, showing the increasing urbanisation in the catchment, the multiple management authorities, and protection policies (Map projection: GDA94). (**a**) Catchment characteristics of Forrestdale Lake; (**b**) Legend; (**c**) Location of Forrestdale Lake.

Climate change, via its impact on rainfall and groundwater recharge, is an important regional driver of wetland hydrology and ecological functions [38,39]. Since the 1970s, this region has experienced a 10–20% decrease in average annual rainfall that resulted in a mean annual rainfall of 775 mm in the period 2004–2014 [40–42]. There is evidence that climate change has been impacting the hydrology of the unconfined aquifer since the 1970s [43–45], leading to less surface water availability [46,47]. Local-scale hydrologic changes associated with land-use change and groundwater abstraction may also impact water levels of wetlands. Although, these changes are considered minimal when compared to region-wide changes in rainfall and consequently recharge of the aquifer [48,49]. A growing population and greater demand for groundwater (Figure 3) is expected to put more stress on the already over-allocated groundwater resources.

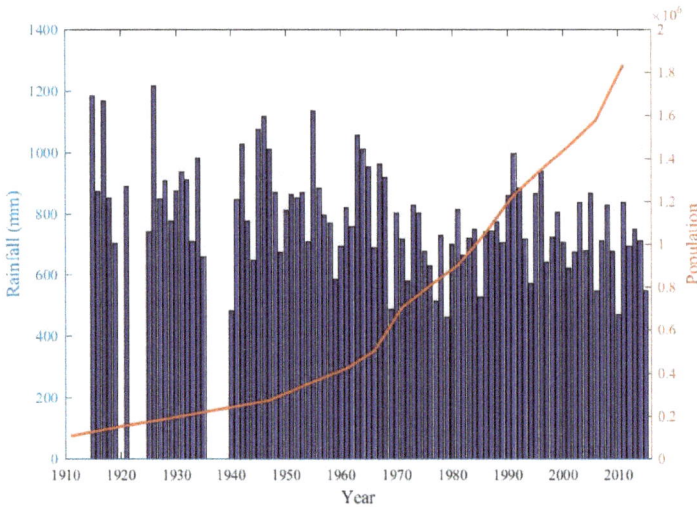

Figure 3. Population growth in Perth between 1910 and 2015 [50] shown against the annual rainfall data over the same period, where available [51]. The decreasing annual rainfall results in reduced water availability.

When considering that the case study area is under multiple stressors, it is expected that current management policies are already inadequate, and that management authorities have the desire to understand past effects of climate change on maintaining individual socio-ecological objectives. With the high likelihood of a management plan that needs to be updated according to new research findings, multi-scale policies requiring review, and limited availability of ecological and hydrological data, the ATP methodology is suitable to apply to this wetland.

2.2. Data Collection and Analyses

Data were collected, and thresholds defined through a literature review and interviews. Hydrological time series data for each socio-ecological objective from the management strategy [52,53], and minimum and maximum water level thresholds were compared with mandated management objectives and policies, respectively [54].

2.2.1. Step 1: Legislative Framework and Impacts of Climate Change–Literature Review

The legislative framework consists of gradually introduced laws and policies first aimed to protect the rights to use groundwater resources, and more recently, to protect the natural resources. In Figure 4, we present a timeline of the legislation framework for Forrestdale Lake and its groundwater catchment area, with key social and environmental events that have occurred. During the time period from colonial settlement until the mid-20th century the wetlands suffered due to negative perceptions of mosquitos, and through degradation due to land use changes. As the degradation of the environmental resources progressed, new knowledge about the ecosystem helped to shift legislation to protect species and ecosystems. Prior to the 1950s, the wetland was classified as a 'groundwater through flow lake', but is now, depending on rainfall and groundwater, considered a 'permanently inundated and perched lake' [55–57]. However, even more recently a combination of disconnection from groundwater and decreasing annual rainfall has resulted in the lake only being seasonally inundated (CCWA 2005); the trend of the drying climate is likely to continue during the 21st century [40,42]. We reviewed all of the policies and legislation that have been introduced to protect the wetlands, including policies that are aimed to protect groundwater resources, species, and connectivity of green zones within urban areas. Legislation and policies have been introduced on both state and national levels; on the local level, statutory documents are produced that provide detailed environmental objectives and an overview of the responsible managing authorities.

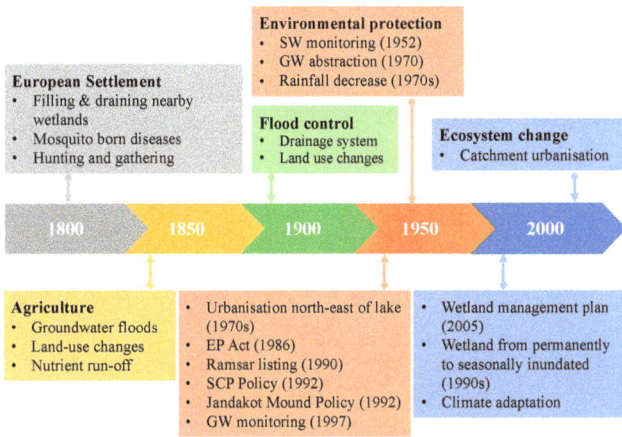

Figure 4. A historical representation of time and scale the traditional human-nature system and water resources system of Forrestdale Lake with indicated key events of the four subsystems: Natural resources, infrastructure, socio-economics and institution.

2.2.2. Step 2: Select Objectives and Quantify Threshold Values–Literature Review

In the second step, we reviewed the current wetland management strategy for policy objectives, indicators, and threshold values of the wetland ecological processes. These functions represent the critical objectives of the wetland management strategy. Certain water depths are needed within a wetland to sustain a variety of ecological processes [36,38,58,59], therefore we used water depth as a proxy to link ecological objectives to mandated policy thresholds [54], shown in Table 1. We identified two pathways within the SES via which water depth may impact on wetland ecological objectives:

i   Water depth may reach levels that are too low to:

- maintain sediment processes;
- provide habitat needed by waterbirds, frogs, freshwater turtles, and macro-invertebrates for survival and reproduction;
- inhibit the growth of mosquitoes and midges.

ii   Water depth may reach levels that are too low or too high, such that they lead to:

- the death of phreatophytic (i.e., groundwater dependent) and fringing vegetation;
- the compromise of the habitat needed for terrestrial birds and mammals; and,
- increased weed invasion and compromise the habitat needed for wading birds.

Table 1. Threshold values for the ecological objectives to determine ATPs for surface water (SW) and groundwater (GW) levels in (non)-consecutive months derived from the state water policy. 21.6 m AHD (mean water level in Australian Height Datum in meters) is the height of the lake bed, which we here denote as zero; all thresholds are defined as water depth with respect to the lake bed.

Ecological Objectives	Water Level (m)	Threshold Definition	Source
1. protect vegetation and mammals; definition of drought	SW < 0	3 consecutive months; 1 in 5 years	[33,54,58]
2. prevent mosquitoes	SW < 0	1 month per year; 1 in 1 year	[33]
3. protect waterbirds	SW < 0	6 consecutive months; 1 in 5 years	[33,54,60]
4. protect frogs	SW < 0	8 months; 1 in 5 years	[58,59]
5. protect tortoises	SW < 0	3 months; 1 in 5 years	[58,59]
6. protect macro-invertebrates	SW < 0.4	3 consecutive months; 1 in 5 years	[58,59]
7. prevent exposure of Acid Sulphate Soils	GW < −0.5	3 consecutive months; 1 in 5 years	[58]
8. maintain sediment processes	GW < −0.5	3 consecutive months; 1 in 5 years	[58]

From the aforementioned pathways, we derived eight critical ecological objectives, as shown in Table 1. The objectives were taken from the Forrestdale Lake wetland management strategy [33]; the Ministerial Water Requirements [54], and from discussion with two experts from different management authorities (the Department of Parks and Wildlife and the Department of Water; since 2017 the Department of Biodiversity, Conservation and Attractions and the Department of Water and Environmental Regulation resp.). For each ecological objective, minimum water depth requirements were obtained (i.e., threshold) using the Ministerial water requirements (Table 1). 21.6 m AHD is the height of the lake bed, which we here denote as zero; all of the thresholds are defined as water depth with respect to the lake bed. The appraisal of the ecological objectives in Table 1 reveals an inundated lake is needed to support the socio-ecological objectives. The minimum water level (depth) for vegetation, mammals, and terrestrial birds is >0 m.; and 0.4 m. to maintain waterbirds, freshwater turtles, frogs, and macro-invertebrates (Table 1). In cases where water level thresholds were not informed by the Ministerial water requirements, we relied on peer-reviewed literature (See 'Source' column, Table 1). A detailed description of each ecological objective were obtained from previous research [32,48,52,60–62]. In addition, two expert interviews were conducted to determine both the accepted exceedance frequency, and to define threshold definitions not previously included in policy or the literature. We also included experts from other government department and actors that are involved in the management of the wetland to discuss the threshold definitions and determine the consensus for using these threshold values. These actors are listed below with their role and tasks:

- the local government (city council, responsible for land division and drainage);
- the State Department of Parks and Wildlife (conservation authority);
- the Department of Water (water regulator, responsible for ground- and surface water allocation and monitoring); and,
- community and local conservation groups (community, involved in monitoring birds, revegetation and rehabilitation of the wetland buffer zone).

2.2.3. Step 3: Determine ATPs—Statistical Analyses

Time series datasets of surface and groundwater depths [63] were sourced from the Department of Water's water information database. The data were divided into two time periods, 1978–1995 and 1996–2012, so that each period reflects a sufficient amount of time for policy implementation and linked to the downward trend in rainfall. To evaluate the ecological resilience of the wetland, we assessed when and for how long the water level in Forrestdale Lake crossed the thresholds. In order to estimate the frequencies of occurrence of threshold exceedance (see thresholds, Table 1) by annual minimum series, we used the observed historical time series of water levels and the following equation proposed by [64]:

$$G(x) = 1 - \left[1 - k \left(\frac{x - x_0}{\alpha} \right) \right]^{\frac{1}{k}} \text{ for } k \neq 0 \qquad (1)$$

where $G(x)$ is the distribution of the magnitude of events (x) smaller than a threshold (x_0) over a (non)-consecutive duration over a period of years (T). Here, α and k are constants derived from the average highest and lowest values in sets of T annual minima, and the minimum value to be expected once in T years. Arrival rate (λ) = the average number of minimum values (x_0) per year. Constant $\alpha = (2 \times 10.88) - \lambda 1 = -0.05$; Lower bound $(\xi) = \lambda - (0.5572 - \alpha) \times 0.5572 = $ constant; Probability value (p) = $(1 - (1/T)$; Expected water levels = $\xi - (\alpha \times LN(-1 \times LN(p)))$.

To interpret the occurrence of ATPs in context with the ecological tipping points; we extended our analyses by comparing the drought frequency, duration, and start month for both the pre- and post-1995 water-level time series. A drought was defined by experts as a dry period when the water depth was zero m. for three consecutive months. We compared the water levels with the available historical ecological data to make an estimation of the trajectories over time.

3. Results

The results are presented in accordance with our methodology, as per Figure 1 (Column ATP assessment). The results of the literature review (Step 1), along with an analysis of the multi-scale legislative framework of the case study area (Step 2) are presented in Section 3.1, while the results from the time series analyses (Step 3) of historical surface and groundwater level data from 1978–2012 are presenting in Section 3.2. The understanding of alternate systems states with the ATP assessment (Step 4A) is presented in Section 3.3.

3.1. Legislative Framework across Scales

The scope of the assessment for Forrestdale Lake was defined as stipulated in existing legislation (Supplementary Materials). In Western Australia, the *Environmental Protection Act* (1986) [65] is the legislative act that underpins the environmental protection of wetlands. According to the *Environmental Protection Act*, the *Ministerial water requirements for the Gnangara Mound and Jandakot wetlands* (1992) [54] mandates ecological water requirements that consist of upper and lower thresholds to maintain ecological processes; the State water regulator holds the responsibility to maintain these water requirements. Protection of biodiversity or conservation values, such as maintaining biodiversity, is included in the *Conservation and Land Management Act* (1984) and the *Wildlife Conservation Act* (1950) [66,67]. Large regional wetlands have also been listed under the Ramsar Convention (e.g., Forrestdale Lake) to protect waterbirds (Ramsar 1994) [68], as well as to protect migratory birds under several international agreements (JAMBA 1981; CAMBA 1988; ROKAMBA 2006) [69–71]. However, the protection of nationally and internationally important flora, fauna, and ecological communities is arranged by the Commonwealth of Australia under the *Environment Protection and Biodiversity Conservation Act* (EPBC 1999) [72]. The above-mentioned Acts and Agreements provide the statutory base to formulate wetland management plans. In contrast to international conventions and Commonwealth legislation, the State government departments and local governments cooperate to maintain the ecological functions, as described in the wetland management plan. A previous wetland management plan from 1993 for Forrestdale Lake was updated in 2005; this now includes the ecological values of the wetland, proposes management actions to control invasive species, and mentions the risks of declining water levels [33]. However, the plan fails to address how to cope with declining water levels.

The literature review revealed that the protection of the regionally important Forrestdale Lake wetland is provided by legislation and policies on different levels and scales (Figure 5). The management of the lake is therefore organised on different levels of government departments that have their own scale of operation (e.g., local council vs. state-wide department). Due to the different institutions and their operational levels, the execution of the wetland management strategy is a shared responsibility of all of the stakeholders. However, the co-ordination of this strategy is the responsibility of a government department with state-wide legislative powers (Department of Parks and Wildlife). System controls (e.g., policy and legislation) are mandated on larger spatial scales, whereas accumulated stressors (e.g., reduced rainfall or lowering groundwater table) have larger impacts on lower spatial scales, such as on the whole ecosystem scale or only on part of it. Drying of the lake and ecological degradation are translated by threshold exceedance of ecological processes. Also, the separation/disconnect of water and ecological policy increases the risk of mismanagement. For example, the Department of Water is responsible for groundwater abstraction and the reporting of threshold exceedance to the environmental regulator (the Environmental Protection Authority). While the State government needs to ensure that the ecological functions of the lake are maintained, the Department of Parks and Wildlife is responsible for the ecological state and not for water related management.

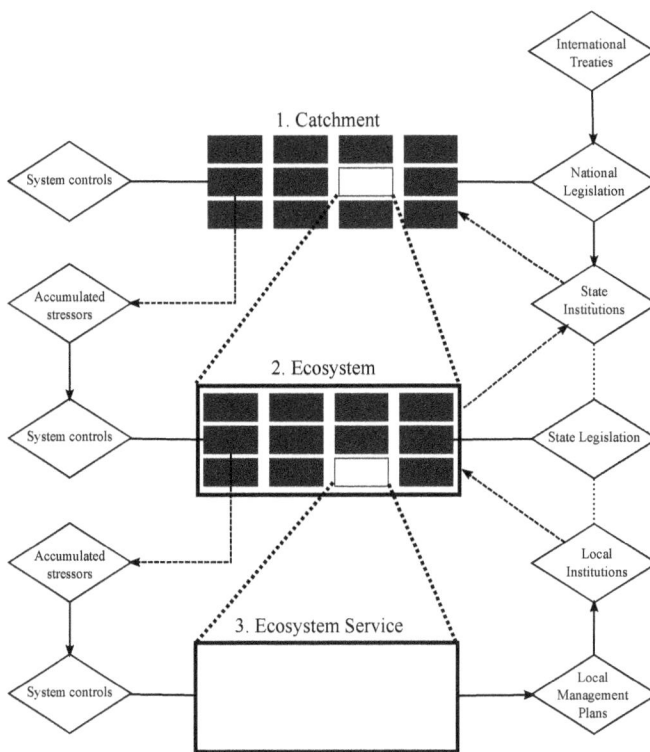

Figure 5. Ecosystem and legislative organisation: across spatial levels of ecosystem organisation stressors are accumulated and trigger a response for system controls in the legislative organisation. Due to fragmented legislative organisation responses are inadequate to maintain ecological resilience.

From the extensive variety of policies and legislation in place to protect the ecological values of the wetland, we were able to derive the important socio-ecological objectives for the wetland. For each objective, we determined the critical water requirement thresholds. However, the water requirement policies did not provide maximum exceedance frequencies (return period) for each objective in our analyses. Where return periods for certain objectives in the management strategy were lacking, stakeholders were able to provide expert knowledge to determine threshold definitions, such as for drought duration, water availability for birds, and exposure of acid sulphate soils.

The findings from the interviews with experts showed that the legislation and policy aims are a good starting point for discussion with stakeholders that operate on a state-wide scale. The experts interviewed represent management authorities that are responsible for the implementation of larger scale (top-down) policies and legislation, and their roles are to build consensus with other governing institutions that contribute to the wetland management plan.

A combination of a review of peer-reviewed literature and government reports provided a comprehensive overview of ecological studies that were undertaken in Forrestdale Lake. Data are predominantly available in government reports rather than in peer-reviewed media. This included data on bird counts, macro-invertebrate species composition, and vegetation transects. Ecological data is often patchy and only available for certain time frames in the 1990s and 2000s for Forrestdale Lake. Bird counts for the lake have been discontinued since 2009 [73] and vegetation transects are not conducted on regular basis as mandated in policy. Groundwater level data was only available from 1997, while surface water levels were recorded from 1952. In addition, surface water level observations from 1952–1978 contained too many data gaps to adequately perform ATP analyses, as consecutive observations up to six months are not available.

3.2. ATPs and Ecological Resilience

ATPs were determined by calculating the re-occurring water level depth using the values from Table 1 with Equation 1. The time series analysis employed here suggests that a drying climate has compromised four ecological objectives of Forrestdale Lake (Table 2). ATPs occurred after 1995 and threshold crossings occurred for vegetation and mammals, waterbirds, turtles, and macro-invertebrates. Water levels for the remaining objectives are close to exceeding thresholds, such as the capacity of the lake to deliver sediment processes and limiting the risk of oxidation of acid sulphate soils in the lake bed.

Table 2. Adaptation tipping points calculated with eq. 1 for each ecological function of Forrestdale Lake. Bold values indicate that the water level is below the threshold value and consequently result in an ATP. The two time periods reflect the timeframe for policy adaption. SW = surface water; GW = groundwater.

Ecological Objective	Water Level (m)		
	Threshold	1978–1995	1996–2012
1. protect vegetation and mammals	SW < 0	0.06	**−0.21**
2. prevent mosquitoes	SW > 0	−0.27	−0.19
3. protect waterbirds	SW < 0	0.24	**−0.16**
4. protect frogs	SW < 0	0.42	0.01
5. protect tortoises	SW < 0	0.06	**−0.21**
6. protect macro-invertebrates	SW < 0.4	0.06	**−0.21**
7. prevent exposure of Acid Sulphate Soils	GW < −0.5	0.06	−0.21
8. maintain sediment processes	GW < −0.5	0.06	−0.21

Figure 6 shows that Forrestdale Lake dried more frequently than the recommended return period of one in five years, and that each dry period exceeded the maximum duration of three consecutive months. Drying is most frequent in summer (December, January, and February) which is in line with regulation that drying of the lake should not occur before April/May, in order to ensure a waterlogged lake bed throughout the year. When the drought frequency and duration are compared for both periods, pre-, and post-1995, no droughts according to the policy definition occur. However, the lake did dry completely for shorter durations during summer. In contrast to the regulation, it is completely logical that drying is more likely to occur over summer, with longer periods per year of limited water availability for species.

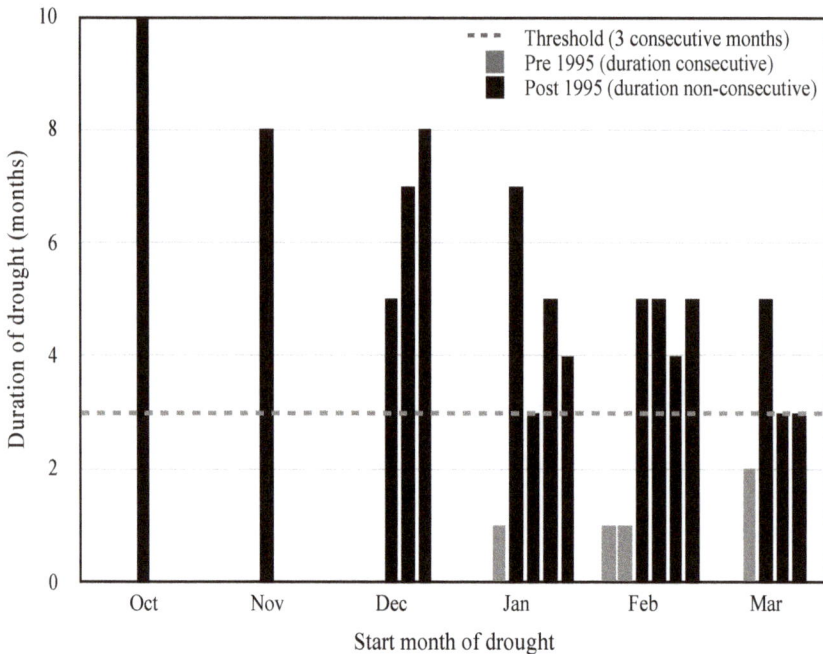

Figure 6. Comparison of the onset and duration of drought during the period 1978–2012 at Forrestdale Lake, shown pre and post 1995. The policy definition of a dry period is ≥3 consecutive dry months which must not start prior to April/May. Each bar represents a dry period and respective start month. Drying of the lake prior to 1995 is added as a reference, as the lake dried in the period, but, according to the policy definition, was not considered as drought.

Although there was not enough data to conduct trend analyses, the frequency of droughts and the duration of each drought has markedly increased since 1995. When we combine the results from our ATP analyses (Table 2) with the drought analyses (Figure 6), we observe a regime shift in the ecosystem from a permanently to seasonally inundated wetland. The effect of this hydrological shift translates into failing to meet the defined threshold level that is enforced in policy and leading to an ATP. In Figure 7 we graphically present the minimum thresholds for all of the objectives, the water levels from 1978–2012 as compared to the initiation of groundwater abstraction, and the implementation of the water policy requirements.

Frequent water level and drought exceedance for objectives only occur in the period after the water policy was implemented in 1992. Between the 1970s and the implementation period of the water policy in 1992, no significant research was conducted on the gradual decline of water levels in the Swan Coastal Plain wetlands. With available quantitative ecological data on ecological responses we base our representation on stylised lines to explain individual ecological responses when compared to declining water levels from the 1970s (Figure 7). This representation is a combination of historical data from previous research and information from the expert interviews (Supplementary Materials). The decline of the ecological processes coincides with the increased duration and frequency of dry periods during the 1990s. After the mid-1990s, we observe that the management of the lake did not respond to maintain declining water levels on the mandated threshold levels; indeed, the minimum water requirements for the wetland were not updated during the period 1992–2005. However, new water level requirements were proposed in 2005 to reflect the current hydrological regime of the wetland.

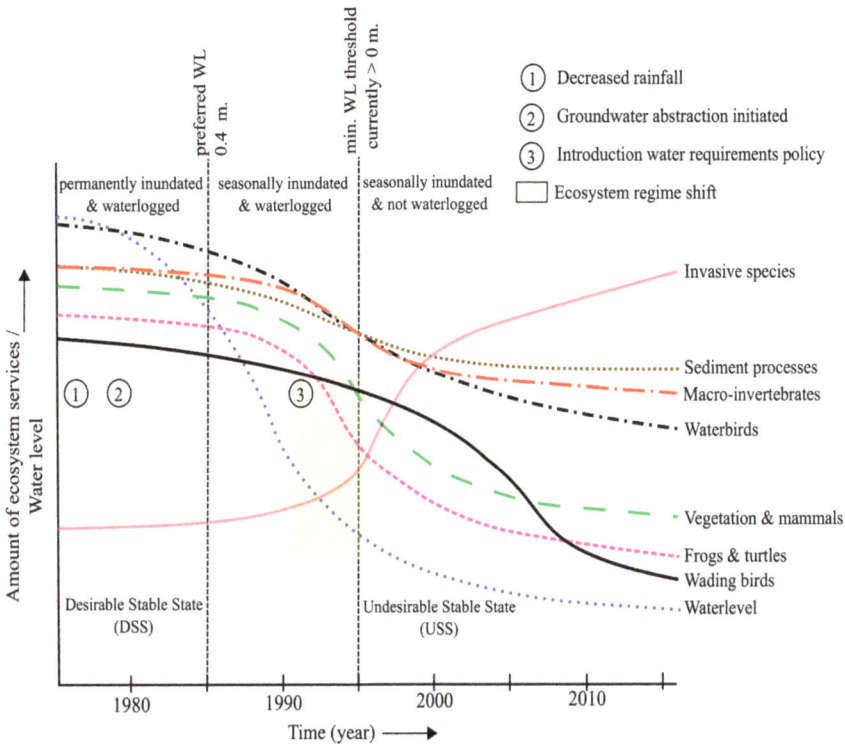

Figure 7. Ecosystem regime shift on the onset of dry periods with declining water levels (WL) and the change of conditions of ecological processes over time. Incremental management and policy compared to non-linear ecosystem responses over time are ineffective when sudden changes occur.

3.3. ATP Assessment and Alternate System States

A major gap in the science-policy interface and socio-hydrologic systems literature is here defined as: (i) the identification of inadequate policy to inform managers or policy makers about the durability of an ecosystem management strategy; or, (ii) the performance of assessments of hydrological variables when data is lacking. With the ATP methodology presented, where possible, we have tried to close the gaps in the literature. The methodology presented assessed whether an existing baseline ecosystem management strategy was sufficient to sustain the ecological resilience of the ecosystem. With the ATP framework, we assessed resilience of the hydrological system across spatial and temporal scales by the: (i) magnitude of the reaction of the ecosystem; (ii) temporal scale and ecosystem responses to increased perturbations; and, (iii) recovery rate or shift from a desirable stable state to an alternate/undesirable stable state with limited ecological processes [74]. We linked eight critical socio-ecological objectives to explain subsystem changes and the implications for decision-making to reach mandated policy thresholds, which was considered a literature gap for ATP assessments [11].

4. Discussion

4.1. Temporal and Spatial Hydrological Responses in Atp Analysis Applied to Ecosystems

The observed climatic shift evident in the late 1960s/early 1970s in south-west Western Australia [75] follows the stepwise decreasing rainfall trend in our hydrological time series. With shorter periods of inundation in the 1990s a hydrological response is evident, and ATPs occur simultaneously in the same time period. The hydrological shift from permanent to intermittent water availability in the lake due decreased surface water availability from lower rainfall is explained in previous studies [38,57,76,77]. The observations of consistent reductions of water levels result in more frequent, prolonged dry periods, and studies confirm that a significant reduction in water levels for consecutive years could threaten the regional function of wetlands to sustain multiple ecological functions [76–78].

The analysis points to an ineffective water requirements policy, as water levels requirements are not met for four of the eight ecological functions; thresholds were crossed in the 1990s, which occurred concurrently with the observed hydrological response. The main ecological processes of the lake depend on waterlogged soils during low water availability, however are at increasing risk when the lake bed dries completely over summer. Early drying of the lake implies a lack of surface water availability for species that have a limited action radius to alternative habitats, such as macrophytes, freshwater tortoises, frogs, and macro-invertebrates. Our study did not include the investigation of ecological responses, however, the hydrological change and ATPs are followed by declining trends in the ecology. Previous studies on this wetland have shown:

- increasing weed invasion and exotic species establishing in the understory, along with deterioration of fringing vegetation [76,78];
- a gradual declining trend in the species numbers and composition of macro-invertebrates; in particular, a reduced number of families was observed (down from 40 in 1987 to 34 in 2009 [61]) due a loss of some species [46,60,77]; and,
- decreasing numbers of birds from over 20.000 birds in the 1980s to just over 10.000 birds in 2009 [79].

The responses of ecosystems after perturbations, and the shifts that could occur from a desirable higher stable state into an undesirable lower stable state with higher resilience and reduced ecological processes are described in the literature [2,3,12]. However, a lack of data makes it difficult to determine shifts between multiple or alternate stable states [80]. From our results, we see that a gradual transition of the boundary condition (reduced rainfall) failed to trigger management interventions to maintain the rapid responses in an ecosystem. Management responses are also absent when, for example, rapid hydrological processes and the slow response of ecological processes, such as vegetation shifts [81], are not detected when monitored at different spatial scales [82]. This mismatch is magnified when different government departments are responsible for monitoring and management responses.

To draw attention to the different responses of ecological processes we started a discussion among management authorities to consider management objectives and threshold values. The management objectives are derived from different sources such as the State-scale water level criteria; the national (Commonwealth) ecological objectives that are linked to the Ramsar guidelines; and, the key socio-ecological objectives from the local management plan. Currently, Ramsar criteria, such as the number of (water) birds is infrequently monitored, and objectives from the local management plan are only partly monitored (vegetation and macro-invertebrates). The ATP analysis and the discussion among the different actors for wetland management showed that the jurisdiction of the government departments in question does not cover the spatial scale of certain ecological processes. Some ecological processes rely on factors which are managed by different institutions. For example, the decline of vegetation quality depends on regional groundwater availability, which is regulated by the water regulator; whereas, protecting flora in the buffer zones is the responsibility of the conservation authority. Despite the strong indication of declining ecological values, national and state level policies are only partly informed by the policies determined at the local scale. Research confirms the need to incorporate all the relevant institutions to achieve institutional-ecosystem function fit [83], while the identification of underlying gaps in multi-sector governance as described, will form the basis to negotiate closing the gaps in governance.

4.2. Informing Ecosystem Management

The ATPs that are presented in this case study area are intended as guiding principles (early-stage) to existing ineffective ecosystem management strategies. The ineffectiveness of other policies has been shown in: Flood risk studies [15,17], flood mitigation under climate change [18], river restoration [26], and the impact of the hydrological regime of a river on salmon re-introduction, and shipping [24]. Central in these studies is to determine when and how much action is needed to determine alternative management strategies [81], but for a SES, when to take action is far more complicated. Due to the jurisdiction of decision makers or managing authorities, ATPs are used as a starting point to explore if and when adaptation measures need to be taken to adequately resolve the critical adaptation tipping point of different ecological processes [23]. When, such as in our case, quantitative data is not readily available to support a complex model, with predicted feedback mechanism, in the socio-environmental system [4,5,84,85], the outcomes of an ATP analyses provide a better understanding of the role of individual processes before making more complex models [86]; highlighting the potential dynamics of scale of legislation and policy, and the interaction of management authorities in the hydrological system. As described previously, management interventions can be considered by different institutions that will provide the appropriate outcome for each ecological process. This requires understanding the scale and level of policy and legislation in the analysis prior to embarking on a process to deliver adaptation measures for the different socio-ecological objectives, such as those that we included in our analysis.

In order to adequately improve existing management practices, we should first consider the whole set of clearly stated objectives in a management strategy without prioritising or aggregating them. As a result, we may then provide the alternate states of ecological processes within the spatial and temporal scales of processes and governance systems [14]. Introducing multiple management aims overcomes a focus on separate ecological objectives, which may lead to a lack of quantitative boundaries or thresholds for acceptable ecological change [11,16,28,87]. Studies have shown that when law or policy enforced threshold levels along an environmental gradient are passed [7], that not all ecological processes will show a direct decline of species or shift in species composition, thus making it more difficult to reverse different conditions of the ecosystem [2]. Therefore, informing decision-makers at an early stage prevents costly measures to reverse undesirable changes to the system.

In the absence of clearly defined thresholds, our framework provides active involvement of the management authorities [10,28] from a multi-purpose perspective [24]. The ATP analyses stimulate stakeholders to look at the resilience of their approach [16]. Continuous improvement in the processes of adaptive management is an ongoing challenge, but studies have demonstrated successful frameworks for collaborative research in the science-policy interface across several scales [88,89]. When management practices need to be updated, the threshold definitions for management approaches should reflect the ideas of multiple management authorities that are involved. In the absence of a combined eco-hydrological and social model, we were able to distinguish the trade-offs between vulnerability (performance) of the ecosystem as compared to thresholds of subsystem processes that were defined by policies and legislation across spatial scales.

4.3. Adapting Management Strategies

For effective governance, developing a better understanding of climate and hydrological impacts is required [89]. With the involvement of stakeholders in our assessment, we can account for the exploration of future hydrological events and provide decision-makers the information on under which conditions the current policies will expire; however, the exact timing of expiry remains problematic due to different timescale of system responses. We aimed to overcome this by including threshold values (only partly available) that represent ecosystem processes across scales. Although the ATP assessment includes some options to identify measures for adequate governance decisions; further exploration for how long these are sufficient under future climate scenarios needs to be investigated [10]. This could include: (1) physical/engineered measures, (2) adoption of new or amended policy instruments,

(3) adoption of policy strategies (combination of options 1 and 2), or (4) implementation of an adaptation strategy [12,16,90]. Successful adaptation requires a critical understanding of the scale and level of implementation of existing policies, legislation, or management strategies, as these are often barriers to local scale adaptation.

Despite the exceedance of critical thresholds, management has not adequately responded to changing hydrological variation in the ecosystem. We assumed climate change to be the main external driver for the ecosystem regime shift, although this does not assume a non-adaptive management strategy. The ATP application is adequate for ecosystems when a clear external driver of change can be determined (e.g., climate change), stakeholders agree on setting thresholds, and expand individual management objectives to objectives across several levels of policies. However, the study of systems becomes complicated when multiple stressors are responsible for subsystem change and stakeholders do not include objectives or thresholds defined by different or new policies. The limitations of system study include the effects of multiple stressors on the system, a limited focus on new strategies, and including objectives or thresholds that change over time due to socio-economic changes. In this paper we have addressed the difficulties to determine ATPs for an ecosystem with respect to existing policies and management objectives. Further collection of ecological data and monitoring ecological responses will be helpful to determine alternative strategies with stakeholders to postpone or eliminate existing ATPs, according to the steps of the original ATP methodology. The dynamic adaptive policy pathways approach could be a useful tool to guide this process [10].

5. Conclusions

The extended ATP method presented in this paper provides a combination of a qualitative and quantitative analysis of datasets of a wetland ecosystem. We applied the concept of 'adaptation tipping points' to identify when management responses became inadequate to prevent decline in ecological integrity. Through a combination of conceptual and visual representation of the ecological processes, we were able to identify major trends and transitions in the system, in the presence of strong drivers of change and variable hydrological conditions. This approach was useful to determine the effectiveness of an ecosystem management strategy when data availability was limited, and where social-ecological dynamic models to fully assess the tipping point and potential points for interventions were absent. This study showed that a lack of data, quantitative boundaries, or thresholds to define acceptable ecological change can be overcome by the inclusion of pre-existing thresholds based on available information about shifts of the wetland's hydrological regime. This information included the importance of reviewing a range of policies to enable discussion among stakeholders to determine existing and new management objectives/thresholds. Through stakeholder discussions, we found unacceptable adverse ecological changes to the unique set of identifiers, and then used the input of expert knowledge to determine the critical wetland objectives and thresholds for wetland management. We showed that informing stakeholders about the effectiveness of existing wetland policy can be used to adapt or accept objectives and thresholds, both seen here in context with ATPs and undesirable ecological changes. ATPs could be established a proxy indicator for lag-responses in the ecology to adapt ecosystem management in a timely manner before ecological processes deteriorate to unaccepted levels.

Supplementary Materials: The following are available online at http://www.mdpi.com/2073-4441/10/2/234/s1, Table S1. Overview legislation framework; Tables S2–S7. Results from stakeholder workshop 1 with the problem statement, objectives, drivers and performance metrics; Table S8. Identified adaptation measures.

Acknowledgments: This research was funded within program B4.2 of the Cooperative Research Centre of Water Sensitive Cities. The authors thank the Department of Parks and Wildlife and the Department of Water for providing the ecological and water level data of Forrestdale Lake. The RStatistics code to compute the water level data was provided by Chrianna Bharat at The University of Western Australia. Liah Coggins provided valuable feedback to improve the structure of the manuscript. Amar Nanda was supported by a Scholarship for International Research Fees (SIRF) funded by The University of Western Australia. We sincerely thank the stakeholder representatives from each of the government departments that participated in the workshop; all subjects gave their informed consent for inclusion before they participated in the study. The research involving human data reported in this study was assessed and approved by The University of Western Australia Human Research Ethics Committee (Approval #: RA/4/1/7999).

Author Contributions: A.N. designed the study, applied for human ethics approval, collected the data, organised the stakeholder workshop and wrote the manuscript under the supervision of B.G., M.R.H. and A.G.; B.G. assisted with the statistical analyses and provided guidance for the stakeholder workshop; L.B. provided feedback on the ecological analyses and introduction; L.L. assisted with understanding the local and regional hydrology of the lake.

Conflicts of Interest: The authors declare no conflict of interest. The funding sponsors had no role in the design of the study; in the collection, analyses, or interpretation of data; in the writing of the manuscript, and in the decision to publish the results.

References

1. Holling, C.S. Resilience and stability of ecological systems. *Annu. Rev. Ecol. Syst.* **1973**, *4*, 1–23. [CrossRef]
2. Scheffer, M.; Carpenter, S.; Foley, J.A.; Folke, C.; Walker, B. Catastrophic shifts in ecosystems. *Nature* **2001**, *413*, 591–596. [CrossRef] [PubMed]
3. Folke, C.; Carpenter, S.; Walker, B.; Scheffer, M.; Elmqvist, T.; Gunderson, L.; Holling, C.S. Regime shifts, resilience, and biodiversity in ecosystem management. *Annu. Rev. Ecol. Evol. Syst.* **2004**, 557–581. [CrossRef]
4. Sivapalan, M.; Savenije, H.H.G.; Blöschl, G. Socio-hydrology: A new science of people and water. *Hydrol. Process.* **2012**, *26*, 1270–1276. [CrossRef]
5. Elshafei, Y.; Sivapalan, M.; Tonts, M.; Hipsey, M.R. A prototype framework for models of socio-hydrology: Identification of key feedback loops and parameterisation approach. *Hydrol. Earth Syst. Sci.* **2014**, *18*, 2141–2166. [CrossRef]
6. Dudgeon, D.; Arthington, A.H.; Gessner, M.O.; Kawabata, Z.-I.; Knowler, D.J.; Lévêque, C.; Naiman, R.J.; Prieur-Richard, A.-H.; Soto, D.; Stiassny, M.L. Freshwater biodiversity: Importance, threats, status and conservation challenges. *Biol. Rev.* **2006**, *81*, 163–182. [CrossRef] [PubMed]
7. Walker, B.; Meyers, J.A. Thresholds in ecological and social–ecological systems: A developing database. *Ecol. Soc.* **2004**, *9*, 3. [CrossRef]
8. Walker, B.; Holling, C.S.; Carpenter, S.; Kinzig, A. Resilience, adaptability and transformability in social–ecological systems. *Ecol. Soc.* **2004**, *9*. [CrossRef]
9. Rijke, J.; Brown, R.; Zevenbergen, C.; Ashley, R.; Farrelly, M.; Morison, P.; van Herk, S. Fit-for-purpose governance: A framework to make adaptive governance operational. *Environ. Sci. Policy* **2012**, *22*, 73–84. [CrossRef]
10. Haasnoot, M.; Kwakkel, J.H.; Walker, W.E.; ter Maat, J. Dynamic adaptive policy pathways: A method for crafting robust decisions for a deeply uncertain world. *Glob. Environ. Chang.* **2013**, *23*, 485–498. [CrossRef]
11. Werners, S.; Pfenninger, S.; van Slobbe, E.; Haasnoot, M.; Kwakkel, J.; Swart, R. Thresholds, tipping and turning points for sustainability under climate change. *Curr. Opin. Environ. Sustain.* **2013**, *5*, 334–340. [CrossRef]
12. Folke, C.; Hahn, T.; Olsson, P.; Norberg, J. Adaptive governance of social-ecological systems. *Annu. Rev. Environ. Resour.* **2005**, *30*, 441–473. [CrossRef]
13. Schlueter, M.; McAllister, R.; Arlinghaus, R.; Bunnefeld, N.; Eisenack, K.; Hoelker, F.; Milner-Gulland, E.; Müller, B.; Nicholson, E.; Quaas, M. New horizons for managing the environment: A review of coupled social-ecological systems modeling. *Nat. Resour. Model.* **2012**, *25*, 219–272. [CrossRef]
14. Niemi, G.J.; McDonald, M.E. Application of ecological indicators. *Annu. Rev. Ecol. Evol. Syst.* **2004**, *35*, 89–111. [CrossRef]
15. Lavery, S.; Donovan, B. Flood risk management in the thames estuary looking ahead 100 years. *Philos. Trans. R. Soc. Lond. A Math. Phys. Eng. Sci.* **2005**, *363*, 1455–1474. [CrossRef] [PubMed]
16. Kwadijk, J.; Haasnoot, M.; Mulder, J.; Hoogvliet, M.; Jeuken, A.; van der Krogt, R.; van Oostrom, N.; Schelfhout, H.; van Velzen, E.; van Waveren, H.; et al. Using adaptation tipping points to prepare for climate change and sea level rise: A case study in the netherlands. *Wiley Interdiscip. Rev. Clim. Chang.* **2010**, *1*, 729–740. [CrossRef]
17. Reeder, T.; Ranger, N. *How do You Adapt in an Uncertain World? Lessons from the Thames Estuary 2100 Project*; World Resources Report Uncertainty Series; World Resources Institute: Washington, DC, USA, 2011.

18. Gersonius, B.; Ashley, R.; Pathirana, A.; Zevenbergen, C. Climate change uncertainty: Building flexibility into water and flood risk infrastructure. *Clim. Chang.* **2012**, *116*, 413–423. [CrossRef]

19. Brown, C.; Werick, W.; Leger, W.; Fay, D. A decision-analytic approach to managing climate risks: Application to the upper great lakes1. *J. Am. Water Resour. Assoc.* **2011**, *47*, 524–534. [CrossRef]

20. Poff, N.L.; Brown, C.M.; Grantham, T.E.; Matthews, J.H.; Palmer, M.A.; Spence, C.M.; Wilby, R.L.; Haasnoot, M.; Mendoza, G.F.; Dominique, K.C.; et al. Sustainable water management under future uncertainty with eco-engineering decision scaling. *Nat. Clim. Chang.* **2016**, *6*, 25. [CrossRef]

21. Lawrence, J.; Sullivan, F.; Lash, A.; Ide, G.; Cameron, C.; McGlinchey, L. Adapting to changing climate risk by local government in new zealand: Institutional practice barriers and enablers. *Local Environ.* **2013**, *20*, 298–320. [CrossRef]

22. Fünfgeld, H. Facilitating local climate change adaptation through transnational municipal networks. *Curr. Opin. Environ. Sustain.* **2015**, *12*, 67–73. [CrossRef]

23. Hanger, S.; Pfenninger, S.; Dreyfus, M.; Patt, A. Knowledge and information needs of adaptation policy-makers: A european study. *Reg. Environ. Chang.* **2013**, *13*, 91–101. [CrossRef]

24. van Slobbe, E.; Werners, S.E.; Riquelme-Solar, M.; Bölscher, T.; van Vliet, M.T.H. The future of the rhine: Stranded ships and no more salmon? *Reg. Environ. Chang.* **2016**, *16*, 31–41. [CrossRef]

25. Werners, S.; Swart, R.; van Slobbe, E.; Bölscher, T. Turning points in climate change adaptation. *Glob. Environ. Chang.* **2013**, *16*, 253–267.

26. Bölscher, T.; van Slobbe, E.; van Vliet, M.T.; Werners, S.E. Adaptation turning points in river restoration? The rhine salmon case. *Sustainability* **2013**, *5*, 2288–2304. [CrossRef]

27. Wardekker, J.A.; de Jong, A.; Knoop, J.M.; van der Sluijs, J.P. Operationalising a resilience approach to adapting an urban delta to uncertain climate changes. *Technol. Forecast. Soc. Chang.* **2010**, *77*, 987–998. [CrossRef]

28. Haasnoot, M.; Middelkoop, H.; Offermans, A.; Beek, E.V.; Deursen, W.P.A.V. Exploring pathways for sustainable water management in river deltas in a changing environment. *Clim. Chang.* **2012**, *115*, 795–819. [CrossRef]

29. Myers, N.; Mittermeier, R.A.; Mittermeier, C.G.; Da Fonseca, G.A.; Kent, J. Biodiversity hotspots for conservation priorities. *Nature* **2000**, *403*, 853–858. [CrossRef] [PubMed]

30. Bekle, H.; Gentilli, J. History of the perth lakes. *R. West. Aust. Hist. Soc.* **1993**, *10*, 441–460.

31. Bekle, H. *The Wetlands Lost: Drainage of the Perth Lake Systems*; Geographical Society of W.A.: Perth, Australia, 1981.

32. Balla, S.A. *Wetlands of the Swan Coastal Plain. Volume 1, Their Nature and Management*; Water Authority of WA: Perth, Australia, 1993.

33. Conservation Commission of Western Australia. *Forrestdale Lake Nature Reserve Management Plan 2005*; Management Plan No. 53; Conservation Commission of Western Australia; Government of Western Australia: Perth, Australia, 2005.

34. Storey, A.W.; Vervest, R.M.; Pearson, G.B.; Halse, S.A. *Wetlands of the Swan Coastal Plain, Volume 7, Waterbird Usage of Wetlands on the Swan Coastal Plain*; Water Authority of WA: Perth, Australia, 1993.

35. Davis, J.A.; Froend, R. Loss and degradation of wetlands in southwestern australia: Underlying causes, consequences and solutions. *Wetl. Ecol. Manag.* **1999**, *7*, 13–23. [CrossRef]

36. Barron, O.; Barr, A.; Donn, M. Effect of urbanisation on the water balance of a catchment with shallow groundwater. *J. Hydrol.* **2013**, *485*, 162–176. [CrossRef]

37. Department of Water (Ed.) *Assessment of the Declining Groundwater Levels in the Gnangara Groundwater Mound, Report hg14*; Hydrogeological Record Series; Department of Water: Perth, Australia, 2008.

38. Eamus, D.; Froend, R. Groundwater-dependent ecosystems: The where, what and why of gdes. *Aust. J. Bot.* **2006**, *54*, 91–96. [CrossRef]

39. Barron, O.; Froend, R.; Hodgson, G.; Ali, R.; Dawes, W.; Davies, P.; McFarlane, D. Projected risks to groundwater-dependent terrestrial vegetation caused by changing climate and groundwater abstraction in the central perth basin, western australia. *Hydrol. Process.* **2013**, *28*, 5513–5529. [CrossRef]

40. Charles, S.; Silberstein, R.; Teng, J.; Fu, G.; Hodgson, G.; Gabrovsek, C.; Crute, J. *Climate Analyses for South-West Western Australia*; A report to the Australian Government from the CSIRO South-West Western Australia Sustainable Yields Project; CSIRO: 92-92; CSIRO: Canberra, Australia, 2010.

41. Petrone, K.C.; Hughes, J.D.; Van Niel, T.G.; Silberstein, R.P. Streamflow decline in southwestern australia, 1950–2008. *Geophys. Res. Lett.* **2010**, *37*. [CrossRef]
42. Smith, I.; Power, S. Past and future changes to inflows into perth (western australia) dams. *J. Hydrol. Reg. Stud.* **2014**, *2*, 84–96. [CrossRef]
43. Froend, R.; Sommer, B. Phreatophytic vegetation response to climatic and abstraction-induced groundwater drawdown: Examples of long-term spatial and temporal variability in community response. *Ecol. Eng.* **2010**, *36*, 1191–1200. [CrossRef]
44. Froend, R.H.; Farrelly, C.F.; Wilkins, C.C.; McComb, A.J. *Wetlands of the Swan Coastal Plain. Volume 4, the Effects of Altered Water Levels on Wetland Plants*; Water Authority of WA: Perth, Australia, 1993.
45. Ali, R.; McFarlane, D.; Varma, S.; Dawes, W.; Emelyanova, I.; Hodgson, G. Potential climate change impacts on the water balance of regional unconfined aquifer systems in south-western australia. *Hydrol. Earth Syst. Sci.* **2012**, *16*, 4581–4601. [CrossRef]
46. Sommer, B.; Horwitz, P. Macroinvertebrate cycles of decline and recovery in swan coastal plain (western australia) wetlands affected by drought-induced acidification. *Hydrobiologia* **2009**, *624*, 191–203. [CrossRef]
47. Sommer, B.; Froend, R. Resilience of phreatophytic vegetation to groundwater drawdown: Is recovery possible under a drying climate? *Ecohydrology* **2011**, *4*, 67–82. [CrossRef]
48. Townley, L.; Turner, J.; Barr, A.D.; Trefry, M. *Wetlands of the Swan Coastal Plain Volume 3: Interaction between Lakes, Wetlands and Unconfined Aquifers*; Education Department of Western Australia: Perth, Australia, 1993.
49. McFarlane, D. *The Effect of Climate Change on South West WA Hydrology*; CSIRO: Canberra, Australia, 2012.
50. Australian Bureau of Statistics. 3105.0.65.001—Australian Historical Population Statistics. 2014. Available online: http://abs.gov.au/ausstats (accessed on 8 November 2016).
51. Bureau of meteorology. Bureau of Meteorology, Monthly Rainfall Midland (Perth) 1886–2015, Station Number 9025. Available online: http://www.bom.gov.au/climate/data/index.shtml (accessed on 8 November 2016).
52. Froend, R.; Loomes, R.; Horwitz, P.; Rogan, R.; Lavery, P.; How, J.; Storey, A.; Bamford, M.; Metcalf, B. *Study of Ecological Water Requirements on the Gnangara and Jandakot Mounds under Section 46 of the Environmental Protection Act, Task 1: Identification and Re-Evaluation of Ecological Values Prepared for: The Water and Rivers Commission*; Water and Rivers Commission, Ed.; Water and Rivers Commission: Perth, Australia, 2004.
53. Eamus, D.; Froend, R.; Loomes, R.; Hose, G.; Murray, B. A functional methodology for determining the groundwater regime needed to maintain the health of groundwater-dependent vegetation. *Aust. J. Bot.* **2006**, *54*, 97–114. [CrossRef]
54. Environmental Protection Authority (Ed.) *Jandakot Mound Groundwater Resources, Bulletin 1155*; Environmental Protection Authority: Perth, Australia, 1992.
55. Semeniuk, C.A. Wetlands of the darling system- a geomorphic approach to habitat classification. *J. R. Soc. West Aust.* **1987**, *69*, 95–112.
56. Hill, A.L.; Australia, W.; Australia, W.; Water and Rivers Commission. *Wetlands of the Swan Coastal Plain: Wetland Mapping, Classification and Evaluation, Main Report. Volume 2a*; Water and Rivers Commission and Department of Environmental Protection: Leederville, WA, USA, 1996.
57. Dawes, W.; Barron, O.; Donn, M.; Pollock, D.; Johnstone, C. *Forrestdale Lake Water Balance; CSIRO Water for a Healthy Country National Research Flagship*; CSIRO: Perth, Australia, 2009.
58. Froend, R.; Loomes, R.; Horwitz, P.; Bertuch, M.; Storey, M.; Bamford, M. *Study of Ecological Water Requirements on the Gnangara and Jandakot Mounds under Section 46 of the Environmental Protection Act, Task 2: Determination of Ecological Water Requirements*; Water and Rivers Commission: Perth, Australia, 2004.
59. Canham, C. The Response of Banksia Roots to Change in Water Table Level in a Mediterranean-Type Environment. Ph.D. Thesis, Edith Cowan University, Joondalup, Australia, 2011.
60. Balla, S.; Davis, J. Seasonal variation in the macroinvertebrate fauna of wetlands of differing water regime and nutrient status on the swan coastal plain, western australia. *Hydrobiologia* **1995**, *299*, 147–161. [CrossRef]
61. Dale, P.; Knight, J. Wetlands and mosquitoes: A review. *Wetl. Ecol. Manag.* **2008**, *16*, 255–276. [CrossRef]
62. Department of Environment and Conservation (Ed.) *Treatment and Management of Soils and Water in Acid Sulfate Soil Landscapes*; Department of Environment and Conservation: Perth, Australia, 2011.
63. Department of Water (Ed.) *Water Information (Win) Database—Time-series Data Site ID 14578 and 12781400*; Department of Water, Water Information Section: Perth, Australia, 2015.
64. Jenkinson, A.F. The frequency distribution of the annual maximum (or minimum) values of meteorological elements. *Q. J. R. Meteorol. Soc.* **1955**, *81*, 158–171. [CrossRef]

65. Environmental Protection Authority. *Environmental protection act 1986*; Environmental Protection Authority: Perth, Australia, 2003.

66. Conservation and Land Management Act. In *Conservation and Land Management Act 1984*; Government of Western Australia 126 of 1984; State Law Publisher: Perth, Australia, 1984.

67. Wildlife Conservation Act. Wildlife Conservation Act. Wildlife conservation act 1950. In *Government of Western Australia 1950, 077 of 1950*; (14 & 15 Geo. VI No. 77); State Law Publisher: Perth, Australia, 1950.

68. Ramsar. *The Convention on Wetlands Text, as Amended in 1982 and 1987: The List of Wetlands of International Importance*; United Nations Educational, Scientific and Cultural Organization (UNESCO): Paris, France, 2014; pp. 1–47.

69. The Japan Australia Migratory Bird Agreement (JAMBA). Agreement between the government of australia and the government of japan for the protection of migratory birds in danger of extinction and their environment. In *Australian Treaty Series*; State Law Publisher: Perth, Australia, 1981; No. 6.

70. The China Australia Migratory Bird Agreement (CAMBA). Agreement between the government of australia and the government of the people's republic of china for the protection of migratory birds and their environment. In *Australian Treaty Series*; State Law Publisher: Perth, Australia, 1988.

71. The Republic of Korea and Australia Migratory Bird Agreement (ROKAMBA). Agreement between the government of australia and the government of the republic of korea on the protection of migratory birds. In *Australian Treaty Series*; State Law Publisher: Perth, Australia, 2007.

72. Environment Protection and Biodiversity Conservation (EBPC). In *Environment Protection and Biodiversity Conservation Act 1999*; State Law Publisher: Perth, Australia, 1999.

73. Department of Water (Ed.) *Environmental Management of Groundwater from the Jandakot Mound; Triennial Compliance Report to the Office of the Environmental Protection Authority*; Department of Water, Government of Western Australia: Perth, Australia, 2012.

74. Zevenbergen, C.; Veerbeek, W.; Gersonius, B.; Van Herk, S. Challenges in urban flood management: Travelling across spatial and temporal scales. *J. Flood Risk Manag.* **2008**, *1*, 81–88. [CrossRef]

75. Verdon-Kidd, D.C.; Kiem, A.S.; Moran, R. Links between the Big Dry in Australia and hemispheric multi-decadal climate variability. *Hydrol. Earth Syst. Sci.* **2014**, *18*, 2235–2256. [CrossRef]

76. Davis, J.; Brock, M. Detecting unacceptable change in the ecological character of ramsar wetlands. *Ecol. Manag. Restor.* **2008**, *9*, 26–32. [CrossRef]

77. Maher, K.; Davis, J. *Ecological Character Description for the Forrestdale and Thomsons Lakes Ramsar Site; a Report to the Department of Environment and Conservation*; Murdoch University: Perth, Australia, 2009.

78. Froend, R.; Rogan, R.; Loomes, R.; Horwitz, P.; Bamford, M.; Storey, A. *Study of Ecological Water Requirements on the Gnangara and Jandakot Mounds under Section 46 of the Environmental Protection Act, Task 3 & 4: Parameter Identification and Monitoring Program Review*; Water and Rivers Commission, Ed.; Water and Rivers Commission: Perth, Australia, 2004.

79. Bamford, M.; Bancroft, W.; Raines, J. *Effects of Remote Rainfall Events on Waterbird Populations on the Jandakot Mound Wetlands*; Department of Water, by Bamford Consulting Ecologists: Perth, Australia, 2010.

80. Capon, S.J.; Lynch, A.J.; Bond, N.; Chessman, B.C.; Davis, J.; Davidson, N.; Finlayson, M.; Gell, P.A.; Hohnberg, D.; Humphrey, C.; et al. Regime shifts, thresholds and multiple stable states in freshwater ecosystems; a critical appraisal of the evidence. *Sci. Total Environ.* **2015**, *534*, 122–130. [CrossRef] [PubMed]

81. Sivapalan, M.; Blöschl, G. Time scale interactions and the coevolution of humans and water. *Water Resour. Res.* **2015**, *51*, 6988–7022. [CrossRef]

82. Elshafei, Y.; Tonts, M.; Sivapalan, M.; Hipsey, M.R. Sensitivity of emergent sociohydrologic dynamics to internal system properties and external sociopolitical factors: Implications for water management. *Water Resour. Res.* **2016**, *52*, 4944–4966. [CrossRef]

83. Ekstrom, J.; Young, O. Evaluating functional fit between a set of institutions and an ecosystem. *Ecol. Soc.* **2009**, *14*, 16. [CrossRef]

84. Di Baldassarre, G.; Viglione, A.; Carr, G.; Kuil, L.; Yan, K.; Brandimarte, L.; Blöschl, G. Debates-perspectives on sociohydrology: Capturing feedbacks between physical and social processes. *Water Resour. Res.* **2015**, *51*, 4770–4781. [CrossRef]

85. Di Baldassarre, G.; Kooy, M.; Kemerink, J.S.; Brandimarte, L. Towards understanding the dynamic behaviour of floodplains as human-water systems. *Hydrol. Earth Syst. Sci.* **2013**, *17*, 3235–3244. [CrossRef]

86. Hipsey, M.R.; Hamilton, D.P.; Hanson, P.C.; Carey, C.C.; Coletti, J.Z.; Read, J.S.; Ibelings, B.W.; Valesini, F.J.; Brookes, J.D. Predicting the resilience and recovery of aquatic systems: A framework for model evolution within environmental observatories. *Water Resour. Res.* **2015**, *51*, 7023–7043. [CrossRef]
87. Hallegatte, S. Strategies to adapt to an uncertain climate change. *Glob.Environ. Chang.* **2009**, *19*, 240–247. [CrossRef]
88. Mitchell, B.; Hollick, M. Integrated catchment management in western australia: Transition from concept to implementation. *Environ. Manag.* **1993**, *17*, 735–743. [CrossRef]
89. Davis, J.; O'Grady, A.P.; Dale, A.; Arthington, A.H.; Gell, P.A.; Driver, P.D.; Bond, N.; Casanova, M.; Finlayson, M.; Watts, R.J.; et al. When trends intersect: The challenge of protecting freshwater ecosystems under multiple land use and hydrological intensification scenarios. *Sci. Total Environ.* **2015**, *534*, 65–78. [CrossRef] [PubMed]
90. Nelson, D.R.; Adger, W.N.; Brown, K. Adaptation to environmental change: Contributions of a resilience framework. *Annu. Rev. Environ. Resour.* **2007**, *32*, 395–419. [CrossRef]

water

MDPI

Article

Urban Surface Water Quality, Flood Water Quality and Human Health Impacts in Chinese Cities. What Do We Know?

Yuhan Rui [1], Dafang Fu [1], Ha Do Minh [2], Mohanasundar Radhakrishnan [2,*], Chris Zevenbergen [2] and Assela Pathirana [2]

[1] School of Civil Engineering, SEU-Monash Joint Research Centre for Water Sensitive Cities, Southeast University (SEU), Nanjing 210096, China; ruiyh1103@163.com (Y.R.); 101002314@seu.edu.cn (D.F.)
[2] Water Science & Engineering Department, IHE Delft Institute for Water Education, #7 Westvest, 2611AX Delft, The Netherlands; h.do@un-ihe.org (H.D.M.); c.zevenbergen@un-ihe.org (C.Z.); a.pathirana@un-ihe.org (A.P.)
* Correspondence: m.radhakrishnan@un-ihe.org; Tel.: +31-647-618-689

Received: 11 January 2018; Accepted: 16 February 2018; Published: 26 February 2018

Abstract: Climate change and urbanization have led to an increase in the frequency of extreme water related events such as flooding, which has negative impacts on the environment, economy and human health. With respect to the latter, our understanding of the interrelationship between flooding, urban surface water and human health is still very limited. More in-depth research in this area is needed to further strengthen the process of planning and implementation of responses to mitigate the negative health impacts of flooding in urban areas. The objective of this paper is to assess the state of the research on the interrelationship between surface water quality, flood water quality and human health in urban areas based on the published literature. These insights will be instrumental in identifying and prioritizing future research needs in this area. In this study, research publications in the domain of urban flooding, surface water quality and human health were collated using keyword searches. A detailed assessment of these publications substantiated the limited number of publications focusing on the link between flooding and human health. There was also an uneven geographical distribution of the study areas, as most of the studies focused on developed countries. A few studies have focused on developing countries, although the severity of water quality issues is higher in these countries. The study also revealed a disparity of research in this field across regions in China as most of the studies focused on the populous south-eastern region of China. The lack of studies in some regions has been attributed to the absence of flood water quality monitoring systems which allow the collection of real-time water quality monitoring data during flooding in urban areas. The widespread implementation of cost effective real-time water quality monitoring systems which are based on the latest remote or mobile phone based data acquisition techniques is recommended. Better appreciation of health risks may lead to better flood risk management. In summary, there is still a limited understanding of the relationship between urban surface water quality, flood water quality and health impacts. This also holds true for Chinese cities. Given the widespread and frequent occurrence of urban flooding, further research into this specific cross-cutting field is mandatory.

Keywords: urban water quality; floodwater; human health; water quality monitoring and assessment system

---

1. Introduction

Urban flooding occurs when surface water cannot be drained quickly through drainage systems, streams and rivers. A reduction in permeability, which is a result of an increase of impervious surface area (i.e., road, building, parking lot, sidewalk), increases run off, peak discharge and flood frequency [1]. The analysis of urban flood risk is mostly based on an assessment of the underlying hydrological and physical features of an urban area. For example, flood risk maps are usually drawn from an assessment of the recorded and/or predicted rainfall intensities and frequencies in conjunction with a detailed land-use analysis revealing relevant spatial information of the urban fabric [2–4]. In these risk analyses, it is common practice to use the estimated damage due to flooding as a proxy [5]. The estimated flood damage includes the direct and ephemeral impact of flooding, and does not include the associated human health influences which are perennial and implicit.

The quality of flood water is generally overlooked, even in comprehensive flood risk assessments such as the European Union directive on the assessment and management of flood risks [6]. For example, in Section 4, Article 6, Chapter III of the EU directive 2007/60/EC [6] on the assessment and management of flood risk, the various aspects of flooding pertain to flood extent, water depth, and flow velocity, whereas there is no mention of flood water quality [6]. This is due to the fact that deterioration of surface water quality is considered acceptable as an exception during flooding [7]. Floods also endanger human health through wound infections, diarrheal illness and post-traumatic stress disorder in evacuees [8]. This paper aims to collate and analyze research papers which address the link between urban flooding, urban water quality and public health.

The paper presents an overview and a comparison of studies on public health, urban flooding and surface water quality from a global perspective, but with emphasis on China. The latter is relevant in the context of contemporary flood resilient city initiatives such as the "Sponge Cities" program, which aims to enhance infiltration, evapotranspiration, and the capture and reuse of storm water in Chinese cities [9]. The general objectives of the Sponge Cities concept are to 'restore' the city's capacity to absorb, infiltrate, store, purify, drain and manage rainwater and 'regulate' the water cycle as much as possible to mimic the natural hydrological cycle. This paper is divided into five sections: (i) background, where an overview of research on urban flooding, surface water quality and public health is presented; (ii) methodology section which elaborates on the literature review procedure; (iii) results section where the assessment outcomes are presented; (iv) discussion section were the results are discussed in relation to the contemporary urban context in China; and (v) conclusion.

2. Background: Connections between Urban Surface Water, Floods and Human Health

A wide spectrum of contaminants such as heavy metal (copper, zinc, lead, and chromium), halogenated aliphatics (gasoline), polycyclic aromatic hydrocarbons (chloroform, benzene, etc.), pesticides, phenol (chlordane, lindane, etc.), and associated macro pollutants (phosphorus, nitrogen, etc.) accumulate in the urban environment (air, land and water bodies) due to local emissions of these contaminants driven by urbanization (c.q. transportation and power production) and industrial development [10]. These contaminants are known to be associated with cancer, cardiovascular, gastrointestinal, kidney, liver and neurological diseases [11]. A significant fraction of these contaminants eventually accumulates in bodies of water, such as lakes, rivers and streams, in the event of runoff or flooding after precipitation [12]. An additional source of contamination of these water bodies are the outflows from sewage treatment plants leading to eutrophication [13].

Fluvial and pluvial flooding can impact the urban surface water quality. However, the effect of inundation on urban surface water quality is usually temporary. Depending upon the type and concentration of the contaminants of the receiving surface waters and flood water and their dilution rate, inundation may improve or deteriorate the water quality of the receiving surface waters [14]. Suspended Solids (SS), Biochemical Oxygen Demand (BOD), Chemical Oxygen Demand (COD) and Dissolved Organic Carbon (DOC) usually tend to increase during the flood phase and then return to pre-flood levels [15]. In Botic Creek, Prague, agglomeration, that is, heavy metal remobilization from sediments, during flooding has been observed, which has led to an increase in the heavy metal concentration in the surface waters. After the flood water receded, the concentration subsequently decreased to a level lower than the former level, probably because of depletion of the adsorbed fraction of heavy metals due to remobilization, followed by removal of these constituents via surface water flow. Though urban flooding may have negative impacts (as it may serve as a source of contamination), it also may have positive impacts on the surface water quality (as it may cause dilution and/or removal of contaminants) [16]. It follows from the above that floodwater may have an impact on the urban surface water quality. These impacts result in temporary changes in contaminant concentration of urban surface waters. Given the dynamic and temporary nature of flood water, continuous monitoring is required to better assess these impacts. This certainly holds for the long-term impacts.

There is a growing body of research directed to assessing the contaminant concentrations of storm water. For example, a nine-year long monitoring program of urban storm water runoff in Vilnius, Lithuania, has revealed that the mean values of suspended solids (SS) and total petroleum hydrocarbons (TPH) are most likely to exceed the permissible limits [17]. The results from an investigation of the Liangshui River in Beijing has indicated that the concentration of NH_4–N increased rapidly during an intense rainfall event [18]. An assessment of the water quality of runoffs from an urban road in in Guangzhou showed NH_4-N concentrations around 30 mg/L [19], which indicates that road runoff is a significant contributor to pollution of urban water bodies [18]. Contaminations associated with traffic, such as toxic metals Zn, Cr, Cu, Hg, Ni, Cd and Pb [20] and polycyclic aromatic hydrocarbons (PAHs), can lead to a deterioration of the water quality of urban rivers [21]. However, biological oxygen demand (BOD) and chemical oxygen demand (COD) from road runoffs does not seem to result in urban surface water concentrations which exceed the acceptable limits [16]. In this literature review no further research has been found which provides additional information on the correlation between the water quality of urban water bodies and runoff [15]. With respect to the sediment quality of urban water bodies, it should be noted here that high concentrations of heavy metals and PAHs concentrations have been found in sediments of urban small streams [14]. These high concentrations may adversely impact the surface water quality in the longer term. However, due to their complexity, our understanding of the underlying desorption and weathering processes causing mobilization of these contaminants is still very limited [14].

Natural disasters, namely floods, droughts and earthquakes often have a dramatic impact on human society, such as loss of lives, and economic and ecological losses [22]. In the past two decades, our understanding of these risks and how to manage them have increased significantly. However, very little is known about the impacts of these disasters on human health. This especially holds true for those associated with flooding. It is still common practice, to express the severity of a flood disaster by the number of deaths and injured [23]. The impacts of flooding on human health encompassing the physical and psychological effects, are excluded from these assessments as they are often remote and difficult to assess.

A review investigating the literature on flooding and human health impacts before 2012, indicated that there has been an increase in disease outbreaks, such as hepatitis E, gastrointestinal disease and leptospirosis, particularly in areas with poor hygiene and displaced populations [11]. For instance, in Bangladesh, with seventy percent of the population living in flood prone low-land regions, diarrhea is the major illness cause, which accounted for thirty-five percent of the 45.000 hospital admissions and twenty-seven percent of 154 reported deaths [23]. The most common pathogens causing waterborne outbreaks that have been reported during extreme water-related weather events, are *Vibrio* spp. (21.6%) and *Leptospira* spp. (12.7%) [23].

Also, Cann et al. (2013) [24] concluded that outbreaks of waterborne infectious disease do occur after an extreme water-related weather event in both developed and developing countries. A study on the quality of urban floodwater in Utrecht, the Netherlands, has shown that the floodwater quality is similar to that of sewage [25]. This is to be expected as some of the old urban areas have a combined sewer system which overflows during periods of excessive rainfall, resulting in a pollution outbreak. Similarly, a study on the quality of floodwater in Can Tho City, Vietnam, has shown that the pathogen and contaminant levels of the flood water are almost as high as that of the sewers [26]. From Figure 1 it can be seen that the Salmonella concentration in flood water are as high as in sewage and hence wading through flood water of this type is likely to affect health [26]. Hence flood water quality is a concern in both developed and developing countries.

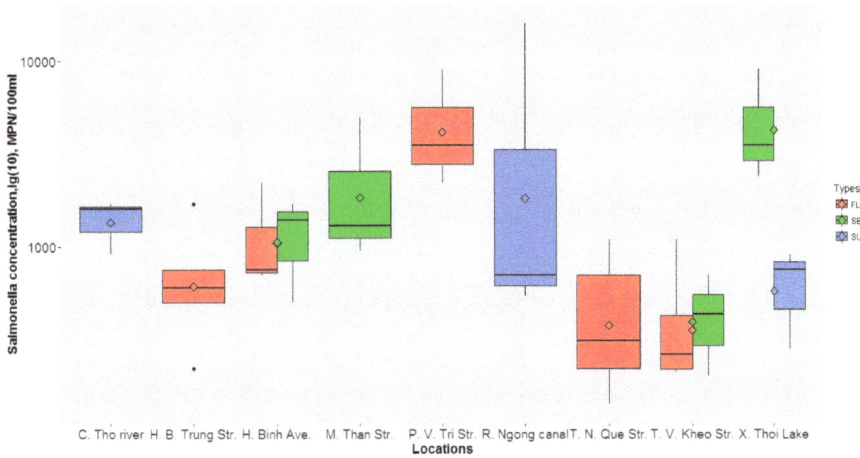

Figure 1. Salmonella concentration in flood water (FL), sewage (SE) and surface water (SU) bodies in Can Tho observed during October 2013 flood event [26].

3. Methodology/Approach

In order to review relevant published literature around the world, especially for China, databases comprising scientific literature on the crossroad of human health, urban flooding and urban water quality were selected and keyword searches were performed. The detailed research methodology is presented in Figure 2.

Figure 2. Review methodology.

Categorization: As the main focus of the paper is to assess flood water impacts on human health, the literature in the following broad categories were identified for further assessment: (i) urban water quality; and (ii) urban flood and human health.

Keyword identification: In order to narrow down the search, specific keywords were used to identify and filter the literature for a thorough assessment. The keywords pertaining to water quality are: "drinking water quality", "potable water quality", "urban water quality", "flood water quality", "urban water quality and health", "flood water quality and health" and "floodwater quality". The keywords pertaining to urban flood and human health are: "health risk", "flood", "urban", "water quality", "waterlogging and flood" and "urban stream (creek) water quality".

Keyword search: A keyword search was executed across a variety of scientific journals and international databases and Chinese libraries to ascertain the global context and the Chinese context, respectively, on public health during flooding. International scientific journal repositories, such as Water Research, Water Science and Technology, Environmental Monitoring and Assessment, Journal of Hydrology and Water, were accessed and searched using the keywords mentioned in Step 2 and the number of identified papers were further downsized using a detailed assessment. Also, Chinese libraries, such as Chinese Science Citation Database (CSCD) and Chinese Core Journal of Peking University [27] were also assessed and the number of papers were narrowed down using the keywords.

Literature analysis: The selected literature (following from the keywords search) between 1999 and 2017 was assessed in detail based on the content and geographical focus of the literature. In this manner, the number and geographical location of the studies reported in the literature on drinking/potable water quality; flood water quality; urban water quality; flood and health; and, urban water and health have been assessed.

Presenting and interpreting results: After assessing the number and location of these studies the papers were analyzed and collated.

This research used the following definitions for urban surface water, flooding, surface water quality, flood water quality, drinking water quality and urban water quality: (i) Urban surface water pertains to the water bodies in urban areas such as ponds, lakes, streams, canals, rivers, swales and wetlands; (ii) flooding is the undesirable consequence that pertains to the covering or submerging of normally dry land with a large amount of water; (iii) surface water quality pertains to the water quality of the surface water bodies in urban areas; (iv) flood water quality pertains to the water quality of flood water in urban areas; (v) drinking water quality pertains to the quality of drinking water in the drinking water distribution systems and in places where drinking water is consumed; and (vi) urban water quality pertains to the quality of drinking water, quality of surface bodies and quality of flood water in urban areas.

4. Results

The keywords used for the literature search and few papers based containing those keywords are presented in Table 1. The number of papers published in each year between the year 1997 and 2017, containing the keywords, are presented in Figure 3. It can be seen from Figure 3 that the number of publications containing the five keywords has increased over time.

Table 1. Select literature based on keyword search.

Keywords	Literature
Drinking/potable water quality	[28–37]
Flood water quality	[38–47]
Urban water quality	[48–57]
Flood water & health	[58–67]
Urban water & health	[68–77]

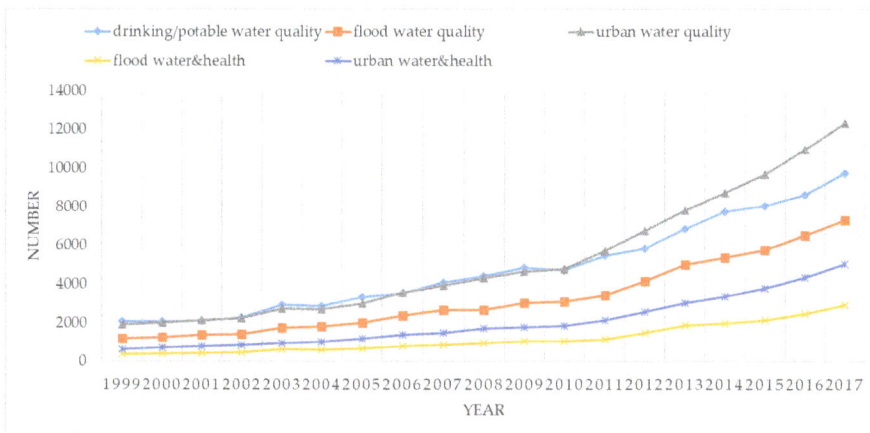

Figure 3. Number of papers published in each year on urban surface water, human health and flooding based on keyword search.

From Figure 3 it can be seen that there is no significant increase in publications related to drinking/potable water quality and urban water quality between 1999 and 2010 with the number of publications containing these two keywords increasing from 2000 to 4000 in this period. Figure 3 also shows that there has been an increase in emphasis on urban water quality related research after 2011 as the number of publications attributed to urban water quality is steadily increasing. The latter can be attributed to the developments in technology and/or a (global) rising concern about water security [78,79]. Also, the number of studies on flood water quality are less than those on urban water quality. This may be due to the following practical reasons: (i) floods are infrequent and short lasting, which complicate the design and management of a flood water quality monitoring system; and (ii) flood events pose a threat to human life, hence it can be a risky undertaking to collect flood water samples in order to assess the flood water quality in real time [26]. As a result, the number of studies comprising both water quality of urban water and health and water quality of flood and human health are much less compared to the number of studies on the water quality of drinking water, flood water and urban surface water.

Research publications from five journals (Water Research, Water Science and Technology, Environmental Monitoring and Assessment, Journal of Hydrology and Water) in the field of human health impact were retrieved, which include direct exposure to flood water (Table 2). Three papers from Water Research and one paper from the Journal of Hydrology have addressed the human risk due to direct exposure to floods [25,58,80,81]. Most papers in these journals concern the health risk from potable water or agricultural water [82,83], which are due to direct and indirect consumption of water and cutaneous absorption. A paper on the analysis of floodwater quality in Utrecht, the Netherlands, indicates that fecal indicator organism concentrations, such as Campylobacter, Cryptosporidium and Giardia, are similar to those found in raw sewage under high-flow conditions [25]. In the Netherlands the risk for children of infection when exposed to flooding originating from combined sewers, storm sewers and rainfall generated surface runoff are 33%, 23% and 3.5% respectively; whereas the risk of infection for adults are 3.9%, 0.58% and 0.039% respectively [58]. A study on water quality from a water plaza in Rotterdam, the Netherlands, shows that the Campylobacter infection risks for children playing in a water plaza are higher than the annual average figure for the general population through all exposure pathways [80]. Mosquito breeding in rain water harvesting systems has been reported as a significant public health threat in Melbourne, Australia [84]. Mosquito breeding in storm water storage systems is also a public health threat in China [85]. All these health risks are likely to increase with an increase in frequency and severity of urban flooding due to an increase of intense rainfall events and temperature driven by climate change [86].

Table 2. Research papers connecting flooding and health risk in select journals.

Journal	Type		
	Health Risk	Flood & Health Risk	Direct Exposure to Flood Water
Water Research	2110	212	3
Water Science and Technology	398	51	0
Environmental Monitoring and Assessment	5706	747	0
Journal of Hydrology	406	224	1
Water	32	3	0

Numerous research papers on floodwater quality in China, with the keywords "urban", "water quality", "urban runoff water quality", "waterlogging" and "flood" were retrieved using the data from CSCD and the Chinese core Journal of Peking University [28]. Three studies were found on urban floodwater quality. Two studies pertain to mosquito breeding in Shanghai and Ningbo and one study is about runoff quality in Xi'an [85,87,88]. The geographical distribution of these studies is presented in Figure 4. There have been 32 studies on urban runoff water quality. None of these pertain to urban flooding situations [89–120]. Although most of these studies focus on cities in the south-eastern part of China, there has been four studies involving cities in other regions of the country, like Lanzhou [111], Baoji [115], Chongqing [109] and Urumqi [117]. The urban water quality monitoring reported by these publications are relevant to the major parameters, SS, Nitrogen (N), Phosphate (P) and COD/BOD, which are associated with environmental problems and not health risk [105–107,110]. Two hundred and two research papers were found using the keywords, "urban", "river (stream, creek)" and "water quality" of which 145 papers included urban water quality. China (Figure 5).

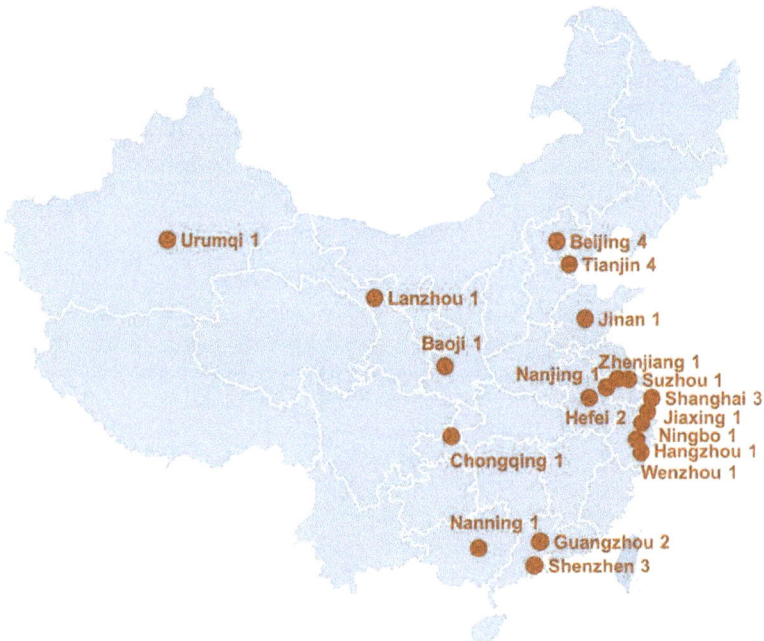

Figure 4. Distribution of urban runoff quality studies in China. The numbers after the city names indicate the number of runoff quality research papers based on the city.

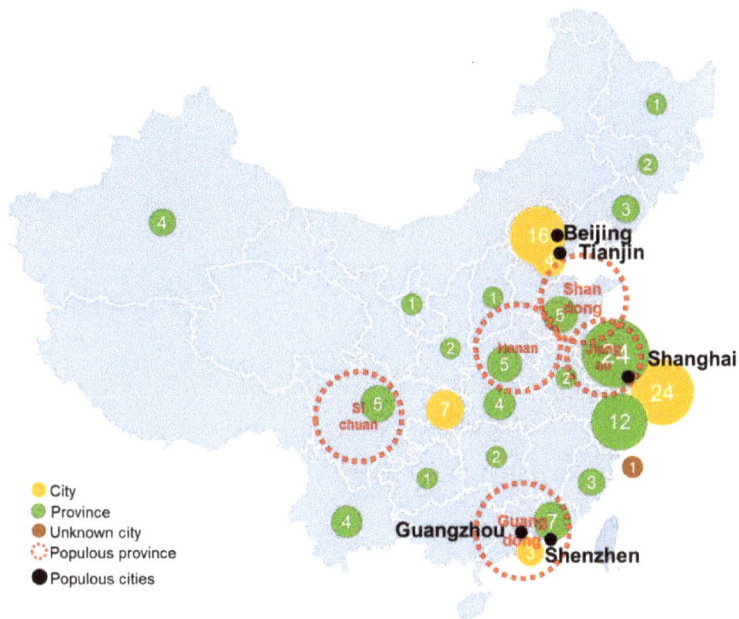

Figure 5. The distribution of urban water quality and population.

From the assessment, it can be seen that the study locations are unevenly distributed, with most of the studies in the densely populated eastern part of China (Figure 5). Most of the studies have a focus on the water quality problems of Shanghai and Beijing, and predominantly include physical and chemical parameters, such as N, P, DO and COD/BOD (Figure 6). These macro-constituents have relevance to environmental health rather than health risk. There are a few studies that use other parameters, i.e., environmental hormones [121], neuroactive substances [122], viruses and Escherichia coli [123].

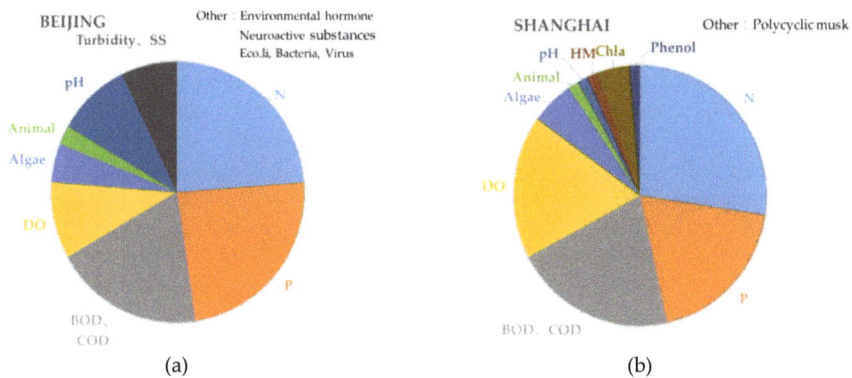

Figure 6. *Cont.*

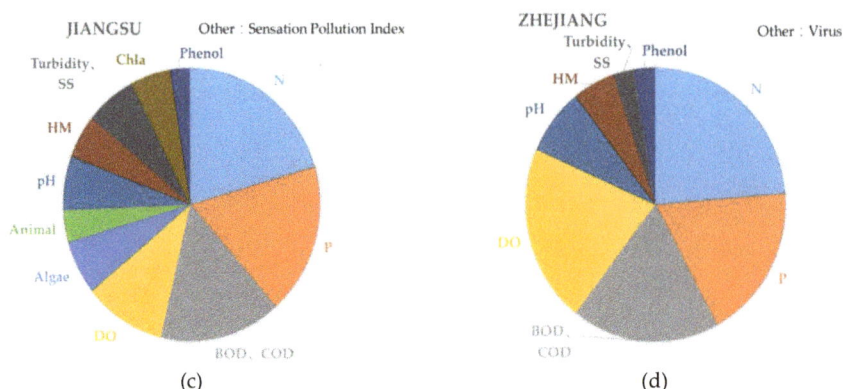

Figure 6. Water quality parameters assessed in four major cities and provinces of China. (**a**) Beijing, (**b**) Shanghai, (**c**) Jiangsu, (**d**)Zhejiang.

Although the most populated provinces are Guangdong, Henan and Shandong, the two provinces where water studies are predominant are Jiangsu and Zhejiang (Figure 6). Though these provinces are relatively less urbanized, they are highly industrialized. This might be the reason for more water quality studies being done in these two provinces.

5. Discussion

Floods are omnipresent but infrequent and short lasting, which may explain the practical constraints to timely collection of samples. This explains in part the lack of flood water quality data. From a global perspective, flood water quality assessments have been carried out only in few countries (Figure 7). In Bangladesh, Vietnam, Sudan, Nigeria, Thailand and Indonesia, which are prone to flooding, the number of studies on floodwater quality is less than in developed countries such as United Kingdom, Australia and the Netherlands [25,38–43,61]. There are studies in Dhaka, Bangladesh which model health risk from flooding exemplified by cholera, based on the mixing of pollutants in flood water using a deterministic hydraulic model and data on human vulnerability using dose response functions [124]. Initiatives such as Preparing for Extreme and Rare events in coastal regions (PEARL) is developing holistic flood risk assessment for coastal communities that also include health risks due to flooding [125]. Also, from Figure 7 it can be seen that there is no flood water quality data publicly available in China. Availability of funding or enforcement of strict water quality regulations may explain why more studies have been conducted in developed countries [126,127]. Whereas a lack of resources, delegation of resources to other needs such as evacuation or emergency management and weak enforcement of water quality monitoring guidelines can be attributed to the observed lack of water quality assessment studies and water quality monitoring during flood events in developing countries [26]. The absence of water quality studies in the western part of China does not mean that there are no water quality problems in this region of China [128,129]. Flooding is a problem in China [130,131] and the lack of flood water quality studies in China is glaring (Figure 7).

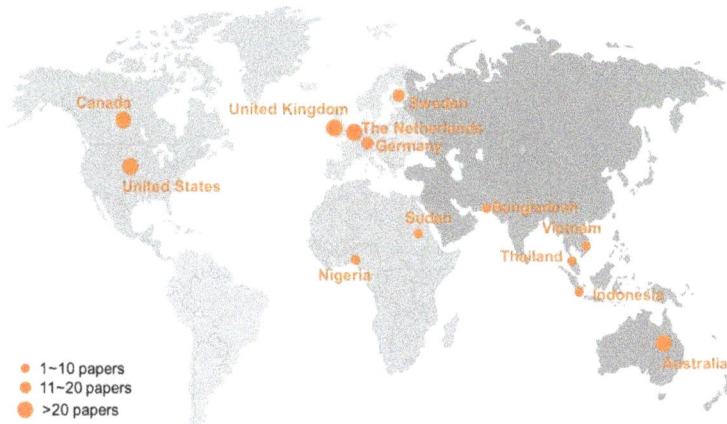

Figure 7. Geographical distribution of flood water quality assessments in the world.

As the central and western parts of China are being industrialized, urban environmental problems might increase in the near future and are likely to be exacerbated by climate change. Hence, it is likely that more research and initiatives will be dedicated to urban water quality in these regions of China in the near future. Urban water quality monitoring assessment and public health assessment can be linked to the Sponge City initiative as this program aims to guide cities to follow a sustainable pathway of urban development and transformation where water quality improvement is an essential component [132]. In optima forma, the water quality component also captures human health aspects during flooding. The cities selected in 2015 and 2016 as pilot Sponge Cities are presented in Figure 8 [132]. Comparison of Figures 6 and 8 reveals that Sponge City pilot studies on urban water quality are concentrated in the Eastern region. These studies are likely to provide a wealth of data which can feed into a knowledge base on urban water quality in the coming years.

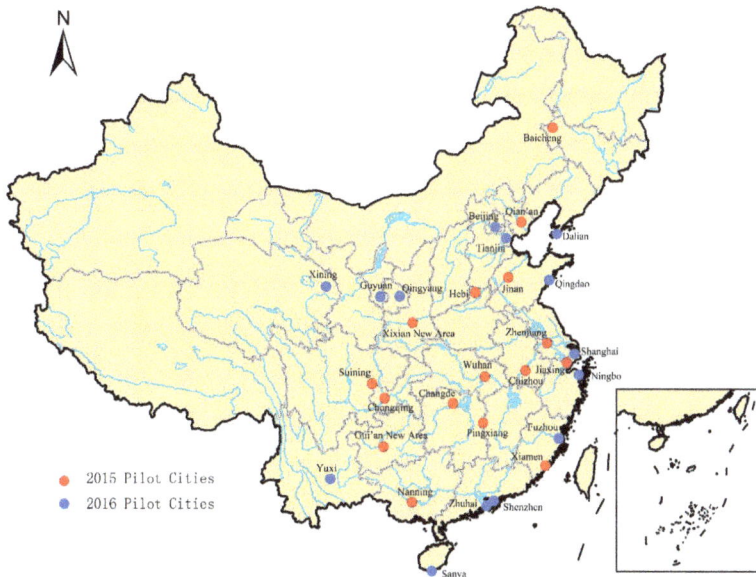

Figure 8. Location of Sponge City pilots in China (Source: Li et al. (2017)) [132].

Another explanation for the current knowledge gap is that there is still little attention to the health impacts associated with floods because these impacts are generally remote and difficult to assess. Monitoring flood water quality in real time is a labor-intensive process which needs elaborate planning arrangements for anticipating flooded areas, collecting samples, transporting them to labs and ensuring the safety of the personnel involved [26]. Some of the provincial capitals in China such as Nanjing, Chongqing, Beijing, Guangzhou, Shanghai and Hangzhou have started extensive water quality monitoring programs [133–138]. Real-time water quality data from 148 locations across China are now available [139].

Recent advances in in situ water testing technology (such as smartphone-based sensors) in conjunction with online data platforms allowing for rapid data analysis, visualisation and decision-making, will further accelerate the deployment of real-time water quality monitoring systems during flood events in Sponge Cities [140]. These developments will ultimately contribute to a rapid increase in our understanding of the complex inter-relationship between water quality, flooding and public health in pilot Sponge Cities. In addition to real-time flood water quality monitoring, flood water quality can be simulated using models such as EPA Storm Water Management Model (SWMM) [141] and Delft-FEWS [142] open data handling platform, which can be used to forecast water quality. The water quality models validated and calibrated using real-time water quality systems can be used in the context of the Sponge Cities. This can be futher extended to assess the microbal risk to human health at a particular location or throughout the city based on methods such as Quantitative Microbial Risk Assessment (QMRA) [143,144] based on hazard, dose-response, exposure, and risk characterization [145].

6. Conclusions

This paper has assessed the scientific literature on the inter-relationship between surface water quality, flood water quality and human health in urban areas around the world, and China in particular. Even though flooding, surface water and human health in urban areas are interconnected, this assessment reveals that the importance of flood risk and health impacts have generally been ignored and underestimated. There is a growing body of evidence that the health impacts due to flooding are becoming more important as more people are being exposed to floodwater every year. From our assessment, it is clear that (i) most of the studies are focused on either water quality or health; (ii) the number of cross-cutting studies on flood water quality and health are limited, but growing; and (iii) the effects of flood water quality on health impacts in urban areas are still largely unknown, but are receiving more attention in assessments of health risks in cities. However, based on this study, it can be assumed that the health impacts in urban areas with high frequency low impact (direct damage) flood events are more severe than those with low frequency high impact events. The first type of event is often ignored in flood risk assessments. This study also reveals that water quality assessment studies in China have focused on the highly populated south-eastern cities, albeit with particular attention to the water quality parameters which have relevance to environmental assessments. These parameters have no or limited value to health impact assessments. Expanding the current water quality monitoring activities with the flood water quality parameters required to assess human health impacts in China is needed and this is in alignment with the ambition of the Sponge Cities program. Hence, flood water quality monitoring should be an integral part of the Sponge City program. Though collection and sharing of flood water quality data are a challenge, the Sponge Cities program provides an opportunity to make a leap forward in the application, testing, rolling out and upscaling of innovative flood water quality monitoring technology and health impact analyses throughout China. Better appreciation of flood induced health risks will lead to better flood risk management.

Acknowledgments: This work was supported by Southeast University (SEU), Nanjing, China; and IHE Delft Institute for Water Education, The Netherlands.

Author Contributions: The paper is based on the results obtained from the research of Yuhan Rui, which was done for the partial fulfilment of requirements for the Master of Science degree at the Southeast University under the supervision of Dafang Fu and Chris Zevenbergen. The first draft was written by Yuhan Rui, Ha Do Minh and Mohanasundar Radhakrishnan. Dafang Fu, Chris Zevenbergen and Assela Pathirana helped to review and improve the paper.

Conflicts of Interest: The authors declare no conflict of interest.

References

1. Konrad, C.P. Effects of Urban Development on Floods. Available online: https://pubs.usgs.gov/fs/fs07603/ (acccessed on 21 February 2018).
2. Zhang, C.-L.; Chen, F.; Miao, S.G.; Li, Q.C.; Xuan, C.Y. Influences of urbanization on precipitation and water resources in the metropolitan Beijing area. In Proceedings of the 21st American Metrological Society Conference on Hydrology, San Antonio, TX, USA, 17 January 2007.
3. Huong, H.T.L.; Pathirana, A. Urbanization and climate change impacts on future urban flood risk in Can Tho city, Vietnam. *Hydrol. Earth Syst. Sci.* **2013**, *17*, 379–394. [CrossRef]
4. Chen, Y.; Zhou, H.; Zhang, H.; Du, G.; Zhou, J. Urban flood risk warning under rapid urbanization. *Environ. Res.* **2015**, *139*, 3–10. [CrossRef] [PubMed]
5. Olsen, A.S.; Zhou, Q.; Linde, J.J.; Arnbjergnielsen, K. Comparing Methods of Calculating Expected Annual Damage in Urban Pluvial Flood Risk Assessments. *Water* **2015**, *7*, 255–270. [CrossRef]
6. European Parliament; Council of the European Union. Directive 2007/60/EC of the European Parliament and of the Council of 23 October 2007 on the Assessment and Management of Flood Risks (Text with EEA Relevance). Available online: http://data.europa.eu/eli/dir/2007/60/oj (accessed on 21 February 2018).
7. European Parliament; Council of the European Union. Directive 2000/60/EC of the European Parliament and of the Council of 23 October 2000 Establishing a Framework for Community Action in the Field of Water Policy. Available online: http://eur-lex.europa.eu/legal-content/EN/TXT/?uri=CELEX:32000L0060 (accessed on 21 February 2018).
8. Horton, R. Extreme rain, flooding, and health. *Lancet* **2017**, *390*, 1005.
9. Liu, D. Water supply: China's sponge cities to soak up rainwater. *Nature* **2016**, *537*, 307. [CrossRef] [PubMed]
10. Novotny, V. *Water Quality: Diffuse Pollution and Watershed Management*; John Wiley & Sons: New York, NY, USA, 2002.
11. Alderman, K.; Turner, L.R.; Tong, S. Floods and human health: A systematic review. *Environ. Int.* **2012**, *47*, 37–47. [CrossRef] [PubMed]
12. Chiew, F.H.S.; Mcmahon, T.A. Modelling runoff and diffuse pollution loads in urban areas. *Water Sci. Technol.* **1999**, *39*, 241–248.
13. Wang, A.L.; Sun, X.; Chen, Q.K.; Yang, L.Y. Effect of ammonia in the tailwater from wastewater treatment plant on the growth of myriophyllum spicatum. *Chin. J. Ecol.* **2015**, *34*, 1367–1372.
14. Nabelkova, J.; Kominkova, D.; Jirak, J. The impact of highway runoff on the chemical status of small urban streams. *Urban Environ.* **2012**, *19*, 297–306.
15. Hayashi, H.; Tasaki, M.; Uchiyama, N.; Morita, M. Water quality and pollution load during flood and non-flood periods in an urban tidal river. *J. Hydraul. Eng.* **2013**, *69*, I_1723–I_1728. [CrossRef]
16. Nábelková, J.; Stastná, G.; Komínková, D. Flood impact on water quality of small urban streams. *Water Sci. Technol.* **2005**, *52*, 267–274. [PubMed]
17. Karlavičienė, V.; Švedienė, S.; Marčiulionienė, D.E.; Randerson, P.; Rimeika, M.; Hogland, W. The impact of storm water runoff on a small urban stream. *J. Soils Sediment.* **2008**, *9*, 6–12. [CrossRef]
18. Yan, L.; Bi, J.L.; Wang, L.S.; Tang, W.Z.; Shan, B.Q.; Yang, L.; Chen, J. Effect of storm runoff on the water quality of urban rivers with unconventional water sources. *Acta Sci. Circumst.* **2015**, *35*, 443–448.
19. Gan, H.Y.; Zhuo, M.N.; Li, D.Q.; Zhou, Y.Z. The water quality characteristics of runoff on urban road in Guangzhou. *Ecol. Environ. Sci.* **2006**, *15*, 969–973.
20. Floresrodrîguez, J.; Bussy, A.L.; Thevenot, D. Toxic metals in urban runoff: Physico-chemical mobility assessment using speciation schemes. *Water Sci. Technol.* **1994**, *29*, 83–93.
21. Zehetner, F.; Rosenfellner, U.; Mentler, A.; Gerzabek, M.H. Distribution of Road Salt Residues, Heavy Metals and Polycyclic Aromatic Hydrocarbons across a Highway-Forest Interface. *Water Air Soil Pollut.* **2009**, *198*, 125–132. [CrossRef]

22. Tai, K.C. Flood plain management models for economic, environmental and ecological impact analysis. In Proceedings of the Conference on Ecological Modelling, Copenhagen, Denmark, 28 August–2 September 1978.
23. Cash, R.A.; Halder, S.R.; Husain, M.; Islam, M.S.; Mallick, F.H.; May, M.A.; Rahman, M.; Rahman, M.A. Reducing the health effect of natural hazards in Bangladesh. *Lancet* **2013**, *382*, 2094–2103. [CrossRef]
24. Cann, K.F.; Thomas, D.R.; Salmon, R.L.; Wyn-Jones, A.P.; Kay, D. Extreme water-related weather events and waterborne disease. *Epidemiol. Infect* **2013**, *141*, 671–686. [CrossRef] [PubMed]
25. ten Veldhuis, J.A.E.; Clemens, F.H.L.R.; Sterk, G.; Berends, B.R. Microbial risks associated with exposure to pathogens in contaminated urban flood water. *Water Res.* **2010**, *44*, 2910–2918. [CrossRef] [PubMed]
26. Nguyen, H.Q.; Radhakrishnan, M.; Huynh, T.T.N.; Baino-Salingay, M.L.; Ho, L.P.; Steen, P.V.D.; Pathirana, A. Water Quality Dynamics of Urban Water Bodies during Flooding in Can. Tho City, Vietnam. *Water* **2017**, *9*, 260. [CrossRef]
27. Zhu, J.; Cai, R.H.; He, J. *A Guide to the Chinese Core Periodical*; Peking University Press: Beijing, China, 2011.
28. Ochoo, B.; Valcour, J.; Sarkar, A. Association between perceptions of public drinking water quality and actual drinking water quality: A community-based exploratory study in newfoundland (Canada). *Environ. Res.* **2017**, *159*, 435–443. [CrossRef] [PubMed]
29. Heibati, M.; Stedmon, C.A.; Stenroth, K.; Rauch, S.; Toljander, J.; Säve-Söderbergh, M.; Murphy, K.R. Assessment of drinking water quality at the tap using fluorescence spectroscopy. *Water Res.* **2017**, *125*, 1–10. [CrossRef] [PubMed]
30. Abera, B.; Bezabih, B.; Hailu, D. Microbial quality of community drinking water supplies: A ten year (2004–2014) analyses in west amhara, ethiopia. *Sustain. Water Qual. Ecol.* **2017**, *9–10*, 22–26. [CrossRef]
31. Bridgeman, J.; Baker, A.; Brown, D.; Boxall, J.B. Portable led fluorescence instrumentation for the rapid assessment of potable water quality. *Sci. Total Environ.* **2015**, *524–525*, 338–346. [CrossRef] [PubMed]
32. Emelko, M.B.; Silins, U.; Bladon, K.D.; Stone, M. Implications of land disturbance on drinking water treatability in a changing climate: Demonstrating the need for "source water supply and protection" strategies. *Water Res.* **2011**, *45*, 461–472. [CrossRef] [PubMed]
33. Storey, M.V.; Van, d.G.B.; Burns, B.P. Advances in on-line drinking water quality monitoring and early warning systems. *Water Res.* **2011**, *45*, 741–747. [CrossRef] [PubMed]
34. Smith, H.G.; Sheridan, G.J.; Lane, P.N.J.; Nyman, P.; Haydon, S. Wildfire effects on water quality in forest catchments: A review with implications for water supply. *J. Hydrol.* **2011**, *396*, 170–192. [CrossRef]
35. Mkandawire, T.; Banda, E. Assessment of drinking water quality of mtopwa village in bangwe township, blantyre. *Desalination* **2009**, *248*, 557–561. [CrossRef]
36. Cidu, R.; Frau, F.; Tore, P. Drinking water quality: Comparing inorganic components in bottled water and italian tap water. *J. Food Compos. Anal.* **2011**, *24*, 184–193. [CrossRef]
37. Macova, M.; Toze, S.; Hodgers, L.; Mueller, J.F.; Bartkow, M.; Escher, B.I. Bioanalytical tools for the evaluation of organic micropollutants during sewage treatment, water recycling and drinking water generation. *Water Res.* **2011**, *45*, 4238–4247. [CrossRef] [PubMed]
38. Trinh, H.T.; Marcussen, H.; Hansen, H.C.; Le, G.T.; Duong, H.T.; Ta, N.T.; Nguyen, T.Q.; Hansen, S.; Strobel, B.W. Screening of inorganic and organic contaminants in floodwater in paddy fields of hue and thanh hoa in vietnam. *Environ. Sci. Pollut. Res. Int.* **2017**, *24*, 7348–7358. [CrossRef] [PubMed]
39. Nwonumara, N.; Okogwu, O. The impact of flooding on water quality, zooplankton composition, density and biomass in lake iyieke, cross river-floodplain, southeastern nigeria. *Acta Zool. Litu.* **2013**, *23*, 138–146. [CrossRef]
40. Mahmood, M.I.; Elagib, N.A.; Horn, F.; Sag, S. Lessons learned from khartoum flash flood impacts: An integrated assessment. *Sci. Total Environ.* **2017**, *601–602*, 1031–1045. [CrossRef] [PubMed]
41. Costa, D.; Burlando, P.; Priadi, C. The importance of integrated solutions to flooding and water quality problems in the tropical megacity of jakarta. *Sustain. Cities Soc.* **2016**, *20*, 199–209. [CrossRef]
42. Miller, J.D.; Hutchins, M. The impacts of urbanisation and climate change on urban flooding and urban water quality: A review of the evidence concerning the United Kingdom. *J. Hydrol. Reg. Stud.* **2017**, *12*, 345–362. [CrossRef]
43. Wallace, J.; Stewart, L.; Hawdon, A.; Keen, R.; Karim, F.; Kemei, J. Flood water quality and marine sediment and nutrient loads from the tully and murray catchments in north queensland, australia. *Mar. Freshw. Res.* **2009**, *60*, 1123–1131. [CrossRef]

44. Mcmillan, M.D.; Rahnema, H.; Romiluy, J.; Kitty, F.J. Effect of exposure time and crude oil composition on low-salinity water flooding. *Fuel* **2016**, *185*, 263–272. [CrossRef]
45. Schuch, G.; Serrao-Neumann, S.; Morgan, E.; Choy, D.L. Water in the city: Green open spaces, land use planning and flood management—An australian case study. *Land Use Policy* **2017**, *63*, 539–550. [CrossRef]
46. Lintern, A.; Leahy, P.J.; Heijnis, H.; Zawadzki, A.; Gadd, P.; Jacobsen, G.; Deletic, A.; Mccarthyad, D.T. Identifying heavy metal levels in historical flood water deposits using sediment cores. *Water Res.* **2016**, *105*, 34–46. [CrossRef] [PubMed]
47. Dortch, M.S.; Zakikhani, M.; Kim, S.C.; Steevens, J.A. Modeling water and sediment contamination of lake pontchartrain following pump-out of hurricane katrina floodwater. *J. Environ. Manag.* **2008**, *87*, 429–442. [CrossRef] [PubMed]
48. Wijesiri, B.; Deilami, K.; Mcgree, J.; Goonetilleke, A. Use of surrogate indicators for the evaluation of potential health risks due to poor urban water quality: A bayesian network approach. *Environ. Pollut.* **2017**, *233*, 655. [CrossRef] [PubMed]
49. Constantine, K.; Massoud, M.; Alameddine, I.; El-Fadel, M. The role of the water tankers market in water stressed semi-arid urban areas: Implications on water quality and economic burden. *J. Environ. Manag.* **2017**, *188*, 85–94. [CrossRef] [PubMed]
50. Feng, T.; Wang, C.; Hou, J.; Wang, P.; Liu, Y.; Dai, Q.; Yang, Y.Y.; You, G.X. Effect of inter-basin water transfer on water quality in an urban lake: A combined water quality index algorithm and biophysical modelling approach. *Ecol. Indic.* **2017**, in press. [CrossRef]
51. Nazemi, A.; Madani, K. Urban water security: Emerging discussion and remaining challenges. *Sustain. Cities Soc.* **2017**, in press. [CrossRef]
52. Minomo, K.; Ohtsuka, N.; Nojiri, K.; Matsumoto, R. Influence of combustion-originated dioxins in atmospheric deposition on water quality of an urban river in japan. *J. Environ. Sci.* **2018**, *64*, 245–251. [CrossRef]
53. Xue, Z.H.; Yin, H.L.; Xie, M. Development of integrated catchment and water quality model for urban rivers. *Chin. J. Hydrodyn. B* **2015**, *27*, 593–603. [CrossRef]
54. Neto, S. Water governance in an urban age. *Util. Policy* **2016**, *43*, 32–41. [CrossRef]
55. Hur, S.; Nam, K.; Kim, J.; Kwak, C. Development of urban runoff model ffc-qual for first-flush water-quality analysis in urban drainage basins. *J. Environ. Manag.* **2017**, *205*, 73–84. [CrossRef] [PubMed]
56. Nosrati, K. Identification of a water quality indicator for urban roof runoff. *Sustain. Water Qual. Ecol.* **2017**, *9–10*, 78–87. [CrossRef]
57. Scott, A.B.; Frost, P.C. Monitoring water quality in toronto's urban stormwater ponds: Assessing participation rates and data quality of water sampling by citizen scientists in the freshwater watch. *Sci. Total Environ.* **2017**, *592*, 738–744. [CrossRef] [PubMed]
58. De Man, H.; van den Berg, H.H.J.L.; Leenen, E.J.T.M.; Schijven, J.F.; Schets, F.M.; van der Vliet, J.C.; van Knapen, F.; de Roda Husman, A.M. Quantitative assessment of infection risk from exposure to waterborne pathogens in urban floodwater. *Water Res.* **2014**, *48*, 90–99. [CrossRef] [PubMed]
59. Vardoulakis, S.; Dimitroulopoulou, C.; Thornes, J.; Lai, K.M.; Taylor, J.; Myers, I.; Heaviside, C.; Mavrogianni, A.; Shrubsole, C.; Chalabi, Z.; et al. Wilkinson, P. Impact of climate change on the domestic indoor environment and associated health risks in the UK. *Environ. Int.* **2015**, *85* (Suppl. 1), 299–313. [CrossRef] [PubMed]
60. Picou, J.S.; Nicholls, K.; Guski, R. Environmental stress and health. *Int. Encycl. Soc. Behav. Sci.* **2015**, *12*, 804–808.
61. Chaturongkasumrit, Y.; Techaruvichit, P.; Takahashi, H.; Kimura, B.; Keeratipibul, S. Microbiological evaluation of water during the 2011 flood crisis in thailand. *Sci. Total Environ.* **2013**, *463–464*, 959–967. [CrossRef] [PubMed]
62. Harris, R.F. Floodwater challenges. *Curr. Biol.* **2005**, *15*, 815–816. [CrossRef]
63. Ashley, N.A.; Valsaraj, K.T.; Thibodeaux, L.J. Elevated in-home sediment contaminant concentrations—The consequence of a particle settling-winnowing process from hurricane katrina floodwaters. *Chemosphere* **2008**, *70*, 833–840. [CrossRef] [PubMed]
64. Grigg, B.C.; Beyrouty, C.A.; Norman, R.J.; Gbur, E.E.; Hanson, M.G.; Wells, B.R. Rice responses to changes in floodwater and n timing in southern usa. *Field Crops Res.* **2000**, *66*, 73–79. [CrossRef]

65. Modlmaier, M.; Kuhn, R.; Kaaden, O.R.; Pfeffer, M. Transmission studies of a european sindbis virus in the floodwater mosquito aedes vexans (diptera: Culicidae). *Int. J. Med. Microbiol.* **2002**, *291* (Suppl. 33), 164. [CrossRef]
66. Abbas, H.B.; Routray, J.K. Assessing factors affecting flood-induced public health risks in kassala state of sudan. *Oper. Res. Health Care* **2014**, *3*, 215–225. [CrossRef]
67. Waroux, O.L.P.D. Floods as Human Health Risks. In *Encyclopedia of Environmental Health*; Elsevier Science: Amsterdam, The Netherlands, 2011.
68. Cui, Q.; Fang, T.; Huang, Y.; Dong, P.; Wang, H. Evaluation of bacterial pathogen diversity, abundance and health risks in urban recreational water by amplicon next-generation sequencing and quantitative pcr. *J. Environ. Sci.* **2017**, *57*, 137–149. [CrossRef] [PubMed]
69. Crocker, W.; Maute, K.; Webb, C.; French, K. Mosquito assemblages associated with urban water bodies; implications for pest and public health threats. *Landsc. Urban Plan.* **2017**, *162*, 115–125. [CrossRef]
70. Renouf, M.A.; Serrao-Neumann, S.; Kenway, S.J.; Morgan, E.A.; Low, C.D. Urban water metabolism indicators derived from a water mass balance—Bridging the gap between visions and performance assessment of urban water resource management. *Water Res.* **2017**, *122*, 669–677. [CrossRef] [PubMed]
71. Nel, J.L.; Maitre, D.C.L.; Roux, D.J.; Colvin, C.; Smith, J.S.; Smith-Adao, L.B.; Maherry, A.; Sitas, N. Strategic water source areas for urban water security: Making the connection between protecting ecosystems and benefiting from their services. *Ecosyst. Serv.* **2017**, *28*, 251–259. [CrossRef]
72. Fang, T.; Cui, Q.; Huang, Y.; Dong, P.; Wang, H.; Liu, W.T.; Ye, Q. Distribution comparison and risk assessment of free-floating and particle-attached bacterial pathogens in urban recreational water: Implications for water quality management. *Sci. Total Environ.* **2017**, *613–614*, 428–438. [CrossRef] [PubMed]
73. Fuhrimann, S.; Pham-Duc, P.; Cissé, G.; Tram, N.T.; Ha, H.T.; Dung, D.T.; Ngoc, P.; Nguyen, H.; Anh Vuong, T.; Utzinger, J.; Schindler, C.; Winkler, M.S. Microbial contamination along the main open wastewater and storm water channel of hanoi, vietnam, and potential health risks for urban farmers. *Sci. Total Environ.* **2016**, *566–567*, 1014–1022. [CrossRef] [PubMed]
74. Wang, T.; Xu, Z.; Li, Y.; Liang, M.; Wang, Z.; Hynds, P. Biofilm growth kinetics and nutrient (n/p) adsorption in an urban lake using reclaimed water: A quantitative baseline for ecological health assessment. *Ecol. Indic.* **2016**, *71*, 598–607. [CrossRef]
75. Völker, S.; Kistemann, T. Developing the urban blue: Comparative health responses to blue and green urban open spaces in germany. *Health Place* **2015**, *35*, 196–205. [CrossRef] [PubMed]
76. Rana, M.D.S. Status of water use sanitation and hygienic condition of urban slums: A study on rupsha ferighat slum, khulna. *Desalination* **2009**, *246*, 322–328. [CrossRef]
77. Chanan, A.; Kandasamy, J.; Vigneswaran, S.; Sharma, D. A gradualist approach to address australia's urban water challenge. *Desalination* **2009**, *249*, 1012–1016. [CrossRef]
78. Chong, M.N.; Jin, B.; Chow, C.W.; Saint, C. Recent developments in photocatalytic water treatment technology: A review. *Water Res.* **2010**, *44*, 2997–3027. [CrossRef] [PubMed]
79. Ellis, J.B.; Marsalek, J.; Chocat, B. *Urban Water Quality*; John Wiley & Sons, Ltd.: New York, NY, USA, 2006; pp. 285–292.
80. Sales-Ortells, H.; Medema, G. Microbial health risks associated with exposure to stormwater in a water plaza. *Water Res.* **2015**, *74*, 34–46. [CrossRef] [PubMed]
81. Ruin, I.; Creutin, J.-D.; Anquetin, S.; Lutoff, C. Human exposure to flash floods–Relation between flood parameters and human vulnerability during a storm of September 2002 in Southern France. *J. Hydrol.* **2008**, *361*, 199–213. [CrossRef]
82. Phan, K.; Sthiannopkao, S.; Kim, K.-W.; Wong, M.H.; Sao, V.; Hashim, J.H.; Yasin, M.S.M.; Aljunid, S.M. Health risk assessment of inorganic arsenic intake of Cambodia residents through groundwater drinking pathway. *Water Res.* **2010**, *44*, 5777–5788. [CrossRef] [PubMed]
83. Murray, A.; Ray, I. Wastewater for agriculture: A reuse-oriented planning model and its application in peri-urban China. *Water Res.* **2010**, *44*, 1667–1679. [CrossRef] [PubMed]
84. Moglia, M.; Gan, K.; Delbridge, N. Exploring methods to minimize the risk of mosquitoes in rainwater harvesting systems. *J. Hydrol.* **2016**, *543*, 324–329. [CrossRef]
85. Yang, S.-J.; Ma, X.; Zhu, G.-F.; Xu, M.; Chen, X.-Y.; Wang, G.-A.; Shi, B.-J. Analysis of the breeding sources of overwintering Aedes albopictus and influencing factors in Ningbo city, 2016. *Chin. J. Vector Biol. Control* **2017**, *28*, 69–71.

86. Intergovernmental Panel on Climate Change (IPCC). Climate Change 2014: Synthesis Report. Available online: http://www.ipcc.ch/report/ar5/syr/ (accessed on 21 February 2018).

87. Xu, Y.X.; Xu, R.Q.; Wang, S.Z. Investigation of the breeding place of mosquito larva in outdoor environmental seeper. *Shanghai J. Prev. Med.* **2004**, *16*, 153–156.

88. Pei, Q.-B.; Liu, W.-J.; Zhang, J.-F.; Liu, Q.; Zhao, X.-Y.; Yu, Q.-F. Characterization of Urban Roadway Runoff in Xi'an City. *Water Sav. Irrig.* **2013**, *4*, 46–49.

89. Yang, B.; Huang, Y.Q.; Wu, T.; Li, J.J.; Xu, M.L. Research on the Characteristics of Urban Runoff in Zhenjiang: Rainwater Utilization Demonstration Project. *Environ. Eng.* **2010**, *28*, 31–35.

90. Li, C.; Tu, X.-J.; Qin, Y.-Q.; Huang, X.-Y.; Hu, Z.-B.; Wei, Q. Analysis of pollutants concentration of initial rainwater in nanning roads. *Environ. Eng.* **2017**, *7*, 70–75.

91. Zou, A.-P. Analysis on stormwater management project in Baoan District, Shenzhen City. *China Water Wastewater* **2010**, *16*, 71–73.

92. Feng, C.M.; Mi, N.; Wang, X.T.; Cai, Z.W.; Di, W.Z. Analysis of Road Runoff Pollutants in Northern City Based on the Typical Rainfall. *Ecol. Environ. Sci.* **2015**, *3*, 418–426.

93. Liu, D.X.; Li, Q.Q.; Li, T.L.; Wang, W.; Jin, C.H.; Yang, Z.Z.; Cao, J.G. Study on the pollution status of rainfall runoff in Tianjin. *China Water Wastewater* **2015**, *11*, 116–119.

94. Deng, F.; Chen, W. Probe to the Rainwater Utilization Scheme in Nanjing City Residential Area. *China Water Wastewater* **2003**, *19*, 95–97.

95. Jin, G.X.; Wang, P.X.; Qiu, W.G. Rainwater quality monitoring and treatment in Shanghai City. *Water Wastewater Eng.* **2007**, *33*, 47–51.

96. Huang, G.R.; Nie, T.F. Characteristics and Load of Non-Point Source Pollution of Urban Rainfall Runoff in Guangzhou, China. *J. South China Univ. Technol. Nat. Sci. Edit.* **2012**, *40*, 142–148.

97. Jin, H.W.; Hua, L.; Chen, Y.Y.; Shan, W.J.; Shi, W.X.; Huang, Z.F.; Jiao, Z.Z. The Characteristics of Pollution in Urban Rainwater Pipe Network and Its Influence on the Water Quality of Receiving Water Body. *Environ. Chem.* **2012**, *31*, 208–215.

98. Sun, H.; Liu, Z.Q.; Liu, H.H.; Tian, Y. Analysis and Research on the Rainwater Utilization on One Campus of Tianjin. *Res. Soil Water Conserv.* **2013**, *2*, 288–292.

99. Shao, H.H.; Li, Y.; Fang, X.J.; Jiang, Z.H. Study on the water quality characteristics of the road stormwater runoff in Beijing City. *Water Wastewater Eng.* **2011**, *37*, 130–133.

100. Liu, J.S.; Guo, L.C.; Luo, X.L.; Chen, F.R.; Zeng, Y.P. The Characteristics of Rainwater-Runoff-Flow Pollution Chain in Guangzhou. *Environ. Chem.* **2014**, *33*, 1040–1041.

101. Wang, J.X.; Wang, Y.J. Water Quality Analysis and Recycling Suggestions for a Campus in Lanzhou. *Water Wastewater Eng.* **2013**, *39*, 151–153.

102. Shen, J.; Li, T.; Qian, J.; Peng, S.H.; Qian, L.P. Types of Detention Tank for Xinghua Drainage System in Hefei City. *China Water Wastewater* **2012**, *28*, 40–43.

103. Lin, L.F.; Li, T.; Li, H. Characteristics of Surface Runoff Pollution of Shanghai Urban Area. *Chin. J. Environ. Sci.* **2007**, *28*, 1430–1434.

104. Li, M.; Yu, X.J. The analysis of change trend of runoff water quality and recycling methods in Jinan. *Environ. Pollut. Control* **2008**, *30*, 98–99.

105. Zhang, N.; Zhao, L.J.; Li, T.L.; Jin, Z.H. Characteristics of pollution and monitoring of water quality in Tianjin. *Ecol. Environ. Sci.* **2009**, *18*, 2127–2131.

106. Hou, P.Q.; Ren, Y.F.; Wang, X.K.; Ouyang, Z.P.; Zhou, X.P. Research on Evaluation of Water Quality of Beijing Urban Stormwater Runoff. *Chin. J. Environ. Sci.* **2012**, *33*, 71–75.

107. Xu, P.; Si, S.; Zhang, J.Q.; Zhang, Y.J.; Zheng, K.B.; Sun, K.P. Study on control effects of permeable asphalt road and road retention on water quality and quantity of runoff. *Water Wastewater Eng.* **2015**, *11*, 64–69.

108. Li, Q.Q.; Li, T.L.; Liu, D.X.; Jin, Z.H. Pollution characteristics of runoff in different function area of Tianjin. *Environ. Pollut. Control* **2011**, *33*, 22–26.

109. Zhang, Q.Q.; Wang, X.K.; Hao, L.L.; Hou, P.Q.; Ouyang, Z.Y. Characterization and Source Apportionment of Pollutants in Urban Roadway Runoff in Chongqing. *Chin. J. Environ. Sci.* **2012**, *33*, 76–82.

110. Zhang, S.F.; Li, T.; Gao, T.Y. Study on Pollution Load of Urban Surface Runoff in Shanghai. *China Water Wastewater* **2006**, *22*, 57–60.

111. Zhu, T.; Zhao, Y.; Che, W.; He, W.H.; Lu, C.Y. Pollution analysis and control countermeasures of rainwater runoff in Hangzhou. *China Water Wastewater* **2015**, *17*, 119–123.

112. Fei, W.; Wang, J.L.; Che, W. Analysis on characteristics of stormwater runoff flush on different land surfaces. *Tech. Equip. Environ. Pollut. Control* **2012**, *6*, 817–822.

113. Xie, W.M.; Zhang, F.; Zhang, J.D.; Lin, H. Study on the Change Regularity and Treatment of Urban Rainwater Runoff. *Environ. Sci. Technol.* **2005**, *28*, 30–31.

114. Wang, X.H.; Lai, Q.Y.; Du, J.Y.; Bao, Y.S.; Zheng, W.W.; Ye, F.X. Analysis of water quality characteristics of different bedding surface rainfall runoff in ningbo city. *Chin. J. Environ. Sci.* **2016**, *S1*, 312–316.

115. Wang, Y.H.; Han, Y.; Peng, D.C. Analysis and characteristics of water quality of urban rain runoff. *Chin. J. Environ. Sci.* **2006**, *24*, 84–85.

116. Tian, Y.J.; Li, T.; Ye, G.J.; Tang, X.Y. Study on the pollution characteristics of surface runoff in fengqiao industry park of Suzhou. *Environ. Pollut. Control* **2009**, *31*, 39–42.

117. Yi, Y.R.; Hai, M.Y.; Zhao, L.L. Analysis on the characteristics of storm runoff water quality of different underlaying surfaces in urumqi. *Res. Soil Water Conserv.* **2010**, *17*, 247–251.

118. Chen, Z.R.; Hu, S.; Huang, H.; Li, H. Study on surface runoff pollution in the baimang river basin of shenzhen city. *China Water Wastewater* **2011**, *S1*, 128–132.

119. Yan, Z.J.; Liu, H.Q.; Sun, H.L.; Zhou, L.Z.; Zhang, M.H. Study on the characteristics of surface runoff pollution in different functional areas in Wenzhou. *Environ. Sci. Technol.* **2012**, *S1*, 203–208.

120. Xu, W.; Gao, H.J.; Li, T. Pollution Characterization of Impermeability Surface Runoff in Typical Urban of Hefei. *Environ. Sci. Technol.* **2013**, *4*, 84–88.

121. Chen, M.; Ren, R.; Wang, Z.J.; Lin, X.T.; Liu, L.L.; Wu, S.H.; Zhang, S.F.; Chen, S. Investigation on Status of Environmental Hormone Pollution in the Industrial Wastewater and Urban Sewage in Beijing. *Res. Environ. Sci.* **2007**, *20*, 1–7.

122. Zhang, Y.; Zhang, T.T.; Guo, C.S.; Hua, Z.D.; Zhang, Y.; Xu, J. Pollution Status and Environmental Risks of Illicit Drugs in the Urban Rivers of Beijing. *Res. Environ. Sci.* **2016**, *29*, 845–853.

123. Yang, Y.; Wei, Y.S.; Zheng, X.; Wang, Y.W.; Yu, M.; Xiao, Q.C.; Yu, D.W.; Sun, C.; Yang, Y.; Gao, L.J.; et al. Investigation of microbial contamination in Wenyu River of Beijing. *Acta Sci. Circumst.* **2012**, *32*, 9–18.

124. Mark, O.; Jørgensen, C.; Hammond, M.; Khan, D.; Tjener, R.; Erichsen, A.; Helwigh, B. A new methodology for modelling of health risk from urban flooding exemplified by cholera—Case Dhaka, Bangladesh. *J. Flood Risk Manag.* **2018**, *11*, S28–S42. [CrossRef]

125. Preparing for Extreme And Rare Events in COASTAL REGIONS (PEARL). Available online: http://www.pearl-fp7.eu/ (accessed on 21 February 2018).

126. United States Geological Survey (USGS) Office of Water Quality. USGS Water-Quality Information. USGS Water-Quality Sampling of Flood Waters. Available online: https://water.usgs.gov/owq/floods/ (accessed on 21 February 2018).

127. Hawdon, A.; Keen, R.; Kemei, J.; Vleeshouwer, J.; Wallace, J. Design and application of automated flood water quality monitoring systems in the wet tropics. *CSIRO Land Water Sci. Rep.* **2007**, *49*, 27.

128. Qiu, X.C.; Zhao, H.X.; Yin, J.; Zhang, W.J. An Analysis of Water Environment Factors and an Evaluation of Water Quality of Aiyi River. *China Rural Water Hydropower* **2014**, *12*, 52–55.

129. Li, H.F.; Zhou, L.R.; He, Q.; Li, Y.W. Biotoxicity Analysis of the Gulin River Quality. *Ecol. Econ.* **2013**, *7*, 182–185.

130. Li, K.; Wu, S.H.; Dai, E.F.; Xu, Z.C. Flood loss analysis and quantitative risk assessment in China. *Nat. Hazards* **2012**, *63*, 737–760. [CrossRef]

131. Yin, J.; Ye, M.W.; Yin, Z.N.; Xu, S.Y. A review of advances in urban flood risk analysis over China. *Stoch. Environ. Res. Risk Assess.* **2015**, *29*, 1063–1070. [CrossRef]

132. Li, H.; Ding, L.Q.; Ren, M.L.; Li, C.Z.; Wang, H. Sponge City Construction in China: A Survey of the Challenges and Opportunities. *Water* **2017**, *9*, 594. [CrossRef]

133. Nanjing Environmental Protection Agency. Available online: http://www.njhb.gov.cn/43462/43466/ (accessed on 21 February 2018).

134. Chongqing Environmental Protection Agency. Available online: www.cepb.gov.cn/ (accessed on 21 February 2018).

135. Beijing Environmental Protection Agency. Available online: www.bjepb.gov.cn/bjhrb/index/index.html (accessed on 21 February 2018).

136. Guangzhou Environmental Protection Agency. Available online: www.gzepb.gov.cn/ (accessed on 21 February 2018).

137. Shanghai Environmental Protection Agency. Available online: www.hzepb.gov.cn/ (accessed on 21 February 2018).
138. Hangzhou Environmental Protection Agency. Available online: www.hzepb.gov.cn/ (accessed on 21 February 2018).
139. National Surface Water Quality Automatic Monitoring and Real-Time Data Release System. Available online: http://123.127.175.45:8082/ (accessed on 21 February 2018).
140. AKVO Organization. Available online: https://akvo.org/blog/akvo-caddisfly-a-water-quality-testing-kit-for-sdg-monitoring/ (accessed on 21 February 2018).
141. Gironás, J.; Roesner, L.A.; Rossman, L.A.; Davis, J. A new applications manual for the Storm Water Management Model (SWMM). *Environ. Model. Softw.* **2010**, *25*, 813–814. [CrossRef]
142. Deltares Institution. Available online: http://oss.deltares.nl/web/delft-fews/about (accessed on 21 February 2018).
143. Juntunen, J.; Merilainen, P.; Simola, A. Public health and economic risk assessment of waterborne contaminants and pathogens in Finland. *Sci. Total Environ.* **2017**, *599*, 873–882. [CrossRef] [PubMed]
144. World Health Organization. *Quantitative Microbial Risk Assessment: Application for Water Safety Management*, 1st ed.; WHO: Geneva, Switzerland, 2016.
145. Haas, C.N.; Rose, J.B.; Gerba, C.P. *Quantitative Microbial Risk Assessment*, 2nd ed.; John Wiley & Sons, Inc.: New York, NY, USA, 2014.

water

MDPI

Article

Hydrological Performance of LECA-Based Roofs in Cold Climates

Vladimír Hamouz [1,*], Jardar Lohne [1], Jaran R. Wood [2] and Tone M. Muthanna [1]

[1]   Department of Civil and Environmental Engineering, Norwegian University of Science and Technology, 7491 Trondheim, Norway; jardar.lohne@ntnu.no (J.L.); tone.muthanna@ntnu.no (T.M.M.)
[2]   Leca International, Årnesvegen 1, 2009 Nordby, Norway; Jaran.Wood@Leca.no
*   Correspondence: vladimir.hamouz@ntnu.no; Tel.: +47-410-70-318

Received: 15 December 2017; Accepted: 28 February 2018; Published: 3 March 2018

Abstract: Rooftops represent a considerable part of the impervious fractions of urban environments. Detaining and retaining runoff from vegetated rooftops can be a significant contribution to reducing the effects of urbanization, with respect to increased runoff peaks and volumes from precipitation events. However, in climates with limited evapotranspiration, a non-vegetated system is a convenient option for stormwater management. A LECA (lightweight expanded clay aggregate)-based roof system was established in the coastal area of Trondheim, Norway in 2016. The roof structure consists of a 200 mm-thick layer of LECA® lightweight aggregate, covered by a concrete pavement. The retention in the LECA-based roof was estimated at 9%, which would be equivalent to 0.27 mm/day for the entire period. The LECA-based configuration provided a detention performance for a peak runoff reduction of 95% (median) and for a peak delay of 1 h and 15 min (median), respectively. The relatively high moisture levels in the LECA-based roof did not affect the detention performance. Rooftop retrofitting as a form of source control may contribute to a change in runoff characteristics from conventional roofs. This study of the LECA-based roof configuration presents data and performance indicators for stormwater urban planners with regard to water detention capability.

Keywords: detention; cold climate; hydrological performance; LECA-based roof; lightweight aggregate; sustainable drainage systems (SuDS); water-detaining non-green roof

1. Introduction

Stormwater management is experiencing raised awareness due to an increased frequency of damaging rain-induced flood events across the world. The existing infrastructure is not typically fit to handle the combined effects of ever-increasing urbanization (including the proliferation of impervious surfaces) and climate change [1]. Densely-urbanized areas have limited space for retrofitting with green solutions, or for the reduction of imperviousness. This encourages communities to seek out new, emerging solutions. One possible way to rethink stormwater management is to focus on building rooftops. In developed cities, rooftops account for almost half of impervious surfaces [2].

New constructions and retrofitting existing buildings with sustainable drainage systems (SuDSs) seem to be efficient measures to counteract the effect of impervious covers in the cityscape [1,3,4]. Additionally, they contribute to the reduction of both sewer overflows and flood risks. Rooftop retrofitting differs from many other SuDS approaches, as it does not require additional land acquisition. Rooftop solutions such as green roofs belong to the first of the so-called three-step stormwater treatment train as a form of source control [5,6]. In 2008, the Norwegian Water Association adopted a national guideline for surface stormwater management that uses a three-step approach, where a source control should be able to collect and infiltrate runoff following small events (the rainfall intensity classification of small is location-specific) [7].

The main drivers behind rooftop source control are detention and retention of runoff. Retention occurs through the combined process of evapotranspiration for vegetated solutions, and its annual runoff reduction has been extensively investigated [8–13]. On the other hand, detention performance indicators are increasingly required by stormwater designers to alleviate urban flooding due to capacity exceedance in sewer systems [14,15]. Green roof performance depends largely on the local climate. Most studies scrutinized within the context of this research have reported limited hydrological performance in cold and wet climates, when evaporation and transpiration is limited due to climatic factors [14–20]. Johannessen et al. [20] investigated potential evapotranspiration in cold and wet regions across 14 locations in northern Europe, and concluded that retention on green roofs varied between 0% and 1% for Nordic countries in the winter period. In order to address the challenges outlined in the literature, it was decided to test the performance of a non-vegetated lightweight filter. For the sake of reference, this paper benchmarks the hydrological performance of the new non-vegetated solution against that of green roofs.

A LECA-based roof system was constructed at Høvringen (Trondheim, Norway), where the testing of roofs for water detention and retention are piloted. The Trondheim region registers an average of 150 days of precipitation a year [21]. Based on research studying the evapotranspiration and evaporation from vegetated and non-vegetated roofs, the water loss was comparable for both roofs for a period of approximately 50 h [22]. The following two weeks of dry period demonstrated that the additional ability of plants to transpire water outperformed the evaporation by more than 60%.

Hydrological performance indicators relevant to this study are peak flow reduction, peak flow delay with an event-based perspective, and retention within a long-term rainfall/runoff water balance perspective.

In order to address the hydrological performance of the LECA (lightweight expanded clay aggregate)-based roof, the following research questions were proposed:

(1) What is the seasonal and annual retention capacity of the LECA-based roof in cold climates?
(2) What is the event-based detention capacity of the LECA-based roof in cold climates?
(3) How do antecedent stormwater events affect the hydrological performance of the examined roof?

Limitations to the Study

Given that this is an in-situ field setup, the study is limited to the actual weather phenomena that occurred during this period. As such, there was only one event with a return period greater than 2 years. Thus, a limited amount of data was collected to investigate the extreme performance of the roof.

2. A Brief Literature Review

A brief literature review was performed to address challenges within stormwater management, more specifically stormwater retention and detention on rooftops. Most of the relevant studies focused on rooftops with vegetation. Thus, water losses due to plant uptake show a clear difference when comparing results between vegetated and non-vegetated solutions. In this study, no transpiration was expected because of the non-vegetated setup. Furthermore, seasonally low evaporation rates were expected due to the cold and wet weather conditions [20,23]. The review identified the requirements of stormwater designers and planners regarding sustainable solutions that enable the reduction of annual runoff, as well as the management of short and large design vents. A dataset based on 18 studies was analyzed. Emphasis was given to studies which focused on non-vegetated roofs (including reference black roofs), as well as cold and wet climates. The majority of the studies were focused on retention performance, rather than detention.

Retention occurs during dry periods, when water is evaporated into the atmosphere. In terms of retention performance, VanWoert et al. [3] studied the total rainfall retention from different media. The retention ranged from 27.2% for gravel ballast to 50.4% for a bare growing media, and 60.6% for a green treatment. Similarly, Mentens et al. [1] compared the annual runoffs from green and gravel-covered roofs. They presented a 25% reduction for the gravel roof, and a 50% reduction for the green roof. Berghage et al. [22] reported the annual rainfall retention from three different setups. The retention ranged from 14.1% for an asphalt roof to 29.7% for a media roof, and 52.6% for a vegetated roof. Comparing LECA-based (non-vegetated) and vegetated setups, higher levels of retention were observed using vegetated beds [10], with an annual volumetric retention of 54.5% for a LECA-based setup and 75.1% for a green roof. Johannessen and Muthanna [24] presented an annual runoff reduction of 17–30% for three coastal cities in Norway. In a study focused on a long dry period, major differences were found in retention through evapotranspiration by vegetated and non-vegetated configurations [25]. Berretta et al. [9] studied moisture loss from a growing medium during a dry period of cold and warm months. They presented a mean moisture loss ranging from 0.34 mm/day to 1.65 mm/day in the period of March through July. Special attention should be given to the regeneration of roof storage capacity, which depends on physical configuration, precipitation patterns, and evaporation during dry periods [17]. Overall, the average retention performance is useful in a context where stormwater discharge to the sewer system is billable.

The detention effect occurs when temporally detained stormwater is subsequently released [14,15,18]. The evapotranspiration effect, which restores storage capacity during dry periods, may be neglected at this time in the interest of detention. Comparing detention performance, Liu et al. [26] and Villarreal et al. [14] presented peak flow reductions and peak delays of an intensive green roof on an event basis. The peak reductions varied between 25% and 65%, and peak delays varied between 20 min and 40 min. Stovin et al. [10] concluded that peak reduction for rainfall larger than 10 mm varied from 29% for a LECA-based bed to 68% for a sedum roof. They also noted that vegetated beds with brick-based substrates offer consistently greater attenuation compared with the LECA-based substrate. Stovin et al. [2] investigated the performance of an extensive green roof subjected to events with a return period of over one year. They presented a per-event peak reduction of 59.22% (mean) and 58.67% (median), and a per-event peak-to-peak delay of 54.16 min (mean) and 18 min (median). Li et al. [27] reviewed the typical hydrological performance of green roofs. It was shown that they attenuate a peak flow of 22% to 93%, and delay a peak flow of 0 to 30 min.

3. Materials and Methods

3.1. Geometrical Description and Structure Composition of the LECA-Based Roof

A full-scale LECA-based setup (Figure 1) was built to monitor the hydrological balance between rainfall and runoff on the roof of a wastewater treatment plant at Høvringen in Trondheim, Norway, approximately 50 m a.s.l. (63°26′47.5″ N; 10°20′11.0″ E). According to the Köppen-Geiger climate classification map (http://koeppen-geiger.vu-wien.ac.at), Trondheim is situated at the interface of oceanic (Cfb) and subarctic (Dfc) climates [23]. Main characteristics are strong seasonality, short summers, and no predominant dry seasons. The Norwegian Meteorological Institute recorded an annual precipitation of 950 mm and an annual average temperature of 3.8 °C in 2016. Cold climate is defined as a climate where the mean temperature of at least one month per year is below +1 °C [28]; in Trondheim, this occurs in January, February, November, and December [21].

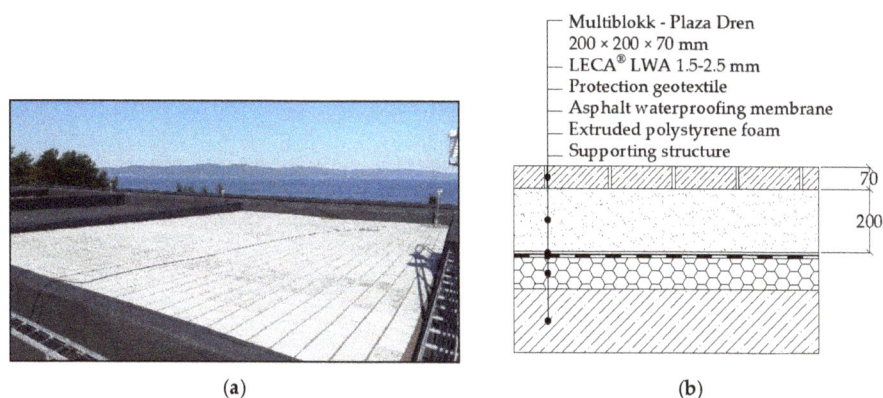

Figure 1. The LECA-based roof with concrete paving stones and its cross-section. LWA: lightweight aggregate. (a) The full-scale LECA-based roof at Høvringen (Trondheim, Norway); (b) The LECA-based roof components in a cross-section

The dimensions of the LECA-based roof-cells are 8 × 11 m, with a longitudinal slope of 2%. The full-scale configuration prevents impact associated with the scaling factor, which is important when accounting for the lateral flow across the roof-to-roof drain. The structure composition is made up of an underlying protection layer, a 200 mm thick layer of LECA® lightweight aggregates (LWA), and covering concrete pavers (200 × 200 × 70 mm). A geotextile is used as a separation layer, and to prevent fine particles from being washed out. LECA® LWA is an expanded lightweight crushed clay aggregate with a bulk density of 500 kg/m^3, a particle density of 1050 kg/m^3, and a particle size range of 1.5–2.5 mm [29]. Laboratory tests were also performed. The specific fraction was found to be ~60% of the proportion of voids in a sample, with a maximum water holding capacity (MWHC) of 26.2%, which is defined as the water content of a substance after two hours draining post-saturation. The tests were performed according to the Guidelines for the Planning, Construction and Maintenance of Green Roofing of the German Landscape Development and Landscaping Research Society [30]. The saturated hydraulic conductivity was measured to be 143.2 cm/h. The weight of the LECA-based roof was calculated at 251 kg/m^2 based on completely dry materials, and 310 kg/m^2 for wet conditions (MWHC). This includes LECA and pavers.

3.2. Data Collection and Event Analysis

Hydrological data were collected for all four seasons, from January 2017 to November 2017. Precipitation was monitored by a heated tipping bucket rain gauge (Lambrecht meteo GmbH 1518 H3, Lambrecht meteo GmbH, Göttingen, Germany) with a resolution of 0.1 mm at 1-min intervals. The runoff collection was measured using a weight-based system with two tanks downstream of the drainage outlets. The collection tanks had two conditions for emptying: they were automatically emptied either every 30 min, or when the weight of the water approached the capacity of the tank (30 kg).

A CR1000 data logger (Campbell Scientific, Inc., Logan, UT, USA) recorded all the parameters at 1-min intervals. Single precipitation events were defined according to a minimum period of 6 h of antecedent dry weather (ADWP), as commonly used by several previous studies, among others [2,3]. A threshold precipitation depth of 0.5 mm was used to exclude insignificant precipitation events. Similarly, a threshold discharge of 0.1 L/min was set to specify the start and end of runoff events. The moisture content in LECA® LWA was recorded using Decagon 5TM soil moisture and temperature sensors, which were delivered at the end of June. The moisture sensors were pre-calibrated in the laboratory for minimum and maximum degrees of saturation (0% and 100% saturation). Events were identified and sorted into five groups based on the type of precipitation: *rain, rain on snow,*

snow, *snowmelt*, and *mixed*. A total of 127 events were registered in the period between January 2017 and November 2017: 94 *rain* events, 12 *rain on snow* events, 9 *snow* events, 4 *snowmelt* events, and 8 *mixed* events. The events which were designated as *mixed* typically had a long duration (several days) and experienced several changes of precipitation type.

3.3. Retention Capacity

Retention was considered as long-term permanent water removal on a monthly basis, and a mean value for the entire studied period. Retention capacity was determined as follows:

$$Ret = P - R,$$
(1)

where *P* is precipitation, *R* is runoff and *Ret* is retention.

The retention at any given time will be the sum of the evapotranspiration and the water currently stored in the LECA medium.

3.4. Detention Capacity

The detention capacity of the LECA-based roof was assessed as the ability to attenuate and delay peak flows compared to the response of the black roof. This analysis was carried out on an event basis. In some cases, several peaks were observed in a single event due to the long duration (several days). In these cases, only the highest peak per event was analyzed. Peak flow reduction (PR) was determined as follows:

$$PR = 1 - \frac{Q_{LR,max}}{Q_{BR,max}},$$
(2)

where $Q_{LR,max}$ is the maximum flow recorded per event from the LECA-based roof (LR), and $Q_{BR,max}$ is the maximum flow recorded per event from the black roof (BR). Peak delay (PD) was determined as follows:

$$PD = T_{LR,max} - T_{BR,max},$$
(3)

where $T_{LR,max}$ is the time of maximum flow recorded per event from the LECA-based roof (LR) and $T_{BR,max}$ is the time of maximum flow recorded per event from the black roof (BR). Additionally, any delays were analyzed as delays of centroid of individual events.

4. Results

4.1. Precipitation Events and Time for Regeneration of the LECA-Based Roof Storage Capacity

The event durations and precipitation depths varied considerably. Of the selected highest-intensity events, the shortest lasted 8 min in July, and the longest lasted 122 h in October. This presents a widespread range; therefore, the median value of 8.2 h might be more representative. In terms of total precipitation depths, the events ranged from 0.5 mm to 85.4 mm. The total duration of precipitation events and dry periods were determined. Assuming the 6 h ADWP, precipitation occurred 22% of the time during these eleven months. This leaves 78% of the time (dry period) for the regeneration of the roof storage capacity, considered as time between events.

The precipitation at Høvringen (11-months dataset) was compared to the Risvollan stations, located 82 m a.s.l, at an areal distance of 7 km, due to the unavailability of Intensity–Duration–Frequency (IDF) curves at Høvringen. For the observation period, there was 8% more precipitation recorded at Høvring than at Risvollan. The higher elevation of Risvollan makes this difference expected. Comparing the observed events at Høvringen with the IDF curves from the Risvollan station, one can see that the all the events fall below a 2-year return period, with the exception of one from August 19, which lasted more than 1 day (Figure 2). The figure also shows a zoom to a 30 min resolution, where eventual rapid storms can be found. However, they registered low precipitation depths.

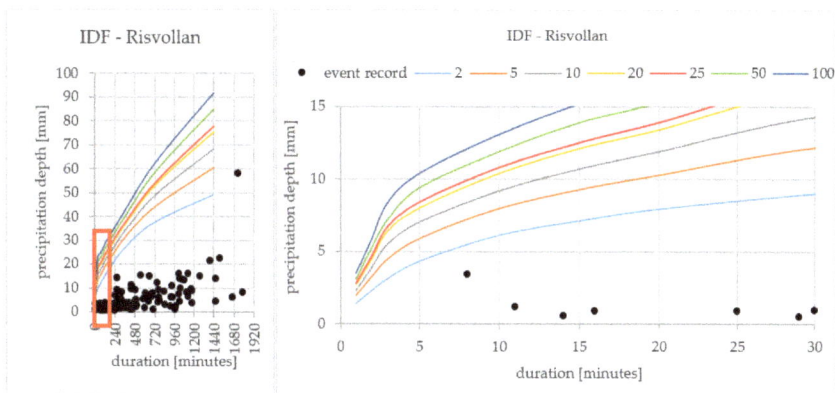

Figure 2. (**left**) The event record comparison with Risvollan Intensity–Duration–Frequency (IDF) curves for different return periods; (**right**) 30-min resolution.

The mean event intensities ranged from 0.1 mm/h to 25.6 mm/h. Particularly, the maximum mean intensity from July 26 that lasted 8 min exceeded the rest of the values; however, it is still below a 2-year return period event.

4.2. Retention Performance

The total rainfall-runoff rate can be observed between 23 January 2017 and 30 November 2017. During this period, the precipitation gauge measured 937.6 mm precipitation. The runoff depths during the examined period were 912.1 mm and 852.4 mm for the black and LECA-based roofs, respectively (Figure 3, Table 1). This indicated a discrepancy of 59.7 mm (black vs. LECA-based roof) and 85.2 mm (precipitation gauge vs. LECA-based roof), which is the evaporated volume. Overall, the difference between the precipitation and runoffs was a 3% volume reduction by the black roof, and 9% for the LECA-based roof.

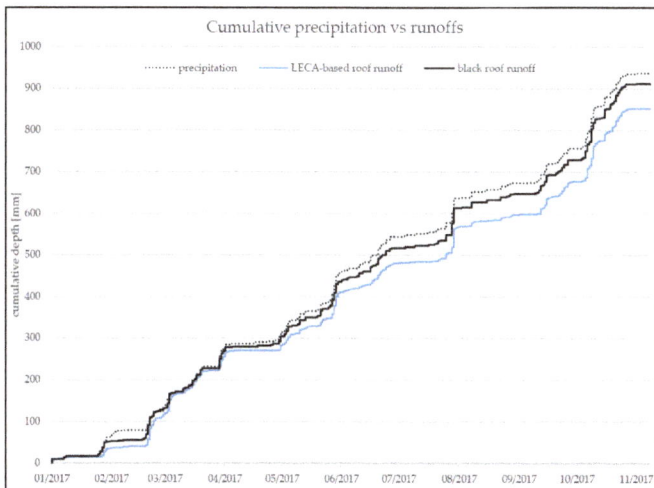

Figure 3. Cumulative hydrograph from the rain gauge and runoffs of the black and the LECA-based roofs.

The seasonal variations are shown in Table 1. The 11 months were divided into four groups: November, January, February, and March represent the winter period; April, May, and June represent the spring period; July and August represent the summer period; and September and October represent the fall period. The winter period confirmed zero evaporation during the cold climate condition. The minor negative difference between the black roof and the precipitation gauge can be attributed to signal noise and measuring uncertainties. The snowfall measurements are most likely the most significant here. Evaporation in the spring season exceeded values from the summer season (Table 1). This could be explained by an alteration of rainfall patterns in the summer season, when higher intensity rainfalls occur, resulting in decreased retention.

Table 1. Seasonal variation in retention performance.

Season	Number of Days	Total Precipitation (mm)	Total Runoff (mm)		Runoff Reduction (%)		Retention (mm)		Normalized Daily Retention (mm/day)	
2017		Rain Gauge	LECA	Black	LECA	Black	LECA	Black	LECA	Black
Winter	97	349.4	343.6	355.3	2%	−2%	5.7	−5.9	0.06	−0.06
Spring	91	301.3	251.9	276.7	16%	8%	49.4	24.7	0.54	0.27
Summer	62	182.7	160.9	179.1	12%	2%	21.8	3.7	0.35	0.06
Fall	61	104.2	96.0	101.1	8%	3%	8.3	3.2	0.14	0.05
Total	311	937.6	852.4	912.1	9%	3%	85.2	25.5	0.27	0.08

The runoff coefficients calculated from total precipitation and runoff records were 0.97 for the black roof and 0.91 for the LECA-based roof. The normalized daily retention estimated from total precipitation and runoff measurements was 0.27 mm per day. This reflects the fact that the climate in the Trondheim region is relatively cold and wet. The detention in the expanded clay aggregate (LECA), followed by a subsequent slow drainage of the system, was much greater than the evaporation loss rate.

4.3. Detention Performance

The detention capacity of the system was evaluated using peak flow delay and peak flow reduction. Evaluating the LECA-based roof using a wide range of performance indicators for all events indicated that the performance was mainly influenced by duration, intensity, moisture content, and ADWP of the individual events. At the same time, the detention indicators were highly sensitive to the chosen subsets of the rainfall dataset which was used in the calculations. Therefore, the eight events with the highest 5-min peak intensity were selected for detailed examination. Figure 4 illustrates the eight largest events, ranging in duration from 8 min to almost 3 days, with depths of 3.2 mm to 59.6 mm (Table 2). Additionally, the largest snowmelt (event 24) and rain-on-snow event (event 27) were included to show the different types of events observed on the roofs. For event 24 (the pure snowmelt event), there was a negative lag time delay between the black and the LECA-based roof. This can be explained by the observed temperatures in the LECA-based roof, which were more stable compared to the black roof. The black roof was typically colder than the LECA-based roof, meaning that higher net radiation was needed to initiate snowmelt. Though this was the largest snowmelt event, it was a relatively small compared to the other events in Tables 2 and 3, making the peak flow reduction less relevant. Event 27 (the rain-on-snow event) is very difficult to evaluate without knowing the mass of snow on the roofs at the onset of rain. It is not possible to compare peak lag times, as it is possible that the snowmelt initiated prior to the precipitation runoff. A complete mass balance would be needed of the initial snowfall until it was completely melted again. Due to these constraints, these two events were excluded from the comparisons in Tables 2 and 3.

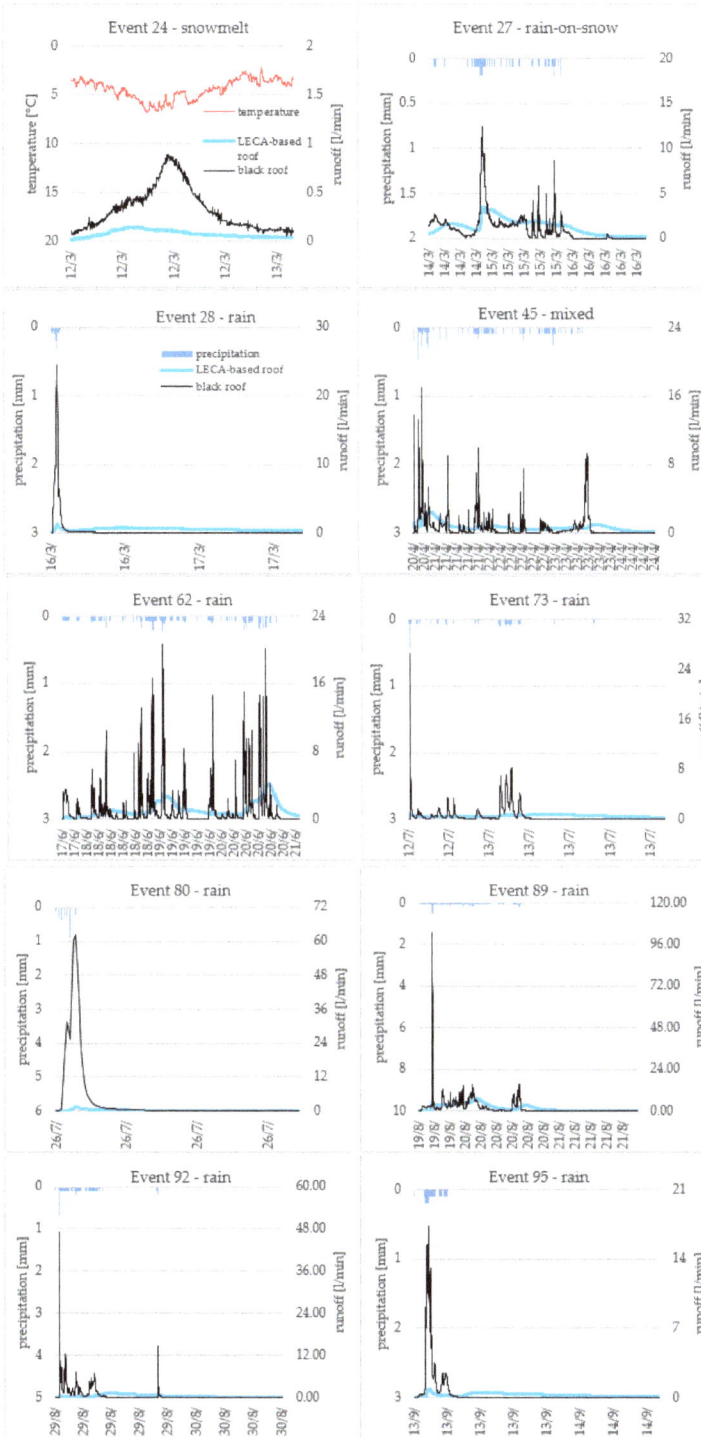

Figure 4. The events with the highest peak 5-min intensity.

Two alternative precipitation intensities were presented. The mean intensity indicates the ratio between the depth and duration, whereas the peak 5-min intensity is the peak intensity measured over 5 min. Event 45 turned to snow at a later stage, followed by a snowmelt in the end, and therefore the event type was classified as mixed. However, this event was used in the final evaluation, since the peak 5-min intensity and corresponding runoff occurred during "rainy" conditions. Event 80 was quite untypical in terms of mean intensity, which was more than ten times higher than the remaining events. With respect to the return period, event 89 was the only one that fell between the 2-year and 5-year return period (Figure 2).

Table 2. Rainfall characteristics for the significant events.

Event	Type	Start	Duration (hh:mm)	Precipitation Depth (mm)	Mean Intensity (mm/h)	Peak 5-min Intensity (mm/h)
28	rain	16.03.2017 19:43	1:38	3.3	2.13	14.40
45	mixed	20.04.2017 19:04	69:34	35.6	0.57	17.64
62	rain	17.06.2017 16:47	74:38	59.6	0.81	11.64
73	rain	12.07.2017 16:53	19:35	8.8	0.46	16.32
80	rain	26.07.2017 16:20	0:08	3.2	25.58	21.60
89	rain	19.08.2017 13:41	28:47	57.7	2.02	40.80
92	rain	29.08.2017 02:52	7:15	11.2	1.57	21.60
95	rain	13.09.2017 03:05	5:25	7.9	1.51	12.00

Table 3 summarizes responses of both black and LECA-based roofs for the eight events. Here, individual events were characterized by comparing the maximum (peak) values registered from the roofs. Overall, the peak delay totaled 1 h and 15 min in median and 7 h and 23 min in mean. A long dry period before the events naturally led to the freeing of the storage capacity. However, this does not necessarily mean ideal conditions for delaying peak runoffs, as can be seen for event 95. On the contrary, the short dry period led to a sufficient delay for event 45. A focus on maximum values was not always the best solution when evaluating the runoff delays. One can see very long delays for the long-duration events 62 and 89. Event 62 even experienced two heavier rainfalls, which obviously led to two responses. Because of this, two alternative solutions of peak delays were suggested (Figures 5 and 6).

In terms of peak reduction (Table 3), the roof demonstrated a high efficiency, with a reduction rate of 80% to 97%, irrespective of the length of the antecedent dry period or the degree of previous saturation. The latter indicates the extent to which the voids in the expanded clay aggregates are filled with water. The saturation measurements ranged between 31% and 61% for the period after which the sensors were installed. In addition to the degree of saturation, the initial runoff may also be used as a performance corrector or predictor. Higher initial runoffs correlate to higher degrees of saturation in the media from previous events.

Table 3. Comparison of runoff characteristics of black and LECA-based roofs for the significant events. ADWP: antecedent dry weather period.

Event	ADWP	Peak Delay	Peak Reduction	Initial Degree of Saturation	Initial Runoff	Runoff Duration	Runoff Duration
		Peak-to-Peak	Peak-to-Peak		LECA	Black	LECA
	(hh:mm)	(hh:mm)	(%)	(%)	(L/min)	(hh:mm)	(hh:mm)
28	23:41	0:00	95%	-	0.27	2:43	14:26
45	7:37	2:30	85%	-	0.7	69:42	93:40
62	17:14	36:42	80%	-	0.08	74:57	81:41
73	8:44	0:01	97%	35.7	0.42	13:00	27:02
80	23:56	0:00	97%	32.1	0	0:51	1:22
89	62:42	12:17	93%	32.5	0.02	29:25	58:04
92	66:36	7:35	97%	32.2	0.01	15:46	33:28
95	165:12	0:00	95%	32.1	0.01	4:56	29:51

There were large differences in the runoff duration between the examined roofs. The median runoff duration of the LECA-based roof (31.5 h) lasted 2.2 times longer than of the black roof (14.5 h). Considering average values, the mean runoff duration of the LECA-based roof (42.5 h) lasted 1.6 times longer than of the black roof (26.5 h).

Figure 5 serves to recognize centroid delays of individual runoffs. The steep rises in cumulative runoffs associated with the black roof response after intense rainfalls may be considered a potential cause of rapid floods. One can see that the LECA-based roof transformed these rises to either flat (effective detention performance) or—in extreme cases—gradual runoffs. The extreme cases may be seen in events 45, 62, and 89. Additionally, Figure 5 shows the largest snowmelt and rain-on-snow events; however, these were different types of events, shown only for comparison. Further calculations include only the eight largest events, as outlined in the methods.

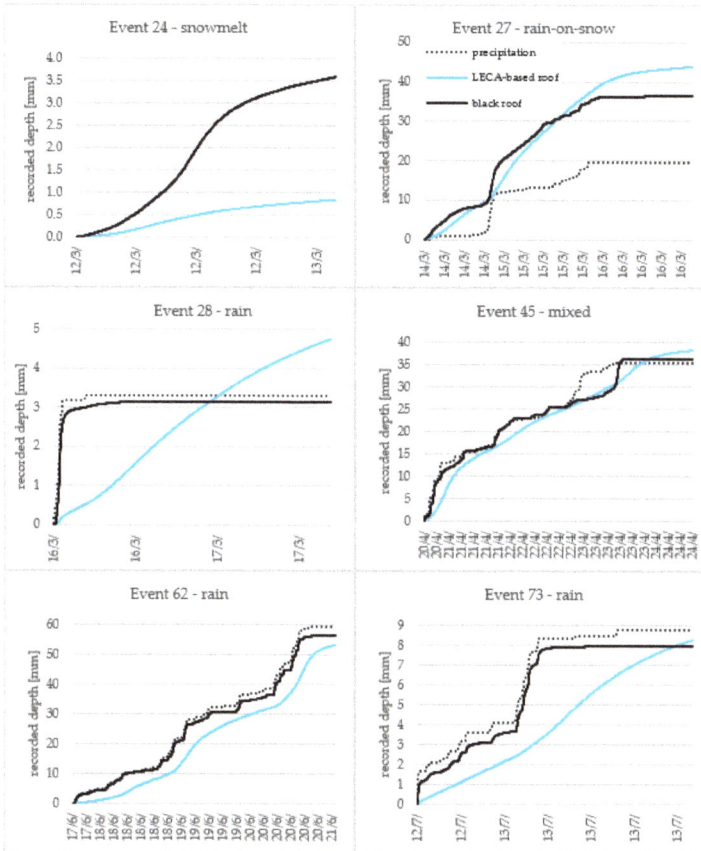

Figure 5. *Cont.*

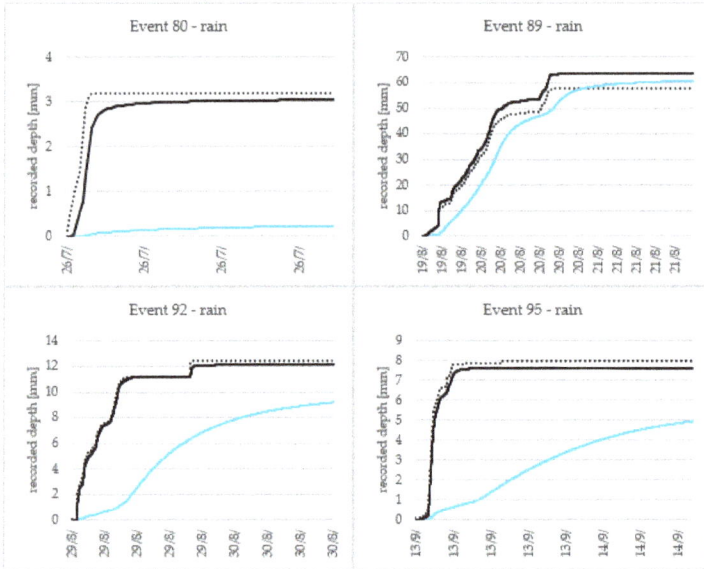

Figure 5. Cumulative runoff responses to the events with the highest peak 5-min intensity.

Considering shorter ADWP between events would result in a larger number of individual events. For instance, event 62 needed a 3 h ADWP to separate the event into two parts. Events 89 and 92, sharing a similar pattern, needed shorter breaks within the rainfall dataset (e.g., approximately 5–10 min), as there was continuous rainfall during this whole event. Using shorter breaks in the rainfall dataset would underestimate the total performance. For instance, the peak delay would change dramatically to 1 min (median) and 28 min (mean). Considering the centroids as the representative values, the peak delay counted 6 h (median) and 5 h and 51 min (mean).

Figure 6 presents a comparison of the alternative methods for runoff delays. The centroid delays were influenced by the short event 80, with small rainfall depth lasting only 8 min; this obviously led to a shortening of the centroid delay as well. During events 28 and 95, the quick responses can be clearly seen by the LECA-based roof runoffs, compared to the black roof peak runoff at the very beginning. This was probably caused by water collecting directly in the outlet (0.25 m^2). Therefore, those peak runoffs should not be considered as real responses of the LECA-based roof. Overall, the peak reduction presents 95% in median and 92% in mean.

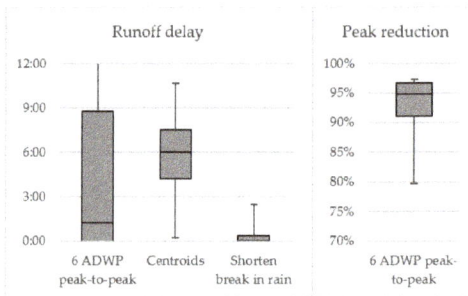

Figure 6. Comparison of the different methods of (**left**) the runoff delay and (**right**) the peak reduction.

5. Discussion

5.1. Retention Capacity

Retention is expected to be highest in the warm season. During the winter months, there was very limited measurable retention observed (2%). This is consistent with research [20] focused on potential evapotranspiration in cold and wet regions. Overall, the difference between the precipitation and runoff resulted in a 3% volume reduction for the black roof and 9% for the LECA-based roof. This is lower than the results presented by previous non-vegetated and green roof studies [1,3,10–13,22,24]. In terms of Norwegian conditions, Braskerud [31] presented a 25% runoff reduction in extensive green roofs in Oslo. This also agrees with the limited retention performance of the LECA-based roof. On the other hand, the result is most comparable with studies conducted in locations with wet and cool conditions [17,24]. The normalized daily retention during spring accounted for 0.54 mm per day, which is comparable with results from Beretta et al. [9]. As for the summer, 0.24 mm/day is about five times higher.

Nevertheless, it is a generally confirmed fact that water losses from vegetated roofs are higher due to evapotranspiration [25]. Berretta et al. [9] previously concluded lower daily moisture loss when using non-vegetated roofs compared to green roofs. Additionally, the reason for lower retention may be attributable to the concrete pavers, which cover and seal much of the surface area.

Average performance indicators may be useful for comparing different systems or even the same system exposed to different climatic conditions and/or for determining annual runoff, which does not have to be treated in a wastewater treatment plant. However, this data is very limited in terms of stormwater management design.

5.2. Detention Capacity

The performance indicators should reflect the performance in the non-daily events. Therefore, the eight largest events were examined. Evidence that the LECA-based roof can reduce peaks up to 95% (median) and 92% (mean) for significant events, as well as delay peaks by 1 h and 15 min (median) and 7 h 23 min (mean), provides support for its use in urban stormwater management strategies. The results of the peak reduction and delay, when neglecting the evapotranspiration effect, show a better performance than in the reviewed literature, for both the non-vegetated [10] and the green roofs [2,14,26,27,31].

The difficulties with defining events using the 6 h ADWP (resulting in some very long events) indicated that it might not be the best-suited time definition for the climate zone. These long duration events also cause decreasing mean intensity of individual events. Considering a maximum runoff as the representative value for an event may overestimate the overall performance in terms of the peak delay, even though it is more natural to compare events according to maximum registered values. The irregularity of natural rainfall patterns, combined with the variability within detention effect in specific events, complicates the identification of peak-to-peak delays. Centroid delays are perceived to be a more robust indicator of the delays in bulk runoff than peak delays [15]. Considering the centroids as the representative values, the peak delays total up to 6 h (median) and 5 h and 51 min (mean). Even though the vertical movement of stormwater (due to high saturated hydraulic conductivity; K_{sat} = 143.2 cm/hour) through the expanded clay aggregate is rather quick, the lateral movement through the media as well as the size of the roof (the distance between the sides to the outlet) and the slope of the roof are decisive and cause the high detention in the LECA-based roof.

5.3. Effect of Antecedent Events

Moisture levels in the expanded clay aggregate were higher than expected during dry periods. The expanded aggregate detains water for a long time, demonstrating why it is used for planting. However, very low saturation could be seen in November, when the moisture sensor registered low or negative temperatures. This can be explained by the inability of the sensor to accurately measure moisture in low temperatures, or a measurement error due to frozen media. Between July and November (Figure 7), every month experienced a long dry period: 156 h in July, 172 h in August, 278 h in September, 166 h in October, and 191 h in November.

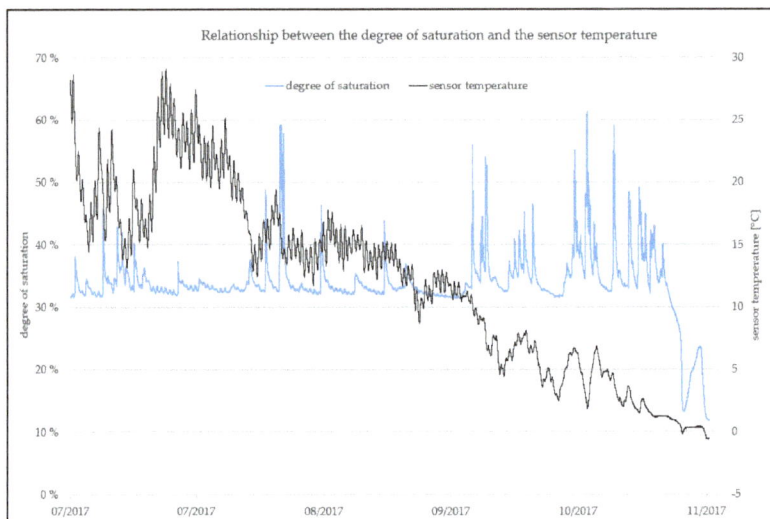

Figure 7. Variation of the degree of saturation within July and November.

Disregarding the long dry periods, the expanded clay aggregate kept a relatively high content of water within the medium. This clearly shows that the expanded clay aggregate was unable to fully regenerate its storage capacity to completely dry conditions, as in the laboratory. Despite the relatively high remaining saturation before an event, the LECA-based roof provided high performance in terms of peak reduction, as well as slowing down and transforming the runoff into a more natural flow for all seasons. Based on the moisture content measurements, a maximum storage capacity of 3.2 mm (0.28 m^3) was determined. This was also validated by a maximum observed precipitation of 2.7 mm, which did not generate runoff. The total available voids space (120 mm) is permanently taken by an inaccessible volume of 31%. This could indicate that there might be some capacity that could be gained by optimizing the size fractions. The void space is a function of this LECA type, making this yet another parameter which could be further investigated for optimal water detention. However, the water is adsorbed by the void spaces in the LECA, which additionally makes the water detention capacity dependent on the rainfall intensity.

5.4. Practical Implications

In light of the results and limitations, the LECA-based retrofitting shows promising results in terms of handling runoff from precipitation depths as a source control. This solution outperformed the conventional green roof solutions [14,18,26,27] in several aspects. The hydrological performance was evaluated for retention and detention, as well as for resistance to different forms of precipitation and durability. The practical implication of the LECA-based roof with respect to the detention performance includes runoff delays and peak flow reductions. Extending the durations of the runoff may significantly decrease the number of combined sewer overflows or the design volume of underground detention basins. The hydrological performance did not decline during the largest events recorded in this study, giving a strong indication of its performance, even during the more intense precipitation events. The LECA-based solution offers a detention capacity on the roof without allowing standing water on the roof membrane, as the water is held back and absorbed into the LECA material. An alternative solution would be an open or closed detention basin on the roof, or a cistern/closed tank system. A detention basin directly on the roof would exert water pressure on the roof membrane, increasing the risk of leaks. Standing water on the roof in a cold climate would freeze and contribute to blocked drains and ice formation on the roof. This would be a significant risk factor, as the ice expansion could also result in frost failure of the drains, which would lead to malfunctioning drains, and in the worst case, leaks into the underlying layers. A cistern-type solution would also experience freezing problems unless it was fully insulated or located indoors. The latter solution would be technically possible, as flat roofs always have internal roof drains in cold climates, again to minimize freezing of standing water. However, in urban areas where space is at a premium, it is more attractive to utilize the currently unused rooftop rather than sacrifice indoor space in buildings. The LECA-based solution is also an attractive solution to retrofit existing roofs, while the cistern solution would only be a feasible option for new buildings, due to the need for indoor space. Though it is not the only non-green alternative possible, the LECA-based system offers rooftop detention without introducing standing water pressure on the roof membrane, and with minimal ice formation risk due to the draining capacity, leaving no standing water on the roof.

In terms of estimated material costs per m^2, the LECA-based roof is approximately 40 €/m^2, whereas a green roof is 70 €/m^2 (following consultations with providers). Since the LECA-based roof performance does not rely on the evapotranspiration effect, it can used worldwide. The findings are specific to the LECA-based roof study and the specific actual rainfall that occurred during the study.

The weak points of the LECA-based roof in comparison to the green roof are in weight and in lower retention ability during warmer months. The LECA-based roof weighs 251 kg/m^2 in dry conditions and 310 kg/m^2 in wet conditions, whereas a green roof weighs 25 kg/m^2 in dry conditions and 50 kg/m^2 in wet conditions [32]. On the other hand, green roofs require irrigation during dry periods, periodical fertilization throughout the year, and have a deteriorated performance in extreme conditions [33].

6. Conclusions

The comparative study of precipitation and runoff data from two parallel rooftops (the LECA-based roof and the referenced "black" roof) was carried out at a coastal part of Trondheim, Norway. Eleven months of data were collected, analyzed, and divided into 127 events (94 rain events) according to a 6 h ADWP.

With respect to the previously reported findings, the retention performance of the LECA-based roof was lower than that of typical green roofs. This is because the water loss is only actuated by evaporation. For the entire studied period, the balance between precipitation and runoff of the LECA-based roof was estimated at 0.27 mm/day; this performance is 0.19 mm/day higher than the normalized daily retention of the black roof.

The runoff characteristics regarding to detention capacity of the LECA-based roof are particularly encouraging, even though the performance of source control systems typically struggle with intense rainfalls of short duration. Overall, the study demonstrated that the LECA-based configuration provided a large improvement compared with the black roof runoffs, with a peak reduction of 95% (median) and 92% (mean), and with a peak delay of 1 h 15 min (median) and 7 h 23 min (mean). This indicates that the LECA-based roof, with its improved detention performance, could be a good solution for the retrofitting of already-existing roof areas.

Acknowledgments: The field station was built with the support and contribution of the Klima2050 Centre for Research-based Innovation (SFI), financed by the Research Council of Norway and its consortium partners. The Klima2050 project aims to reduce the societal risks associated with climate change and the enhanced precipitation and floodwater exposure within urban areas [34].

Author Contributions: Vladimír Hamouz and Tone M. Muthanna conceived and designed the experiment. Vladimír Hamouz and Tone M. Muthanna performed the experiment and analyzed the data. Vladimír Hamouz and Jaran R. Wood interpreted the results with a general discussion with all authors. Jardar Lohne helped with structure of the manuscript and language correction. Vladimír Hamouz wrote the majority of the text, with feedback and input from Tone M. Muthanna, Jardar Lohne and Jaran R. Wood.

Conflicts of Interest: The authors declare no conflict of interest.

References

1. Mentens, J.; Raes, D.; Hermy, M. Green roofs as a tool for solving the rainwater runoff problem in the urbanized 21st century? *Landsc. Urban Plan.* **2006**, *77*, 217–226. [CrossRef]
2. Stovin, V.; Vesuviano, G.; Kasmin, H. The hydrological performance of a green roof test bed under UK climatic conditions. *J. Hydrol.* **2012**, *414*, 148–161. [CrossRef]
3. VanWoert, N.D.; Rowe, D.B.; Andresen, J.A.; Rugh, C.L.; Fernandez, R.T.; Xiao, L. Green roof stormwater retention. *J. Environ. Qual.* **2005**, *34*, 1036–1044. [CrossRef] [PubMed]
4. Torgersen, G.; Bjerkholt, J.T.; Lindholm, O.G. Addressing flooding and suds when improving drainage and sewerage systems—A comparative study of selected Scandinavian cities. *Water* **2014**, *6*, 839–857. [CrossRef]
5. Woods-Ballard, B.; Kellagher, R.; Martin, P.; Jefferies, C.; Bray, R.; Shaffer, P. *The Suds Manual*; CIRIA: London, UK, 2007; Volume 697.
6. Fletcher, T.D.; Shuster, W.; Hunt, W.F.; Ashley, R.; Butler, D.; Arthur, S.; Trowsdale, S.; Barraud, S.; Semadeni-Davies, A.; Bertrand-Krajewski, J.-L. SUDS, LID, BMPS, WSUD and more–the evolution and application of terminology surrounding urban drainage. *Urban Water J.* **2015**, *12*, 525–542. [CrossRef]
7. Lindholm, O.G.; Endresen, S.; Thorolfsson, S.T.; Sægrov, S.; Jakobsen, G.; Aaby, L. Veiledning i klimatilpasset overvannshåndtering. *Rapport* **2008**, *162*, 2008.
8. Berndtsson, J.C. Green roof performance towards management of runoff water quantity and quality: A review. *Ecol. Eng.* **2010**, *36*, 351–360. [CrossRef]
9. Berretta, C.; Poë, S.; Stovin, V. Moisture content behaviour in extensive green roofs during dry periods: The influence of vegetation and substrate characteristics. *J. Hydrol.* **2014**, *511*, 374–386. [CrossRef]
10. Stovin, V.; Poë, S.; De-Ville, S.; Berretta, C. The influence of substrate and vegetation configuration on green roof hydrological performance. *Ecol. Eng.* **2015**, *85*, 159–172. [CrossRef]
11. Carson, T.; Marasco, D.; Culligan, P.; McGillis, W. Hydrological performance of extensive green roofs in New York City: Observations and multi-year modeling of three full-scale systems. *Environ. Res. Lett.* **2013**, *8*, 024036. [CrossRef]
12. Gregoire, B.G.; Clausen, J.C. Effect of a modular extensive green roof on stormwater runoff and water quality. *Ecol. Eng.* **2011**, *37*, 963–969. [CrossRef]
13. Garofalo, G.; Palermo, S.; Principato, F.; Theodosiou, T.; Piro, P. The influence of hydrologic parameters on the hydraulic efficiency of an extensive green roof in mediterranean area. *Water* **2016**, *8*. [CrossRef]
14. Villarreal, E.L. Runoff detention effect of a sedum green-roof. *Hydrol. Res.* **2007**, *38*, 99–105. [CrossRef]
15. Stovin, V.; Vesuviano, G.; De-Ville, S. Defining green roof detention performance. *Urban Water J.* **2017**, *14*, 574–588. [CrossRef]
16. Bengtsson, L.; Grahn, L.; Olsson, J. Hydrological function of a thin extensive green roof in Southern Sweden. *Hydrol. Res.* **2005**, *36*, 259–268.
17. Stovin, V.; Poë, S.; Berretta, C. A modelling study of long term green roof retention performance. *J. Environ. Manag.* **2013**, *131*, 206–215. [CrossRef] [PubMed]
18. Locatelli, L.; Mark, O.; Mikkelsen, P.S.; Arnbjerg-Nielsen, K.; Jensen, M.B.; Binning, P.J. Modelling of green roof hydrological performance for urban drainage applications. *J. Hydrol.* **2014**, *519*, 3237–3248. [CrossRef]
19. Vanuytrecht, E.; Van Mechelen, C.; Van Meerbeek, K.; Willems, P.; Hermy, M.; Raes, D. Runoff and vegetation stress of green roofs under different climate change scenarios. *Landsc. Urban Plan.* **2014**, *122*, 68–77. [CrossRef]
20. Johannessen, B.G.; Hanslin, H.M.; Muthanna, T.M. Green roof performance potential in cold and wet regions. *Ecol. Eng.* **2017**, *106*, 436–447. [CrossRef]
21. Norwegian Meterological Insitute (MET). Available online: www.met.no (accessed on 23 January 2018).
22. Berghage, R.D.; Beattie, D.; Jarrett, A.R.; Thuring, C.; Razaei, F.; O'Connor, T.P. Green roofs for stormwater runoff control. 2009. Available online: http://nepis.epa.gov/exe/zypurl.cgi?dockey=p1003704.txt (accessed on 23 January 2018).

23. Peel, M.C.; Finlayson, B.L.; McMahon, T.A. Updated world map of the köppen-geiger climate classification. *Hydrol. Earth Syst. Sci. Discuss.* **2007**, *4*, 439–473. [CrossRef]

24. Johannessen, B.G.; Muthanna, T. Hydraulic performance of extensive green roofs in cold climate. *Contrôle À La Source/Source Control.* 2016. Available online: http://documents.irevues.inist.fr/bitstream/handle/2042/60499/1B1P04-134JOH.pdf (accessed on 2 March 2018).

25. Poë, S.; Stovin, V.; Berretta, C. Parameters influencing the regeneration of a green roof's retention capacity via evapotranspiration. *J. Hydrol.* **2015**, *523*, 356–367. [CrossRef]

26. Liu, K.; Minor, J. Performance evaluation of an extensive green roof. Presented at Green Rooftops for Sustainable Communities, Washington, DC, USA, 5–6 May 2005; Volume 1, pp. 1–11.

27. Li, Y.; Babcock, R.W. Green roof hydrologic performance and modeling: A review. *Water Sci. Technol.* **2014**, *69*, 727–738. [CrossRef] [PubMed]

28. Smith, D.W. *Cold Regions Utilities Monograph*; American Society of Civil Engineers: Reston, VA, USA, 1996.

29. Filtralite. Available online: http://www.filtralite.com/ (accessed on 23 January 2018).

30. Forschungsgesellschaft Landschaftsentwickung Landschaftsbau (FLL). *Guidelines for the Planning, Construction and Maintenance of Green Roofing: Green Roofing Guideline*; Forschungsgesellschaft Landschaftsentwickung Landschaftsbau: Bonn, Germany, 2008.

31. Braskerud, B.C. *Grønne Tak Og Styrtregn*; NVE: Oslo, Norway, 2014.

32. Bergknapp. Available online: http://www.bergknapp.no/hjem (accessed on 23 January 2018).

33. Committee, M.S.S. *The Minnesota Stormwater Manual*; Minnesota Pollution Control Agency: St.Paul, MN, USA, 2005.

34. Klima2050. Available online: http://www.klima2050.no/ (accessed on 23 January 2018).

water

MDPI

Article

Study on the Influence of Clogging on the Cooling Performance of Permeable Pavement

Jianguang Xie *, Sicheng Jia, Hua Li and Lei Gao

Department of Civil Engineering, Nanjing University of Aeronautics and Astronautics, Nanjing 210016, China; jiasc1993@163.com (S.J.); lihua112358@gmail.com (H.L.); glzjy@nuaa.edu.cn (L.G)
* Correspondence: xiejg@nuaa.edu.cn

Received: 6 February 2018; Accepted: 5 March 2018; Published: 10 March 2018

Abstract: Permeable pavement is often known as "cool road". However, the cooling performance will be weakened due to clogging. In this paper, the temperature field distribution model of asphalt pavement was obtained by Green's function. Gradations of porous asphalt mixture were designed to obtain different porosities, and the thermal properties of specimens with different porosities were tested and calculated. The simulation test was carried out to obtain the heating curves, which were used to verify the accuracy of the temperature model by comparing the results of the theoretical calculation to results of the test. The daily solar radiation intensity and air temperature changing functions were plugged into the model to calculate the temperature at the bottom of the middle surface. In this way, the simplified model of void fraction and cooling performance of the porous asphalt pavement was obtained. The results showed that the temperature at the bottom of the middle surface for permeable pavement was lower than that for traditional asphalt pavements. The gap was between 0.29 to 2.75 °C and it increased as the porosity of permeable pavement increased.

Keywords: sponge city; permeable pavement; Green's function; cooling performance

---

1. Introduction

As an important part of a "sponge city", permeable pavement plays a significant role in regulating local climate, accelerating groundwater recycling and reducing runoff on road surface. Permeable pavement is also known as "cool road". Due to its porous structure, the maximum temperature in the permeable pavement is lower than that in traditional asphalt pavement in summer [1]. The maximum temperature of asphalt pavement affects the service and safety performance. Problems such as rutting and deflection are more likely to occur as the maximum temperature rises, leading to instability failures [2–5]. The difference between the maximum temperatures of the permeable asphalt pavement and traditional asphalt pavement is defined as the cooling performance of permeable asphalt pavement. In the long-term use of permeable pavement, the study found that it was easily affected by the environment and the compaction conditions, leading to void clogging [6–8], weakening its permeability characteristics, and also affecting the cooling performance.

The numerical analysis method is often used to predict the temperature distribution internal to the pavement [9–12]. This method is based on finite element or finite difference method. In these numerical models, pavement structures are discretized into elements for solving governing equations. However, the accuracy of the calculation results is highly dependent on the density of the mesh and total number of elements in the model.

Different from the numerical analysis method, the analytical method can be used to obtain the rigorous solutions of pavement temperature fields with appropriate assumptions. Additionally, various transfer methods were used in the analytical method. With this approach, each layer of the pavement was treated as isotropic [13]. Based on the Fourier–Biot heat conduction equation and transform methods such as Laplace transform method and Green Function method, an analytical solution to heat conduction can be obtained very fast by solving the initial boundary value problems. The solution is usually an equation of instant temperature, thermal properties of materials in each layer and the heat flux on the surface of pavement [14–16]. Comparing with the numerical analysis method, the analytical method is more convenient and accurate. So the analytical method was adopted in the paper.

According to the analytical method, the changing rule between thermal properties and porosity of the asphalt mixture should be studied first. Then the maximum internal temperatures of the pavement can be obtained, based on which the changing rule of cooling performance with void clogging can also be obtained.

2. Objective

This paper aims to establish a prediction model of the cooling performance decline for layered permeable pavements due to clogging problems. In order to achieve the objective, both theoretical analysis and laboratory test were performed. The Green's function method was used in the prediction of pavement temperature fields. The thermal properties of porous asphalt mixture with different porosities were tested in laboratory. Finally, the cooling performance was represented by the difference between the maximum temperatures in the permeable asphalt pavement and traditional pavement.

3. Pavement Temperature Field with Green's Function Method

3.1. Heat Conduction Model of Multilayer Pavement Structure

Because the horizontal dimension of each layer of the pavement is much larger than the vertical direction, the heat conduction problem of the pavement can be regarded as the heat conduction problem of the large flat plate. Therefore, in the prediction of the temperature field, the method of one-dimensional heat conduction was adopted [17–19]. The pavement structure is shown in Figure 1. At the surface of the pavement $z = z_0 = 0$. The surface is considered as a mixed boundary where the pavement receives the solar radiation, radiates part of the heat into the atmosphere and exchanges heat with the air. When the depth is infinitely deep, the temperature can be considered as a constant.

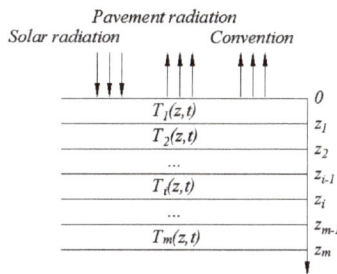

Figure 1. Heat transfer diagram of multilayered pavement system.

According to the theory of heat transfer, heat conduction and heat convection that occurred due to direct contact between fluid and solid are called convention heat transfer, which can be calculated according to Newton cooling formula:

$$q_r = h_r(T_{surf} - T_{air}) \tag{1}$$

where q_r is the heat flux when convective heat transfer occurs, $W \cdot m^{-2} \cdot s^{-1}$; h_r is convective heat transfer coefficient, $W/(m^2 \cdot K)$. T_{surf} is the temperature of the surface, K; and T_{air} is the temperature of the air, K.

The energy entering the road consists of short wave radiation and long wave radiation. Meanwhile, the road will radiate outward in a long wave way. The total amount of long wave radiation energy can be expressed as the following formula:

$$E_L = \varepsilon_a \sigma_{sb} T_{air}^4 \tag{2}$$

where E_L is the amount of long wave radiation energy, $W \cdot m^{-2} \cdot s^{-1}$; σ_{sb} is Stefan–Boltzmann constant, $5.67 \times 10^{-8} \ W \cdot m^{-2} \cdot K^{-4}$; and ε_a is atmospheric long wave emissivity.

The long wave radiation [9] of asphalt pavement emission can be calculated according to the following formula:

$$q_L = \varepsilon \sigma_{sb} T_{surf}^4 \tag{3}$$

where q_L is the outward amount of pavement radiation, $W \cdot m^{-2} \cdot s^{-1}$; and ε is pavement emissivity.

So the heat flux into the pavement system can be expressed as:

$$q = \varepsilon_e E_g + \varepsilon_f E_L - q_L - q_r \tag{4}$$

where q is the heat flux into the pavement; E_g is the total solar shortwave radiation, $W \cdot m^{-2} \cdot s^{-1}$; ε_e is the absorptivity of solar shortwave radiation; and ε_f is the absorptivity of atmospheric long wave radiation.

3.2. Model of Temperature Field in Asphalt Pavement Based on Green's Function

Green's function method is widely used to solve heat conduction problems, especially when the boundary conditions are nonhomogeneous. It can be used to solve single dielectric problems, as well as inhomogeneous problems for composite dielectrics. The heat conduction problem can be expressed as follows:

$$\alpha_i \frac{\partial^2 T_i(z,t)}{\partial z^2} = \frac{\partial T_i(z,t)}{\partial t} \qquad z_{i-1} < z < z_i, \ t > 0 \tag{5}$$

where $T_i(z,t)$ is the distribution of temperature in i-th layer, K; z is the depth, m; and α_i is the thermal diffusivity of i-th layer, $m^2 \cdot s^{-1}$.

The boundary conditions can be expressed as follows:

$$-k_1 \frac{\partial T_1(0,t)}{\partial z} + h_1 T_1(0,t) = h_1 f_1(t) \tag{6}$$

$$k_i \frac{\partial T_i(z,t)}{\partial z}\Big|_{z=z_i} = k_{i+1} \frac{\partial T_{i+1}(z,t)}{\partial z}\Big|_{z=z_i} \tag{7}$$

$$T_i(z,t)|_{z=z_i} = T_{i+1}(z,t)|_{z=z_i} \tag{8}$$

$$T_m(z,t)|_{z=z_m} = f_2(t) \tag{9}$$

$$T_i(z,0) = I_i(z) \tag{10}$$

where k_i is the thermal conductivity of i-th layer, $W \cdot m^{-1} \cdot K^{-1}$; $f_1(t)$ is the heat flux into the pavement; $f_2(t)$ is the distribution of temperature as $z = z_m$.

Construct the solution of $T_i(z,t)$ as the form as follows:

$$T_i(z,t) = \varphi_i(z) f_1(t) + \psi_i(z) f_2(t) + \theta_i(z,t) \tag{11}$$

where $\varphi_i(z)$ and $\psi_i(z)$ are functions of z; $f_1(t)$ and $f_2(t)$ are functions of t; and $\theta_i(z,t)$ is function of z and t.

$\varphi_i(z)$ should satisfy the steady state heat conduction problem given as Equation (12):

$$\frac{d^2 \varphi_i(t)}{dz^2} = 0 \qquad z_{i-1} < z < z_i \tag{12}$$

The boundary conditions of Equation (12) are shown as Equations (13)–(16).

$$\varphi_m(z)|_{z=z_m} = 0 \tag{13}$$

$$\varphi_i(z)|_{z=z_i} = \varphi_{i+1}(z)|_{z=z_i} \tag{14}$$

$$k_i\frac{d\varphi_i(z)}{dz}|_{z=z_i} = k_{i+1}\frac{d\varphi_{i+1}(z)}{dz}|_{z=z_i} \tag{15}$$

$$-k_1\frac{d\varphi_1(z)}{dz}|_{z=0} + h_1\varphi_1(z)|_{z=0} = 0 \tag{16}$$

In addition, $\psi_i(z)$ should satisfy the steady state heat conduction problem given as Equation (17), subjected to the boundary conditions Equations (18)–(21).

$$\frac{d^2\psi_i(t)}{dz^2} = 0 \qquad z_{i-1} < z < z_i \tag{17}$$

$$\psi_m(z)|_{z=z_m} = 1 \tag{18}$$

$$\psi_i(z)|_{z=z_i} = \psi_{i+1}(z)|_{z=z_i} \tag{19}$$

$$k_i\frac{d\psi_i(z)}{dz}|_{z=z_i} = k_{i+1}\frac{d\psi_{i+1}(z)}{dz}|_{z=z_i} \tag{20}$$

$$-k_1\frac{d\psi_1(z)}{dz}|_{z=0} + h_1\psi_1(0) = h_1 f_1(t)|_{z=0} \tag{21}$$

$\theta_i(z,t)$ should satisfy the transient state heat conduction problem given as Equations (22) and (23), subjected to the boundary conditions Equations (24)–(28).

$$\alpha_i\frac{\partial^2\theta_i(z,t)}{\partial z^2} + g_i(z,t) = \frac{\partial T_i(z,t)}{\partial t} \qquad z_{i-1} < z < z_i, \ t > 0 \tag{22}$$

$$g_i(z,t) = \varphi(z)\frac{df_1(t)}{dt} + \psi(z)\frac{df_2(t)}{dt} \tag{23}$$

$$-k_1\frac{\partial\theta_1(z,t)}{\partial z}|_{z=0} + h_1\theta_1(z,t)|_{z=0} = 0 \tag{24}$$

$$k_i\frac{\partial\theta_i(z,t)}{\partial z}|_{z=z_i} = k_{i+1}\frac{\partial\theta_{i+1}(z,t)}{\partial z}|_{z=z_i} \tag{25}$$

$$\theta_i(z,t)|_{z=z_i} = \theta_{i+1}(z,t)|_{z=z_i} \tag{26}$$

$$\theta_m(z,t)|_{z=z_m} = 0 \tag{27}$$

$$\theta_i(z,0) = I_i(z) - f_1(0)\varphi_1(x) - f_2(0)\psi_1(x) \equiv I_i^*(x) \tag{28}$$

The solution of $\phi_i(z)$ and $\psi_i(z)$ can be constructed as Equations (29) and (30):

$$\varphi_i(z) = A_i + B_i z \tag{29}$$

$$\psi_i(z) = C_i + D_i z \tag{30}$$

The unknown coefficients in Equations (29) and (30) can be determined by Equations (12)–(21).

Because there is no heat source inside the pavement, the solution of $\theta_i(z,t)$ can be constructed as $\theta_i(z,t) = Z(z)\,\Gamma(t)$; Equations (31) and (32) can be obtained as $\theta_i(z,t)$ substituted into Equations (22) and (23).

$$\frac{d\Gamma(t)}{dt} + \beta_n^2\Gamma(t) = 0 \qquad t > 0 \tag{31}$$

$$\frac{d^2 Z(z)}{dz^2} + \frac{\beta_n^2}{\alpha_i}Z(z) = 0 \qquad z_{i-1} < z < z_i \tag{32}$$

where β_n is the eigenvalue, which can be determined by Equations (22) and (23). Equations (33) and (34) can be obtained by equations above.

$$\Gamma(t) = e^{-\beta_n^2 t} \tag{33}$$

$$Z(z) = A_{i,n} \sin(\frac{\beta_n}{\sqrt{\alpha_i}} z) + B_{i,n} \cos(\frac{\beta_n}{\sqrt{\alpha_i}} z) \tag{34}$$

So $\theta_i(z,t)$ can be expressed as follows:

$$\theta_i(z,t) = \sum_{n=1}^{\infty} c_n e^{-\beta_n^2 t} Z(z) \tag{35}$$

$$c_n = \frac{1}{N(\beta_n)} \sum_{i=1}^{N} \frac{k_i}{\alpha_i} \int_{z_i}^{z_{i+1}} Z_i(\beta_n, z) I_i^*(z) dz \tag{36}$$

$$N(\beta_n) = \sum_{i=1}^{N} \frac{k_i}{\alpha_i} \int_{z_i}^{z_{i+1}} Z_i^2(\beta_n, z) dz \tag{37}$$

Equation (22) can be deduced as:

$$\begin{aligned}
\theta_i(z,t) &= \sum_{j=1}^{N} \frac{k_j}{\alpha_j} \int_{z_j}^{z_{j+1}} [\sum_{n=1}^{\infty} \frac{1}{N(\beta_n)} e^{-\beta_n^2 t} Z_i(\beta_n, z) Z_i(\beta_n, z') \cdot I_j^*(z') dz' \\
&= \sum_{j=1}^{N} \int_{z_j}^{z_{j+1}} G_{ij}(z, t|z', \tau)|_{\tau=0} I_j^*(z') dz'
\end{aligned} \tag{38}$$

where

$$G_{ij}(z, t|z', \tau) = \sum_{n=1}^{\infty} \frac{1}{N(\beta_n)} e^{-\beta_n^2 (t-\tau)} Z_i(\beta_n, z) Z_j(\beta_n, z') \tag{39}$$

According to the heat transfer theory, once the Green's function is determined, the temperature field of the pavement can be determined directly as:

$$\begin{aligned}
T_i(z,t) &= (A_i + B_i z) f_1(t) + (C_i + D_i z) f_2(t) \\
&+ \sum_{j=1}^{N} [\frac{k_i}{\alpha_i} \int_{z_j}^{z_{j+1}} G_{ij}(z, t|z', \tau)|_{\tau=0} I_j^*(z') dz' \qquad z_{i-1} < z < z_i, t > 0 \\
&+ \int_0^t G_{i1}|_{z'=0} f_1(\tau)
\end{aligned} \tag{40}$$

4. Test Process

4.1. High Viscosity Modified Asphalt

Different from other studies, the high viscosity additive (HVA) and styrene-butadiene-styrene (SBS) block copolymer modified asphalt were used in this test. The indexes of asphalt and HVA were measured by the methods in Chinese standards. The results of SBS modified asphalt test are shown in Table 1. The test results of HVA are shown in Table 2 and the indexes of high viscosity modified asphalt modified by high viscosity additives are shown in Table 3.

Table 1. Properties of SBS modified asphalt.

Test	Value	Specification Limits
Penetration 25 °C, 100 g, 5 s (0.1 mm)	54	40–60
Softening point (°C)	88.0	≥75
Ductility, 5 °C, 5 cm/min (cm)	28	≥20
Density, 25 °C (g/cm³)	1.031	
After aging in rolling thin film oven		
Mass change (%)	+0.045	±1.0
Retained penetration, 25 °C (%)	83	≥65
Retained ductility, 5 °C (cm)	19	≥15

Table 2. Properties of high viscosity additive.

Index	Value	Specification Limits
Mass of single particle (g)	0.022	≤0.03
Density (g/cm³)	0.978	0.90–1.00
Appearance	Granular, uniform and plump	-

Table 3. Properties of high viscosity modified asphalt.

Index	Value	Specification Limits
Penetration 25 °C, 100 g, 5 s (0.1 mm)	44	40–60
Softening point (°C)	98.0	≥90
Ductility, 5 °C, 5 cm/min (cm)	35	≥30
Dynamic viscosity, 60 °C(Pa·s)	440,806	≥400,000
Density, 25 °C	1.027	-
After aging in rolling thin film oven		
Mass change (%)	−0.023	±0.6
Retained penetration, 25 °C (%)	82.4	≥65
Retained ductility, 5 °C (cm)	25	≥20

4.2. Mix and Structure Design

Asphalt concrete (AC) is widely used in the traditional pavement. Different from the traditional pavement, the concrete used in the permeable pavement is often called porous asphalt concrete (PAC). Besides, the maximum particle size of upper-surface is usually 13 mm in China, and for the mid-surface, it is usually 20 mm. In addition, for better simulation effect, the test specimens were all made up of two layers. Traditional pavement specimen was made up of 4 cm thick AC13 mixture and 6 cm thick AC20 mixture. The permeable pavement specimens were made up of 4 cm thick PAC13 mixture and 6 cm thick AC20 mixture. Additionally, the different clogging situations were simulated by porosities ranged from 15% to 24%.

4.2.1. Mix Design

The present study shows that the porosity of PAC13 is greatly influenced by the aggregate passing proportion of 4.75 mm ($P_{4.75}$) and 2.36 mm ($P_{2.36}$) [20]. So the porosity of PAC13 mixture was mainly adjusted by controlling $P_{4.75}$ and $P_{2.36}$. The gradation is shown in Table 4 and the gradation curves are shown in Figure 2.

Table 4. Gradation of aggregate blends for asphalt mixture PAC13.

Sieve Size		16	13.2	9.5	4.75	2.36	1.18	0.6	0.3	0.15	0.075	Porosity
Upper limit		100	100	71	30	22	18	14.0	12	9	7	
Lower limit		100	90	40	10	8	6	4	3	3	3	
	1	100	92.7	56.8	16.7	10.4	7.9	6.6	5.2	4.5	3.8	20.79%
	2	100	95.0	65.0	25.0	16	14.0	11.0	9.0	6.0	4.0	19.66%
Gradation	3	100	95.0	67.5	26.5	18.5	14.3	10.0	8.0	6.0	4.0	18.11%
	4	100	95.0	70.0	28.0	21	14.5	10.5	9.0	6.0	4.0	16.36%
	5	100	95.0	45.0	10.0	8.0	7.0	6.0	5.0	4.0	3.0	23.05%

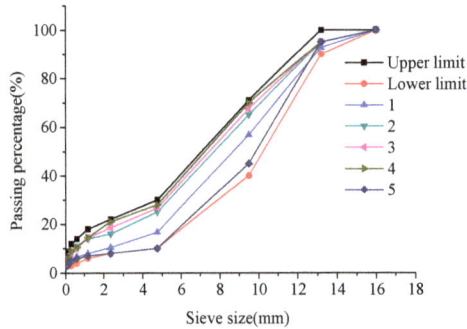

Figure 2. Gradation of PAC13 asphalt mixture.

In addition, the gradations of aggregate blends for asphalt mixture AC13 and AC20 are listed in Tables 5 and 6.

Table 5. Gradation AC13 asphalt mixture.

Sieve Size	16	13.2	9.5	4.75	2.36	1.18	0.6	0.3	0.15	0.075
Upper limit	100	100	85	68	50	38	28	20	15	8
Lower limit	100	90	68	38	24	15	10	7	5	4
Gradation	100	96.9	70.2	41.8	29.1	19.9	14.4	10.5	8.2	5

Table 6. Gradation AC20 asphalt mixture.

Sieve Size	26.5	19.0	16.0	13.2	9.5	4.75	2.36	1.18	0.6	0.3	0.15	0.075
Upper limit	100	100	95	86	70	48	33	23	16	11	9	6
Lower limit	100	90	83	73	56	35	22	15	10	6	5	4
Gradation	100	96.8	89.4	78.9	60.9	42.8	29.3	21.1	14.6	10.7	8.3	5.5

4.2.2. Structure Design

A total of 6 experimental groups were set up in the test. The size of the specimens was 30 cm × 30 cm × 10 cm. They were divided into 2 types: traditional group (AC) and permeable group (S1–S5). The specific combination form is shown in Table 7. Additionally, in order to facilitate later experiment, every gradation of asphalt mixture was also made into rutting plate specimens in the size of 30 cm × 30 cm × 5 cm (Labeling AC13, AC20 and PAC13-1–PAC13-5).

Table 7. Structure design of specimens.

Type	Porosity/Material (Upper Layer)	Porosity/Material (Lower Layer)	Label
Traditional	AC13	AC20	AC
Permeable	16.68%/PAC13	AC20	S1
	18.11%/PAC13	AC20	S2
	19.66%/PAC13	AC20	S3
	20.79%/PAC13	AC20	S4
	23.05%/PAC13	AC20	S5

4.2.3. Specimen Preparation

The temperature field in the pavement needed to be monitored in the later tests. Thus, the thermocouple sensors should be embedded in the specimen preparation process. The length of the thermocouples inside the specimen was 15 cm, and the depth of 2 cm, 4 cm, 7 cm and 10 cm respectively, which was shown in Figure 3. Two thermocouples were embedded in each depth, and the temperature of each layer is expressed as the average value of the measured values of the two sensors. In order to facilitate the embedment of the sensor, the customized model in Figure 4a was made. The specimens were prepared by the method of layered compaction. After the lower layer was cooled, the upper layer was added and compacted.

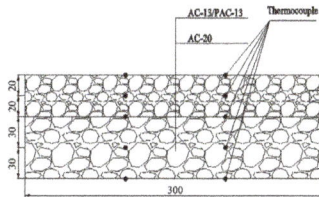

Figure 3. Location of thermocouple embedded.

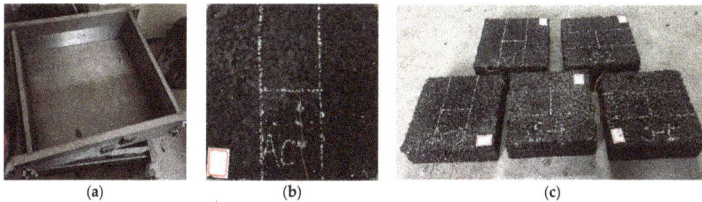

Figure 4. (**a**) Customized models; (**b**) AC specimen; (**c**) S1–S5 specimens.

4.3. Determination of Thermal Properties

Parameters in the heat conduction model included mass density, specific heat and thermal conductivity of the mixture. The mass density of the mixture was expressed by the bulk density. In addition, in the study of the thermally physical properties of asphalt mixtures, Zou [21] proposed that the specific heat can be expressed in a parallel model, which is shown as follows:

$$A_c = \sum A_i B_i \tag{41}$$

where A_c is the specific heat of asphalt mixtures; A_i is the specific heat of the *i*-th component; B_i is mass fraction of the *i*-th component.

The mass densities and specific heat of each gradation are shown in Table 8.

Table 8. Specific heat and density of 5 cm thick specimens.

Numbering	Porosity (%)	Specific Heat (J/(kg·K))	Density (kg/m^3)
PAC13-1	16.68	926.88	2229.52
PAC13-2	18.11	926.09	2194.62
PAC13-3	19.66	926.09	2142.77
PAC13-4	20.79	925.30	2116.84
PAC13-5	23.05	915.82	2090.92
AC13	-	922.51	2420.96
AC20	-	920.60	2381.07

The thermal conductivity of the mixture was related to the conductivity of aggregate, asphalt, mineral powder and porosity. Williamson [22] once put forward a formula:

$$k_m = (k_a)^g \cdot (k_b)^h \cdot (k_v)^i \cdot (k_w)^j \tag{42}$$

where k_a, k_b, k_v and k_w are the thermal conductivity of the aggregate, asphalt, mineral powder and air respectively. G, h, i and j are volume fraction of each component. The thermal conductivities [23] of the components were listed in Table 9.

Table 9. Thermal conductivities of the components.

Index	Aggregate	Asphalt	Mineral Powder	Air
Thermal conductivity (W/(m·K))	2.18	0.66	0.2	0.026

In addition, the thermal conductivities of the specimens were also tested by heat flow meter method, the results of the test and Williamson's formula were shown in Table 10.

Table 10. Thermal conductivities of 5 cm thick specimens.

Numbering	Porosity (%)	Thermal Conductivity (W/(m·K))	
		Williamson's Formula	Test
PAC13-1	16.68	0.82	1.03
PAC13-2	18.11	0.78	0.97
PAC13-3	19.66	0.73	0.93
PAC13-4	20.79	0.70	0.88
PAC13-5	23.05	0.67	0.8
AC13	-	1.07	1.15
AC20	-	1.16	1.38

From Table 10, for PAC13 mixtures, the calculation results of Williamson formula and the measured results decreased with the increase of porosity. The increasing trend is consistent, so it was considered whether there was linear relationship between them. To facilitate the analysis and comparison, the data of Table 10 is shown in Figure 5, and the relationship between the tested and calculated values was obtained by linear fitting.

Can be seen from Figure 5, it can be linear fitted well between test value and the theoretical value of the thermal conductivity of PAC mixture. So in this paper, a new model was proposed based on the Williamson formula to explain the relationship between thermal conductivity of the mixture and that of the components, which is shown as Equation (43):

$$k = -0.365 + 1.808(k_a)^g \cdot (k_b)^h \cdot (k_v)^i \cdot (k_w)^j \tag{43}$$

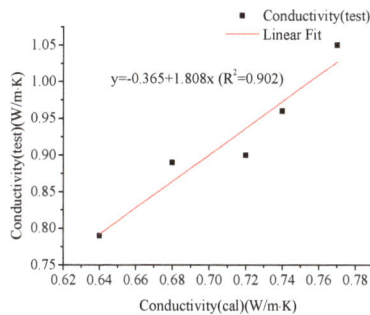

Figure 5. Thermal conductivities of 5 cm thick specimens.

4.4. Test Method

The infrared lamp was used to simulate the solar radiation in summer. Considering the size ratio of the vertical and horizontal directions of the road surface, the sides and the bottom of the rutting plate specimens can be set as an adiabatic surface while the upper surface was set as a mixed surface to exchange heat with the outside.

As the temperature of the road surface in summer was about 65–70 °C, the target of the equilibrium temperature of the heating test was set in the gap. The AC specimen was used to adjust the height of the infrared light. When the specimen temperature reaches the equilibrium range, the height of the infrared lamp would be fixed. The result showed that when the height of the infrared lamp was controlled at 28 cm above the surface and the test lasting time was 60 min, the temperature of the specimen surface would reach the target range and kept stable.

The heating tests were conducted after the height was determined. The infrared lamp should be cooled after each heating test to avoid the influence of the remaining warmth on the later test. The experimental devices were shown in Figure 6a. Additionally, in order to prevent the influence of wind, the enclosure was installed around the specimen, shown in Figure 6b.

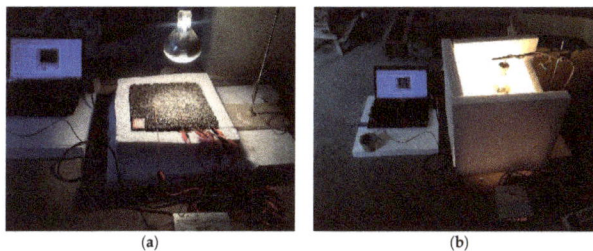

Figure 6. (a) The experimental devices; (b) The enclosure around the specimen.

4.5. Data Comparison

Several hypotheses were adopted in the theoretical calculation.

(1) The heating power on the infrared light remained constant during the test;
(2) The sides and the bottom of the specimen were adiabatic boundaries, and the upper surface of the specimen is a mixed boundary of fluid and solid;
(3) Each layer of asphalt mixture was isotropic materials;
(4) The thermal properties of asphalt mixture remained constant during the test;
(5) There was no thermal resistance between two linked layers.

The AC specimen was set as the matched group. With the thermally physical properties of the AC specimen plugged into Equation (40) and surface temperature curve as the reference, the power of the infrared radiation was determined to be 0.157 W/mm². Then the thermal properties of the remaining specimens were also plugged into Equation (40). The results of theoretical calculation and tests are shown in the following figures.

Firstly, it can be found that the theoretical calculation results are similar with the actual results. So it can be determined that the model was suitable with the heat transfer of porous asphalt mixture, and it can be used to predict the temperature of the whole pavement.

In addition, it can be seen from Figures 7 and 8 that the temperature on the surface of the specimen rose fastest. As the test went on, the temperature rising curve of the specimen gradually became gentle and tended to be stable. In addition, by comparing the temperature among different depths, it can be concluded that the temperature conduction in the asphalt pavement had the hysteresis in the vertical direction. With the increase in depth, the hysteresis phenomenon became more obvious. Taking AC specimen as an example, the surface temperature changes rapidly in the first 5 min. But at the bottom, the temperature had almost no change in 30 min.

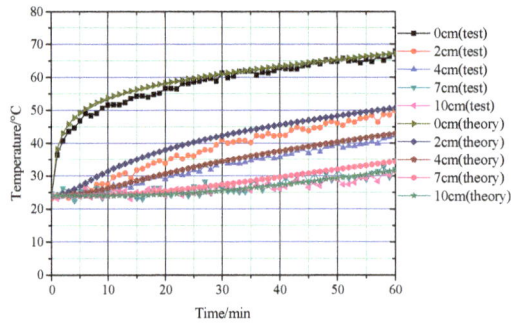

Figure 7. Results of theoretical calculation and tests of AC.

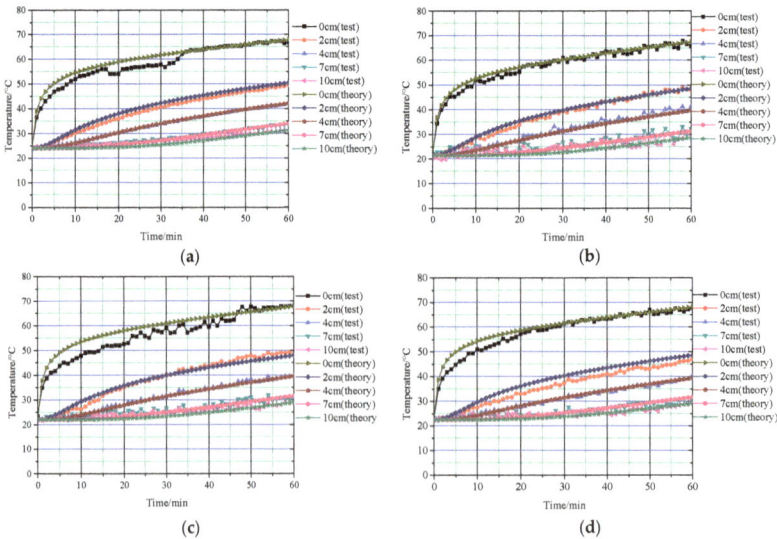

Figure 8. *Cont.*

(e)

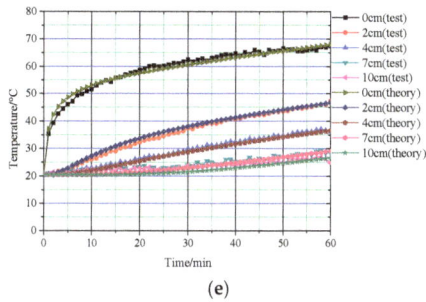

Figure 8. (a) Results of theoretical calculation and tests of S1; (b) Results of theoretical calculation and tests of S2; (c) Results of theoretical calculation and tests of S3; (d) Results of theoretical calculation and tests of S4; (e) Results of theoretical calculation and tests of S5.

The temperature of each depth of the specimens can also be obtained by the curves shown in the figures above. The temperature in the depth of 4 cm and 10 cm of each specimen was extracted and is listed in Table 11.

Table 11. Temperature in the specimens.

Specimen	Porosity/Material (Upper)	Porosity/Material (Lower)	Temperature in 4 cm (°C)	Temperature in 10 cm (°C)
AC	AC13	AC20	42.4	31.4
S1	16.68%/PAC13	AC20	42.0	31.2
S2	18.11%/PAC13	AC20	41.4	30.3
S3	19.66%/PAC13	AC20	40.8	29.5
S4	20.79%/PAC13	AC20	40.5	29.4
S5	23.05%/PAC13	AC20	40.0	28.9

It can be visually seen from Table 11 that in both depth of 4 cm and 10 cm, the temperatures in permeable pavements were lower than the traditional pavement. It showed that the cooling effect still existed in the permeable pavement under the extreme conditions of no wind on the surface or water in the void. Through the comparison of single layered PAC asphalt specimens, it can be found that with the increase of porosity, both of the temperature values in the two depths were showing a downward trend. The results showed that the cooling effect of drainage asphalt increased with the increase of void fraction. As the porosity of PAC13 layer changed from 16.68% to 23.05%, the maximum temperature of permeable asphalt pavement at the depth of 4 cm and 10 cm were 0.4–2.4 °C and 0.2–2.5 °C lower than the traditional pavement respectively.

5. The Cooling Performance of the Temperature Field with Different Porosities

The pavement was usually divided into six layers. The upper surface was AC13/PAC13, the middle surface was AC20, and the lower surface was AC25. The order and thickness of each layer was shown in Figure 9, where CTB stands for cement treated base, LS stands for lime-stabilized soil and SG stands for subgrade.

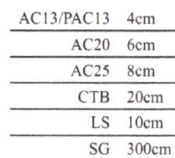

AC13/PAC13	4cm
AC20	6cm
AC25	8cm
CTB	20cm
LS	10cm
SG	300cm

Figure 9. Structure of the pavement. CTB: cement treated base; LS: lime-stabilized soil; SG: subgrade.

The thermally physical properties of each layer were shown in Table 12.

Table 12. Thermally physical properties of each layer.

Layer	Density (kg/m^3)	Specific Heat (J·kg^{-1}·K^{-1})	Conductivity (J·m^{-1}·h^{-1}·K^{-1})
AC25	2300	924.9	1.3
CTB	2200	911.7	1.56
LS	2100	942.9	1.43
SG	1800	1040	1.56

After the structural form was determined, it was necessary to ascertain the thermal physical parameters of each layer, the solar radiation intensity and the variation law of the daily air temperature. The amount of solar radiation [12] in the model was expressed as the form of $q(t)$ by Equation (44).

$$q(t) = \begin{cases} 0 & 0 \le t \le 12 - \frac{c}{2} \\ q_0 \cos m\omega(t - 12) & 12 + \frac{c}{2} \le t \le 12 + \frac{c}{2} \\ 0 & 12 + \frac{c}{2} \le t \le 24 \end{cases} \tag{44}$$

where q_0 is the maximum radiation intensity at midday, $q_0 = 0.131\,mQ$, $m = 12/c$; Q is the total amount of solar radiation in the day, J/m^2; c is the actual effective sunshine time, h; and ω is angular frequency, $\omega = 2\pi/24$, rad.

The changing rule of the daily air temperature [12] was shown in Equation (27)

$$T_a = \overline{T}_a + T_m[0.96\sin\omega(t - t_0) + 0.14\sin 2\omega(t - t_0)] \tag{45}$$

where T_a is the air temperature, °C; $\overline{T}_a$ is the average value of the temperature in a day, $\overline{T}_a = (T_{max} + T_{min})/2$, °C; T_m is the daily temperature variation, $T_m = (T_{max} - T_{min})/2$, °C; t_0 is the initial phase; and $T_{max} = 40$ °C, $T_{min} = 26$ °C, $t_0 = 3$.

Because the temperature at the bottom of the middle surface has the greatest influence on rutting, the change of temperature in the depth of 10 cm within 24 h was studied. The calculation results of AC and S5 were shown in Figure 10.

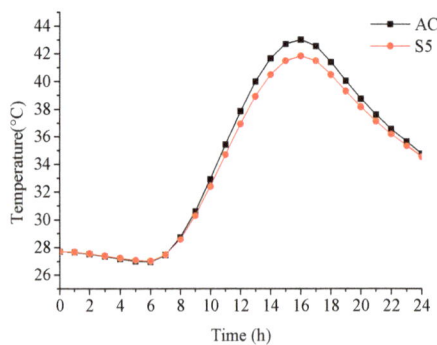

Figure 10. Temperature at the bottom of the middle surface.

It can be seen from Figure 10 that the temperature would be more stable and the maximum temperature would lower in the permeable pavement. It indicated that the permeable pavement could reduce the internal maximum temperature and be more environmental-friendly. The maximum temperatures at the bottom of middle surface of each pavement type were calculated and shown in

Table 13. Additionally, the cooling performance was represented by the reduced temperature between AC and permeable pavements.

Table 13. Cooling performance of all structures.

Type	Maximum Temperature (°C)	Reduced Temperature (°C)
AC	43.00	0.00
S1	42.71	0.29
S2	42.51	0.49
S3	42.37	0.63
S4	42.18	0.82
S5	41.82	1.18

From the theoretical calculation results, it can be seen that the cooling performance would be better with the increase of porosity. The maximum value of cooling performance could reach 1.18 °C. So it was an effective method to contrast the urban heat island effect by adopting permeable pavement in the building of sponge cities. The results in Table 13 were shown in Figure 11.

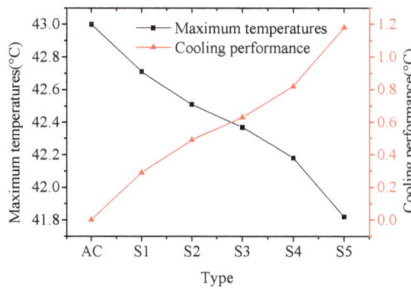

Figure 11. Cooling performance of the structures.

In addition, for the permeable pavement, as the porosity was 16.68%, the maximum temperature of the bottom of the middle surface was only 0.29 °C lower than that of traditional pavement. There was almost no cooling performance, and the porosity of permeable pavement was usually 20%, whose cooling performance was about 0.63 °C. So it can be judged that the cooling performance of the permeable pavement would reduce by over 50% as the porosity changed from 20% to 16.68%. Additionally, there was a linear relationship between the cooling performance and porosity in the single layered permeable pavements, which was shown in Figure 12.

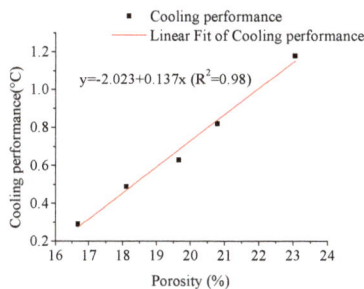

Figure 12. Relationship between the cooling performance and porosity.

The simplified model of cooling performance and porosity can be used to judge the attenuation of cooling effect of permeable asphalt pavement quickly in practical engineering and determine the clean cycle of the pavement.

6. Conclusions

In this paper, the temperature field distribution model of a layered pavement system was obtained based on Green's function. The internal temperature of PAC specimens with different porosities was studied by comparing theory calculations to test results. The temperature field model was applied to evaluate cooling performance of the permeable pavement under different clogging conditions. Through the analysis of the results, the following conclusions were obtained:

(1) The prediction model of the temperature field of permeable pavement was obtained based on Green's function, and the model was verified by the experimental results. The values of theoretical calculation were close to the experimental results. This indicated that the model had a wide applicability, which could be applied to the theoretical analysis of heat conduction problem for asphalt pavement.
(2) The linear fitted model was proposed based on the Williamson formula and the results of the test. The model could explain the relationship between thermal conductivity of the mixture and that of the components well.
(3) According to the results of test, the cooling performance of pavement became worse with the attenuation of porosity. When the porosity of permeable asphalt pavement reaches 23.05%, the cooling performance at the depth of 10 cm could reach 1.18 °C. When the porosity reached 16.68%, the cooling effect declined to 0.29 °C.
(4) Void clogging has a great influence on the cooling effect of drainage pavement. At present, the porosity of single layered drainage pavement is about 20%. The cooling effect of the pavement under this porosity was about 0.63 °C. If the porosity declined by about three percent, the cooling performance would be less than half of the original.
(5) Through the regression analysis of the relationship between cooling performance and porosity in the permeable pavement, a linear model was set up. The model could be used as a reference for rapid judgment of pavement cooling performance in the field, so as to determine the cleaning cycle of permeable pavement.

Acknowledgments: The authors would like to acknowledge Technological Innovation Project of Ministry of Transport of the People's Republic of China (2015315Q11020), Jiangsu Scientific and Technological Development Program (BE2015349), National Key Technology Research and Development Program of the Ministry of Science and Technology of China (2015BAL02B00) and the Fundamental Research Funds for the Central Universities (NQ2018001) for its financial support in this project.

Author Contributions: In this article, Jianguang Xie conceived and designed the experiments; Sicheng Jia performed the experiments; Lei Gao, Sicheng Jia and Hua Li analyzed the data; Jianguang Xie contributed analysis tools; Sicheng Jia wrote the paper.

Conflicts of Interest: The authors declare no conflict of interest.

References

1. Buyung, N.R.; Ghani, A.N.A. Permeable pavements and its contribution to cooling effect of surrounding temperature. In Proceedings of the International Conference of Global Network for Innovative Technology and Awam International Conference in Civil Engineering, Bukit Jambul, Malaysia, 8–9 August 2017; p. 170003.
2. Van Thanh, D.; Feng, C.P. Study on Marshall and Rutting test of SMA at abnormally high temperature. *Constr. Build. Mater.* **2013**, *47*, 1337–1341. [CrossRef]
3. Walubita, L.F.; Faruk, A.N.M.; Zhang, J.; Hu, X.; Lee, S.I. The Hamburg rutting test—Effects of HMA sample sitting time and test temperature variation. *Constr. Build. Mater.* **2016**, *108*, 22–28. [CrossRef]

4. Javilla, B.; Mo, L.; Hao, F.; Shu, B.; Wu, S. Multi-stress loading effect on rutting performance of asphalt mixtures based on wheel tracking testing. *Constr. Build. Mater.* **2017**, *148*, 1–9. [CrossRef]

5. Zheng, Y.; Zhang, P.; Liu, H. Correlation between pavement temperature and deflection basin form factors of asphalt pavement. *Int. J. Pavement Eng.* **2017**, 1–10. [CrossRef]

6. Al-Rubaei, A.M.; Stenglein, A.L.; Viklander, M.; Blecken, G.T. Long-Term Hydraulic Performance of Porous Asphalt Pavements in Northern Sweden. *J. Irrig. Drain. Eng.* **2013**, *139*, 499–505. [CrossRef]

7. Coleri, E.; Kayhanian, M.; Harvey, J.T.; Yang, K.; Boone, J.M. Clogging evaluation of open graded friction course pavements tested under rainfall and heavy vehicle simulators. *J. Environ. Manag.* **2013**, *129*, 164–172. [CrossRef] [PubMed]

8. Cantisani, G.; D'Andrea, A.; Di Mascio, P.; Loprencipe, G. Reliance of Pavement Texture Characteristics on Mix-Design and Compaction Process. In *8th RILEM International Symposium on Testing and Characterization of Sustainable and Innovative Bituminous Materials*; Springer: Dordrecht, The Netherlands, 2016; pp. 271–281.

9. Hermansson, A. Simulation model for calculating pavement temperature including maximum temperature. *Transp. Res. Rec.* **2000**, *1699*, 134–141. [CrossRef]

10. Hermansson, A. A mathematical model for calculating pavement temperatures, comparisons between calculated and measured temperatures. *Transp. Res. Rec. J. Transp. Res. Board* **2001**, *1746*, 180–188. [CrossRef]

11. Barber, E.S. Calculation of maximum pavement temperatures from weather reports. *Highw. Res. Board Bull.* **1957**, *168*, 1–8.

12. Straub, A.; Schenck, H.N., Jr.; Przbycien, F.E. Bituminous Pavement Temperature Related to Climate. *Highw. Res. Rec.* **1968**, *256*, 53–77.

13. Yan, Z. Analysis of the Temperature Field in Layerd Pavement System. *J. Tongji Univ.* **1984**, *3*, 76–85. (In Chinese)

14. Liu, C.; Yuan, D. Temperature distribution in layered road structures. *J. Transp. Eng.* **2000**, *126*, 93–95. [CrossRef]

15. Gao, L.; Ni, F.; Charmot, S. High-temperature performance of multilayer pavement with cold in-place recycling mixtures. *Road Mater. Pavement Des.* **2014**, *15*, 804–819. [CrossRef]

16. Wang, D. Analytical approach to predict temperature profile in a multilayered pavement system based on measured surface temperature data. *J. Transp. Eng.* **2012**, *138*, 674–679. [CrossRef]

17. Chen, J.; Li, L.; Zhao, L.; Dan, H.-C.; Yao, H. Solution of pavement temperature field in "Environment-Surface" system through Green's function. *J. Central South Univ.* **2014**, *21*, 2108–2116. [CrossRef]

18. Chen, J.; Wang, H.; Zhu, H. Analytical approach for evaluating temperature field of thermal modified asphalt pavement and urban heat island effect. *Appl. Therm. Eng.* **2017**, *113*, 739–748. [CrossRef]

19. Sreedhar, S.; Biligiri, K.P. Development of pavement temperature predictive models using thermophysical properties to assess urban climates in the built environment. *Sustain. Cities Soc.* **2016**, *22*, 78–85. [CrossRef]

20. Xing, M. Research on Composition Design and Performance of Pervious Asphalt Mixture. Master's Thesis, Chang'an University, Xi'an, China, 2007. (In Chinese)

21. Zou, L. Research on Thermal Physical Parameters of Asphalt Mixture. Master's Thesis, Chang'an University, Xi'an China, 2011. (In Chinese)

22. Williamson, R.H. Effects of environment on pavement temperature. *Intl. Conf Structural Design Proc.* **1972**, *9*, 144–158.

23. Li, X. Research on Thermophysical Characteristic of Asphalt Mixture. Master's Thesis, Harbin Institute of Technology, Harbin, China, 2007. (In Chinese)

![water logo] *water*

MDPI

Article

A Semi Risk-Based Approach for Managing Urban Drainage Systems under Extreme Rainfall

Carlos Salinas-Rodriguez [1,2,*], Berry Gersonius [1,2], Chris Zevenbergen [1], David Serrano [1] and Richard Ashley [1,2]

[1] Flood Resilience Chair Group, Water Science and Engineering Department, IHE-Delft Institute for Water Education, 2611 AX Delft, The Netherlands; b.gersonius@un-ihe.org (B.G.); c.zevenbergen@un-ihe.org (C.Z.); serra13@un-ihe.org (D.S.); r.ashley@sheffield.ac.uk (R.A.)

[2] Cooperative Research Centre for Water Sensitive Cities, Monash University, Clayton, VIC 3800, Australia

* Correspondence: c.salinas@un-ihe.org or csalirod@gmail.com; Tel.: +31-614-900-976

Received: 4 January 2018; Accepted: 21 March 2018; Published: 26 March 2018

Abstract: Conventional design standards for urban drainage systems are not set to deal with extreme rainfall events. As these events are becoming more frequent, there is room for proposing new planning approaches and standards that are flexible enough to cope with a wide range of rainfall events. In this paper, a semi risk-based approach is presented as a simple and practical way for the analysis and management of rainfall flooding at the precinct scale. This approach uses various rainfall events as input parameters for the analysis of the flood hazard and impacts, and categorises the flood risk in different levels, ranging from very low to very high risk. When visualised on a map, the insight into the risk levels across the precinct will enable engineers and spatial planners to identify and prioritise interventions to manage the flood risk. The approach is demonstrated for a sewer district in the city of Rotterdam, the Netherlands, using a one-dimensional (1D)/two-dimensional (2D) flood model. The risk level of this area is classified as being predominantly very low or low, with a couple of locations with high and very high risk. For these locations interventions, such as disconnection and lowering street profiles, have been proposed and analysed with the 1D/2D flood model. The interventions were shown to be effective in reducing the risk levels from very high/high risk to medium/low risk.

Keywords: extreme rainfall; risk; quick scan; urban drainage management

1. Introduction

Over the last decades, changes in frequency and intensity of severe rainfall events have been consistently associated with changes in weather and climate [1–3]. Urbanisation will amplify the projected changes in heavy rainfall due to the urban heat island effect [4–6]. Understanding the changes in severe rainfall events is of importance, particularly for the estimation of impacts on society, economics, and the environment [7]. Among other sectors, this poses a challenge to urban drainage management, when considering that drainage systems will have to deal with increased frequency and volume of stormwater flows [8–12]. Traditionally, drainage systems have been designed as a collection of grey infrastructure that, in most cases, will not be able to cope with such changes in stormwater flows. The main purpose of these collector networks may not be flood protection; it is to capture the runoff from the early part of rainfall events with highest pollutant loads, and conveying this to wastewater treatment plants. It is common for collector networks to overflow during episodes of heavy rainfall and for shallow water to flow along the street for a short duration (few minutes to hours). One of the main causes for the occurrence of urban flooding is the limited capacity of the drainage system, which, under extreme rainfall conditions, may result in flow being discharged to the catchment surface where it interacts with the incoming overland flow. In such cases, water runs to natural or constructed storage or runs along the surface using the major system components (i.e., above ground drainage components) [13].

Design standards for drainage systems specify the performance that is expected in terms of capturing and conveying runoff water. Although the performance of drainage systems varies between countries, there is a common agreement as to how the required performance should be defined, and some actions have been taken to formulate general design guidelines to serve to the international community (e.g., ISO/CD 20325 [14]). The conventional design of urban drainage systems is often based on the definition of a return period, T, which corresponds to the frequency with which an event occurs [15], and assuming a certain probability of occurrence of the event of 1/T per year [16]. Design standards for drainage systems rarely include severe rainfall events, other than considering what happens when the drainage system is exceeded (e.g., BS EN752 [16]).

Planning collector networks (i.e., the minor drainage system) in order to manage severe rainfall events may result in having uneconomical infrastructure, and most of the times water will flow above ground as a result of the limited storage/conveyance capacity of the system when compared with the flows during severe rainfall events. The major drainage system typically comprises roads, footpaths, and natural ground depressions, as well as smaller water courses. This system can convey flood water over significant distances causing flooding at locations away from the point where the network capacity is exceeded [15]. This means that intra-urban flooding is a complex aggregate of different components, like the physical process of flooding, the use of floodplains, and the people and infrastructures that influence or are subject to flooding and its impacts [17]. This, in turn, adds considerable complexity to the analysis and management of flood risk in urban areas [18].

The need for managing the drainage system (as a whole) by taking into consideration a broader range of rainfall events has become more evident. While design events for collector networks are typically in the range of 1 to 10 years return period, exceedance events may be in the range of 50 to 100 years return period, with extreme events being even higher, in the range of 100 to 1000 years return period. Contrary to the design events, managing exceedance flows and flood waters for a certain frequency of occurrence (or magnitude of rainfall) is not straightforward, and setting an acceptable threshold value has proven to be difficult [19,20]. This is because the adoption of a rigid threshold/criterion might not be feasible in practice, especially when it has to be implemented for the city or drainage area/catchment as a whole. This particularly applies to existing urban areas, where exceedance flows have not yet been managed proactively. In these areas, the management of exceedance using a rigid threshold/criterion could lead to excessive adaptation cost, for example, when streets or building blocks have to be adapted in order to meet the proposed threshold/criterion. For new developments it is common practice to establish management criteria for building in low-risk/safe areas. However, when there are large catchments, these can generate hazardous flooding, for which structural works, such as protection channels or embankments are being built, associated with higher design standards (e.g., 100 years return period).

In spite of the practical difficulties in managing exceedance flows and flood waters, there is a call for adaptation of infrastructure and the spatial layout of built up areas to reduce economic losses, particularly in relation to extreme events [10,21–25]. Various approaches are available for understanding (the need for) and assessing adaptation actions, with many of these requiring an analysis of the impacts of rainfall (or other climate-related) events. Risk assessment is a standard approach that can be defined as the process of identifying, evaluating, selecting, and implementing adaptation actions to reduce risk to human health and ecosystems [12,26,27]. As explained in Zhou et al. [27], performing a risk assessment relies on the identification of the hazards, exposure, and vulnerabilities. Generally, hazards describe the external loadings in terms of return period, and exposure and vulnerabilities describe the spatial distribution of social groups and properties that are susceptible to impacts [28,29]. Difficulties in performing a risk assessment relate to the high level of uncertainties associated with its constituent components, assigning probabilities to different climate change and socio-economic scenarios, and valuing the impacts and costing the various adaptation actions [30,31].

As the standards-based approach (i.e., set return periods) has limitations in dealing with severe rainfall events and the risk-based approach is mainly applied for large scale adaptation projects (e.g., to manage coastal or river flooding), there is an opportunity to suggest a complementary approach. The objective of this paper is to demonstrate an approach that uses a risk matrix, and thereby takes advantage of some of the strengths of both the standards-based and risk-based approaches. From this, it is possible to introduce a guidance tool for defining the priority in taking adaptation actions in order to reduce the risk of intra-urban flooding. In the following sections, first the rationale behind the approach is described through a comparison of the standards-based and risk-based approach (Section 2). Then, the semi risk-based approach, including the concept of a risk matrix, is explained in a step-by-step manner (Section 3). The characteristics of the case study and model set up (Spaanse Polder in Rotterdam, The Netherlands) are presented in Section 4, followed by the application of the approach to this case study in Section 5. The paper ends with a reflection on both the approach and the case study, and conclusions.

2. Rationale

This section provides an overview of two well-established approaches for the design and analysis of drainage systems: the standards-based and the risk-based approach.

The standards-based approach is focused on keeping the system functioning for a certain design value, as the magnitude of the load, usually with small return periods. The design value (e.g., for rainfall intensity) is, in many cases, prescribed in regulations or design guidelines, which means that the need for stakeholder involvement (e.g., in setting acceptable threshold values) is limited. The analysis of the system functioning is done in a deterministic way, and for this analysis, the single design value is used. The approach aims to optimise the performance of the (minor) drainage system for a single objective (e.g., no water on streets), without considering the efficiency of investments to manage the potential impacts; as it does not weigh the benefits of interventions, in terms of reduced impacts, against the investment costs [32].

The risk-based approach incorporates the outcomes of both flood hazard and impact assessments, which allows for the comparison of different interventions based on their expected impacts [16]. Therefore, a risk-based approach helps to make informed choices after the analysis of benefit and costs of several alternative interventions. Usually, the benefit is the reduction of the expected annual damage (EAD). This corresponds to the average annual consequences that can be expressed in monetary terms representing the expected persons or properties affected by flooding. In practice, flood risk reduction is also used to refer to a reduction in water level, to a decrease in flood probability, to a reduced area at risk, or to fewer potential flood losses [33]. Contrary to standard-based approaches, the risk-based approach requires the participation of experienced stakeholders for configuring the probabilities and impacts to be included in the analysis. This approach is analytically costly and therefore not standard practice in analysis of the performance of drainage systems. There may also be a lack of information regarding the impact data and its relationship to various geographical scales (from catchment to precinct). Although some countries like The Netherlands, are developing standardized information for this.

Both of the approaches, based on standards and the analysis of risk, provide some advantages and disadvantages when undertaking an analysis of system performance. The aforementioned characteristics are considered for comparison in Table 1: type of analysis; the objective; the requirement of investment; the economic guidance; the involvement of stakeholders; and, the extent of practice.

Table 1. Comparison between standard and risk-based approaches.

Characteristic	Standard-Based Approach	Risk-Based Approach
Type of analysis	Deterministic: single design value	Probabilistic: values with different probability
Objective	Functioning/performance of (minor) drainage system	Optimization of benefits/costs
Economic guidance	No insight into efficiency (i.e., into benefits/costs)	Looks for the economic optimum (i.e., optimum of benefits/costs)
Stakeholder involvement	Limited need for stakeholder involvement	Needs experienced stakeholders
Use in practice for urban drainage	Standard practice for small scale (precinct) drainage systems	Not common for small scale (precinct) drainage systems

This paper introduces a complementary approach to those above for working with a range of rainfall events, including extreme events. As the proposed approach builds upon the characteristics of both the standards-based and the risk-based approaches, it is termed the semi risk-based approach. This approach is intended to facilitate good housekeeping practices, which is aimed at tackling higher risk areas in the short term, while postponing the consideration of adaptation of medium risk areas to a later stage. This should be informed by the process of taking advantage of opportunities for mainstreaming adaptation, which will emerge from the interaction of investment cycles for different urban systems (roads, green spaces, social systems) [34,35]. The semi risk-based approach aims to provide decision support to the planning and urban design process and guidance for identifying adaptation actions, in terms of (interventions for) critical locations (i.e., hotspots).

The rationale for the semi risk-based approach can be derived from the mathematical relationship between risk, probability and consequences. This relationship has been explained from three perspectives by Klijn et al. [36], as shown in Figure 1. These authors represent risk as a function of the probability of flooding and the consequences, which are (in turn) determined by the exposure and vulnerability of the receptors (e.g., buildings and roads). A matrix representation, as proposed by the semi risk-based approach, is a straightforward way to describe this [37]. In the so-called risk matrix, the rows represent the probability (i.e., the frequency of occurrence) of flooding and the columns represent the severity of the consequences. The approach is coined semi-quantitative when one axis (usually probability) is based on a qualitative scale, while the other axis (usually consequences) is based on a qualitative scale [38]. This approach requires relative judgments to categorise the consequences, according to their (perceived) severity. This is often done by considering object-based vulnerabilities in terms of the type of objects or functions exposed to flooding. This is the most significant simplification of the (strict) risk-based approach, which quantifies the values at risk by using depth-damage curves [39]. The justification of this simplification is commonly the lack of quantitative information, such as about the flood damages resulting from rainfall flooding [40].

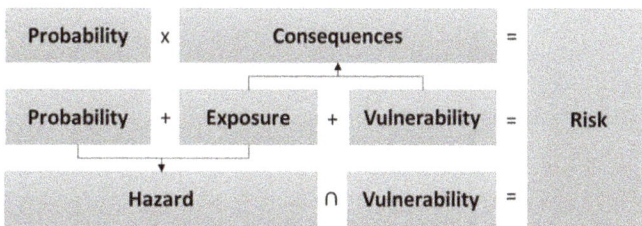

Figure 1. Risk-based approach from three perspectives: "flooding probability, exposure determinants and vulnerability of receptors" Source: Klijn et al., 2015 [36].

As discussed by Wall [37], the two parameters (probability and consequences) must be complemented by a third one, being the decision maker preference. This parameter describes how the decision maker values the possible outcomes representing the consequences. Risk cannot be quantified/qualified without the knowledge of how much the decision maker desires to avoid the different outcomes. This points to the need for a third dimension (or axis) to what is a two-dimensional matrix. In practice, this is accounted for by color-coding and/or scoring the cells of the risk matrix. Wall [37] demonstrates with mathematical relationships that if the color-codes and/or scores are not determined by the decision maker, then these do not represent risk and are meaningless. Furthermore, engagement with the decision maker and other stakeholders helps to increase their risk awareness. This also supports the process to engage more widely to access potentially less costly adaptation actions, as well as adaptation opportunities that are provided by urban (re)development and infrastructure investments.

In recent years, the understanding of the best ways to manage urban drainage has advanced in understanding and perspectives with different approaches emerging from research (e.g., [34,35]). However, this knowledge has not been taken up in practice and implementation of these good practices has been slow [41].

3. Semi Risk-Based Approach

The semi risk-based approach provides information and guidance for engineers and urban planners for defining the priority in which to take adaptation actions in anticipation of severe rainfall events. The proposed approach is based on the hydrological and hydraulic modelling of drainage systems using various rainfall events. The approach supports the identification of (cost-) effective interventions based upon the adaptation opportunities provided at specific locations that are at risk of flooding. In doing so, it promotes good housekeeping on the part of the drainage authority and/or municipality, along with other relevant stakeholders. The approach is set out below in a number of sequential steps.

3.1. Step 1. Selection of Events

The drainage system is subject to stress by hydrological drivers, being the range of possible rainfall events. Determining the probability/magnitude of rainfall events to be included in the analysis is the starting point of the semi risk-based approach. This should be done by the relevant stakeholders, such as the drainage authority, municipality, and others, like critical infrastructure providers, using a bottom-up assessment (also stress test). As stated by AGWA, the purpose of a stress test is to define the vulnerabilities of the system and to evaluate their occurrence within a certain range of magnitude of drivers, such as rainfall probability and/or intensity [42]. The selection of rainfall events, or classes of events, has to be made by the stakeholders by considering the local climate and other conditions, which is not usually straightforward. When historical records are available, it is possible to know how the system has previously responded to rainfall events, or at least the design event. This provides some indication of how to set the magnitude of rainfall relevant for the location. As a consequence of climate change, however, severe rainfall events will become increasingly common as well as increasingly intense. To reduce the significance of climate change uncertainty, the analysis can be executed by using fixed magnitudes of rainfall events. These represent a certain range of probabilities now and in the future, and should be defined according to local conditions. These pre-defined events will be used in the next step to model and map potential flooding.

3.2. Step 2. Modelling and Hazard Mapping

Modelling has an important role in understanding the performance of the minor drainage system, when considering its interaction with the urban terrain. It also provides understanding of where and how floodwaters are likely to collect and flow across the urban landscape, how fast they might rise and fall, and how frequent [43]. The role of modelling comprises the transition from raw data to analysed data and modelling outcomes, such as hazard maps [44]. Urban drainage modelling specifically involves the simulation of hydrological conditions, which is used for the analysis of the whole of drainage system response under these conditions and subsequently for analysing interventions. Therefore, the different rainfall events defined in the previous step need to be modelled and analysed to obtain improved knowledge about the functioning of the drainage system. Input data for this step include the temporal description of the rainfall event (i.e., hyetograph), a hydrological and hydraulic model of the drainage system, a digital elevation model (DEM), and parameter descriptions for the water exchange between the one-dimensional (1D) and two-dimensional (2D) model. The outcomes comprise a range of flood hazard maps that show the locations of inundation and simulated maximum water depths that are associated with the defined rainfall events. These hazard maps are used in the next step to develop risk maps, which form the baseline for the identification and assessment of interventions (step 4 and 5).

3.3. Step 3. Risk Categorization and Mapping

The flood hazard maps provide information on the areas susceptible to flooding based on the frequency of occurrence of the defined rainfall events. In this step, their frequency of occurrence is combined with the severity of flood damages, which can also be considered in different classes, such as: nuisance, minor damage, moderate damage, and major damage (adapted from [45]). The classification scheme for severity should be developed by the decision maker/stakeholders, and should preferably consider object-based vulnerability. Here, the classes could vary according to the type of object (building, road, etc.) and/or the function of the object (house, metro station, hospital, etc.). The classes might also take into account thresholds for the depth of flooding. This type of information is important, because e.g., 10 cm of water depths may be tolerable for some houses, but critical for metro stations or hospitals. However, the classification scheme will not typically consider vulnerability in terms of the values at risk and does not need detailed data about the impacts. The latter sets the semi risk-based approach apart from the strictly risk-based approach, and hence it might be considered to be a more (but not entirely) qualitative approach. The classification of severity requires inherently subjective judgments and/or arbitrary decisions about how to group the range of consequences into classes. Wall [37] notes that labels for the classes of probability and consequences use terms that vary from one application to the next, and that the meaning of these labels is open to interpretation. The need for subjective judgments, and the potential for inconsistencies in making such judgments, implies that there may be no entirely objective way to fill out a risk matrix [46].

The combination of frequency and severity gives an indication of the flood risk of an area. Different combinations of frequency and severity can be presented in a risk matrix. This is a matrix with several classes of frequency for its columns and several classes of severity or impact for its rows (see Figure 2). This associates a level of risk and/or some risk management action (based upon the tolerability of risk) with each row-column pair [46]. As for the frequency and severity classes, it is the role of the stakeholders to decide upon the definition of tolerable risk levels, and most importantly, upon the need for risk management actions. The risk matrix helps the stakeholders in identifying how a particular level of risk can be managed. For example, a (very) low risk, e.g., related to water on streets, would likely be tolerated by the stakeholders, whereas a (very) high risk, e.g., related to flooding of critical infrastructure, could be unacceptable. It is preferable to set up a stakeholder engagement process when deciding upon the tolerable levels of risk, and how this should be defined. This process should bring together the key stakeholders with a responsibility in managing flood risk, including the private sector, such as critical infrastructure providers. Through the engagement process, the negotiated risk matrix becomes meaningful for the stakeholders and local conditions. It also increases commitment for the implementation of risk management actions. The stakeholder process could be designed based on different approaches that are described elsewhere (e.g., [47]).

3.4. Step 4. Selection of Interventions

The levels of risk are the starting point for the selection of interventions. This is part of the engagement process with stakeholders. Stakeholder involvement is essential to better capture the specifics of the local context and it will also expand the available range of interventions to include options such as lowering street profiles or disconnecting areas from the collector network. The risk matrix and risk map provide the stakeholders with the information and guidance that is needed to define a timeline for interventions based on the tolerability of a level of risk. Specifically, it enables them to take account of the development dynamics in urban areas, such as the maintenance and renewal of infrastructure, buildings and public spaces [32]. In areas with (very) high risk (red colour), interventions are required most urgently, given the occurrence of severe damages. This indicates that these interventions have, in many cases, to be implemented in the short-term, often independent of any adaptation opportunity (i.e., mainstreaming). This is because adaptation opportunities will mostly occur in the medium term (next 5 to 30 years). The selection of interventions for areas with a medium risk (orange colour) is more challenging. In these areas, the damage that is caused is not so severe that it demands urgent action, and for that same reason the available budget is constrained. Here, a more detailed analysis may be required for the selection process, including the specific expertise of stakeholders (e.g., spatial planners). By sharing investment agendas, the stakeholders involved should aim to take advantage of adaptation opportunities that are provided by the dynamics of other urban systems, such as road maintenance. This could lead to cost savings on the implementation of interventions. In this case, there may be evidence from the analysis allowing for the postponement of the design and selection of interventions (to deal with a medium risk) to a later stage when the adaptation opportunity arises.

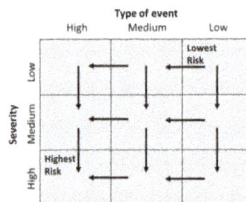

Figure 2. Risk matrix, general example. The axes are related with the consequence severity (y), and probability/type of event (x). The arrows represent the directions from lower risk regions to higher risk regions. Adapted from Elmontsri [48].

3.5. Step 5. Repetition and Refining

The semi risk-based approach is iterative in that it involves the repetition of steps 2 to 4 until the level of risk is lowered to a tolerable level. By doing this, the design or modifications of the drainage system (as a whole) can be undertaken in the most affordable manner. The involvement of stakeholders representing different sectors is desirable, particularly during steps 3 and 4 of the approach. Their involvement throughout the iterative process increases awareness among them about the effectiveness of possible interventions, and, in turn, secures feedback on the design that may result in innovative ways to manage flooding. This is because stakeholders have a better understanding of the area being studied and may think of interventions that are outside the engineer's normal practices [49].

As part of the ongoing research collaboration with the City of Rotterdam, a drainage system was selected as a way to test the proposed approach. Rotterdam is a city with particular interest on developing new ways of managing urban flooding, and therefore it makes available relevant information for undertaking case studies.

4. Case Study and Model Set Up

4.1. Case Study and Stakeholder Involvement

Spaanse Polder is one of the 42 districts of Rotterdam, located in the northwest of the city. The district comprises a low-lying polder which covers a drainage area of 190 hectares with twenty-seven types of land use. This area is an industrial park that has been developed in the 1920s and further densified in the 1980s and 1990s. Almost 30% of the land use is unpaved surfaces covered by green space, whereas paved areas comprise both open paved and closed paved covered 70% of land use. A few green spaces are situated adjacent to open watercourses and along linear infrastructure like roads [50]. There is only one major road in the area running in the North-South direction. Figure 3 shows the location of the case study area.

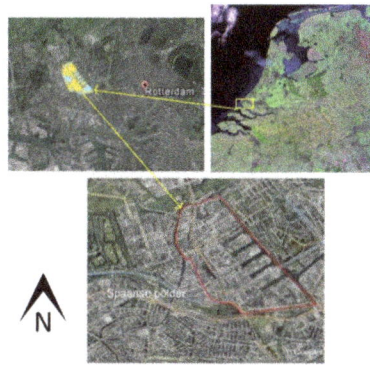

Figure 3. Location of Spaanse Polder. Source: Google Maps, 2016 [51].

Over the past two decades severe rainfall events have occurred in the area, which have led to water on the streets with the capacity of the drainage system being exceeded. In July of 2005, the area was subject to a severe rainstorm causing around 70 cm of water on the streets, which led to the saturation of the drainage system [50]. There were in total 170 affected buildings, including basements, households, and commercial buildings. As for the flooding mechanisms, pluvial flooding has been found as the main source of the problems. The following factors have worsened the impacts of these heavy rainfall events:

- The district has a high percentage of imperviousness (around 70% of the area) due to the dense urbanization. As a consequence, the hydrologic response to these heavy rainfall events is very fast and translates almost immediately into high runoff peaks.
- There are no storage facilities in which the rainwater can be collected, so surface runoff drains directly into the drainage system, causing flooding when extreme rainfall occurs.
- Under design rainfall conditions, rainwater is transported to the wastewater treatment plant by means of a pumping system. Under more severe rainfall conditions, an excess of stormwater is pumped out of the system towards the river Maas. However, the pump capacity is limited, and when the water volumes exceed this capacity, water is drained out of the system to surface water bodies.

The City of Rotterdam was involved in the case study as a partner of the Cooperative Research Centre for Water Sensitive Cities. Their involvement consisted mainly of the participation of two experts (a program manager and a policy advisor) in various research meetings. They provided advice on the development of the approach, as well as feedback on the results of its application. They also assisted in collecting relevant data, the 1D flood model and grey literature. A comprehensive process for stakeholder involvement was beyond the scope of the case study, which was the demonstration of the semi risk-based approach. It is recommended, however, that such a process is set-up within the proposed policy development trajectory on managing flood impacts.

4.2. Model Set Up

Simulation of the urban drainage system was performed using PCSWMM, which is an urban drainage modelling package with a GIS system providing a user-friendly interface [52]. The 1D drainage model simulates the pipe flow and the 2D overland flow model simulates the surface flow. The interaction between the two models occurs via connection nodes or structures, like manholes and open channels. The flood hazard extent has been assessed for the selected rainfall events, using this 1D-2D coupled model. The DEM was taken from the Dutch National System AHN2 with 0.5 m × 0.5 m of resolution. Using such resolution helped identify the occurrence of surface water in detail. In addition, the quality of the model is represented by the physical characteristics of e.g., land cover map and 1D sewerage network model developed from GIS database of sewer assets. Further data collection (e.g., water levels in an actual event in the channel) is required to improve predictions and even then, unless the event is of sufficient rarity, its validity for model verification could be questioned.

4.2.1. Overland Flow (2D)

PCSWMM performs 2D calculations by discretizing the 2D domain into a mesh consisting of various cells. In this case, the grid is adapted to specific details. In general, a grid size of 5 m for the main blocks, 1 m use for streets, and 0.5 m for small alleys or corridors. Each cell is represented by a 2D node, whose invert elevation is represented by the average bottom (or ground) elevation of that cell. All of the 2D nodes are connected to their adjacent nodes by 2D conduits or rectangular open channels. In order to preserve continuity, the surface areas of the nodes are usually small and the surface area in each cell is assigned to the 2D conduits connected to the node. The overland flow is spread according to the Shallow Water Equations (SWE) in two dimensions, including continuity and momentum. The routing step use in the model is 6 s, and the manning range from 0.012 to 0.025, according to the type of area.

4.2.2. Sewer Network (1D)

The 1D drainage model consisted of 637 manhole nodes, seven external weir nodes, and two boundary nodes (which serve as drainage outlets) connected by 691 pipes. The discharge into the system is computed as a function of rainfall and runoff factors. The runoff from built-up areas come into the (minor) drainage system collected through sub-catchments, assigned for each of the nodes, and generated in the 1D drainage model. As flow increases, water can flow out to the surface through the connections. Depending on the flow conditions, the 1D-2D coupled model allows water to flow back into the (minor) drainage system.

5. Results

The semi risk-based approach has been applied to Spaanse Polder in Rotterdam, the Netherlands. The focus of this application was on the demonstration of the approach as a guidance tool for defining the priority in taking adaptation actions to reduce the risk of intra-urban flooding.

5.1. Step 1. Selection of Events

For the case study (Spaanse Polder), the following intensities of rainfall were selected in consultation with the City of Rotterdam: 20, 40, 60, and 80 mm per hour. This broad range of events has been selected to obtain an understanding of how the drainage system responds to different pressures (i.e., rainfall intensities). According to the Dutch National Standards, the design event corresponds to 19.6 mm per hour. Here, the (design) performance of the minor drainage system was assessed with the 20 mm/h event. From this assessment, it was found that the minor drainage system has insufficient conveyance capacity in a number of locations, leading to water on the streets. This gave an initial indication of where flooding might be expected to occur for the more severe rainfall events.

5.2. Step 2. Modelling and Hazard Mapping

Simulation of events provided the water depths and velocities for the study area. Figure 4a shows the flood hazard extent for the design rainfall (i.e., 20 mm). It shows that a few areas have water on the streets. Whereas, the hazard map for the extreme rainfall (i.e., 80 mm) indicates that water on the streets occurs all across the sewer district (Figure 4b). This could be expected as the minor drainage system has not been designed to cope with such extreme rainfall. When combining the results of the model with land use information, the impacts have been initially categorized using a binary approach for determining inundation conditions (i.e., flooded or not flooded). However, the selection of the risk category requires consideration of the water depth, as the limits between different categories are defined by water depth (i.e., 5 and 10 cm for the case study), as can be seen in the following section. Results from the simulation are used for the risk categorization in the next step.

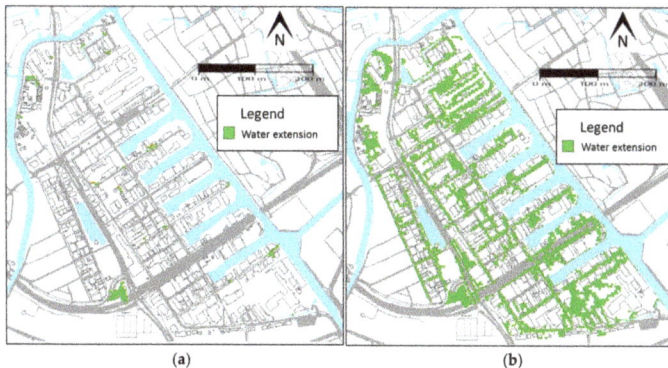

Figure 4. Flood map for Spaanse Polder for (a) the 20 mm event and (b) the 80 mm event.

5.3. Step 3. Risk Categorization and Mapping

The hazard maps for the four rainfall events were combined with the impacts that were derived from a land use map, to obtain a single map with areas of (in) tolerable flood risk, using the risk matrix from Figure 4. As a first (sub) step, a limited number of objects/functions were considered for the severity classes: buildings, roads/access and power supply. This narrow focus resulted from the homogeneous land use of the case study area, being an industrial area. Almost all of the buildings in Spaanse Polder have an industrial function, like a garage, loading area, or storage area. For more heterogeneous areas, it is recommended that differentiation of the types of objects or functions is important, like houses, metro stations, and hospitals.

For the three objects/functions, a description of the consequences was developed together with the City of Rotterdam. For this case study, four classes of severity of the consequences were considered to be sufficient for the objects and functions being studied. The labels for these classes were: nuisance, minor damage, moderate damage, and major damage. Table 2 gives the descriptions per severity class and object/function. As can be observed from Table 2, criteria related to both exposure (being the depth of flooding) and vulnerability (being the type of object/function at risk, together with some critical threshold for damage or outage) have been used for the qualification of the severity. The velocity and duration of flooding were such that these criteria (for exposure) were not deemed to be relevant for the case study. That is the flow velocities were less than 1.1 m/s and the duration of flooding was less than

60 min. The descriptions per object/function have also been aggregated into general descriptions, as given in Figure 5 and below. In the development of the descriptions, the City of Rotterdam has learned from the good practices of the City of Melbourne [45], which is also a partner in of the CRC for Water Sensitive Cities.

Table 2. Description of the consequences per object/function.

Severity Class	Buildings	Roads/Access	Power Supply
Nuisance	Low lying areas (e.g., parking lots or back-yards) are affected: up to 5 cm of water	Low lying areas (e.g., green spaces) next to roads are affected: up to 5 cm of water	Not applicable
Minor damage	Buildings are affected below the floor level	Minor roads are affected: more than 5 cm of water	Transmission stations are affected: up to 10 cm (no outage)
Moderate damage	Buildings are affected at the floor level: up to 10 cm of water	Main roads are affected: more than 5 cm of water	Transmission stations are affected above 10 cm (outage)
Major damage	Buildings are affected above 10 cm	Major roads are affected: more than 5 cm of water	Substations are affected above 25 cm (outage)

The classes of rainfall intensities and severity were subsequently combined to obtain different subjective levels of risk, ranging from very low to very high risk. The lowest class of severity corresponds to nuisance, and the levels of risk that have been associated with this class were low risk and very low risk. Here, shallow water is allowed to flow across land surfaces; this means up to 5 cm of water and up to 60 min in duration. These flows will cause nuisance for the occupants, but all the services and business activities will continue to function. The next class corresponds to minor damage, and three levels of risk have been assigned to this class: very low risk for the extreme event, low risk for the major and moderate events, and medium risk for the minor event. This severity class indicates that low-lying areas are inundated, with some minor roads potentially being closed and flood water flowing into some backyards. The third category corresponds to moderate damage, and has been assigned three levels of risk: low risk for the extreme event, medium risk for the major and moderate events and high risk for the minor event. Within this class, main traffic routes may be affected as well as some buildings above the ground floor level. The most severe class has been described as major damage, where buildings may be inundated by more than 10 cm of water. In addition, some main routes and utility services may be impacted, which could lead to a closure to traffic or utility service interruptions. This class has been associated with four levels of risk, ranging from a low risk for the extreme event to a very high risk for the minor event.

The levels of risk translated to a risk map of the sewer district in accordance with the combinations of frequency and severity are given in Figure 5. This was done by overlapping the four hazard maps with one another in GIS, placing the map for the extreme event at the bottom. The resulting risk map, as shown in Figure 5, has a few areas with high and very high risk, requiring attention for identifying and selecting interventions to reduce the flood risk to a tolerable level, being either medium or (very) low risk. These areas are marked with A, B, and C in Figure 6. The main traffic route is affected, but most of the areas of Spaanse Polder are classified as very low or low risk. As a rough validation step, a site visit has been made to the areas with the highest risk to obtain an understanding of the flow of water on streets. In this regard, it is noted that the area located in the northern most corner is a private precinct, which was not accessible for a site visit. Through the research meeting, the experts from the City of Rotterdam furthermore confirmed the identified areas as being exposed to flooding.

	Type of event			
	Minor event	**Moderate event**	**Major event**	**Extreme event**
Magnitude	20 mm	40 mm	60 mm	80 mm
Description	Smaller/often			Bigger/less often
Nuisance (areas are affected with shallow water. Up to 5 cm and 60 min duration)	Low risk	Low risk	Very low risk	Very low risk
Minor damage (Low-lying areas are inundated, minor roads may be closed)	Medium risk	Low risk	Low risk	Very low risk
Moderate damage (Main routes may be affected, some building may be affected above floor level)	High risk	Medium risk	Medium risk	Low risk
Major damage (Main routes are affected and buildings are affected above floor level > 10cm)	Very high risk	High risk	Medium risk	Low risk

Figure 5. Risk matrix, developed for the case study of Spaanse Polder.

5.4. Step 4. Selection of Interventions

Following the risk categorization and mapping, a short list of interventions was proposed as a way to improve the current situation for the areas that are most at risk. This has considered the characteristics of the topography, type of construction, type of property, and the connection (or rather, distance) to the open water system. The aim for selecting the interventions was their potential in lowering the risk level to a tolerable level. The following interventions were considered:

- Disconnection from the drainage system. This can be done in the areas near to open water bodies, such as canals.
- Lowering the profile of streets. By lowering specific streets more water can be stored within the street profile or can be diverted to a temporal storage, such as a green space. This appears most favourable in (large) open spaces, like parking areas.
- Flood proofing of buildings. Placing physical barriers in front of a building for preventing water flowing into the structure. This intervention can be realised at the building scale and with a low investment cost.

The interventions were selected as the result of a consultation with experienced professionals and stakeholders, who helped to define promising actions to be taken.

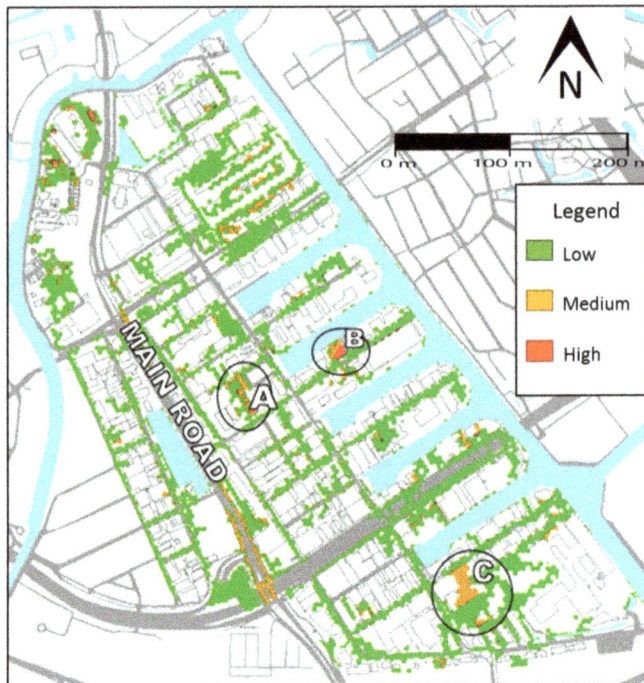

Figure 6. Risk map of Spaanse Polder. A, B, and C are medium to high risk locations.

5.5. Step 5. Repetition of Step 2 to 4 and Refinement

The proposed interventions were first modelled individually to check their performance and effects on reducing flooding or its impacts. The interventions that were included in the simulation/analysis were: changes in the street profile (in zone C), disconnection of areas (in zone B), and flood-proofing buildings (in zone A). As the affected areas are small and relatively far from each other, the effect of the interventions proved to be very local, but nonetheless positive in reducing the risk level. Subsequently, all of the interventions were combined and their effect was analysed and translated into a risk map with the interventions in place (Figure 7).

In zone A, flood proofing of buildings reduced the level of risk from very high to low, although there is a small area where the risk has not changed (medium risk). In zone B, the disconnection of the area has been effective as the level of risk has been reduced from high to low. By this intervention, water is no longer running into the (minor) drainage system, but directly to the open water system that was located at the border of the neighbourhood. The intervention proposed in zone C shows that lowering the profile of streets is effective in reducing the level of risk from medium to low. In this area, the change in street profile was modelled in combination with lowering the parking area, so that it can retain flood water.

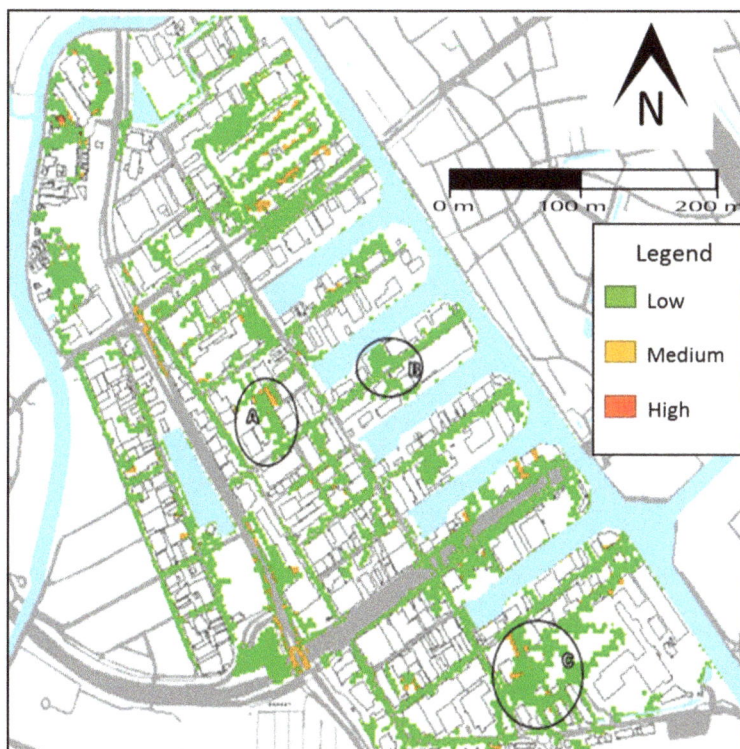

Figure 7. Effect of actions taken in most affected areas in Spaanse Polder. A, B and C are the locations were the risk has been reduced low risk after interventions.

6. Discussion

The discussion is presented with respect to the model, the approach, and the application using the case study.

6.1. Reflections on the Model

The computational domain of the model presents limitations that can be analysed. Some of the aspects include: the boundaries, friction map, grid resolution, the equations, the program, and physical domain. In this case, different options were selected. One change in the settings in PCSWMM (e.g., dynamic wave to diffusion wave equations) would slightly vary the results in the flood extent and depth. First, boundaries of the model are chosen according to the area of study. There is no initial level and discharge exchange. As a result, the depth of water remains the same inside and outside the area. Second, the friction map is selected according to the surfaces: paved, unpaved, sloped, and flat roofs, the Manning values are averaged for each type of surface. Third, the irregular grid uses different sizes of polygons according to the specific area, for instance, roads are smaller than buildings. After manipulation of the DEM, it takes an average value for each polygon, which can generate some changes in the surface. Fourth, the equations used to solve the water movement in the surface are the Shallow Water Equations in two dimensions, which means that the analysis excludes the acceleration term in the vertical axis and it is considered as a full dynamic model. It is convenient because the water level (z) is smaller when compared with the flood extent area (x, y). [53]. It is also possible to simulate using the diffusion wave equations to reduce the computational time of the model, but this would not then use the acceleration and pressure terms [54]. Each of these factors are assumptions that set limits to the model.

In the physical domain, the velocity is interpreted according to the characteristics of the study area: terrain and type of flood. First, the terrain is low-lying land, with elevations varying from −0.5 m to 1 m above Dutch National Datum (i.e., NAP). This is one of the main reasons for simplifying the hazard maps to exhibit water depth only. Second, another reason is related to the type of flood. In this case, pluvial flood, which starts from rainfall events. Then, the velocities are lower when compared to the velocities from a fluvial (river) flood. Although, it is advisable to include also fluvial and coastal flood in the model simulation for a complete representation.

In terms of the impacts, it is possible to consider the buildings, pedestrians, and/or vehicles. First, the representation of the ground levels without the physical representation of the buildings suggests which structures will have problems of water coming into properties. The walls or hard boundaries of the buildings could modify the path of the flood. However, due to the low velocities that remain in a range of 0 to 1.1 m/s, and shallow depths, this is not considered. In order to improve the accuracy of the simulation it would be useful to include in the DEM the elevation of the property thresholds (e.g., 10 cm). Second, the pedestrians. If the study area had velocities in the range 1.17 to 3.17 m/s, when a person feels unsafe, the vulnerability for the pedestrians will increase [55,56]. Third, the vehicles. In Figure 6, there are some places that could present difficulties for driving such as A, B or the main road. However, the multiplication of velocity and water depth does not represent a stability problem according to the proposed AR&R draft stability criteria for stationary vehicles (e.g., equation of stability for small vehicles $v \times y \leq 0.3$ m^2/s) [57].

Additionally, there is lack of field data to calibrate and validate the flood extension of the model. In the absence of data for the study area (e.g., satellite images, eye-witness map of actual flood events), it is advisable for stakeholders to recognize and verify the places flooded. However, this is not physical validation of the water extent and depth, which should be determined, for example, by varying the computational domain of the model (e.g., friction coefficient), to compare it statistically with a historical event, and changing it until there is the best approximation to observations [58]. Nevertheless, this 1D-2D model could be used as a reference for 1D/1D a fast assessment model [59] [60]. The calibration coefficient of the weirs for the connection in 1D to 2D is another factor that can modify the flood map with steady or unsteady flow [61]. In the same way, the process to convert rainfall to runoff can be further elaborated [62], for example the connections of the roofs that go directly to the sewer network can be included. Regarding the storage capacity of the sewer network, the manholes use a maximum volume of 1 m^3.

6.2. Reflections on the Approach

Standards-based approaches are most common practice in designing drainage systems. These approaches require the definition of specific levels of service in order to assure the serviceability of the system under such conditions, generally linked to a single objective (e.g., conveyance of runoff). In contrast, risk-based approaches have been extensively used in flood and coastal risk management. A risk-based approach requires extensive work on the impacts/consequences, focused on economic values of vulnerable objects [21]. For flood or coastal risk analyses, typically significant resources are made available for the analysis as well as data from studies that authorities have already conducted. These approaches have not often been used for drainage systems [21], mainly due to a lack of information regarding the consequences for the objects/functions affected and the scale of any affected areas. This points to the need to incorporate (into practice) new ways of planning and designing drainage systems that facilitate the analysis of the impacts from rainfall events in a simple and practical way. A potential way to do this is the adoption of the semi risk-based approach. This approach overcomes the resource and data constraints of the strict risk-based approach by qualitatively describing object-based vulnerabilities (instead of quantifying the values at risk). A qualitative scale is used to distinguish between the most severe and less severe consequences for the objects/functions. A quantitative scale is applied only to the probabilities. Although the risk matrix is only an approximate tool, the analysis and interpretations to underpin management decisions are a significant improvement over the standard-based approach as the impacts are taken into account.

Although it requires fewer resources and data, the semi risk-based approach still requires a rigorous analysis. The involvement of the decision maker/stakeholders in the analysis is crucial for the outcomes (i.e., the levels of risk) to be meaningful. This is because the risk is determined by the decision maker preferences over the possible values of the outcomes, and these preferences should be incorporated in the development of the risk matrix [37]. In addition, the process to involve stakeholders will help to increase their risk awareness. The semi risk-based approach visualizes the level of risk in an area with a risk matrix and risk map. This mapping is helpful when involving stakeholders, other than engineers, because they can observe where and how the flood impacts occur. The risk matrix and risk map can also be useful instruments for spatial planning, such as for the zoning of different land uses according to the (potential) level of risk. This is particularly relevant for decision makers (e.g., urban planners) who need to take decisions on how to deal with flood risk in new or redevelopments.

An advantage of the semi risk based approach is that it facilitates the connection of interventions with adaptation opportunities. By defining the tolerable risk levels in a more flexible way, the management of flood risk in an area can be connected with the adaptation opportunities [32]. These are opportunities that may occur at a localised and small scale in order to intervene in the development process. Such fit-for-purpose interventions can facilitate the transformation of an entire area at higher scale level, by the aggregation of many smaller interventions. The connection with adaptation opportunities is also facilitated because the tolerable levels of risk give an indication about the timing (i.e., prioritization) of interventions. Those interventions for dealing with (very) high risk areas require implementation in the short-term, while other interventions (i.e., for medium risk areas) could be postponed to the mid-term or even longer. Such interventions, in particular, can be timed to coincide with other investments in an area. In contrast to this, establishing rigid thresholds (tolerable levels) for the probability or risk, such as set return periods, may lead to over-dimensioned interventions that will require excessive, stand-alone investments.

A further advantage of the semi risk-based approach is that it does not consider monetary values at risk. As shown by Thaler et al. [63], risk-based approaches tend to contribute to distribute the investments to locations with high-value properties, leaving the poor unprotected or less protected from the same hazard. This is because investments are prioritized based on their benefit-cost ratio (from highest to lowest). As the semi risk-based approach does not use a benefit-cost rationale, it is less likely to give rise to social injustice.

The limitations of the use of risk matrices have been identified by Cox [46], and these also apply for its use for drainage systems. This includes the following limitations. Risk matrices can assign the same level of risk to quantitatively very different risks. They can even wrongly assign higher qualitative ratings to quantitatively smaller risks. This is particularly the case if some of the consequence severities are associated with large variances in severity. Thirdly, the classes provided by risk matrices do not necessarily support efficient decision making on investments in risk management measures. This is because the approach does quantify/monetise the impacts or the potential benefits (being the reduction in the impacts). Lastly, the classifications of severity cannot be made objectively and these classifications, along with the outputs (i.e., levels of risk) require subjective interpretation.

Regarding its use in urban drainage, we have identified the time needed for the flood modelling as a potential limitation. In particular, this may be substantial for the 2D model. Furthermore, there is potential for the improvement of the approach by considering all types of flooding in the analysis. This may prove beneficial for urban areas located in coastal zones, for example, because the effects of the interaction of different sources happen at the same time, when considering that there may be consequences from extensive fluvial floods, the breaching of coastal defenses without good warning, urban areas flooded by intensive and sudden rainstorm events, leading to rapid runoff and high flood water velocities, and floods in small steep catchments [33]. This is especially important in contexts where different sources of flooding are present, like in Rotterdam.

6.3. Reflections on the Case Study

Having most of its area covered by pervious surfaces, Spaanse Polder does not offer much room for implementing alternative solutions without changing the nature of the surfaces extensively [46]. However, by tackling the problem in a localised way, first at urban catchment level and then at a precinct level, the neighbourhood problems from overflowing water from the drainage system on to the streets can be tackled. Considering that there are three main areas where the biggest damages occur (zones A, B and C in Figure 4), and that these are not directly connected, the interventions can be implemented independently. Moreover, as the interventions are proposed at small scale, they are likely to be affordable, especially if they are associated with some ongoing works in other urban systems, like the renovation of street surface or changes in land use or building situation, which can result in reducing the cost of the interventions [33,64].

As shown in Figure 4a, the interventions needed for the standards-based approach have a different focus to that identified using the range of rainfall events used for the semi risk-based approach. The results from the standards-based approach do not always necessitate the need to intervene, whereas the results from the semi risk-based approach highlight the localised hot spots. This may lead to investment in areas that are not those most affected by extreme events. In some areas, the effect of the proposed interventions has been shown to provide limited reduction of water on the streets (i.e., flood proofing buildings in Area A). As urban drainage systems degrade with time, this type of analysis should be undertaken recurrently, so that results can be updated for the capacity of coping with impacts, and to propose new interventions when needed.

Financing of the interventions shown to be effective in the analysis process illustrated here is often difficult due to the allocation of responsibilities and the willingness or ability to pay. Some interventions, like changing the profile of streets, is the responsibility of the municipality or the competent authority, whereas flood proofing or disconnection responsibilities are not as clear, especially in an area that is mainly industrial and for which the benefits of the interventions are not necessarily evident to the residents. The City of Rotterdam has defined a new strategy for opportunistic investment, meaning that they are able to defer the interventions that are proposed in this work until a mainstreaming opportunity appears in the area. This also means that this approach may not be adopted as regulation, but rather be kept as one of the ways for assessing adaptation needs.

Some drawbacks of the approach may be identified. Firstly, the approach does not provide any indication about the economic benefit of the selected interventions. This is because the priorities are assigned by the defined risk levels (the matrix in Figure 2) that focus the investments to the high-risk locations. Secondly, the approach is dependent on the time that is needed for the modelling of the drainage system. Finally, the involvement of stakeholders depends on the context of application. As explained before, the definition of magnitude of events and risk tolerance is mainly decided in consultation with stakeholders, and therefore their participation becomes very important for the proper understanding of the context and occurrence of flood risk.

7. Conclusions

A semi risk-based approach for urban flood assessment and response has been presented as a simple way of supporting the process of decision making about the location for implementing the possible interventions according to the risk level defined in consultation with relevant stakeholders. The approach defines priority of actions depending upon the urgency highlighted in the analysis (Step 3 and Step 4). The approach has been shown here to be feasible and simple to apply without the need for the extensive analysis of the economics behind the selection of interventions.

The semi-quantitative risk-based approach helps to structure the analysis across a broad range of rainfall events in a simple way and to comprehensively illustrate the results for the stakeholders.

Water **2018**, *10*, 384

The latter helps to facilitate engagement of stakeholders in the design process. It is helpful for raising the awareness about the occurrence of extreme rainfall events among non-technical stakeholders. Although extreme events are usually considered to be a threat, they may also represent an opportunity to implement change in practice, encouraging decision makers to take urgent actions and to adopt a long-term perspective [65]. This, however, requires a change in the current practice of managing urban drainage systems by incorporating the innovation that is needed to deal with extreme rainfall and the exceedance flows that are generated as the result of such events [19].

Although the case study presented here does not demand extensive research of possible interventions, mainly because of its industrial nature, it has shown the usefulness of the approach. Further use and evaluation of the approach is needed in order to gain more evidence of its application when dealing with extreme rainfall and/or other sources of flooding.

Acknowledgments: This paper has been written in the framework of the CRC Water Sensitive Cities, in which the authors participate in the project 'B4.2: Socio-Technical Resilience in Water Sensitive Cities—Adaptation across Spatial and Temporal Scales'. The authors thank the collaboration and support provided by the City of Rotterdam for this research. PCSWMM v5.4 software used for 1D-2D modelling was provided by Computational Hydraulics Education through an educational grant. The authors thank Zohre Naderi and Virdiyana Yuser for their contribution during their M.Sc. studies.

Author Contributions: The paper is based on the results from the research of Carlos Salinas-Rodriguez. Carlos Salinas-Rodriguez, Berry Gersonius, Chris Zevenbergen, and Richard Ashley contributed equally in the production of the present manuscript; David Serrano helped to improve the paper.

Conflicts of Interest: The authors declare no conflict of interest.

References

1. IPCC. *Climate Change 2013: The Physical Science Basis. Contribution of Working Group I to the Fifth Assessment Report of the Intergovernmental Panel on Climate Change*; Cambridge University Press: Cambridge, UK, 2013; p. 1535.
2. Fischer, E.M.; Knutti, R. Observed heavy precipitation increase confirms theory and early models. *Nat. Clim. Chang.* **2016**, *6*, 986–991. [CrossRef]
3. Willems, P.; Olsson, J.; Arnbjerg-Nielsen, K.; Beecham, S.; Pathirana, A.; Gregersen, I.B.; Madsen, H. *Impacts of Climate Change on Rainfall Extremes and Urban Drainage Systems*; IWA Publishing: London, UK, 2012.
4. Westra, S.; Fowler, H.J.; Evans, J.P.; Alexander, L.V.; Berg, P.; Johnson, F.; Kendon, E.J.; Lenderink, G.; Roberts, N.M. Future changes to the intensity and frequency of short-duration extreme rainfall. *Rev. Geophys.* **2014**, *52*, 522–555. [CrossRef]
5. Willems, P.; Olsson, J.; Arnbjerg-Nielsen, K.; Beecham, S.; Pathirana, A.; Gregersen, I.B.; Madsen, H. Impacts of climate change on rainfall extremes and urban drainage systems: A review. *Water Sci. Technol.* **2013**, *68*, 16–28. [CrossRef]
6. Pathirana, A.; Denekew, H.B.; Veerbeek, W.; Zevenbergen, C.; Banda, A.T. Impact of urban growth-driven landuse change on microclimate and extreme precipitation—A sensitivity study. *Atmos. Res.* **2014**, *138*, 59–72. [CrossRef]
7. Shinyie, W.L.; Ismail, N.; Jemain, A.A. Semi-parametric Estimation for Selecting Optimal Threshold of Extreme Rainfall Events. *Water Res. Manag.* **2013**, *27*, 2325–2352. [CrossRef]
8. Arnbjerg-Nielsen, K. Quantification of climate change effects on extreme precipitation used for high resolution hydrologic design. *Urban Water J.* **2012**, *9*, 57–65. [CrossRef]
9. Butler, D.; McEntee, B.; Onof, C.; Hagger, A. Sewer storage tank performance under climate change. *Water Sci. Technol.* **2007**, *56*, 29–35. [CrossRef] [PubMed]
10. Zhou, Q.; Panduro, T.E.; Thorsen, B.J.; Arnbjerg-Nielsen, K. Adaption to Extreme Rainfall with Open Urban Drainage System: An Integrated Hydrological Cost-Benefit Analysis. *Environ. Manag.* **2013**, *51*, 586–601. [CrossRef] [PubMed]
11. Olsson, J.; Amaguchi, H.; Alsterhag, E.; Dåverhög, M.; Adrian, P.E.; Kawamura, A. Adaptation to climate change impacts on urban storm water: A case study in Arvika, Sweden. *Clim. Chang.* **2013**, *116*, 231–247. [CrossRef]
12. Mailhot, A.; Duchesne, S. Design Criteria of Urban Drainage Infrastructures under Climate Change. *J. Water Res. Plan. Manag.* **2010**, *136*, 201–208. [CrossRef]

13. Maksimović, Č.; Prodanović, D.; Boonya-Aroonnet, S.; Leitão, J.P.; Djordjević, S.; Allitt, R. Overland flow and pathway analysis for modelling of urban pluvial flooding. *J. Hydraul. Res.* **2009**, *47*, 512–523. [CrossRef]

14. ISO. *ISO/CD 20325 Service Activities Relating to Drinking Water Supply and Wastewater Systems—Guidelines for Stormwater Management in Urban Areas*; ISO: Geneva, Switzerland, 2009.

15. CIRIA. *C635 Designing for Exceedance in Urban Drainage—Good Practice*; CIRIA: London, UK, 2014.

16. BS EN 752. *Drain and Sewer Systems Outside Buildings*; NBS: Newcastle, UK, 2008.

17. Sayers, P.; Meadowcroft, I. *RASP-A Hierarchy of Risk-Based Methods and Their Application*; HR Wallingford: Wallingford, UK, 2005.

18. Kellagher, R.; Sayers, P.; Counsell, C. Developing a risk-based approach to urban flood analysis. In Proceedings of the 11th International Conference on Urban Drainage, Edinburgh, UK, 31 August–5 September 2008.

19. CIRIA. *Managing Urban Flooding from Heavy Rainfall (Encouraging the Uptake of Designing for Exceedance)—Literature Review*; CIRIA: London, UK, 2014.

20. Mugume, S.N.; Butler, D. Evaluation of functional resilience in urban drainage and flood management systems using a global analysis approach. *Urban Water J.* **2017**, *14*, 727–736. [CrossRef]

21. Kirshen, P.; Caputo, L.; Vogel, R.M.; Mathisen, P.; Rosner, A.; Renaud, T. Adapting Urban Infrastructure to Climate Change: A Drainage Case Study. *J. Water Res. Plan. Manag.* **2015**, *141*, 04014064. [CrossRef]

22. Kleidorfer, M.; Mikovits, C.; Jasper-Toennies, A.; Huttenlau, M.; Einfalt, T.; Rauch, W. Impact of a Changing Environment on Drainage System Performance. *Procedia Eng.* **2014**, *70*, 943–950. [CrossRef]

23. Pregnolato, M.; Ford, A.; Glenis, V.; Wilkinson, S.; Dawson, R. Impact of Climate Change on Disruption to Urban Transport Networks from Pluvial Flooding. *J. Infrastruct. Syst.* **2017**, *23*, 04017015. [CrossRef]

24. Rosbjerg, D. Optimal adaptation to extreme rainfalls in current and future climate. *Water Res. Res.* **2017**, *53*, 535–543. [CrossRef]

25. Löwe, R.; Urich, C.; Domingo, N.S.; Mark, O.; Deletic, A.; Arnbjerg-Nielsen, K. Assessment of urban pluvial flood risk and efficiency of adaptation options through simulations—A new generation of urban planning tools. *J. Hydrol.* **2017**, *550*, 355–367. [CrossRef]

26. Jones, R.N. An environmental risk assessment/management framework for climate change impact assessments. *Nat. Hazards* **2001**, *23*, 197–230. [CrossRef]

27. Zhou, Q. A Review of Sustainable Urban Drainage Systems Considering the Climate Change and Urbanization Impacts. *Water* **2014**, *6*, 976–992. [CrossRef]

28. The World Bank. *Risk and Opportunity. Managing Risk for Development*; The World Bank: Washington, DC, USA, 2013.

29. UNISDR. *UNISDR Terminology on Disaster Risk Reduction*; UNISDR: Geneva, Switzerland, 2009.

30. Dessai, S.; Hulme, M. Does climate adaptation policy need probabilities? *Clim. Policy* **2004**, *4*, 107–128. [CrossRef]

31. Olsen, A.S.; Zhou, Q.; Linde, J.J.; Arnbjerg-Nielsen, K. Comparing Methods of Calculating Expected Annual Damage in Urban Pluvial Flood Risk Assessments. *Water* **2015**, *7*, 255–270. [CrossRef]

32. Gersonius, B.; Nasruddin, F.; Ashley, R.; Jeuken, A.; Pathirana, A.; Zevenbergen, C. Developing the evidence base for mainstreaming adaptation of stormwater systems to climate change. *Water Res.* **2012**, *46*, 6824–6835. [CrossRef] [PubMed]

33. Salinas Rodriguez, C.N.; Ashley, R.; Gersonius, B.; Rijke, J.; Pathirana, A.; Zevenbergen, C. Incorporation and application of resilience in the context of water-sensitive urban design: Linking European and Australian perspectives. *Wiley Interdiscip. Rev. Water* **2014**, *1*, 173–186. [CrossRef]

34. Haasnoot, M.; Kwakkel, J.H.; Walker, W.E.; ter Maat, J. Dynamic adaptive policy pathways: A method for crafting robust decisions for a deeply uncertain world. *Glob. Environ. Chang.* **2013**, *23*, 485–498. [CrossRef]

35. Lansey, K. Sustainable, robust, resilient, water distribution systems. In Proceedings of the WDSA 2012: 14th Water Distribution Systems Analysis Conference, Adelaide, Australia, 24–27 September 2012.

36. Klijn, F.; Kreibich, H.; De Moel, H.; Penning-Rowsell, E. Adaptive flood risk management planning based on a comprehensive flood risk conceptualisation. *Mitig. Adapt. Strateg. Glob. Chang.* **2015**, *20*, 845–864. [CrossRef]

37. Wall, K.D. *The Trouble with Risk Matrices*; DRMI Working Papers Ongoing Research; Naval Postgraduate School: Monterey, CA, USA, 2011.

38. Emblemsvåg, J.; Endre Kjølstad, L. Qualitative risk analysis: Some problems and remedies. *Manag. Decis.* **2006**, *44*, 395–408. [CrossRef]

39. Martins, R.; Leandro, J.; Djordjević, S. Influence of sewer network models on urban flood damage assessment based on coupled 1D/2D models. *J. Flood Risk Manag.* **2016**. [CrossRef]

40. Spekkers, M.H.; Clemens, F.H.L.R.; Ten Veldhuis, J.A.E. On the occurrence of rainstorm damage based on home insurance and weather data. *Nat. Hazards Earth Syst. Sci.* **2015**, *15*, 261–272. [CrossRef]

41. Butler, D.; Ward, S.; Sweetapple, C.; Astaraie-Imani, M.; Diao, K.; Farmani, R.; Fu, G. Reliable, resilient and sustainable water management: The Safe & SuRe approach. *Glob. Chall.* **2017**, *1*, 63–77.

42. Alliance for Global Water Adaptation. *Climate Resilient Investment/Climate Risk Informed Decision Analysis*; AGWA: Amsterdam, The Netherlands, 2016.

43. Melbourne Water. *Flood Management Strategy-Port Phillip and Westernport*; Melbourne Water: Melbourne, Australia, 2015.

44. Price, R.K.; Vojinovic, Z. Urban flood disaster management. *Urban Water J.* **2008**, *5*, 259–276. [CrossRef]

45. Melbourne Water. *Draft Flood Management Strategy Port Phillip and Westernport*; Melbourne Water: Melbourne, Australia, 2015.

46. Cox, L.A.T. What's wrong with risk matrices? *Risk Anal.* **2008**, *28*, 497–512. [PubMed]

47. Van Herk, S.; Zevenbergen, C.; Ashley, R.; Rijke, J. Learning and Action Alliances for the integration of flood risk management into urban planning: A new framework from empirical evidence from The Netherlands. *Environ. Sci. Policy* **2011**, *14*, 543–554. [CrossRef]

48. Elmontsri, M. Review of the strengths and weaknesses of risk matrices. *J. Risk Anal. Crisis Response* **2014**, *4*, 49–57. [CrossRef]

49. Arnbjerg-Nielsen, K. Past, present, and future design of urban drainage systems with focus on Danish experiences. *Water Sci. Technol.* **2011**, *63*, 527–535. [CrossRef] [PubMed]

50. Rain Gain Project. Fine-Scale Rainfall Measurement and Prediction to Enhance Urban Pluvial Flood Management. 2013. Available online: http://www.raingain.eu/sites/default/files/fs2-gen-spaansepolder.pdf (accessed on 30 January 2016).

51. GoogleMaps. Available online: https://www.google.nl/maps/place/Spaanse+Polder,+Rotterdam (accessed on 25 January 2016).

52. *PCSWMM*, version 5.4; Computational Hydraulics International: Guelph, ON, Canada, 2013.

53. Martins, R.; Leandro, J.; Chen, A.S.; Djordjević, S. A comparison of three dual drainage models: Shallow water vs local inertial vs diffusive wave. *J. Hydroinform.* **2017**, *19*, 331–348. [CrossRef]

54. Costabile, P.; Costanzo, C.; Macchione, F. Performances and limitations of the diffusive approximation of the 2-d shallow water equations for flood simulation in urban and rural areas. *Appl. Numer. Math.* **2017**, *116*, 141–156. [CrossRef]

55. Martínez-Gomariz, E.; Gómez, M.; Russo, B. Experimental study of the stability of pedestrians exposed to urban pluvial flooding. *Nat. Hazards* **2016**, *82*, 1259–1278. [CrossRef]

56. Martínez-Gomariz, E.; Gómez, M.; Russo, B.; Djordjević, S. Stability criteria for flooded vehicles: A state-of-the-art review. *J. Flood Risk Manag.* **2016**, *11*, S817–S826. [CrossRef]

57. Cox, R.J.; Shand, T.D.; Blacka, M.J. *Revision Project 10: Appropriate Safety Criteria for Vehicles*; Report Number: P10/S2/020; Australian Rainfall and Runoff (AR&R): Sydney, Australia, 2011.

58. Bennett, N.D.; Croke, B.F.; Guariso, G.; Guillaume, J.H.; Hamilton, S.H.; Jakeman, A.J.; Pierce, S.A. Characterising performance of environmental models. *Environ. Model. Softw.* **2013**, *40*, 1–20. [CrossRef]

59. Leandro, J.; Djordjević, S.; Chen, A.S.; Savić, D.A.; Stanić, M. Calibration of a 1D/1D urban flood model using 1D/2D model results in the absence of field data. *Water Sci. Technol.* **2011**, *64*, 1016–1024. [CrossRef] [PubMed]

60. Leandro, J.; Chen, A.S.; Djordjević, S.; Savić, D.A. Comparison of 1D/1D and 1D/2D coupled (sewer/surface) hydraulic models for urban flood simulation. *J. Hydraul. Eng.* **2009**, *135*, 495–504. [CrossRef]

61. Rubinato, M.; Martins, R.; Kesserwani, G.; Leandro, J.; Djordjević, S.; Shucksmith, J. Experimental calibration and validation of sewer/surface flow exchange equations in steady and unsteady flow conditions. *J. Hydrol.* **2017**, *552*, 421–432. [CrossRef]

62. Chang, T.J.; Wang, C.H.; Chen, A.S. A novel approach to model dynamic flow interactions between storm sewer system and overland surface for different land covers in urban areas. *J. Hydrol.* **2015**, *524*, 662–679. [CrossRef]

63. Thaler, T.; Fuchs, S.; Priest, S.; Doorn, N. Social justice in the context of adaptation to climate change—Reflecting on different policy approaches to distribute and allocate flood risk management. *Reg. Environ. Chang.* **2017**, *18*, 305–309. [CrossRef]

64. Veerbeek, W.; Ashley, R.M.; Zevenbergen, C.; Rijke, J.; Gersonius, B. Building Adaptive Capacity For Flood Proofing In Urban Areas Through Synergistic Interventions. In *Proceedings of the WSUD 2012: Water Sensitive Urban Design; Building the Water Sensiitve Community; 7th International Conference on Water Sensitive Urban Design*; Engineers Australia: West Perth, Australia, 2012; p. 127.

65. Keath, N.A.; Brown, R.R. Extreme events: Being prepared for the pitfalls with progressing sustainable urban water management. *Water Sci. Technol.* **2009**, *59*, 1271–1280. [CrossRef] [PubMed]

water

MDPI

Article

Application of CityDrain3 in Flood Simulation of Sponge Polders: A Case Study of Kunshan, China

Dingbing Wei [1,2,*], Christian Urich [2,3], Shuci Liu [4] and Sheng Gu [5,*]

1   School of Civil Engineering, Southeast University, 2 Sipai Lou, Nanjing 210096, China
2   Southeast University–Monash University Joint Research Centre for Water Sensitive Cities,
    Suzhou 215123, China; christian.urich@monash.edu
3   Department of Civil Engineering, Monash University, Clayton, VIC 3168, Australia
4   Department of Infrastructure Engineering, The University of Melbourne, Parkville, VIC 3010, Australia;
    shucil@student.unimelb.edu.au
5   Kunshan Construct Engineering Quality Testing Center, Kunshan 215337, China
*   Correspondence: kevin.wei@seu.edu.cn (D.W.); shenggu2018@163.com (S.G.); Tel.: +86-25-8379-3223 (D.W.)

Received: 2 March 2018; Accepted: 16 April 2018; Published: 19 April 2018

Abstract: The selection of sponge city facilities (e.g., pump, storage tank, wetland, or bioretention pond) to mitigate urban floods has been a crucial issue in China. This study aims to develop a conceptual flood-simulation model, which can take into account the effects of such facilities of a sponge city. Taking Jiangpu polder in Kunshan City as a case study, CityDrain3 was implemented to develop a baseline model and another three sponge polder models (pump only, storage tank only, pump, and storage tank). A sensitivity analysis was carried out to guarantee the robustness of the newly developed model. In the model application part, firstly, one-hour rainfall scenarios with different return periods (2a, 5a, 10a, 20a, 50a, 100a, with 'a' referring to a year) were employed as inputs to the conceptual baseline model. The growing trend of flood depth (from 12.69 mm to 17.16 mm) simulated by the baseline model under increased return periods (from 3a to 100a) demonstrated the feasibility of polder flood simulations using CityDrain3. Secondly, a one-hour rainfall scenario with a 10-year return period was employed on the baseline model and the three sponge polder models. The results showed that the effect rankings of the control strategies on the total flood volume, peak flow, flood yielding time, and the peak-flow occurrence time were comparable—combined strategies (pump and storage tank) > storage tank only > pump only. The conceptual, and hydrological model developed in this study can serve as a simulation tool for implementing a real-time urban storm water drainage control system in the Jiangpu polder.

Keywords: CityDrain3 model; sponge city; polder; urban flood; sensitivity analysis

---

1. Introduction

A polder is defined as an area where catchments are surrounded by outer rivers [1]. In the Chinese Yangtze River Delta, there are many existing polders. A common feature of polders is that their elevations are lower than the water level in the surrounding outer river [2], resulting in high flooding risk during monsoon seasons. This is particularly an issue when a polder's storage capacity is insufficient, or a pumping station does not work properly. Flooding is a natural disaster that causes damage to human lives, as well as the economy [3]. However, compared with plain catchments [4–8], less effort has been made in mitigating the flooding risks of polders [9–11]. Therefore, investigating measures to control polder flooding problems is of a significant value.

Since 2013, the 'sponge city' concept has become a hot topic in China [12]. It was proposed as an effort to increase city resilience when dealing with storm water disasters. Similar to Best Management Practices (BMPs) [13], Low Impact Development (LID) [14], Water Sensitive Urban Design (WSUD) [15], and Sustainable Urban Design System (SUDS) [16], the Chinese sponge city concept has also adopted several facilities, such as wetlands, permeable pavements, ponds, bioretention ponds, storage tanks, and pumps. In Chinese sponge city practices, Jinan City has constructed a sponge city from four systems—water, road, grass, and architecture. Zhenjiang City has also implemented sponge city facilities with a focus on a river system [17]. These projects have demonstrated the effectiveness of the sponge city concept.

To evaluate the effectiveness of these sponge city strategies, special attention should be paid to the research related to urban storm water drainage systems. The urban storm water drainage system is an integrated infrastructure composed of various components (i.e., storm drainage, pumping stations, inner-rivers, out-rivers, and gates) [18]. To study the flooding performance of an urban area under a rainfall scenario, numerous software packages [19–21] have been developed to simulate urban storm water drainage systems.

Software packages used to simulate urban drainage systems can be classified into two types–commercial software and open source [22]. The commercial software is easy for engineers to implement and test. Some of the popular commercial software on the market include Mouse, Infoworks, Coral, Csoft, XPstorm, Hystem-Extran, and WEST [23,24]. In addition to commercial software, there are open-source tools which are sufficiently flexible to modify and add to the simulation algorithm, which in turn allows the integration of unique features for each drainage system. Two of the most commonly-used open-source tools are SWMM [22,25], and CityDrain [26,27]. The research presented in this paper investigates how CityDrain3 [28] can be implemented to simulate the flood volume of a polder in the Chinese Yangtze River Delta.

In this study, we introduced the concept of the sponge polder, and selected pumping stations and storage tanks as the two sponge facilities to control polder flood. Four control strategies (baseline, pumps only, storage tanks only, pumps, and storage tanks) were proposed to investigate the impact of different scenarios. This forms the basis for the later implementation of real-time controls on the urban storm water drainage system in the Jiangpu polder.

This paper is organized as follows. Section 2 provides a detailed explanation of the flow routing algorithm and the concept of integrated software CityDrain3. Section 3 shows the case study of the Jiangpu polder in China, using the simulation software CityDrain3. The last section draws some conclusions and provides the scope for future work.

2. Modelling Framework

2.1. Flow Routing Models

The urban drainage model is often implemented by using either a hydrological model or a hydrodynamic model. The hydrological model is a conceptualized model, which does not consider detailed information (e.g., detailed elevation data or sewer data). As a result, it is less dependent on computing time. However, the computational result is only appropriate where the influence of the sewer system can be neglected. The hydrodynamic model is often referred to as a detailed, physically-based model, which takes into account more detailed information of the urban catchments. This type of model can calculate the flow regime in the sewer pipes, including backwater effects at the expense of long computing times.

In our research, we aim to simulate the flood volume of the whole polder during specific rainfall scenarios. Because runoff is conveyed into the inner-river either by drainage sewer pipes or through catchment surfaces, we do not need to consider detailed urban drainage networks-thus selecting the hydrological model. Among all the hydrological models, we use CityDrain3, which is a well-known conceptual model adopting the Muskingum method to solve flow routing.

2.2. Modelling Software: CityDrain3

The modelling framework applied in this study is CityDrain3, a conceptual integrated modelling tool for urban drainage systems, which uses the top-down modelling method [29]. The modelling tool, CityDrain3 was developed at the Unit of Environmental Engineering at the University of Innsbruck by Gregor Burger and his colleagues [28]. As an open source simulation tool, CityDrain3 allows users to implement modular designs to solve different research questions, while coding in the C++/Python environment.

In CityDrain3, every part of the urban drainage system (catchment, inner-river, storage facilities, pumping stations, outer-river, wastewater treatment plant, etc.) is described by a corresponding subsystem block. Each subsystem block is made up of in-ports, out-ports, underlying modelling algorithms, as well as physical properties like catchment area, impervious surface ratio, pumping station efficiency curve, river length, etc.

To initiate the calculation at any time step n, the data was imported from the previous $n - 1$ time step and parameter data existing in the underlying blocks in CityDrain3 (i.e., catchment and inner-river physical properties). Each block uses its in-port data, as well as underlying physical properties of the research area to calculate the out-port values with the underlying modelling algorithms. The calculation began with the first rainfall input block, and the result was passed onto the downstream block, and so forth [29]. When the last block finished its calculation, the results of the nth time step were saved in the computer's memory, and the calculation for $n + 1$ time step was initiated. Finally, the result calculated during the whole process was stored into a txt. file.

3. Case Study of Jiangpu Polder

3.1. Site Description

The model was applied at Jiangpu polder, Kunshan City, China (see Figure 1). Kunshan is a well-developed satellite city in the greater Suzhou region, located in the southeastern part of the Jiangsu province, adjacent to the Shanghai Municipality. The area where the case study was conducted is in the northwestern part of Kunshan city, and there were eight pumping stations, gates, and weirs. Moreover, the drainage system in this study area was a separate sewer system; storm water was conveyed to the nearest inner-river through stormwater pipes or flows into the inner-river in the form of catchment surface runoff. For the simulation of this site, four models were considered:

(1) Baseline model: The above described polder was assumed to have an absence of any storm water exchange (pumping stations) between the inner-river and outer-river. Meanwhile, there were no sponge facilities, such as storage tanks, wetlands, bioretention ponds, ponds, etc.

(2) Sponge polder model 1 (Pump only model): The pumping stations were located at the end of each inner-river to pump water from the inner-river to the outer-river, in case of the possibility of flooding. Water in the outer-river of Jiangpu polder then flowed into the Yangtze River and finally flowed into the East China Sea. According to the pumping control strategy, the pumping was started when the water level of the inner-river became higher than 2.6 m and the pumping was stopped when the water level of the inner-river became lower than 2.2 m.

(3) Sponge polder model 2 (storage tank only model): Storage tanks were located at the downstream point of each sub-catchment. Whenever there was sub-catchment surface runoff over the storage tank, a part of the runoff flowed into the storage tank at a fixed flow speed.

(4) Sponge polder model 3 (pump and storage tank): Pumping stations and storage tanks worked at the same time when their working conditions were triggered.

3.2. Application and Data Analysis

In this study, the storm water system of Jiangpu polder was stimulated to calculate the flood volume of the whole area. Different catchments were divided into several parts, with the runoff from each part of the catchment draining into the inner-river section according to the terrain.

The Muskingum model was implemented as the underlying algorithm in the catchment and inner-river blocks. The system simulation scheme is represented in Figure 2.

As evident from Figure 2, when rain fell onto the catchment area, through the Muskingum algorithm, the flow routing was calculated to generate a runoff that flowed into the inner-river section. Usually, the runoff from the several catchments drained into the same inner-river section. The pumping station was installed at each end of an inner-river section to pump water from the inner-river to outer-river. The rainfall and pump working situation were regarded as the input to the system, with the water level of the whole inner-river system being treated as the final output (which could later be used to calculate the flood volume of the whole system).

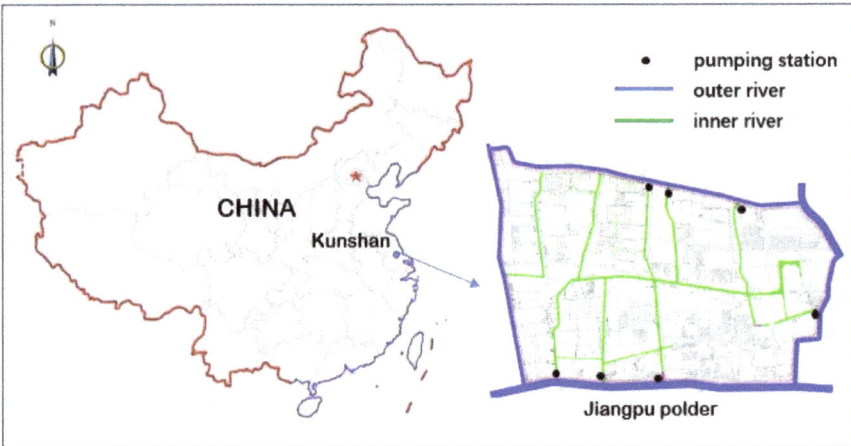

Figure 1. Simulation site: Jiangpu polder of Kunshan City.

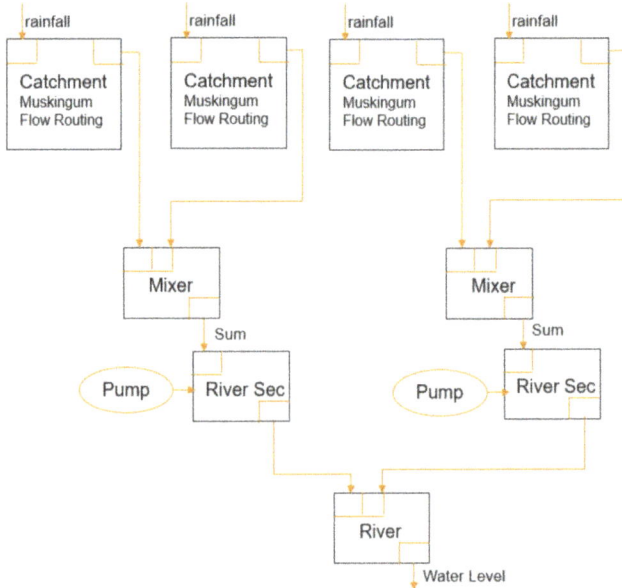

Figure 2. Jiangpu polder system simulation scheme.

3.2.1. Rainfall Data

Rainfall design is an important basis for urban drainage system design and sponge city construction. If the rainfall data far exceed reality, this could result in the construction of a massive drainage system and significant economic burden.

The Chicago Method is widely used in China to design the rainfall process on the basis of the rainstorm intensity equation. The Kunshan government implemented the Chicago Method and deduced the rainstorm intensity equation in 2017 [30] as:

$$i = \frac{9.5336 \times (1 + 0.5917 \times \log T_M)}{(t + 5.9828)^{0.6383}} \tag{1}$$

The rainfall process hydrograph can be described in Equations (5) and (6),

$$i_{before} = \frac{A \times (1 + c \times \log T_M)[\frac{1-n}{r} \times t_1 + b]}{(\frac{t_1}{r} + b)^{n+1}} \tag{2}$$

$$i_{after} = \frac{A \times (1 + c \times \log T_M)[\frac{1-n}{1-r} \times t_2 + b]}{(\frac{t_2}{1-r} + b)^{n+1}} \tag{3}$$

where T_M, A, b, n are the parameters in the rainstorm intensity equation, $b = 5.9828$, $n = 0.6383$; r is the peak coefficient which expresses the ratio between peak flow time point and rain duration. In the Kunshan case, $r = 0.438$; t_1 and t_2 are the time space before and after peak flow time point; i_{before} and i_{after} describe the rainfall intensity before and after peak flow, respectively.

In this study, rainfall duration time was set as one hour, and rainfall return period was set as, 2a, 5a, 10a, 20a, 50a, and 100a. The rainfall hydrograph can be referred to in Figure 3.

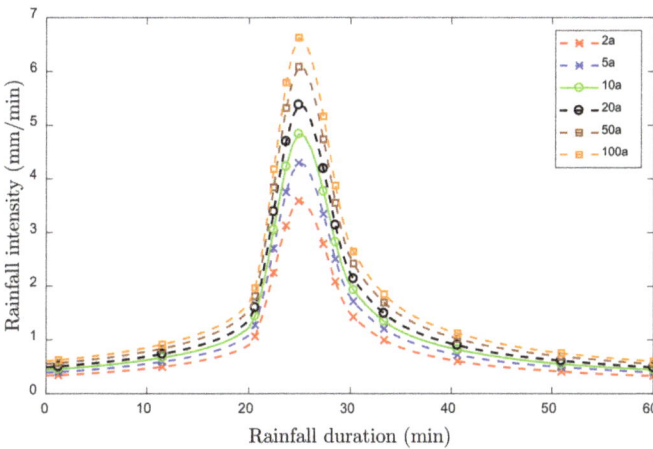

Figure 3. One-hour rainfall hydrograph with different return periods.

3.2.2. Sub-Catchment Physical Properties

Sub-catchment parameters were fixed according to the physical properties of the Jiangpu polder area. The area of each sub-catchment was measured with an electronic ruler on the map. The runoff coefficient was decided as the Chinese government's statistical result, with forest park areas as 0.15, urban areas with intensive buildings as 0.55, and urban areas with sparse buildings as 0.4. Data from the pumping station were also very important, which gave detailed information of each pumping station (Table 1) (Note: WL$_{ON}$ = water level for pump start; WL$_{OFF}$ = water level for pump stop).

Table 1. Pumping station data.

ID	Name	Q (m^3/s)	WL$_{ON}$ (m)	WL$_{OFF}$ (m)
1	GONGYUAN	4.8		
2	DONGDANG	3.2		
3	HONGQIAO	2.85		
4	YUEHE	6.5	2.6	2.2
5	SICHANGGANG	3.0		
6	BAITA	3.2		
7	GONGQING	3.7		
8	XIDANG	7.0		

3.3. Results and Discussion

3.3.1. Sensitivity Analysis

The sensitivity analysis was to understand how each parameter affected the simulation output, and to assess which parameters should be concentrated on. The sensitivity analysis techniques have been developed and categorized into different classes, including derivative analysis, algebraic analysis of model equations, sparse sampling, variance decomposition, Fourier analysis, binary classification, and many more. The first two kinds of sensitivity analysis techniques could provide detailed insight, while sampling-based sensitivity analysis techniques could handle more complex models to obtain the necessary information at a relatively low computational cost [31].

In this study, we implemented two steps in sensitivity analysis techniques. In the first step, we calculated the total flood volume using different parameter values (i.e., changed only one parameter at one time), demonstrating how sensitive the model performance is to the change in individual parameters. In the second step, a traditional sensitivity analysis technique-the Morris screening method [32]—was adopted to further evaluate the impact of the parameter on the model output.

Preliminary Calculation for Sensitivity Analysis

In the above conceptual model, several parameters were considered in the sensitivity analysis. From Figure 2, the polder is substituted by several blocks, e.g., rainfall block, catchment block, river block, mixer block, and pump block. Through experience screening, the parameters that should be included in the sensitivity analysis are highlighted in the catchment, and river blocks, outlined in Table 2.

Table 2. Screening parameters of sensitivity analysis.

Model Components	Parameter	Value Range	Best Fit
	K [s]	100~300, delta 50	300
Catchment	N [-]	3~15, delta 3	3
	X [-]	0~0.5, delta 0.1	0
	K [s]	100~300, delta 50	300
River	N [-]	7~15, delta 2	11
	X [-]	0~0.5, delta 0.1	0

Figure 4 shows the sensitivity analysis result of the Muskingum parameters *K*, *N*, *X* in the catchment and river blocks. Results were calculated by running the CityDrain3 models manually. In comparison, we preliminarily find that *CN*, *RK*, *RN* and *RX* are the sensitive factors to the output (Note: CK = Catchment K; CN = Catchment N; CX = Catchment X; RK = River K; RN = River N; RX = River X; FV = Flood Volume).

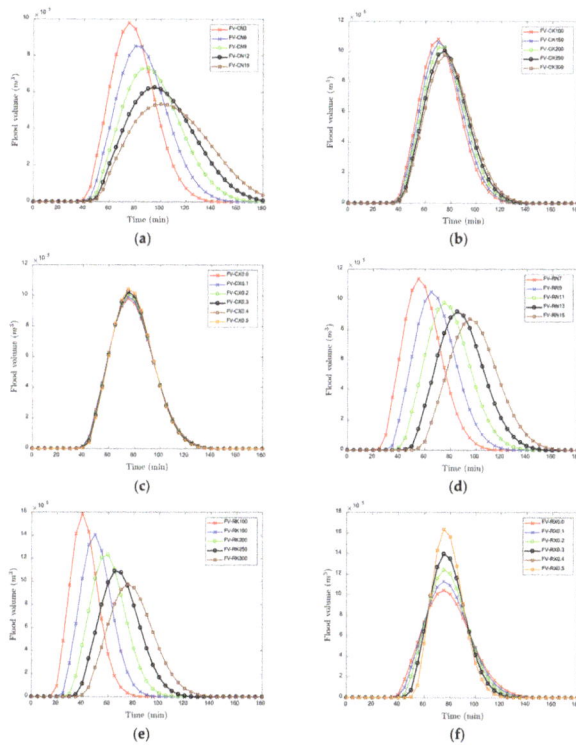

Figure 4. Preliminary sensitivity analysis result of parameters in the catchment and river blocks: (**a**) catchment K, (**b**) catchment N, (**c**) catchment X, (**d**) river K, (**e**) river N, and (**f**) river X.

The Morris Screening Method for Sensitivity Analysis

The Morris screening [33] proposed an elementary effects method to assess which factors may have (a) negligible; (b) linear; (c) nonlinear interactions with other factors [34]. After implementing a number of model runs, the influence of parameters on output can be ranked. This will discard those lower-ranked parameters to avoid expensive computational time cost and allow researchers to focus on the parameters that have an obvious effect on output [31].

In the next step, the Morris screening method was applied to different parameters in the catchment and river blocks (Table 3 and Figure 5). The results of sensitivity measures (μ and σ) for different parameters influencing output (total flood volume of the Jiangpu polder) (Table 3) and the graphical results of each parameter in Morris screening analysis (Figure 5) are represented. As we can see from Table 3 and Figure 5, and the preliminary findings (Figure 4), *CN* showed the most sensitive effect on the overall flood volume result. Therefore, special attention was paid to the study of *CN*. After comparing with the overall flood volume when *N* equals 3–15, it was found that when *N* = 3, the overall flood volume was equivalent to the one that was calculated with different values of *CK* and *CX*.

Table 3. Morris screening sensitivity analysis result of total-flood volume.

Parameter	μ (Mean)	σ (Sigma)	σ/μ
CK	−2458.263608	4498.145451	−1.83
CN	14,947.945884	456,305.8701	30.53
CX	−46,902.31404	39,686.59956	−0.85
RK	−2314.093176	1529.955954	−0.66
RN	−22,624.78659	15,590.01461	−0.69
RX	−103.258607	41.1759731	−0.40

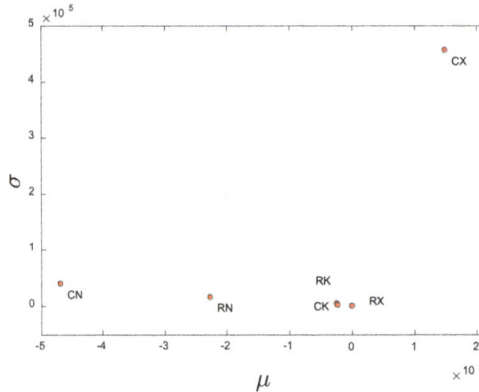

Figure 5. Morris sensitivity analysis result of Muskingum parameters in Jiangpu polder model.

3.3.2. Model Calibration

In order to develop a conceptual model that is sufficiently accurate in simulating the real urban drainage system, a calibration of sub-catchment physical parameters and Muskingum parameters K, N and X in the catchment blocks, as well as in the river blocks in the future should be employed. The results of the Nash-Sutcliffe efficiency coefficient (NSE) [35] should show that the conceptual model can represent the real circumstance well and the conceptual model developed within this procedure is acceptable.

However, in this research, the aim was to compare the effects of different control strategies (baseline, with pump only, with storage tank only, and with both pump and storage tank). The model calibration work will be carried out in the near future as an extension of this work.

3.3.3. Application of the Model

Storms with different return periods were designed as inputs into the developed model to study the effects of sponge polder facilities. In the first step, a preliminary application of several rainfall scenarios was employed to evaluate the relationship between rainfall depth and flood depth. Then, a comparison of the baseline and sponge polder models was carried out to study the effects of different sponge facilities.

Flood Calculation under Different Rainfall Return Periods

For the application of the earlier developed model, precipitation of one-hour duration with different return periods (2a, 5a, 10a, 20a, 50a, and 100a) was implemented as an input into the baseline model. After flow routing with the CityDrain3 model, the rainfall depth and flood depth under every scenario could be compared (Figure 6).

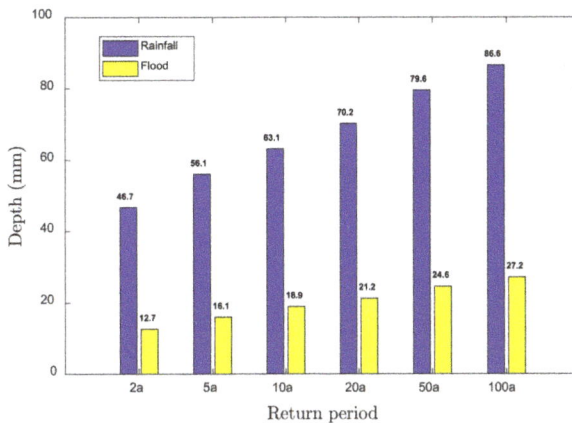

Figure 6. Simulated flood depth in Jiangpu polder with different return period precipitations.

When the rainfall duration was one hour, with return periods of 2a to 100a, the total rainfall depth ranged from 46.7 mm to 86.6 mm (Figure 6). The calculated flood depth ranged from 12.7 mm to 27.2 mm. The growing trend of flood depth under increased precipitation showed the feasibility of polder flood simulation with CityDrain3. This was the basis for the next comparison of baseline and sponge polder models.

Because the developed model was a conceptual system, the flood depth shown in Figure 6 was the average value of the whole polder area. In practice, different sub-catchments have different elevations, so the flood volume would flow to the sub-catchment areas with low elevation. This causes severe flood problems, including loss of human life and economic damage.

Comparison of Baseline and Sponge Polder Models

To investigate the effects of different sponge polder facilities, a one-hour rainfall with return period of 10 years was employed in the developed baseline model and the three sponge polder models. The simulation results are shown in Figure 7, with detailed information listed in Table 4.

Results (Figure 7) indicated that the pump and storage tank could lead to a reduction in total flood volume, flood yielding time, peak flow, and peak-flow occurrence time. Moreover, the effects of combined strategies (pump and storage tank) showed a linear relationship with the effect of each individual control strategy (pump only or storage tank only). The reason was that the both of the individual control strategies chosen in this research had a linear effect on the water volume. It should be noted that there was no control strategy with the infiltration effect in the model.

The effect of control strategies on total flood volume was ranked as—combined strategies (pump and storage tank) > storage tank only > pump only (Table 4 and Figure 7). The effect on peak flow was ranked as—combined strategies (pump and storage tank) > storage tank only > pump only. The effect on the flood yielding time was ranked as—combined strategies (pump and storage tank) > storage tank only > pump only. The effect on the peak-flow occurrence time was ranked as: Combined strategies (pump and storage tank) > storage tank only > pump only. The results were attributed to the linear relationship of the two individual control strategies chosen in this research.

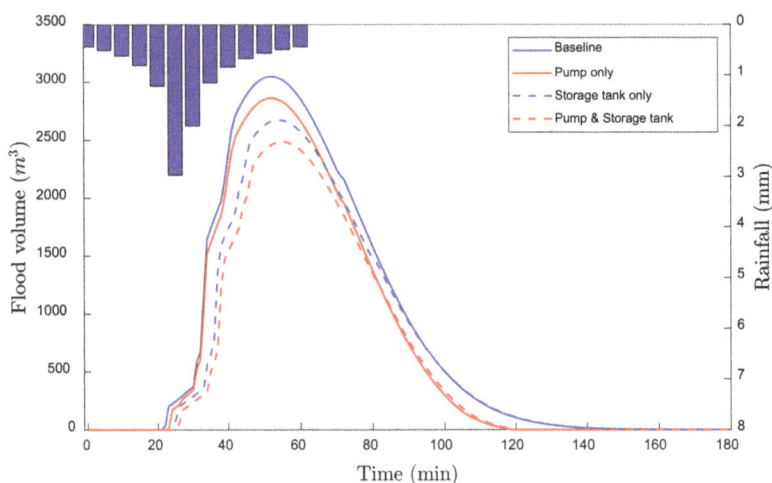

Figure 7. Comparison of total flood volume in the Jiangpu polder area during a rainfall with one-hour duration and 10-year return period under different flood control strategies.

Table 4. Detailed simulation result of polder area under different control strategies.

Type of Models	Baseline	Pump Only	Storage Tank Only	Combination of Pump and Storage Tank
Total rainfall (m^3)			496,005	
Total flood volume (m^3)	146,235	128,533	126,580	108,913
Reduction of flood volume (m^3)		17,702	19,655	37,322
Reduction percentage		12.1%	13.4%	25.5%
Peak flow (m^3/min)	3050	2862	2674	2486
Reduction of peak flow (m^3/min)		188	376	564
Reduction percentage		6.16%	12.33%	18.5%
Flood yielding time (min)	22	23	25	26
Reduction of Flood yielding time (min)		1	3	4
Peak-flow occurrence time (min)	52	52	54	55
Reduction of Peak-flow occurrence time (min)		0	2	3

4. Conclusions and Future Work

In this paper, a conceptual flood-simulation model was developed using CityDrain3. The developed model was applied on a storm water management program at Jiangpu polder, Kunshan, China. The baseline model was first implemented to calculate flood depth under different rainfall scenarios. Later, the baseline model and three sponge polder models were used to estimate the storm water flood result. According to the results and discussions, the following key findings could be concluded:

(1) CityDrain3 can be applied to conceptualize the urban flood performance of polder areas in the Yangtze River Delta area in China. The CityDrain3 model developed in this paper was set up according to the Jiangpu polder land surface information. Sensitivity analysis showed the robustness of the key parameters in each sub-catchment of the model. A hydrological model was used to calculate the flow routing in the polder system.

(2) Preliminary application of the developed baseline model was used to calculate flood depth of the Jiangpu polder under one-hour rainfall with return periods of 2a, 5a, 10a, 20a, 50a, and 100a. It is apparent that the growing trend of flood depth under increased precipitation showed the feasibility of the polder flood simulation using CityDrain3.

(3) A comparative study of the strategies was used to investigate the flood results of the baseline model and three sponge polder models. The effect of control strategies on total flood volume was ranked as—combined strategies (25.5%) > storage tank only (13.4%) > pump only (12.1%). The effect on peak flow was ranked as—combined strategies (18.5%) > storage tank only (12.33%) > pump only (6.16%). The effect on the flood yielding time was ranked as—combined strategies (4 min) > storage tank only (3 min) > pump only (1 min). The effect on the peak-flow occurrence time was ranked as—combined strategies (3 min) > storage tank only (2 min) > pump only (0 min).

(4) The results obtained using the CityDrain3 model can help the engineers and stakeholders to understand and select among the sponge polder facilities.

For future research, there should be more blocks coded to represent sponge polder facilities, e.g., wetlands, bioretention ponds, and grassed swales. This would give CityDrain3 model the ability to realize the real-time control of a polder flood.

Further, there should be flow meters and water-level sensors installed to capture in-situ measurements. The practical measurement data could then be used to verify the developed simulation model. Selecting the optimal control strategy and realizing the real-time control of urban floods could be the main extension of the model developed in this paper.

Acknowledgments: This study was funded by the National key Technologies R&D Program (No. 2015BAL02B05), the Priority Academic Program Development of Jiangsu Higher Education Institutions (No. 1105007002), the Fundamental Research Funds for Southeast University (No. KYLX15_0093) and China Scholarship Council.

Author Contributions: Christian Urich conceived and designed the study topic, imported the simulation software CityDrain3 and did preliminary research with this software; Dingbing Wei chose the case study area, did the main simulation work and analyzed the data; Shuci Liu checked the data and drew figures; Sheng Gu contributed in writing the paper.

Conflicts of Interest: The authors declare no conflicts of interest.

References

1. Deng, Y. Hydrological characteristics analysis and water exchange modelling for polders in plain river network area. In *College of Geographical Science*; Nanjing Normal University: Nanjing, China, 2014. (In Chinese)
2. Jingsen, L. Optimization methods for conventional scheduling of drainage pumps in plain polders. In *Agricultural Soil and Water Engineering*; Yangzhou University: Yangzhou, China, 2014. (In Chinese)
3. Vermuyten, E.; van den Zegel, B.; Wolfs, V.; Meert, P.; Willems, P. Real-time flood control by means of an improved MPC-GA algorithm and a fast conceptual river model for the demer basin in Belgium. In Proceedings of the 6th International Conference on Flood Managament, São Paulo, Brazil, 16–18 September 2014.
4. Maksimović, Č.; Prodanović, D.; Boonya-Aroonnet, S.; Leitao, J.P.; Djordjević, S.; Allitt, R. Overland flow and pathway analysis for modelling of urban pluvial flooding. *J. Hydraul. Res.* **2009**, *47*, 512–523. [CrossRef]
5. Vezzaro, L.; Grum, M. A generalised Dynamic Overflow Risk Assessment (DORA) for Real Time Control of urban drainage systems. *J. Hydrol.* **2014**, *515*, 292–303. [CrossRef]
6. Qin, H.-P.; Li, Z.-X.; Fu, G. The effects of low impact development on urban flooding under different rainfall characteristics. *J. Environ. Manag.* **2013**, *129*, 577–585. [CrossRef] [PubMed]
7. Chang, F.-J.; Chen, P.-A.; Lu, Y.-R.; Huang, E.; Chang, K.-Y. Real-time multi-step-ahead water level forecasting by recurrent neural networks for urban flood control. *J. Hydrol.* **2014**, *517*, 836–846. [CrossRef]
8. Domingo, N.S.; Refsgaard, A.; Mark, O.; Paludan, B. Flood analysis in mixed-urban areas reflecting interactions with the complete water cycle through coupled hydrologic-hydraulic modelling. *Water Sci. Technol.* **2010**, *62*, 1386–1392. [CrossRef] [PubMed]
9. Wei, H. Study on simulation and operation of flooding prevention system in plain river-net region. In *Hydrology and Water Resources*; Hohai University: Nanjing, China, 2007. (In Chinese)
10. Dazhou, X. Research and implementation of automation and dispatch management system of pumping group station in Langxia polder areas. In *Agricultural Automation and Electrization*; Yangzhou University: Yangzhou, China, 2017. (In Chinese)

11. Jianye, X. Effect of polder management methods on flooding prevention. In *Agricultural Water and Soild Engineering*; Yangzhou University: Yangzhou, China, 2009. (In Chinese)

12. Li, X.; Li, J.; Fang, X.; Gong, Y.; Wang, W. Case Studies of the Sponge City Program in China. In Proceedings of the World Environmental and Water Resources Congress 2016, West Palm Beach, FL, USA, 22–26 May 2016.

13. Villarreal, E.L.; Semadeni-Davies, A.; Bengtsson, L. Inner city stormwater control using a combination of best management practices. *Ecol. Eng.* **2004**, *22*, 279–298. [CrossRef]

14. Dietz, M.E. Low impact development practices: A review of current research and recommendations for future directions. *Water Air Soil Pollut.* **2007**, *186*, 351–363. [CrossRef]

15. Wong, T.H. An overview of water sensitive urban design practices in Australia. *Water Pract. Technol.* **2006**, *1*, wpt2006018. [CrossRef]

16. Fletcher, T.D.; Shuster, W.; Hunt, W.F.; Ashley, R.; Butler, D.; Arthur, S.; Trowsdale, S.; Barraud, S.; Semadeni-Davies, A.; Bertrand-Krajewski, J.-L. SUDS, LID, BMPs, WSUD and more—The evolution and application of terminology surrounding urban drainage. *Urban Water J.* **2015**, *12*, 525–542. [CrossRef]

17. Cui, G.; Zhang, Q.; Zhan, Z.; Chen, Y. Research progress and discussion of sponge city construction. *Water Resour. Prot.* **2016**, *32*, 1–4. (In Chinese)

18. Bach, P.M.; Rauch, W.; Mikkelsen, P.S.; McCarthy, D.T.; Deletic, A. A critical review of integrated urban water modelling—Urban drainage and beyond. *Environ. Mode. Softw.* **2014**, *54*, 88–107. [CrossRef]

19. Figueras, J.; Cembrano, G.; Puig, V.; Quevedo, J.; Salamero, M.; Martí, J. Coral off-line: An object-oriented tool for optimal control of sewer networks. In Proceedings of the 2002 IEEE International Symposium on Computer Aided Control System Design, Glasgow, UK, 20 September 2002.

20. Burger, G.; Sitzenfrei, R.; Kleidorfer, M.; Rauch, W. Parallel flow routing in SWMM 5. *Environ. Model. Softw.* **2014**, *53*, 27–34. [CrossRef]

21. Puig, V.; Cembrano, G.; Romera, J.; Quevedo, J.; Aznar, B.; Ramon, G.; Cabot, J. Predictive optimal control of sewer networks using CORAL tool: Application to Riera Blanca catchment in Barcelona. *Water Sci. Technol.* **2009**, *60*, 869–878. [CrossRef] [PubMed]

22. Riaño-Briceño, G.; Barreiro-Gomez, J.; Ramirez-Jaime, A.; Quijano, N.; Ocampo-Martinez, C. MatSWMM—An open-source toolbox for designing real-time control of urban drainage systems. *Environ. Model. Softw.* **2016**, *83*, 143–154. [CrossRef]

23. García, L.; Barreiro-Gomez, J.; Escobar, E.; Téllez, D.; Quijano, N.; Ocampo-Martinez, C. Modeling and real-time control of urban drainage systems: A review. *Adv. Water Resour.* **2015**, *85*, 120–132. [CrossRef]

24. Amdisen, L.K.; Gavranovic, N.; Yde, L. Model-based control-a hydroinformatics approach to real-time control of urban drainage systems. *J. Hydraul. Res.* **1994**, *32*, 35–43. [CrossRef]

25. Muschalla, D.; Vallet, B.; Anctil, F.; Lessard, P.; Pelletier, G.; Vanrolleghem, P.A. Ecohydraulic-driven real-time control of stormwater basins. *J. Hydrol.* **2014**, *511*, 82–91. [CrossRef]

26. Achleitner, S.; Möderl, M.; Rauch, W. CITY DRAIN©–An open source approach for simulation of integrated urban drainage systems. *Environ. Model. Softw.* **2007**, *22*, 1184–1195. [CrossRef]

27. Achleitner, S. *Modular Conceptual Modelling in Urban Drainage Development and Application of City Drain*; IUP-Innsbruck University Press: Innsbruck, Austria, 2008.

28. Burger, G.; Bach, P.M.; Urich, C.; Leonhardt, G.; Kleidorfer, M.; Rauch, W. Designing and implementing a multi-core capable integrated urban drainage modelling Toolkit: Lessons from CityDrain3. *Adv. Eng. Softw.* **2016**, *100*, 277–289. [CrossRef]

29. Forster, C. Urban Water Cycle Modelling with CityDrain3—Water Balance Improvements and a Demonstration Case Study. In *Institute for Urban Water Management*; Dresden University of Technology: Dresden, Germany, 2016.

30. Raoming, S.; Shao, D.; Jin, J.; Zhu, H. *Study on Rainstorm Intensity Equation and Design of Rainstorm in Kunshan City*; H.W.W.S.A.D. Company: Kunshan, China, 2016.

31. Norton, J. An introduction to sensitivity assessment of simulation models. *Environ. Model. Softw.* **2015**, *69*, 166–174. [CrossRef]

32. Ruano, M.; Ribes, J.; Ferrer, J.; Sin, G. Application of the Morris method for screening the influential parameters of fuzzy controllers applied to wastewater treatment plants. *Water Sci. Technol.* **2011**, *63*, 2199–2206. [CrossRef] [PubMed]

33. Morris, M.D. Factorial sampling plans for preliminary computational experiments. *Technometrics* **1991**, *33*, 161–174. [CrossRef]

34. Campolongo, F.; Cariboni, J.; Saltelli, A. An effective screening design for sensitivity analysis of large models. *Environ. Model. Softw.* **2007**, *22*, 1509–1518. [CrossRef]

35. Hwang, S.H.; Ham, D.H.; Kim, J.H. A new measure for assessing the efficiency of hydrological data-driven forecasting models. *Hydrol. Sci. J.* **2012**, *57*, 1257–1274. [CrossRef]

water

MDPI

Article

Rebuild by Design in Hoboken: A Design Competition as a Means for Achieving Flood Resilience of Urban Areas through the Implementation of Green Infrastructure

Robert Šakić Trogrlić [1,*]**, Jeroen Rijke** [2,3]**, Nanco Dolman** [4] **and Chris Zevenbergen** [5]

[1]   School of Energy, Geoscience, Infrastructure and Environment, Institute for Infrastructure and Environment, Heriot-Watt University, Edinburgh EH14 4AS, UK
[2]   Knowledge Centre Engineering and Society, HAN University of Applied Sciences, 6826 CC Arnhem, The Netherlands; j.rijke@han.nl
[3]   Applied Research Centre Delta Areas and Resources, VHL University of Applied Sciences, #26a Larensteinselaan, 6882 CT Velp, The Netherlands
[4]   Royal HaskoningDHV, #47 Contactweg, 1090 GE Amsterdam, The Netherlands; nanco.dolman@rhdhv.com
[5]   Water Science & Engineering Department, IHE Delft Institute for Water Education, #7 Westvest, 2611 AX Delft, The Netherlands; c.zevenbergen@un-ihe.org
*   Correspondence: RS36@HW.AC.UK

Received: 14 January 2018; Accepted: 19 April 2018; Published: 25 April 2018

Abstract: The Rebuild by Design (RBD) competition was launched after the devastating impact of Hurricane Sandy, and the winning designs have put a significant emphasis on green infrastructure (GI) as a means of achieving flood resilience in urban areas. Previous research in the field of urban stormwater management indicates that wide-spread implementation of GI remains a challenge, largely due to a lack of understanding of the required governance approaches. Therefore, by using a case study of Hoboken, for which the winning design was developed, this paper explores whether RBD provides governance structures and processes needed for the uptake of GI. Semi-structured interviews and desk study provided the data for an analysis of the presence of factors for supporting the transformative governance needed to facilitate the uptake of innovative solutions. Results indicate that RBD brought a greater change in terms of governance processes when compared to governance structures. In Hoboken, RBD created a narrative for long-term change, put GI as a preferred solution for tackling multiple challenges, and strengthened the local political buy-in. However, pitfalls were observed, such as limited funding provision, lack of regulatory compliance, economic justification and large investments required from public and private parties. The absence of these factors can hinder the overall uptake of the GI solution. Even though the design competition presents a novel approach to the field of resilience development, further steps should be made in understanding how the RBD methodology can be adjusted to provide results of equal quality in different settings (e.g., less developed regions, different governance contexts).

Keywords: urban resilience; flooding; governance; green infrastructure; Hoboken; Rebuild by Design

---

1. Introduction

In the past, urban areas have been perceived as "safe havens" providing shelter from the impact of natural hazards. However, the modern recognition is that cities are seen as hotspots of disasters and risks [1]. Climate change and its impact will significantly affect cities and their populations, infrastructure and economic activities [2]. The devastating impacts of Hurricane Katrina (2005) in New Orleans and Hurricane Sandy (2012) in the greater New York Metropolitan Area, exemplified to what extent cities of today are vulnerable and susceptible to losses. More recently, in 2017, the same message has been conveyed through Hurricane Harvey and Hurricane Irma.

This paper focuses on urban water systems to manage pluvial flooding due to excessive rainfall, insufficient drainage capacity, inadequate planning and how these systems may impact the characteristics of the built environment and vice versa. It is widely acknowledged that these systems are under increasing threat due to several factors, such as climate change, urban population growth, pollution, limited resources and aging infrastructure [3,4]. These pressures are expected to increase flood risk in urban areas and further degrade the health of open water ways around the globe [5]. Resilience has emerged as an approach to deal with such pressures. In urban areas, it is often envisioned as the ability to absorb, adapt and respond to changes in order to provide a predictable performance of an urban system under a wide range of often unpredictable circumstances [6]. With respect to urban flood risk management, resilience refers to the ability to recover from flooding and to adapt to changing probabilities and consequences of flooding as well as to seize opportunities that may emerge over time. Liao [7] stated that the resilience *"can be conceptualized as the capacity to remain in a desirable regime while experiencing a flood"*, thus adding the characteristic of withstanding the disturbance. Salinas Rodriguez et al. [8] added that the term resilience to flooding also incorporates social, institutional and economic aspects. Thus, the operationalization of resilience in urban areas is not easy to achieve [9,10].

After the devastating impact of Hurricane Sandy on the greater New York City area in 2012, the Rebuild by Design (RBD) competition was launched. Recognized as the first among Cable News Networks' (CNN's) Top 10 Ideas of 2013 [11] and awarded the "Most Groundbreaking Federal Challenge or Prize Competition" by the General Services Administration in 2015 [12], the competition delivered innovative designs to enhance flood resilience in the region. RBD has been replicated in several similar initiatives worldwide, such as in the National Disaster Resilience Competition in the United States, the Global Resilience Partnership in areas of Sahel, Horn of Africa, Southeast Asia and Living with Water in Boston [11,13]. Most recently, the call for the "Bay Area Design Challenge" was released, replicating the original RBD competition for the San Francisco Area [14].

One of the projects that came out as a winning proposal in the competition was developed for the highly urbanized parts of the New York metropolitan area Hoboken, Weehawken and Jersey City, New Jersey. There, green infrastructure (GI; e.g., rain gardens and permeable pavements in new parks) was adopted as a key element of a strategy to achieve flood resilience at a district level.

Since the mid-1990s, significant improvement has been made in the availability of advanced technologies for sustainable urban water management (e.g., GI), but still, the transition to actual implementation is hindered [15]. Even though many examples from Australia, the United States, Singapore, and United Kingdom indicate that implementation of GI is possible with strong political commitment, the implementation on a larger scale is still slow and not yet mainstream in urban planning [16,17]. Previous research recognized that the adoption of sustainable urban water systems (i.e., GI) calls for new governance structures and processes [18–20].

With a current movement towards using the RBD model in other regions, it is important to take stock of the lessons from the original competition in terms of enhancing urban flood resilience through GI. Therefore, the aim of this paper, building on the previous recognition of the GI uptake being primarily a governance issue, is to understand whether RBD provided governance structures and processes needed for the wide-spread implementation of GI, using a case study of GI in Hoboken, New Jersey.

2. Literature Review

2.1. The Process of Rebuild by Design

After the devastating impacts of Hurricane Sandy (October 2012), the disaster relief effort started. Under President Obama, the Hurricane Sandy Rebuilding Task Force was established, led by the

U.S Department of Housing and Urban Development (HUD). In June 2013, this Task Force launched the RBD competition under support by the Rockefeller Foundation and philanthropists. One of the major sources of inspiration for RBD were the "Dutch Dialogues", a series of conferences with local and international experts, initiated in 2006 as a response to the impacts of Hurricane Katrina on New Orleans. These dialogues emphasized the need for a multidisciplinary approach to rebuild Greater New Orleans into a more resilient and attractive city [21].

RBD gathered international expertise to generate strategies to address the resilience challenges in the region through offering innovative designs and presenting a novel approach to policy development. It presented a collaborative effort of government, the private sector, academia, philanthropists and community members. To motivate proposals for development, the federal government allocated the funds for the implementation of the winning proposals by securing $930 million, and private sector secured funds for awarding the teams that developed the winning proposals [11]. The list of the proposals that received funding for implementation is presented in Table 1.

As a novel approach and first design competition on this scale (both spatially and by funding available for implementation), RBD applied an innovative methodological framework for enhancing regional long-term resilience. The competition was organized with the vision of catalyzing the transformation of the whole affected region towards being flood resilient [22].

Table 1. List of projects that received funding for implementation.

Project	Location	Team	Allocated Federal Funds ($ million)	Green Infrastructure Presence in Winning Projects
BIG U	New York, New York	BIG TEAM	335	Yes (e.g., resilience parks)
Living with the Bay: A Comprehensive Regional Resiliency Plan for Nassau County's South Shore	Long Island, New York	Interboro Team	125	Yes (e.g., wetlands, stormwater detention)
New Meadowlands: Productive City + Regional Park	The Meadowlands, New Jersey	MIT CAU + ZUS + URBANISTEN	150	Yes (e.g., resilience parks)
Resist, Delay, Store, Discharge: A Comprehensive Strategy for Hoboken	Hoboken, New Jersey	OMA	230	Yes (e.g., green roofs, permeable pavements)
Hunts Point Lifelines	Bronx, New York	PennDesign/OLIN	20	Yes (e.g., green roofs)
Living Breakwaters	Staten Island, New York	Scape/Landscape Architecture	60	Yes (e.g., stormwater ponds, wetlands)
Resilient Bridgeport	Bridgeport, Connecticut	WB unabridged with Yale ARCADIS	10	Yes (e.g., vegetative buffers)

2.2. The Rise of Green Infrastructure

Over the past few decades, it is internationally increasingly being recognized that there is a need for change in the way urban water systems are traditionally managed [23–25]. Hence, there is a strong movement in transitioning towards more sustainable urban water systems [18,26]. As argued by Rijke et al. [24], sustainable urban water management systems deliver water resource management systems that are adaptive to change and resilient to extremes. Whilst acknowledging the use of different terms related to sustainable stormwater management in the literature (e.g., green infrastructure, low impact development, sustainable urban drainage systems, and water sensitive urban design) [27], due to the location of the case study, this paper adopts the term GI.

If combined appropriately, GI measures contribute to the resilience of urban areas [4,28]. In the domain of sustainable urban stormwater management, GI provides a wide range of social, economic and environmental benefits [29,30]. Switching from grey to green infrastructure in urban areas is generally associated with additional social, educational, economic and environmental improvements [31]. Hence, GI serves larger sustainability goals rather than stormwater management alone. GI plays an important role in the field of flood risk management in urban areas, and its potential for flood risk reduction has been demonstrated in numerous studies. For instance, Liu et al. [32] have investigated the impact of GI on a typical neighborhood in Beijing and have found that by implementing integrated GI, runoff of a five-year return period storm has been 100% reduced. In Augustenborg and Malmo in Sweden, GI has reduced stormwater runoff by 50% [33]. A recent study by the European Environment Agency [34], based on several European case studies, shows that GI can mitigate floods in a cost-efficient way, when compared to traditional solutions. The effectiveness of GI in managing floods in urban areas is amongst others strongly dependent on the local geophysical and hydrological factors [30].

In the United States, the concept of GI as a way to manage stormwater and reduce water pollution has been applied since the early 1990s [35], and nowadays, the Environmental Protection Agency is increasingly promoting GI through regulation [36]. Cities such as Philadelphia (through the Green Cities and Clean Waters) and Oregon (e.g., Green Streets Programme) serve as examples of city-wide scale application of GI. Their success has largely been achieved through the introduction of several initiatives, such as stormwater charge discount programmes, demonstration projects, the requirement of on-site stormwater management for new development and redevelopment, financial incentives, and a progressive stormwater regulation at municipal level [29].

While the body of scientific knowledge of advanced GI technologies is growing, the development of an evidence base for deploying GI is lagging behind [37]. There is an urgent need to have access to information on examples of local success which demonstrate stormwater practices [4,38,39]. Even though the need for a new paradigm is increasingly being advocated, decision-makers across the globe continue to support traditional ways of dealing with stormwater management, rather than promoting innovative technologies and approaches [23]. Still, the favored approach to urban drainage remains pure conveyance to the nearest water body or pond [16,39]. An overview of barriers to implementing GI is presented in Table 2. The identified barriers are numerous and partly of technical origin, but the common denominator in publications on GI centers around socio-institutional aspects [18–20], since innovative stormwater management calls for an alternative to the traditional way of managing and governance of these systems. A typical feature of GI is also that they are highly context specific, as mentioned above, which implies that every case study is unique and cannot be transferred to others since function is dependent on local characteristics.

Table 2. Barriers to green infrastructure implementation as identified by various authors.

Author	Barriers
Roy et al. [40]	(1) lack of empirical data on performance and costs of measures; (2) deficiency in technical standardization and guidances; (3) unclear and fragmented responsibilities; (4) lack of institutional capacity; (5) lack of legislative mandate; (6) lack of funding and effective market incentives; and (7) resistance to change
Brown et al. [25]	(1) uncoordinated institutional framework; (2) limited community engagement, empowerment and participation; (3) limits of regulatory framework; (4) insufficient resources (capital and human); (5) unclear, fragmented roles and responsibilities; (6) poor organizational commitment; (7) lack of information, knowledge and understanding in applying integrated, adaptive forms of management; (8) poor communication; (9) no long term vision- strategy; (10) technocratic path dependencies; (11) little or no monitoring and evaluation; and (12) lack of political and public will
Lee at al. [16]	(1) lack of understanding among stakeholders; (2) lack of common standards, guidelines and technical skills; (3) limited research and knowledge; (4) fragmented stormwater management institutions; (5) lack of institutional provision; and (6) economic cost
Abhold et al. [37]	(1) technical and physical; (2) legal and regulatory; (3) financial; and (4) community and institutional barriers

Table 2. *Cont.*

Author	Barriers
Cettner et al. [19]	(1) insufficient practical knowledge; (2) missing support (organisational, scientific, local community); (3) lack of resources and knowledge; (4) ineffective relations and networks; and (5) discrepancies between interest groups
Thorne et al. [20]	(1) Community perceptions, buy-in, ownership and understanding of GI; (2) willingness to pay/sell; (3) how to change policy support for GI; (4) future governance of GI; (5) keeping GI on the agenda and promoting interagency working; and (6) including climate change in policy/design standards
O'Donnel et al. [17]	(1)Negative past experiences; (2) low priority and/or competing priorities; (3) future land use and climate; (4) political leadership and champions; (5) lack of available space; (6) responsibilities and ownership;(6) institutional capacity and expertise; (7) behaviours and cultural; (8) physical science/engineering uncertainities; (9) legislations, regulations and governance; (10) monetizing the multiple benefits; (11) maintenance and adoption; (12) issues with partnership working; (13) ineffective/lack of communication; (14) funding and costs; (14) lack of knowledge, education and awareness; and (15) reluctance to support new practices

2.3. Green Infrastructure Implementation as a Governance Issue

In the field of urban water governance, three dominant governance approaches are recognized: hierarchical, market and network governance [40]. Hierarchical governance is characterized by formal institutional patterns, lack of stakeholder participation and inflexibility in terms of learning [40,41]. Market governance approaches, by translating private sector management principles to public sector, focus on the efficient allocation of resources, empower citizens and introduction of competition and privatization [42,43]. Network governance is a form of governance where multiple stakeholder groups (public, private and civil actors) interact and as an outcome, develop self-organizing tendencies [43]. This approach is increasingly being advocated as a dominant approach in delivering sustainable water management [41]. RBD can be considered an outing of network governance because it was set up for knowledge exchange amongst multiple "expert" and community actors to develop strategies for urban flood resilience.

Whilst in governance literature importance is placed on facilitating learning through a different approach to institutional design [44], some have seen an opportunity in creating a hybrid governance approach as a combination of standard hierarchical, market and network governance approaches [40], where governance arrangements are a mixture between network and hierarchical approaches and market governance instruments. An analysis of the uptake of stormwater harvesting and reuse schemes in Australia demonstrated how multi-level governance approaches comprised a mix between centralized-decentralized and formal-informal governance, proved to be effective in closing the strategic policy planning-implementation gap [24]. Van de Meene et al. [40] argued that current scholarly work lacks understanding of a setup of governance approaches that facilitates delivery and practical implementation of innovative urban water management solutions (such as GI). The dynamics of governance systems and continuous interactions between different actors create uncertainties that can be addressed through active learning in the process, stakeholder participation and self-organization of governance systems [45].

Led by the idea of a lack of prescriptive approaches for delivering adaptive governance in practice, Farrelly et al. [46] developed practical guidance for supporting transformative governance by identifying eight enabling socio-institutional factors, namely: (1) narrative, metaphor and image; (2) regulatory and compliance agenda; (3) economic justification; (4) policy and planning frameworks and institutional design; (5) leadership; (6) capacity building and demonstration; (7) public engagement and behavior change; and (8) research and partnership with policy and practice. Factors 1–4 refer to governance structures, whereas factors 5–8 refer to governance processes. Governance structures describe different arrangements of institutional design, whereas governance processes are concerned with managing those [47]. Transformative governance refers to a governance approach that has *"the capacity to shape non-linear change in complex systems of people and nature. (...). The goal of transformative governance is to actively shift a socio-ecological system to an alternative and more desirable regime by altering the structures and processes that define the system*" [47]. Even though this presents an attempt at developing a set of operational components of governance practices needed to support innovation, there is a knowledge gap on how (if at all) these elements can be combined in one approach. This analysis aims to explore to which extent the elements identified by Farrelly et al. [46] are contained in the RBD planning approach for GI as an innovative approach to stormwater management in Hoboken.

3. Methodology

3.1. Case Study Research Design

The paper adopted a case study research design [48]. Even though RBD resulted in seven winning case studies, this research focuses exclusively on the case study of GI in Hoboken. Hoboken was chosen for three main reasons. Firstly, the scope of the plans that were developed through RBD went beyond mitigation of hurricane-induced risks. Instead, a more comprehensive strategy for flood resilience was developed by Team OMA (OMA, Royal HaskoningDHV, Balmori Associates and HR&R). The name of the strategy is "Resist, Delay, Store, Discharge: A Comprehensive Urban Water Strategy for Hoboken" (RDSD). The terms "Resist, Delay, Store, Discharge" indicate (respectively): (R) hard infrastructure for protection against storm surges; (D) series of GI for the delay in the runoff; (S) storage areas for excessive rainfall across the City; and (D) discharge pumps. In this paper, the Rebuild by Design (RBD) is used to reflect to a process of the RBD design competition, whereas Resist, Delay, Store, Discharge (RDSD) is used to reflect to the outcomes of the actual winning RBD project in Hoboken. Secondly, the project team had good existing contact among stakeholders in Hoboken. It should be noted that one of the co-authors was a member of Team OMA on behalf of Royal HaskoningDHV. His contribution to this paper was to provide access to the initial set of interviewees and project documentation and to validate the findings from his personal experience in the project. Thirdly, in-depth analysis of one case study was suited to the project logistics (e.g., time and funding available for research, and ease of access to stakeholders).

3.2. Case Study Area

In Hoboken, two main sources of flooding are coastal flooding from the Hudson River and pluvial flash flooding when excessive rainfall coincides with a high tide on the Hudson River [49,50]. The flooding can be attributed to several factors, the main factor being a high level of imperviousness, low lying topography, and a combined sewer system with insufficient capacity during wet weather. Modern Hoboken is built on land reclaimed from the Hudson River. During rainfall episodes, there are approximately 100 combined sewer overflows (CSOs) per year discharged into the Hudson River through seven outfalls. This represents a serious environmental issue and, in the light of upcoming federal regulations, the number of CSOs are expected to be decreased to four per year [50]. The flooding problem, coupled with an overtaxed combined sewer system and lack of open space in the city, asks for a new planning approach for the uncertain future. Prior to RDSD and Hurricane Sandy, resilience was not as high on planning agendas of local decision makers and little significant emphasis was given to GI as a mean of providing multifunctional solutions to a series of challenges.

3.3. Data Collection and Analysis

The analysis is based on the analysis of interview data and document review, thus creating a mixed research approach. Twenty-one semi-structured interviews were conducted with urban water management stakeholders in Hoboken. Semi-structured interviews were a preferred technique because they allow flexibility by discussing specific aspects and provide the interviewer with new knowledge through analysis of the interviewee's point of view [48]. The interview questions were designed to cover the following topics: (1) local context in Hoboken (governance, planning, institutions, budgets, regulation, projects); (2) drivers for GI implementation in Hoboken; (3) hindering factors in implementation process; and (4) the role of RDSD in delivering flood resilience in Hoboken.

The initial list of contacts was developed based on the connections made by Team OMA during the RBD competition. After the initial contacts, additional interviewees were identified based on a snowballing technique, essentially meaning that initial participants proposed interviewees that have experience and knowledge relevant to the project [48]. Interviewees represented a range of different backgrounds and involvement in the process and covered local decision makers, state and federal representatives, developers, academia, non-profit organizations, drainage utility owners, transportation officers, and engineering and planning consultants. In addition, an extensive document review was performed to analyze the local context and the RBD process. To ensure high quality of data, desired information was extracted from peer-reviewed scientific literature, existing legislative documents, guidelines and documents provided by United States Environmental Protection Agency, existing development strategies in Hoboken (e.g., Hoboken Master Plan), and information published by local authorities, the State, project developers and online media. In addition, the review included a detailed examination of the RBD and RDSD documentation.

Coding was used to analyze the collected data, by assigning a code to each portion of text to detect similar information provided through the interviews. The codes used were both analytical (i.e., related to the specific research aims investigated, e.g., outcomes of RDSD in Hoboken) and descriptive (i.e., related to the themes identified, e.g., drivers for GI implementation). After the generation of initial codes and sub-codes, a list of "master" codes (i.e., themes) has been developed. It is important to notice that coding is an iterative process; thus, codes have been refined during the analysis. An identical coding strategy was imposed both for the primary data sources (i.e., interviews) and secondary data (i.e., desk document review). To secure the contextual meaning of the interviews, the authors coded sentences and portions of text. The employed coding strategy allowed for an analysis process that resulted in the in-depth analysis of the existing flood resilience planning in Hoboken, the position of GI, and the role played by RDSD. Research findings were also validated by representatives from the RBD team.

4. Results

RBD, with its focus on innovation and nature-based solutions, emphasized GI as a preferred option for stormwater management from the outset. In order to understand governance structures and processes, this section looks into the existing planning for flood resilience in Hoboken and details the outcomes delivered by RBD in Hoboken. Based on understanding the change brought to the status-quo, outcomes are discussed through the application of the framework developed by Farrelly et al. [46].

4.1. Planning for Flood Resilience in Hoboken

According to the interviewees, prior to RDSD, flood risk management in Hoboken was based on planned construction of "wet weather" discharge pumps to alleviate inundation in the streets and emergency management. The funds for construction of wet weather pumps were secured by the City, whereas operation and maintenance is carried out by the private owner of the drainage system: North Hudson Sewerage Authority (NHSA). This indicates the cooperative relation between core urban water management stakeholders in the City, further reinforced by the overlap between the City's interest in reducing flooding and the NHSA's interest in reducing CSO events.

Even prior to Hurricane Sandy, GI was seen as an inherent part of the City's planning strategy. For instance, Hoboken's Master Plan recognizes the flooding issue and is the first mention of GI as a way of securing more green space in the City together with providing additional benefits (e.g., flood alleviation, urban heat island mitigation) [51]. This idea was further developed in the Master Plan Reexamination Report in 2010, where greening the City was seen as the Mayor's priority and the creation of the Green Element of the Master Plan was advised. However, both planning documents offered only narratives, without detailed operationalizing specifications and instruments. Since a large part of Hoboken is designated as a redevelopment zone, this land is seen as a window of opportunity for incorporating GI.

In the aftermath of Hurricane Sandy, several important initiatives for addressing the flooding issues were created. Firstly, a process of developing a Community Resilience Plan by the local government, aimed at addressing vulnerabilities and developing long term community resilience to disastrous events. Secondly, the Green Infrastructure Strategic Plan proposed a conceptual framework for a GI network [52]. However, according to the interviewees, the most important step forward in planning for GI was Hoboken being heavily impacted by Hurricane Sandy and being a case study in one of the six winning proposals of RBD—the RDSD project developed by Team OMA.

Even though it is evident that there was awareness and initial political support for mitigating the flooding effects in Hoboken, it was not until Hurricane Sandy that a recognition of a need for change fully emerged. Sandy and its devastation acted as a trigger and accelerator for finding a solution. As examples from this section indicate, Hoboken has put a particular focus on recovery and resilience planning since Hurricane Sandy.

4.2. Outcomes of Rebuild by Design in Hoboken

The interview analysis indicates that direct contributions of RBD to enhancing resilience to flooding consist of impulses for ongoing planning processes and the allocation of funds for implementing a part of RDSD.

During the competition, a diagnostic approach was taken to identify all the challenges in the area and to propose innovative solutions. The outcomes created long-term goals and visions, provided a new visual identity and enhanced flood risk management planning. The new approach to achieving resilience to flooding integrated engineering, urban planning, and policy development, and revealed the need for new financial mechanisms to fund projects. The designed flood risk management strategy embraced the vision of accepting water as a resource and an integral part of urban livability, rather than relying on traditional solutions. Hoboken of the future is envisioned as a greener city whose flood protection is multifunctional, serving its original purpose but also providing additional amenities (e.g., recreational opportunities, decrease in CSOs per year, enhanced social capital). According to the local decision-makers and planners, a big step forward was in understanding the potential of GI as a cost-effective alternative to an aged combined sewer system, as *"an alternative that is more financially feasible"* (interviewee from the state government). An interviewee from the federal government explained:

> *"Looking at the cost savings, the co-benefits of putting in green infrastructure as opposed to just doing grey infrastructure, it is really important.(...) New York City, Philadelphia, a lot of municipalities have documented costs savings by replacing some of their grey with green".*

Even though all the proposed solutions are on the *"landscape architecture level of design"* (interviewee from a consultancy) and more detailed planning will be done in the next phase of the project, the first decisive step towards resilient Hoboken has been taken with RDSD. In the meantime, interviewees stated that the municipality is working on delivering the first Resilience Park and incorporating the findings from RDSD, the Green Infrastructure Strategic Plan, plans for designated redevelopment areas and the master plan. As previously mentioned, prior to RBD planning of resilience to flooding was encouraged through a narrative, whereas RDSD provided a more structured vision.

U.S. Department of Housing and Urban Development awarded $230 million to the State of New Jersey through the "Community Development Block Grant Disaster Recovery Funds" to implement the Strategy. The availability of these funds presents a unique opportunity to upgrade the coastal defense, build protection against a once per 500-year storm surge event and create a more attractive coastline for the local residents. As such, the current focus is on implementing the Resist part of the solution. As explained by the interviewee from the local government:

> *"... we have a one-time opportunity to build an enormous levee along our coastline to protect the city from the future storm surges and sea level rise, so we want to invest that 230 million into the Resist strategy first. If there are any funds left, we would invest it in other parts".*

The funding for the other parts (Delay, Store and Discharge) is not secured, and the financial resources are to be found in other streams. Thus far, Hoboken received significant resources for planning and delivery of the projects through grants and loans. An interviewee from a non-governmental sector complimented the local government for "*seeing advantage of the grants and competitions that were available and putting resources in to do that'*, and another interviewee states that the mayor '*directed her staff to spend a lot of time getting all these grants like Rebuild by Design (...)"*.

In addition to enhanced planning and funding for the implementation, having a comprehensive strategy and ambition created an image of the city as a frontrunner in flood resilience efforts. In March 2015, it was named as one of the United Nations Office for Disaster Risk Reduction (UNISDR) Role Model Cities for its flood management practices and RDSD urban water strategy, being only the second city in the US (after San Francisco) to receive this recognition [53].

The interviewees unanimously agreed that RDSD created several important drivers for further development. The main driver for enabling city-wide implementation of GI is the presence of strong political buy-in, leadership, and commitment at the local level. Interviewees believe that the current governing body is seeing GI as a real opportunity and is allocating significant financial and human resources to the implementation. A local government representative stated:

> "*The Mayor of Hoboken is 110% behind green infrastructure and she wants it, she wants to add open space and parks in the city, she wants to reduce flooding and improve environmental quality, she wants to improve property values. So she is in favor of it.*"

Furthermore, some of the interviewees believe this is distinguishing Hoboken from most of the communities in New Jersey, where political commitment is not present. As a representative from the local non-governmental organization pointed out:

> "*In other towns, the leadership is not there. Mayor is not driving it and you are trying to get the community to help push the mayor along because it (GI) is not mandated under regulation or law that they have to do this.*"

Further roots of this driver are the needs of the community for more open space, and localized flooding. Interviewees pointed out that the strong political buy-in is exemplified through approvals of the City Council for applications for low-interest loans from the New Jersey Infrastructure Trust, land acquisition in redevelopment zones, grant applications and cooperation demonstrated while strategies were being developed. However, next to these drivers created after RDSD, interviewees see an important external driver that can move the process forward: a need to comply with the federal regulations and decrease the yearly number of combined sewer overflows.

Despite the recognition that RDSD created a strong momentum for solving the problem in Hoboken in the aftermath of Hurricane Sandy, the most challenging part is yet to start: wide-spread implementation. Interviewees stated that the proposed strategy is comprehensive and on the scale of an urban district, requiring significant investments and future dedication. Even though significant progress (previously presented) has been made, the interviewed stakeholders have identified a range of barriers to the actual realization of the strategy, as presented in Figure 1. It is important to notice that the identified drivers and barriers are related to the implementation of Delay and Store parts of the RDSD (i.e., GI), since these are a new and not previously applied approaches in Hoboken, as well as a part of the RDSD that will not be funded through the prize funds.

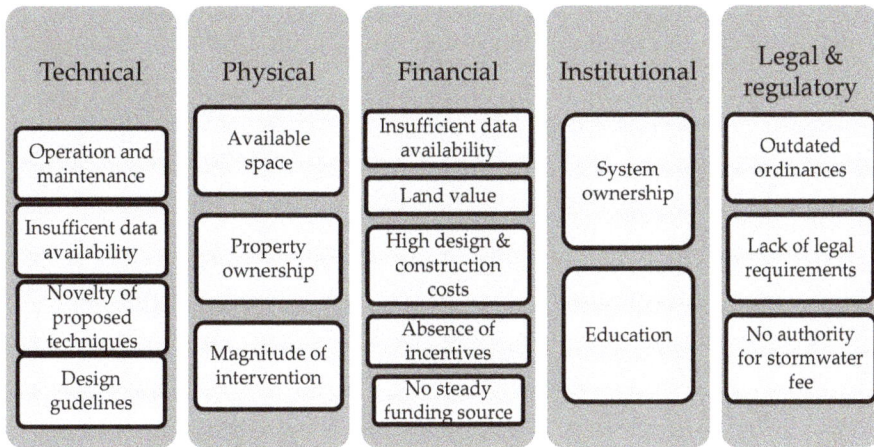

Technical	Physical	Financial	Institutional	Legal & regulatory
Operation and maintenance	Available space	Insufficient data availability	System ownership	Outdated ordinances
Insufficient data availability	Property ownership	Land value		Lack of legal requirements
Novelty of proposed techniques		High design & construction costs	Education	No authority for stormwater fee
Design gudelines	Magnitude of intervention	Absence of incentives		
		No steady funding source		

Figure 1. Barriers for the implementation of the Delay and Store parts of the RDSD, as identified by interviewees

4.3. Rebuild by Design, Governance and the Uptake of Green Infrastructure

In the Literature Review, the identified eight needed factors for enabling and facilitating change and innovation uptake by Farrelly et al. [46] were presented. This section provides critical analysis of the coded interviews on how these structures and processes governance factors were delivered through GI component of RDSD in Hoboken.

4.3.1. Governance Structures

Even though Rebuild by Design was triggered by Hurricane Sandy, the approach taken resulted in a state of the art strategy that went beyond simple rebuilding and increasing the level of protection for the future hurricanes. It provided a platform for designing a solution that addresses multiple challenges in a comprehensive manner. Farrelly et al. [46] argued that a clear vision is important to engage the stakeholders. However, interviewees stated that the RBD approach went a step further by undertaking serious efforts to involve the community while developing the designs. Thereby, it was ensured that the developed RDSD Strategy is representative of the needs of local community, thus creating the ownership of the solution. RDSD created a clear vision and demonstrated a transformational potential of Hoboken.

Even though the RDSD created a long-term vision for Hoboken, it did not contribute to the regulatory and compliance agenda nor provided sound economic justification for GI when compared to traditional infrastructure. However, the project is just entering its second stage, through which more detailed planning and research will be done, so advancement in the development of these factors can be expected. During the interviews, stakeholders were repetitively mentioning a need for a more detailed cost-benefit analysis including additional benefits provided by GI. Such an analysis would prove a business case for GI when compared to traditional infrastructure. As an interviewee pointed out, there is a need for an analysis that would clearly identify where 'the highest impact by lowest cost' is and where 'the highest value of implementation is'. Another suggestion was setting a clear performance target that would allow for constant adjustment in managerial action. The RBD approach, by requiring extensive collaboration, created a nexus between the key urban water management stakeholders in Hoboken.

The approach delivered through RBD and RDSD influenced planning practice and policy creation in Hoboken to a great extent. As a direct consequence of Hurricane Sandy and a direct/indirect consequence of RDSD, several planning initiatives were undertaken. These resulted in a set of policy recommendations (e.g., retention standards for stormwater, and zoning codes update) that could promote GI implementation and contribute to overcoming some of the barriers (Figure 1).

4.3.2. Governance Processes

The fact that one of the key actors for the initiation of RBD was from the federal government level assured strong leadership in the process. The intensive cooperation between the design team and local government in Hoboken led to the legitimacy of RDSD. As stated before, stakeholders were identifying strong leadership from the local mayor as a key force in moving forward with the GI implementation. The competition attracted leaders from the industry and served as a raising awareness activity in the community.

GI, as a preferred option for stormwater management in RDSD, is an innovative approach in Hoboken. RDSD emphasized the importance of demonstrating the concept through the delivery of demonstration projects [50] and this was further confirmed in the interviews. However, the question of whether there is an established approach to understanding where the demonstration projects could deliver most benefit and whether there is a professional capacity for delivering and maintaining these projects arises. In addition, the RDSD did not foresee any funding for demonstration. As stated earlier, partial funding is available only for the coastal defense. To implement the other parts of RDSD, significant public and private resources will be needed. Hence, there is a realistic threat that RDSD will not be fully implemented unless the innovative funding mechanisms are created and/or private sector funds leveraged.

By involving the public while creating the designs, community drivers were considered (e.g., less flooding, and more green space). The evaluation of the RBD competition done by Urban Institute stated that the community involvement increased the awareness of the problem and emphasized the needs for a solution [54]. RDSD places many GI measures on private properties. Thus, having involved and aware residents (i.e., private property owners) can contribute to the rate of implementation of GI in the future and will require a behavioral change.

From the very idea of establishing the competition, the importance of research and partnerships with policy and practice has been identified. All of the design opportunities have been developed as a result of the extensive collaborative research process [11] and by working closely with local decision-makers, the connection with policy was established. Even so, some of the interviewees proposed the necessity for establishing a partnership with academic institutions (e.g., Stevens Institute of Technology, Rutgers University) that would move the process of implementation forward.

4.3.3. RBD, Governance and Wide-Spread GI Implementation

Whilst some of the discussed factors identified by Farrelly et al. [46] in the previous section were a constitutive part of the RBD approach and RDSD in Hoboken (e.g., clear vision, leadership, and research-policy partnerships), others were partially or completely lacking (e.g., regulatory and compliance agenda, and economic justification). The presence of any additional factors, other than those proposed by Farrelly et al. [46] was not observed. It is apparent that the RBD approach contained more factors that fall under the term of governance processes than governance structures. However, as argued by Farrelly et al. [46] all the proposed enabling factors need to be aligned to enable widespread change in the system (i.e., implementation of GI on a city-wide scale). By creating a narrative and long-term vision, RDSD set a foundation for a paradigm change in the way the system was managed in Hoboken, which can be seen as crucial for setting a directional path for change [55]. The results indicate that governance processes go ahead of reform of governance structures. This is in line with findings by Rijke et al. [24] on the emergence of stormwater harvesting and reuse in Australia. In that study, it was demonstrated that in the earlier transformation stages informal networks and innovation testing are facilitated by decentralized and informal governance approaches, whereas in the later stages of transformation, centralized governance approaches secure development of legislative frameworks and stabilization of a new state. The study presented in this paper is, to the best of authors knowledge, the first study analyzing the changes to the status-quo brought to urban stormwater management through a design competition.

The study reveals that design competitions present a novel approach to strategic policy development and have a potential to create a platform for GI, but the translation of the strategies into reality requires significant changes in policy, politics, regulations and financial approaches. The approach developed and applied through RBD challenges the path dependency of delivering urban stormwater management with a single objective in mind. For Hoboken, even though there are many barriers to implementation identified by the stakeholders (Figure 1), detailed planning done through the next phase of the RDSD project and further commitment from the local leaders will decrease the uncertainties connected with GI and help facilitate wider implementation. In addition, learning from successful examples and best practices of GI implementation from the 'champion cities' across the US (e.g., Philadelphia, Portland and Seattle) can provide valuable inputs for reducing stakeholders' perceived barriers for GI implementation and foster the concept of the "city-to-city" learning. For instance, experiences exemplify introduction of the stormwater fee as a steady funding source [56], and introduction of incentives as a way to motivate GI implementation on private properties (e.g., stormwater fee discount, zoning upgrades, awards and recognition programmes) [57]. Furthermore, technical barriers can be overcome with the continued support towards demonstration projects and development of local design guidelines.

5. Conclusions

Based on a series of interviews with local experts and stakeholders, the process of RBD was discussed for the City of Hoboken through a lens of changes to governance structures and processes needed for the wide-spread uptake of GI. The results indicate a more significant change in terms of governance processes when compared to governance structures. RBD brought significant enhancement in the planning of flood resilience by setting a stage for GI and creating a long-term vision, secured a political commitment that lead to increased efforts bringing international recognition to Hoboken. The collaboration between the authorities, the private sector and philanthropists created conditions for innovation through the creation and exchange of ideas between international experts and design teams. The present work unveiled that RBD contains some of the components needed to support transformative governance and facilitate implementation of innovative urban stormwater management at the larger scale.

The study revealed that RBD offered a directional path for change in Hoboken and opened up room for innovation. There were several strong points (e.g., intensive public participation, leadership, research policy interface, capacity building and improvement of long-term planning). However, a lack of a regulatory and compliance agenda and economic justification have been observed. The absence of these factors can hinder the overall uptake of the GI solution and desired transformation. Despite the positive impacts of RBD on Hoboken, the implementation of RDSD has only been partly funded. Thus, without innovative financial mechanisms and involvement of private sector, there is a threat that city-wide implementation of GI will be a difficult goal to accomplish.

The findings from this research, indicating the strong aspects and weaknesses of design competitions as a means of achieving flood resilience in urban areas, are applicable for planners and practitioners involved in the current movement of replicating the RBD model in other settings and offer lessons for decision makers interested in GI. The presented research offers a stockpiling of lessons from the original competition and reveals the areas in which improvements are needed to fully facilitate design competitions as a means of delivering transformative change. Even though design competitions present a novel approach to the field of resilience development, further steps should be made in understanding how the RBD methodology can be adjusted to provide results of equal quality in different settings (e.g., less developed regions, different governance contexts). Further research efforts can be tailored towards a full-scale (i.e., including all winning case studies) evaluation of the effectiveness of RBD to deliver a wider scale implementation of GI, through investigating the shifts and changes brought to governance regimes.

Author Contributions: The paper was written by Robert Šakić Trogrlić based on the result obtained from his research which was done under the close supervision of Jeroen Rijke, Nanco Dolman and Chris Zevenbergen. All co-authors gave ideas for the specific parts of the manuscript and helped to review and improve the paper.

Acknowledgments: This research was financially supported by Royal HaskoningDHV, CRC for Water Sensitive Cities and the European Commission, as part of the Erasmus Mundus MSc Master Programme in Flood Risk Management, all of whom are hereby acknowledged. We thank the Rebuild by Design team, and particularly Tara Eisenberg for their review of a first draft of the paper. Further thanks are directed towards those who offered their precious time to be interviewed. Finally, we thank Annie Visser and Joanne Craven for proofreading the manuscript.

Conflicts of Interest: The authors declare no conflict of interest.

References

1. Wamsler, C.; Brink, E.; Rivera, C. Planning for climate change in urban areas: From theory to practice. *J. Clean. Prod.* **2013**, *50*, 68–81. [CrossRef]

2. Rosenzweig, C.; Solecki, W. Hurricane Sandy and adaptation pathways in New York: Lessons from a first-responder city. *Glob. Environ. Chang.* **2014**, *28*, 395–408. [CrossRef]

3. Ferguson, B.C.; Brown, R.R.; Deletić, A. Diagnosing transformative change in urban water systems: Theories and frameworks. *Glob. Environ. Chang.* **2013**, *23*, 264–280. [CrossRef]

4. Perales-Momparler, S.; Andrés-Doménech, I.; Andreu, J.; Escuder-Bueno, I. A regenerative urban stormwater management methodology: The journey of a Mediterranean city. *J. Clean. Prod.* **2015**, *109*, 174–189. [CrossRef]

5. Pahl-Wostl, C.; Holtz, G.; Kastens, B.; Knieper, C. Analyzing complex water governance regimes: The Management and Transition Framework. *Environ. Sci. Policy* **2010**, *13*, 571–581. [CrossRef]

6. Desouza, K.C.; Flanery, T.H. Designing, planning, and managing resilient cities: A conceptual framework. *Cities* **2013**, *35*, 89–99. [CrossRef]

7. Liao, K. A Theory on Urban Resilience to Floods—A Basis for Alternative Planning Practices. *Ecol. Soc.* **2012**, *17*, 48. [CrossRef]

8. Salinas Rodriguez, C.N.A.; Ashley, R.; Gersonius, B.; Rijke, J.; Pathirana, A.; Zevenbergen, C. Incorporation and application of resilience in the context of water-sensitive urban design: Linking European and Australian perspectives. *WIREs Water* **2014**, *1*, 173–186. [CrossRef]

9. Rijke, J.; van Herk, S.; Zevenbergen, C.; Ashley, R.; Hertogh, M.; ten Heuvelhof, E. Adaptive programme management through a balanced performance/strategy oriented focus. *Int. J. Proj. Manag.* **2014**, *32*, 1197–1209. [CrossRef]

10. Van Herk, S.; Zevenbergen, C.; Gersonius, B.; Waals, H.; Kelder, E. Process design and management for integrated flood risk management: Exploring the multi-layer safety approach for Dordrecht, The Netherlands. *J. Water Clim. Chang.* **2014**, *5*, 100–115. [CrossRef]

11. Gendall, J.; Bisker, J.; Chester, A.; Eisenberg, T.; Davis, S.; Ovink, H. *Rebuild by Design*; Rebuild by Design: New York, NY, USA, 2015.

12. Rebuild by Design. Available online: http://www.rebuildbydesign.org/rebuild-awarded-most-groundbreaking-federal-challenge-or-prize-competition/ (accessed on 11 October 2015).

13. Sutton Grier, A.E.; Wowk, K.; Bamford, K. Future of our coasts: The potential for natural and hybrid infrastructure to enhance the resilience of our coastal communities, economies and ecosystems. *Environ. Sci. Policy* **2015**, *51*, 137–148. [CrossRef]

14. Rebuild by Design. Available online: http://www.rebuildbydesign.org/our-work/city-initiatives/san-francisco-bay-area (accessed on 14 January 2017).

15. Drafting Team of the WFD CIS Working Group Programme of Measures. *EU Policy Document on Natural Water Retention Measures*; Techical Report-2014-2082; Directorate-General for Environment (European Commission): Brussels, Belgium, 2015. [CrossRef]

16. Lee, S.; Yigitcanlar, T. Sustainable urban stormwater management: Water sensitive urban design perceptions, drivers and barriers. In *Rethinking Sustainable Development: Urban Management, Engineering and Design*; Yigitcanlar, T., Ed.; Engineering Science Reference; IGI Global: Hershey, PA, USA, 2010; pp. 22–37.

17. O'Donnell, E.C.; Lamond, J.E.; Thorne, C.R. Recognising barriers to implementation of Blue-Green Infrastructure: A Newcastle case study. *Urban Water J.* **2017**, 1–11. [CrossRef]

18. Brown, R.R.; Farrelly, M.A. Delivering sustainable urban water management: A review of the hurdles we face. *Water Sci. Technol.* **2009**, *59*, 839–846. [CrossRef] [PubMed]

19. Cettner, A.; Ashley, R.; Hedström, A.; Viklander, M. Assessing receptivity for change in urban stormwater management and contexts for action. *J. Environ. Manag.* **2014**, *146*, 29–41. [CrossRef] [PubMed]

20. Thorne, C.R.; Lawson, E.C.; Ozawa, C.; Hamlin, S.L.; Smith, L.A. Overcoming uncertainty and barriers to adoption of Blue-Green Infrastructure for urban flood risk management. *J. Flood Risk Manag.* **2015**. [CrossRef]

21. USGBC. New Orleans Architects Look to the Netherlands for Ideas on Living with Water. Available online: http://plus.usgbc.org/dutch-dialogues/ (accessed on 15 November 2015).

22. Ovink, H.; Boeijenga, J. *Too Big: Rebuild by Design's Transformative Response to Climate Change*; Nai010 Publishers: Rotterdam, The Netherlands, 2018; ISBN 978-94-6208-315-8.

23. Farrelly, M.; Brown, R. Rethinking urban water management: Experimentation as a way forward? *Glob. Environ. Chang.* **2011**, *21*, 721–732. [CrossRef]

24. Rijke, J.; Farrelly, M.; Brown, R.; Zevenbergen, C. Configuring transformative governance to enhance resilient urban water systems. *Environ. Sci. Policy* **2013**, *25*, 62–72. [CrossRef]

25. Brown, R.R.; Sharp, L.; Ashley, R.M. Implementation impediments to institutionalising the practice of sustainable urban water management. *Water Sci. Technol.* **2006**, *54*, 415–422. [CrossRef] [PubMed]

26. Olorunkiya, J.; Fassman, E.; Wilkinson, S. Risk: A Fundamental Barrier to the Implementation of Low Impact Design Infrastructure for Urban Stormwater Control. *J. Sustain. Dev.* **2012**, *5*, 27. [CrossRef]

27. Fletcher, T.D.; Shuster, W.; Hunt, W.F.; Ashley, R.; Butler, D.; Arthur, S.; Trowsdale, S.; Barraud, S.; Semadeni-Davies, A.; Bertrand-Krajewski, J.-L.; et al. SUDS, LID, BMPs, WSUD and more—The evolution and application of terminology surrounding urban drainage. *Urban Water J.* **2014**, *12*, 525–542. [CrossRef]

28. Voskamp, I.M.; Van de Ven, F.H.M. Planning support system for climate adaptation: Composing effective sets of blue-green measures to reduce urban vulnerability to extreme weather events. *Build. Environ.* **2014**, *83*, 159–167. [CrossRef]

29. USEPA. Green Infrastructure Case Studies: Municipal Policies for Managing Stormwater with Green Infrastructure. 2010. Available online: http://www.sustainablecitiesinstitute.org/topics/water-and-green-infrastructure/stormwater-management/green-infrastructure-case-studies-municipal-policies-for-managing-stormwater-with-green-infrastructure (accessed on 17 May 2015).

30. Demuzere, M.; Orru, K.; Heidrich, O.; Olazabal, E.; Geneletti, D.; Orru, H.; Bhave, A.G.; Mittal, N.; Feliu, E.; Faehnle, M. Mitigating and adapting to climate change: Multi-functional and multi-scale assessment of green urban infrastructure. *J. Environ. Manag.* **2014**, *146*, 107–115. [CrossRef] [PubMed]

31. Barbosa, A.E.; Fernandes, J.N.; David, L.M. Key issues for sustainable urban stormwater management. *Water Res.* **2012**, *46*, 6787–6798. [CrossRef] [PubMed]

32. Liu, W.; Chen, W.; Peng, C. Assessing the effectiveness of green infrastructures on urban flooding reduction: A community scale study. *Ecol. Model.* **2015**, *318*, 236–244. [CrossRef]

33. European Commission. Towards an EU Research and Innovation Policy Agenda for Nature-Based Solution & Re-Naturing Cities. 2015. Available online: https://ec.europa.eu/research/environment/index.cfm?pg=nbs (accessed on 22 February 2018).

34. European Environmental Agency. Green Infrastructure and Flood Management: Promoting Cost-Effecient Flood Risk Reduction via Green Infrastructure Solutions. 2017. Available online: https://www.eea.europa.eu/publications/green-infrastructure-and-flood-management (accessed on 22 February 2018).

35. LaBadie, K. Identifying Barriers to Low-Impact Development and Green Infrastructure in Albuquerque Area. Master's Thesis, The University of New Mexico, Albuquerque, NM, USA, 2010.

36. Abhold, K.; Lorraine, L.; Grumbles, B. Barriers and Gateways to Green Infrastructure. Available online: https://issuu.com/savetherain/docs/barriers-and-gateways-to-gi_cwaa (accessed on 22 February 2018).

37. Faivre, N.; Fritz, M.; Freitas, T.; de Boissezon, B.; Vandewoestijne, S. Nature-based solutions in the EU: Innovating with nature to address social, economic and environmental challenges. *Environ. Res.* **2017**, *159*, 509–518. [CrossRef]

38. Brown, R.R. 2005 Impediments to integrated urban stormwater management: The need for institutional reform. *Environ. Manag.* **2005**, *36*, 455–468. [CrossRef] [PubMed]

39. Roy, A.H.; Wenger, S.J.; Fletcher, T.D.; Walsh, C.J.; Ladson, A.R.; Shuster, W.D.; Thurston, H.W.; Brown, R.R. Impediments and solutions to sustainable, watershed-scale urban stormwater management: Lessons from Australia and the United States. *Environ. Manag.* **2008**, *42*, 344–359. [CrossRef] [PubMed]

40. Van de Meene, S.J.; Brown, R.R.; Farrelly, M.A. Towards understanding governance for sustainable urban water management. *Glob. Environ. Chang.* **2011**, *21*, 1117–1127. [CrossRef]

41. Pahl-Wostl, C. Transitions towards adaptive management of water facing climate and global change. *Water Resour. Manag.* **2007**, *21*, 49–62. [CrossRef]

42. Bakker, K. From State to Market: Water Mercantilización in Spain. *Environ. Plan. A* **2002**, *34*, 767–790. [CrossRef]

43. Rijke, J. Delivering Change: Towards Fit-for-Purpose Governance of Adaptation to Flooding and Drought. Ph.D. Thesis, Department of Civil Engineering, Technical University Delft, Delft, The Netherlands, 2014.

44. Huntjens, P.; Lebel, L.; Pahl-Wostl, C.; Camkin, J.; Schulze, R.; Kranz, N. Institutional design propositions for the governance of adaptation to climate change in the water sector. *Glob. Environ. Chang.* **2012**, *22*, 67–81. [CrossRef]

45. Rijke, J.; Brown, R.; Zevenbergen, C.; Ashley, R.; Farrelly, M.; Morison, P.; van Herk, S. Fit-for-purpose governance: A framework to make adaptive governance operational. *Environ. Sci. Policy* **2012**, *22*, 73–84. [CrossRef]

46. Farrelly, M.; Rijke, J.; Brown, R. Exploring Operational Attributes of Governance for Change. In Proceedings of the 7th International Conference on Water Sensitive Urban Design, Melbourne, Australia, 2–7 April 2006.

47. Chaffin, B.C.; Garmestani, A.S.; Gunderson, L.H.; Harm Benson, M.; Angeler, D.G.; Arnold, C.A.; Cosens, B.; Kundis Craig, R.; Ruhl, J.B.; Allen, C.R. Transformative Environmental Governance. *Annu. Rev. Environ. Resour.* **2016**, *41*, 399–423. [CrossRef]

48. Bryman, A. *Social Research Methods*, 4th ed.; Oxford University Press: Oxford, UK, 2012; ISBN 978-0199588053.

49. Cruijsen, A.C. Design Opportunities for Flash Flood Reduction by Improving the Quality of the Living Environment: A Hoboken City Case Study of Environmental Driven Urban Water Management. Master's Thesis, Department of Civil Engineering, Technical University Delft, Delft, The Netherlands, 2015.

50. Team OMA. *Resist, Delay, Store, Discharge: A Comprehensive Urban Water Strategy*; Team OMA: New York, NY, USA, 2013.

51. Hoboken Planning Board. *Hoboken Master Plan*; Hoboken Planning Board: Hoboken, NJ, USA, 2004.

52. Together North Jersey. Available online: http://www.hobokennj.org/docs/communitydev/Hoboken-Green-Infrastructure-Strategic-Plan.pdf (accessed on 11 November 2017).

53. UNISDR. Available online: https://www.unisdr.org/archive/42762 (accessed on 11 November 2017).

54. Urban Institute. Available online: https://assets.rockefellerfoundation.org/app/uploads/20140610165452/The-Evaluation-of-the-Design-Competition-of-Rebuild-by-Design.pdf (accessed on 11 September 2015).

55. Wong, T.; Brown, R.R. Transitioning to Water Sensitive Cities: Ensuring Resilience through a new Hydro-Social Contract. In Proceedings of the 11th International Conference on Urban Drainage, Edinburgh, UK, 31 August–5 September 2008.

56. Campbell, C.W. Western Kentucky University Stormwater Utility Survey. 2014. Available online: http://www.circleofblue.org/waternews/wp-content/uploads/2015/04/wku_stormwater_survey_2014_finalversion.pdf (accessed on 14 March 2015).

57. USEPA. Managing Wet Weather with Green Infrastructure Municipal Handbook: Incentive Mechanisms. 2009. Available online: https://www.epa.gov/green-infrastructure (accessed on 17 April 2015).

![water logo] *water*

MDPI

Article

Objectives and Indexes for Implementation of Sponge Cities—A Case Study of Changzhou City, China

Zhengzhao Li [1], Mingjing Dong [2,3], Tony Wong [4], Jianbing Wang [4], Alagarasan Jagadeesh Kumar [2,3] and Rajendra Prasad Singh [2,3,*]

[1] Changzhou City Planning and Design Institute, Changzhou 213003, China; lzzcool@sohu.com
[2] School of Civil Engineering, Southeast University, Nanjing 210096, China; 220151121@seu.edu.cn (M.D.); jaga.jagadeesh1987@gmail.com (A.J.K.)
[3] Southeast University-Monash University Joint Research Centre for Water Sensitive Cities, Nanjing 210096, China
[4] CRC for Water Sensitive Cities, Clayton, VIC 3800, Australia; tony.wong@crcwsc.org.au (T.W.); wang.jianbin@crcwsc.org.au (J.W.)
*   Correspondence: rajupsc@seu.edu.cn; Tel.: +86-25-83993223

Received: 12 January 2018; Accepted: 19 March 2018; Published: 10 May 2018

Abstract: This paper presents a framework of objectives and indexes for sponge cities implementation in China. The proposed objectives and indexes aims to reflect whether the city is in accord with the sponge city. Different cities have different objectives and indexes as each city has its own geologic and hydrogeological conditions. Therefore, the main problems (e.g., water security and flood risks) in the central urban area of Changzhou city, China were evaluated scientifically. According to the local conditions, four objectives and eleven indexes have been made as a standard to estimate the sponge city and set a goal for the city development to reach the goal of sustainable urban development. The strategy of process control was implemented to improve the standard of urban drainage and flood control facilities, regulate total runoff and reduce storm peak flow, and the ecological monitoring of the function of the rivers and lakes. The objectives of sponge cities include water security, water quality improvement, healthy water ecosystems, and water utilization efficiency. Urban flood prevention capacity, river and lake water quality compliance, and annual runoff control are the key objectives to encourage the use of non-conventional water resources.

Keywords: Yangtze river delta; sponge cities; water ecology; water resources; urban flood risk

1. Introduction

Rapid economic development and urbanization in many developed and developing countries have created numerous environmental as well as developmental challenges. For example, in China more than 600 cities are exposed to frequent flooding, which has a huge social, economic and environmental impact. As waterlogging is considered as one of the major underlying causes of these impacts, the management of urban drainage is a big challenge for both researchers and government authorities [1,2]. Globally, flooding is one of the most common and destructive natural perils causing an average USD 200 billion annually in damages [3]. Most densely populated areas are in coastal zones and in river catchments prone to flooding. Sea-level rise, economic development, and increased frequencies and intensities of storms will require that we continuously have to invest in adapting our flood risk management (FRM) systems, including flood protection infrastructure such as levees, dams and urban drainage systems [3–7]. Water pollution in developed coastal regions due to the higher industrialization and rapid urbanization has become a very critical environmental issue and require proper scientific and effective measures to solve this problem [8]. Dealing with this crisis calls for a multi-stakeholder approach (involving governments, local councils, and citizens). Urban drainage and urban water cycle management have seen a significant change in recent decades leading to an integrative approach to reduce flooding in which multiple objectives in the design and decision making process have been incorporated [9–12].

Various integrated sustainable urban development practices have been developed and adopted in recent years. Many researchers are engaged in research and development of sustainable urban water management approaches such as the concept of sponge cities in China [10–16]. A sponge city refers to sustainable urban development including flood control, water conservation, water quality improvement and natural ecosystem protection. It envisions a city with a water system which operates like a sponge to absorb, store, infiltrate and purify rainwater and releases it for reuse when needed [8]. The sponge city program, launched by the Chinese government in 2014, takes inspiration from the low impact development (LID), best management practices (BMPs) and best planning practices (BPPs) and green infrastructure (GI) in the US [17,18] and Canada [19,20] sustainable drainage systems or sustainable urban drainage systems (Sus Drain/SUDSs) and integrated urban water management (IUWM) in the UK [21,22], alternative techniques (ATs) in France [23] and other European countries [24] and water sensitive urban design (WSUD) in Australia [25] and New Zealand [26]. It promotes natural and semi-natural measures in managing urban storm water and wastewater as well as other water cycles. Since the launch of the program, China has been implementing the sponge city construction initiative, which represents an enormous and unprecedented effort to achieving urban sustainability. A total of 16 cities have been nominated in the first batch as model sponge cities in 2015 and 14 model sponge cities have been nominated in 2016 and received financial aid from the central government of China (Figure 1). According to preliminary estimations, the total investment on the Sponge city construction (SCC) will be roughly 100 to 150 million Yuan (RMB) (US $15 to US $22.5 million) average per square kilometer or 10 Trillion Yuan (RMB) (US $1.5 Trillion) for the 657 cities nationwide.

(a) (b)

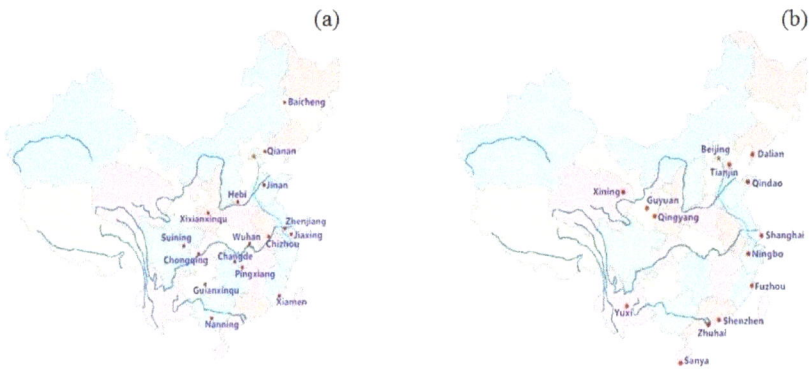

Figure 1. The Geographical distribution of model sponge cities in china; (**a**) Phase 1, (**b**) Phase 2.

To achieve a sustainable urban water ecosystem, it is important to align the interventions related to sponge facilities (referred to as technologies required to develop/transform a city into a Sponge City), ecological treatment of polluted water, non-point source pollution reduce, and rainwater utilization. This alignment should be based on an assessment of the effectiveness of these interventions and their interactions. For example, to reducing the urban heat island effect and improving the ecological environment, an understanding is needed of the relationship between sponge facilities, improvement of micro-climate, and biodiversity. In this paper we are presenting a rational approach to conducting a rapid assessment to improve the livability and sustainability of cities. This approach consists of a framework of objectives and an index system for a sponge city [27–30]. Index is the term generally used for scientific indicators. Based on the regional characteristics of Changzhou city, objectives and indexes related to the current situation have been identified and classified in order to scientifically define and guide the future direction towards new urban developments which comply with the Sponge City Concept (SCC). Paying heed to local conditions is the best way to promote SCC, using an evaluation function, prediction function and guiding function of the objectives and index system. To evaluate whether the objectives can be met and to index them scientifically, based on an objective-oriented approach, relevant urban policy documents, planning, and public demands need to be analyzed to understand the scientific basis and internal driving power to achieve established and periodical objectives and indexes. As well, it would be of great significance to construct sponge cities where stormwater can be naturally stored, purified, and reused [31]. Therefore, the overall objectives and indexes for construction of future sponge cities have been evaluated in the current work. These objectives include, water security, water quality improvement, water discharge management, healthy water ecosystem, water use efficiency, urban flood control capacity, river and lake water quality compliance rate, and the rate of annual runoff control. These are the key objectives to encourage the use of non-conventional water resources [32]. This paper shows that the proposed objectives and index system can provide a guideline for sponge city construction and urban design and planning for future cities. It can also provide a standard to estimate the effects of sponge city construction and set goals for government departments for urban planning, design and construction. Moreover, it is an effective means to ensure that the related sponge facilities can be implemented in the process of urban planning construction.

2. The Sponge City Concept

The concept of "Sponge city" emphasizes the use of natural resources such as soil and vegetation as part of the urban runoff control strategy, which is similar to the low impact development (LID) and green infrastructure (GI) practices being promoted in many parts of the world. The sponge city construction goals not only affect urban flood control but also rainwater harvesting, water quality improvement, natural water discharge and ecological restoration [27]. The primary goals for SCC in China includes on-site retention of 70–90% of average annual rainwater by applying the GI concept and using LID measures, preventing urban flooding, improving urban water quality, mitigating impacts on natural ecosystems, and alleviating the urban heat island impacts (UHI) [28]. Particularly in the Yangtze river delta most of the cities are facing urban flooding problems and runoff pollution is also increasingly affecting the water quality of urban water bodies. Therefore, implementation of sponge cities in this delta is very imminent and as it aims to reduce flood risk and enhances the urban water ecology. The Sponge city construction will also create investment opportunities, infrastructure upgrading, engineering products and new green technology [29].

In the Yangtze river delta, local governments such as the provincial and city governments have to implement the sponge city construction in new development areas. The water system of these areas consists of a dense river network with a slow flow rate of the rivers [8,14]. These newly developed urban areas are receiving irregular precipitation, and have a weak water environmental buffering capacity and thus comprise a vulnerable ecosystem. Therefore, it is critical to have an understanding of the impact of sponge city construction on the hydrological and ecological processes which are as part of the river network. For collection and storage of rainwater effectively and for full rain water utilization, it is imperative to properly understand interventions to establish a sponge city (functions of rain stagnation, purification, infiltration, slow release, and transfusion, such as bio-retention cell, artificial wetland, permeable pavement, existing drainage facilities and the network of rivers and lakes. It can also help to effectively improve the water quality and to make the water cycle in the city more sustainable.

3. Materials and Methods

3.1. Study Area

Changzhou city is located on the south bank of the Yangtze river which is a typical delta region in the eastern coastal area of China with plentiful water resources (Figure 2). This region lies at the heart of China and is also a globally prominent economic sector. There is a dense network of waterways, such as canals and rivers in this region. Most of the cities in this region have a subtropical monsoon climate with high rainfall in summer which often results in severe urban flooding problems [17,28,30]. According to the present status and characteristics of the ecological environment, urban lake and river systems, drainage facilities, land use types etc., Changzhou city is facing two outstanding problems of urban flooding and poor quality of the water environment.

Figure 2. Geographical location of Changzhou city in the Yangtze river delta region, China.

3.2. Technological Pathways for Construction of Sponge Cities

The sponge city concept is based on four target categories: (1) water security, (2) water environment, (3) water ecology, and (4) water resources, to improve urban livability and sustainability comprehensively and determine the framework of objectives and indexes [11]. According to the national development plan for achieving the sustainable urban development goal in China, there are 15 target objectives and indexes for construction/implementation of sponge cities [11]. Based on an problem-oriented to analyze existing problems, background advantage and confronted difficulties and an objective-oriented approach to analyze relevant policy documents, relevant planning, and public demands, overall objectives, building index and control index on the spatial scale of a unit (district) were made for the construction of sponge city by scientifically evaluating their feasibility according to existing advantage and internal driving power to achieve established and periodical objectives and indexes. These objectives and indexes will be assessed separately as qualitatively or quantitatively based on existing problems and challenges, background advantages, relevant policies and public views to achieve the target standards provided in the master plan, regulatory detailed plan and control detailed plan. Background advantages are generally referred to local advantage of building the sponge city, such as rich ecological resources, good natural conditions and favorable hydrological conditions etc. Changzhou city will serve as a case study for the sponge city evaluation process. The technological pathways of objectives and indexes for construction of sponge facilities in Changzhou city are presented in Figure 3.

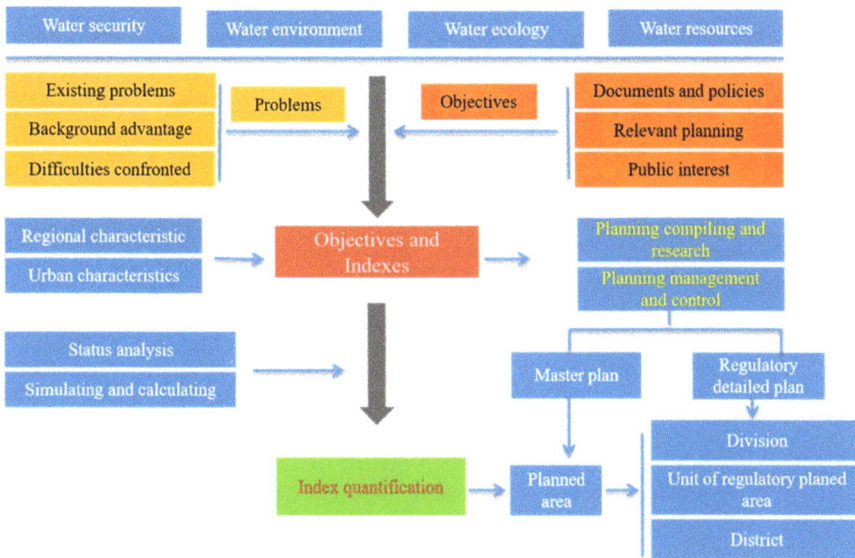

Figure 3. Technological pathways of objectives and indexes for the construction of a sponge city.

The different objectives and indexes will be classified as constraining or enabling local conditions/factors. Based on the knowledge about the local conditions and requirement of the sponge city, we have to determine the objectives and indexes which are in accord with the local problems of water security, water environment, water ecology and water resources. The best way to promote sponge city construction is to develop and implement the construction of sponge cities in local conditions on different areas [33]. The overall objectives of sponge cities are to encourage the use of non-conventional water resources including water security, water environment quality improvement, water ecosystem health, water use efficiency, urban flood control capacity, river and lake water quality compliance rate, and annual runoff control [25,26,33]. The specific site indicators and detailed objectives for sponge cities are quantified by determining and implementing building indexes vis-a-vis overall objectives. The urban water runoff peak flow could be effectively reduced by fences and, green coverage and bioretention systems. Urban runoff quality could also be effectively increased by implementation of bioretention systems [33]. Factors such as, the degree of slopes alongside the river and lake ecosystems, runoff quality, groundwater table, and the permeable paving in urban areas were included as key parameters in the current work. The specific objectives of the development of sponge city in Changzhou city were to achieve water security assurance, water quality improvement, healthy aquatic ecosystems and effective utilization of water resources. The realization of these objectives are rest with indexes to further refine and quantify them. The relationship between objective and index is both mutually independent and unified. Framework objectives and indexes for sponge cities are presented in Figure 4.

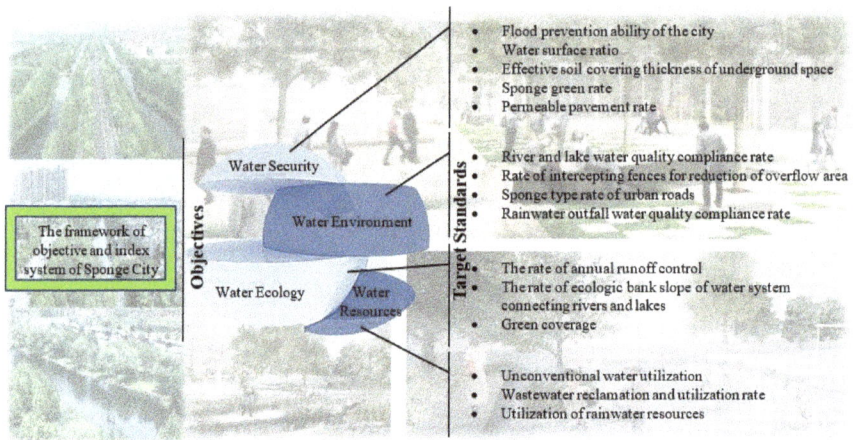

Figure 4. The framework of objectives and index systems for sponge cities.

4. Results and Discussion

Current study has been divided into four categories viz. water security, water environment, water ecology, and water resources based on the existing key problems (e.g., water security and flood risks) occurred in the central urban area of Changzhou city, China. Furthermore, four objectives and eleven indexes have been made as a standard to estimate the sponge city and set a goal for the city development to reach the goal of sustainable urban development.

4.1. Water Security

The main problem of water security in Changzhou city is due to the low water to land surface ratio and the length of the river network. The high amount of hard surface areas is a serious challenge, and the lowland area was existed for a long time and the frequency of heavy precipitation events has increased. Therefore, water security issues and the risk of flooding and its ensuing problems are increasing and have been evaluated scientifically in the central urban area of Changzhou city to find out the most effective strategy to solve these problems (Figure 5).

The strategy involves the improvement of the standard for urban drainage and flood protection facilities, regulation of total runoff and reduce storm flood peak, and the ecological regulation function of rivers and lakes [34,35]. Based on existing urban drainage facilities, we have also considered in the strategy to make full use of an urban green system which has the function of rainwater retention with the use of permeable materials to let rainfall locally infiltrate. Therefore, the objective of the this study was to develop indexes which include water surface ratio, effective soil covering thickness of underground space, sponge green rate, permeable pavement rate, intercepting fence, bioretention to enhance the city's ability to prevent the flooding problem [18,25,26,35].

4.1.1. The City Flood Prevention Ability

The frequency of urban water disasters caused by heavy rain or continuous rainfall can be reduced to lower design rainfall return periods by improving the capacity of inadequate urban drainage facilities [36]. Flooding associated with return period exceeds the storm sewer capacity which is different from what is needed for drainage water; this puts emphasis on utilizing a variety of measures to cope with rainstorm of a great return period. Flood return period and storm sewer capacity should to be comply with the drainage facilities [37]. By accounting for the urban water system, drainage pipe network, runoff control, low impact development construction and other measures, sponge city construction is the best way to reduce frequency of flood disaster. Changzhou city has an area with a dense water network, low-lying terrain and poor connectivity of water networks. The water-surface ratio in the center of the city is 4.4%. The pressure in terms of peak volumes of storm water entering the existing drainage infrastructure is very high and results in a higher risk of system failure than before. Therefore, the primary task of urban water safety is to improve the capacity of the drainage system in order to reduce the risk of flooding. According to the present situation of the central city in Changzhou, the existing standards of flood control and flooding are very low [38].

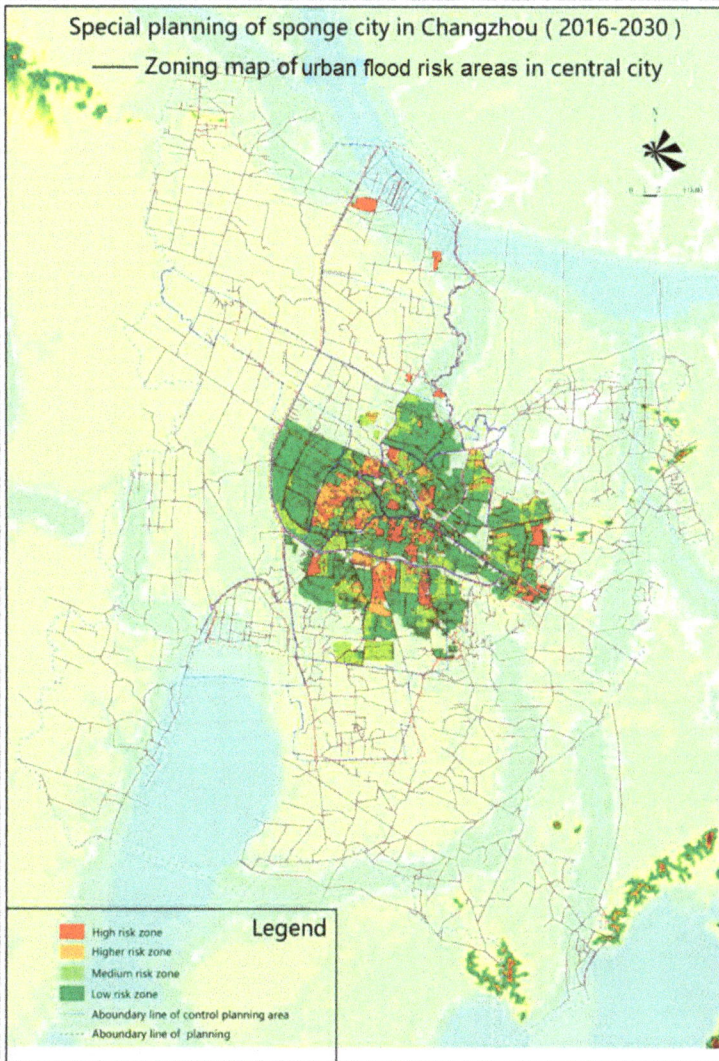

Figure 5. Map displaying the flood risk spots in the central urban area in Changzhou.

4.1.2. Effective Soil Covering Thickness for Underground Spaces

Effective soil covering thickness of underground space is refer to the distance between the top of facilities of underground space and the datum grade of ground. The available "distance" affects the feasibility of building sponge facilities. Often enough room can be designed to construct sponge city practices such as permeable pavement or shallow bioretention.

4.1.3. Sponge Green Factor

Sponge green factor refers to the percentage of total green space area which has the ability to store, convey and treat the surface runoff. Green space vegetation has the functions of water purification and the provision of other eco-system services. Green sponge facilities which have effects on hydrology and water quality by reducing non-point source pollution load, thus improving the quality of rivers and lakes. The type of the land-use planning unit is not used correctly, and therefore, this indicator needs to be delineated according to the planning unit.

4.2. Water Environment

The main problems of water environment in Changzhou city are as follows: high pollution load discharged from sewage treatment plants, high volumes of combined sewage overflow (CSOs), heavy non-point source pollution in runoff and inadequate facilities for the monitoring and thus managing hydrology and water quality based on the water environment and the characteristics of the Yangtze River delta region. The objective and indexes for improving the water quality in the central city of Changzhou have been evaluated and presented in Sections 4.2.1 and 4.2.2.

4.2.1. The Reduction of CSO

Combined sewer outflow is a condition where sometimes, combined sewers receive higher than normal flows during heavy rainfall and sewage treatment plants are unable to handle flows that are more than the designed capacity [38]. Due to this, a mix of excess storm water and untreated wastewater discharges directly into the city's waterways at certain outfalls [38]. According to "Urban Drainage Plan in Changzhou (2011–2020)" [29], the daily average river discharge in 2010 was 72,000 m^3/day, but the actual average river discharge in 2015 increased to 140,200 m^3/day, a 14% annual increase from 2010 to 2015. If the annual rate of increase will gradually drop, the average rate could become 8% in 2015–2020 and even as low as 4% in 2021–2030.

4.2.2. Sponge City Road Rate

Sponge city road rate refers to the proportion of the area of the road which is used in sponge facility to control the runoff. The total area of the present roads and the squares in built-up areas of the central city is about 48.9km^2 and the occupancy rate is about 12.7%. The total area of planned roads and squares in built-up area of the central city is 139.8km^2, with occupancy rate of about 28.5%. The road system collects and transports runoff, and it is also one of the main sources of urban non-point source pollution. Newly built roads have to fully comply with the sponge principles, and improvements to the existing roads will promote implementation of more sponge-type roads in the near future [39].

4.3. Water Ecology

Because ecological protection was neglected in the early stage of urbanization in Changzhou city, some ecosystems have been damaged; while others, such as wetlands, lakes, and urban green areas are continuously shrinking. The deterioration of the ecological function of these systems have become an important factor that restricts sustainable development. As the main issues are related to water ecology, the objectives and indexes have been determined to improve the health of water ecosystems in the central area of Changzhou city. These include: (1) annual runoff control, and (2) the indexes of the river and lake ecological slope rate and green coverage [40].

4.3.1. The Annual Runoff Control Rate

The annual runoff control rate is defined as the proportion of the rainfall which is controlled (not discharged) from the total rainfall in a year. It was analyzed and calculated according to statistical analysis of yearly data related to the natural and artificial enhanced way of penetration, storage, evaporation (transpiration), etc. [41]. According to the relationship between the annual runoff control rate, designed rainfall and rainfall frequency, the rainfall characteristics of Changzhou city and the rainfall data of the past 30 years were analyzed to determine the value of annual runoff control rate [42]. The relationship between the annual runoff control rate and the design rainfall is illustrated in Table 1 and Figure 6.

Table 1. Comparison of annual runoff control rate, the design rainfall and rainfall frequency in Changzhou city.

Annual Runoff Control Rate (%)	50	60	65	70	75	80	85	90
Design rainfall (mm)	10.1	14	16.5	19.5	23.2	28.2	35.2	45.7
Rainfall frequency (%)	72.9	80.0	83.3	86.4	89.5	92.0	94.6	96.4

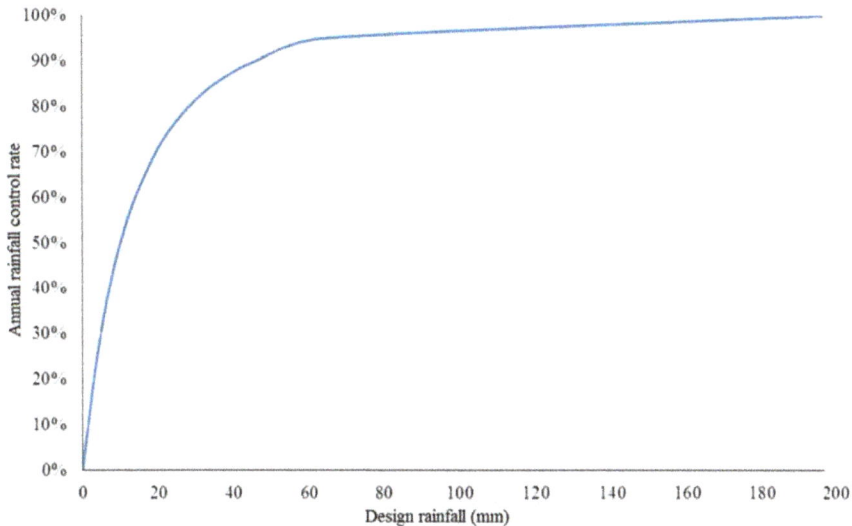

Figure 6. Corresponding relationship between the annual runoff control rate and the design rainfall.

4.3.2. The River and Lake Ecological Revetment Rate

The total length of the ecological revetments rate (excluding shipping river bank and the mean flood discharge channel) analyzed according to the "action plan for construction of sponge city in Changzhou city" [17,29], which reveals that the construction of ecological revetment ratio for sponge city required that, by 2016, must exceed 75% in the central city. Whereas, the "Index system of water ecological civilization in Jiangsu" required that the ratio of ecological revetment must exceed 70%. Changzhou city, with a dense river network has a higher proportion of hard riverbanks and lakes. The rigidity of the revetment obstructs material exchanges between soil and water, which can destroy the ecological function of the natural embankment and reduce the capacity of water self-purification. These factors may lead to water quality deterioration. Therefore, the recovery of ecological function of river banks and lake shores is one of the effective water quality improvement measures [42].

4.3.3. Green Coverage Ratio

The green coverage ratio is the vertical projection of plants/shrubs to the total area. In the "overall planning of Changzhou city" (2015), the green coverage ratio was 39%, increase to 43% in the central city. According to the "Sponge city construction action plan for Changzhou city", the green ratio in built-up areas will be 40% and green coverage rate is predicted to be 45% by 2020. The green coverage ratio will gradually increase taking into account the natural growth of trees/plants. Therefore, the basis for improving the green ration is to increase the proportion of trees/plants, which can effectively alleviate the urban heat island effect [43].

4.3.4. Sponge City Construction

The flood control capacity of Changzhou city should be able to effectively carry the 24hrs storm events for the next 30yrs, and by doing so, the storm water accumulated areas in central urban areas should be basically eliminated. By 2030, current central urban areas which are prone to storm flooding will no longer exist for the next 50yrs. The river drainage capacity should ensure that the storm events will not be overflowing in 20yrs. The black and odorous water bodies in Changzhou city should be basically revived by 2020, and 85% of rivers and lakes in the central urban area should meet the water quality standards. By 2030, 100% of the rivers and lakes will meet the water quality standards. Annual runoff control ratio of sponge city overall objectives in Changzhou city are divided into control units and land-use types. Annual runoff control ratio will vary because the proportions of land types in control units are different; as such annual runoff control ratio should be calculated in control units [44]. Based on the overall planning for goals for Changzhou city, the unconventional water resource utilization rate in the central urban area should be more than 20% by 2020, and should be more than 25% by 2030.

4.4. Water Resource

There are abundant water resources in Changzhou city, and yet its water supply capacity almost completely depends on the Yangtze River. In addition, rainwater and sewage collection and treatment systems are insufficient. Therefore, the ecological function of an urban area as a watershed and water-supply area is difficult to fully utilize. The main problems of water security in Changzhou city are as follows: the large dependency on "conventional" water resources such as river and lakes, pollution-induced water shortages and limited utilization of "unconventional" water resources. Thus, the objectives and indexes were determined to enhance the water utilization rate in Changzhou city, which included the usage of unconventional water resource.

4.4.1. The Rate of Unconventional Water Utilization

The unconventional water utilization is the proportion of the water consumption which is not supplied by water treatment plants, and includes rainwater, sewage treatment plant effluent, gray water, and industrial enterprises that reuse water. Changzhou city initially had no water shortages but recently it has suffered from pollution-induced water shortages. The total water consumption in Changzhou city in 2015 was 45746.69 million m^3 (not including the water consumption of thermal power industry), of which urban recycled water and rainwater utilization contributes 7665 and 3.6 million m^3. Therefore, a major opportunity exists to increase the proportion of unconventional water (e.g., rain water and recycled water), to improve the quality of effluent from sewage treatment plants, and to augment the available quantity of unconventional water resources. These waste water treated through constructed wetlands and ecological treatment (reclaimed water); both can meet the demand for some municipal, industrial, agricultural and landscape irrigation. The ratio of reclaimed water, which meets or surpasses the Chinese standards for surface water class IV requirements, describes the total volume of wastewater discharged from sewage treatment plants. The present ecological processing capacity, sewage load and ecological treatment rate is 10.52 million tons, 69.57 million tons and 15.1%, respectively. At present, the centralized sewage treatment annual growth rate is 17.5% in Changzhou city. It is estimated that the annual volume of sewage treatment will increase at a rate of 8% from 2015 to 2020. The biological treatment rate of effluent from sewage treatment plants will reach to 22% by 2020. It is estimated that the annual volume of sewage undergoing biological treatment will be 5% from 2015 to 2020 [17,29].

4.4.2. The Rainwater Resource Utilization Rate

The rainwater resource utilization rate refers to the proportion of tap water substituted for rainwater resource of the total amount of tap water. It is the severe reality in Changzhou city that the water environment is poor, thus making water resources crisis more prominent. The abundant rainwater resources in Changzhou city are not efficiently used under the traditional approach of urban drainage systems. A vast majority of rainwater runoff is directly discharged into the rivers and lakes, increasing the burden on urban drainage infrastructure. Therefore, an increase of the volume of harvested rainwater provides an ecological function of urban water supply, lessening the need to supply high quality water for the city.

4.5. Major Challenges

Technical Challenges

The original goal of sponge city construction was defined as the introduction of runoff-volume-focused LID techniques to retain 60–90% runoff onsite [17,29]. One year later (in 2016), the goals were expanded to a full array of urban sustainability goals by adding restoration of ecosystem, improving deteriorated urban water bodies, reducing urban heat island impacts, and building smart urban water cycles [17,29]. The concept and practice of LID techniques were introduced into China more than a decade ago [45]. Recent research has been carried out on sustainable urban stormwater management, but the research foundation for sponge city construction on such a large scale is rather limited. The rapid implementation of sponge city measures with such ambitious goals is largely based on very little domestic research [46,47]. A sponge city is an integrated approach that involves a broad range of concepts such as multi-scale conservation and water system management, multi-functional ecological systems, urban hydrology and runoff control frameworks, and impacts of urbanization and human activities on the natural environment. Lacking a sound research foundation can unnecessarily restrict the potentially positive impacts of this new urban water cycle management approach. To implement sponge cities successfully, an appropriate definition of goals and adequate research to understand this new approach (along with sufficient knowledge) is very crucial [31,47].

5. Conclusions

A framework of objectives and indexes has been proposed in this study to providing guidelines to transform the Changzhou city into a sponge city. The application of the framework in this study indicated that: 1. the water surface ratio in the central area of Changzhou city is projected to exceed 4.7% by 2020 and 5% by 2030, 2. the effective soil cover thickness of underground spaces of sponge green lands should not be less than 1.5 meters, and the thickness of soil covering in the upper part of the underground space should not be less than 1.8m ideally, 3. the ratio of sponge green spaces in each control unit consists of different land types, therefore this indicator needs to be delineated according to control units, 4. water quality compliance rate should increase to 85% in the central urban districts of Changzhou city with separated systems, reaching 100% by 2030. 5. the river and lake ecological bank fraction in the central urban districts of Changzhou city should exceed 50% by 2020, and 70% by 2030. 6. rainwater resource utilization ratio should not be less than 1.5% by 2020, and 2% by 2030. Sponge city can be described as a vision that people, the city and water coexist harmoniously in the future. It is a new solution to the problems produced in the city development process in municipalities of the Yangtze River Delta and is the best way for cities to sustainably develop.

Author Contributions: Z.L., T.W., J.W. and R.P.S. conceived and designed the experiments; R.P.S., M.D., A.J.K. performed the experiments and analyzed the data; and R.P.S. wrote the paper.

Funding: This study was funded by the National Key Technologies Research & Development Program, China (No. 2015BAL02B05).

Acknowledgments: Authors would like to thank Willium Hunt, North Carolina State University for his valuable suggestions.

Conflicts of Interest: The authors declare no conflict of interest.

References

1. Chocat, B.; Krebs, P.; Marsalek, J.; Rauch, W.; Schilling, W. Urban drainage redefined; from stormwater removal to integrated management. *Water Sci. Technol.* **2001**, *43*, 61–68. [PubMed]
2. Fltcher, T.D.; Andrieu, H.; Hamel, P. Understanding management and modelling of urban hydrology and its consequences for receiving waters; a state of the art. *Adv. Water Resour.* **2013**, *51*, 261–279. [CrossRef]
3. Boquet, Y. Metro Manila's challenges: Flooding, housing and mobility. In *Urban Development Challenges, Risks and Resilience in Asian Mega Cities*; Springer: Tokyo, Japan, 2014; pp. 447–468.
4. Jiang, Y.; Zevenbergen, C.; Ma, Y. Urban pluvial flooding and stormwater management: A contemporary review of China's challenges and "sponge cities" strategy. *Environ. Sci. Policy* **2018**, *80*, 132–143. [CrossRef]
5. Wang, C.; Wang, Y.; Geng, Y.; Wang, R.; Zhang, J. Measuring regional sustainability with an integrated social-economic-natural approach: A case study of the Yellow River Delta region of China. *J. Clean. Prod.* **2016**, *114*, 189–198. [CrossRef]
6. Shao, W.; Zhang, H.; Liu, J.; Yang, G.; Chen, X.; Yang, Z.; Huang, H. Data integration and its application in the sponge city construction of China. *Proc. Engi.* **2016**, *154*, 779–786. [CrossRef]
7. Feng, H.; Cheng, G.; Liang, J.; Gao, F. Inspiration of rainwater utilization plan in Denmark for the construction of sponge city in China. *DEStech Trans. Environ. Energy Earth Sci.* **2017**, 131–134. [CrossRef]
8. Rui, Y.H.; Fu, D.F.; Do Minh, H.; Radhakrishnan, M.; Zevenbergen, C.; Pathirana, A. Urban surface water quality, flood water quality and human health impacts in Chinese cities: What do we know? *Water* **2018**, *10*, 240. [CrossRef]
9. Fratini, C.; Geldof, G.D.; Kluck, J.; Mikkelsen, P.S. Three points approach (3PA) for urban flood risk management: A tool to support climate change adaptation through transdisciplinarity and multifunctionality. *Urban Water J.* **2012**, *9*, 317–331. [CrossRef]
10. Marsalek, J.; Chocat, B. International report: Stormwater management. *Water Sci. Technol.* **2002**, *46*, 1–17. [PubMed]
11. Wong, T.H.F. Water sensitive urban design—A journey thus far. *Aust. J. Water Resour.* **2007**, *110*, 213–222. [CrossRef]
12. Ashley, R.; Lundy, L.; Ward, S.; Shaffer, P.P.; Walker, L.; Morgan, C.; Saul, A.; Wong, T.; Moore, S. Water sensitive urban design: Opportunities for the UK. *Proc. ICE-Munic. Eng.* **2013**, *166*, 65–76. [CrossRef]
13. Zevenbergen, C.; Herk, S.V.; Rijke, J. Future proofing flood risk management: Setting the stage for an integrative, synergistic and adaptive approach. *Public Works Manag. Policy* **2016**, *22*, 49–54. [CrossRef]
14. Wei, D.; Urich, C.; Liu, S.; Gu, S. Application of CityDrain3 in flood simulation of sponge polders: A case study of Kunshan, China. *Water* **2018**, *10*, 507. [CrossRef]
15. Salinas-Rodriguez, C.; Gersonius, B.; Zevenbergen, C.; Serrano, D.; Ashley, R. A semi risk-based approach for managing urban drainage systems under extreme rainfall. *Water* **2018**, *10*, 384. [CrossRef]
16. Fenner, R. Spatial evaluation of multiple benefits to encourage multi-functional design of sustainable drainage in blue-green cities. *Water* **2017**, *9*, 953. [CrossRef]
17. Ministry of Housing and Urban-Rural Development, China (MHURD-China). Technical Guide for Sponge Cities—Water System Construction with Low Impact Development Techniques. Available online: http://www.mohurd.gov.cn/zcfg/jsbwj_0/jsbwjcsjs/201411/W020141102041225.pdf (accessed on 22 October 2016).
18. United States Environmental Protection Agency (US EPA). *Low-Impact Development Design Strategies: An Integrated Design Approach*; EPA 841-B-00003; US EPA: Washington, DC, USA, 1999.
19. Benedict, M.; McMahon, E. Green infrastructure: Smart conservation for the 21st century. Renew. *Resour. J.* **2002**, *20*, 12–17.

20. British Columbia Ministry of Environment (BCME). Stormwater Planning: A Guidebook for British Columbia. Available online: http://www.env.gov.bc.ca/epd/mun-waste/waste-liquid/stormwater/ (accessed on 4 September 2016).

21. Olewiler, N. *The Value of Natural Capital in Settled Areas of Canada*; Ducks Unlimited Canada and The Nature Conservancy of Canada: Toronto, ON, Canada, 2004.

22. Alexander, D.; Tomalty, R. Smart growth and sustainable development: Challenges, solutions and policy directions. *Local Environ.* **2002**, *7*, 397–409. [CrossRef]

23. STU. *La Maı̂trise du des eaux Pluviales: Quelques Solutions pour L'ame´Lioration du cadre de vie (The Management of Urban Stormwater: Solutions for Environmental Improvement)*; Ministe're de l'Urbanisme et du Logement, Direction de l'Urbanisme et des Paysages, Service Technique de l'Urbanisme: Paris, France, 1982. (In French).

24. Lehmann, S. UNESCO chair in sustainable urban development. In *The Principles of Green Urbanism*; Earth Scan: London, UK, 2010.

25. Beatly, T. *Green Urbanism: Learning from European Cities*; Island Press: Washington, DC, USA, 1999.

26. Sharma, A.K.; Pezzaniti, D.; Myers, B.; Cook, S.; Tjandraatmadja, G.; Chacko, P.; Chavoshi, S.; Kemp, D.; Leonard, R.; Koth, B. Water sensitive urban design: An investigation of current systems, implementation drivers, community perceptions and potential to supplement urban water services. *Water* **2016**, *8*, 272. [CrossRef]

27. Jenkins, S. Towards Regenerative Development. Available online: www.planning.nz (accessed on 10 September 2016).

28. Jia, H.; Wang, Z.; Zhen, X.; Clar, M.; Yu, S.L. China's sponge city construction: A discussion on technical approaches. *Front. Environ. Sci. Eng.* **2017**, *11*, 18. [CrossRef]

29. General Office of the State Council (GOSC). Guideline to Promote Building Sponge Cities. Available online: http://www.gov.cn/zhengce/content/2015-10/16/content_10228.html (accessed on 16 October 2015).

30. Li, H.; Ding, L.; Ren, M.; Li, C.; Wang, H. Sponge city construction in China: A survey of the challenges and opportunities. *Water.* **2017**, *9*, 594. [CrossRef]

31. Marshall, W.E. An evaluation of livability in creating transit-enriched communities for improved regional benefits. *Res. Transp. Bus. Manag.* **2013**, *7*, 54–68. [CrossRef]

32. Zheng, Y.; Yong, X. Unconventional water resources. *Water Purif. Technol.* **2003**, *6*, 38–40. (In Chinese)

33. Zhou, Q.; Mikkelsen, P.S.; Halsnaes, K.; Arnbjerg-Nielsen, K. Framework for economic pluvial flood risk assessment considering climate change effects and adaptation benefits. *J. Hydrol.* **2012**, *414*, 539–549. [CrossRef]

34. Foster, S.B.; Allen, D.M. Ground water—Surface water interactions in a mountain-to-coast watershed: Effects of climate change and human stressors. *Adv. Meteorol.* **2015**, *22*. [CrossRef]

35. Li, C. Ecohydrology and good urban design for urban storm water-logging in Beijing, China. *Ecohydrol. Hydrobiol.* **2012**, *12*, 287–300. [CrossRef]

36. Kang, N.; Kim, S.; Kim, Y.; Noh, H.; Hong, S.J.; Kim, H.S. Urban drainage system improvement for climate change adaptation. *Water* **2016**, *8*, 268. [CrossRef]

37. Torgersen, G.; Bjerkholt, J.T.; Lindholm, O.G. Addressing flooding and SuDS when improving drainage and sewerage systems—A comparative study of selected Scandinavian cities. *Water* **2014**, *6*, 839–857. [CrossRef]

38. United States Environmental Protection Agency (US EPA). *Combined Sewer Overflow Control Manual*; US EPA: Washington, DC, USA, 1984.

39. Wu, J. Analysis on the current situation and development problem of China's sponge city construction—Case study on Ningbo Yaojiang-Cicheng pilot area. *Saudi J. Hum. Soc. Sci.* **2017**, *2*, 572–577.

40. Xiao, L.; Yangn, X.; Cai, H. Responses of sediment yield to vegetation cover changes in the Poyang Lake drainage area, China. *Water* **2016**, *8*, 114. [CrossRef]

41. Xin, R.; Weizhen, T. Application of capture ratio of total annual runoff volume in sponge city. *China Water Wastewater* **2015**, *31*, 105–109. (In Chinese)

42. Wu, Y.; Dai, H.; Wu, J. Comparative study on influences of bank slope ecological revetments on water quality purification pretreating low-polluted waters. *Water* **2017**, *9*, 636. [CrossRef]

43. Shishegar, N. The impact of green areas on mitigating urban heat island effect: A review. *J. Environ. Sustain.* **2014**, *9*, 119–130. [CrossRef]

44. McDonald, R.I.; Green, P.; Balk, D.; Fekete, B.M.; Revenga, C.; Todd, M.; Montgomery, M. Urban growth, climate change, and freshwater availability. *Proc. Natl. Acad. Sci. USA* **2011**, *108*, 6312–6317. [CrossRef] [PubMed]
45. Ries, F.; Schmidt, S.; Sauter, M.; Lange, J. Controls on runoff generation along a steep climatic gradient in the Eastern Mediterranean. *J. Hydrol. Reg. Stud.* **2017**, *9*, 18–33. [CrossRef]
46. Wu, C.; Li, Z. The current situation and future trend of urban rain water harvesting. *Water Wastewater Eng.* **2002**, *28*, 12–14.
47. Wang, H.; Cheng, X.; Li, C. Quantitative analysis of storm water management strategies in the process of watershed urbanization. *J. Hydraul. Eng.* **2015**, *46*, 19–27.

water

MDPI

Article

An Investigation on Performance and Structure of Ecological Revetment in a Sub-Tropical Area: A Case Study on Cuatien River, Vinh City, Vietnam

Van Tai Tang [1,2,3], Dafang Fu [1,2], Tran Ngoc Binh [4], Eldon R. Rene [5], Tang Thi Thanh Sang [6] and Rajendra Prasad Singh [1,2,*]

1   School of Civil Engineering, Southeast University, Nanjing 210096, China; tangvantai@tdt.edu.vn (V.T.T.); fdf@seu.edu.cn (D.F.)
2   Southeast University-Monash University Joint Research Centre for Water Sensitive Cities, Nanjing 210096, China
3   Green Processing, Bioremediation and Alternative Energies Research Group, Faculty of Environment and Labour Safety, Ton Duc Thang University, Ho Chi Minh City 700000, Vietnam
4   Faculty of Public Health, Vinh Medical University, Vinh City 460000, Vietnam; tranngocbinh_tnb@yahoo.com
5   Department of Environmental Engineering and Water Technology, IHE-Delft Institute for Water Education, IHE-Delft, 2601 DA Delft, The Netherlands; e.raj@un-ihe.org
6   Department of Law, Vinh University, Vinh City 461010, Vietnam; sangluat@gmail.com
*   Correspondence: rajupsc@seu.edu.cn

Received: 2 April 2018; Accepted: 6 May 2018; Published: 14 May 2018

Abstract: The current study was performed with an aim to investigate the performance of ecological revetments implemented on the bank of the Cuatien River in Vinh city, Vietnam. Based on the ecological, topographical, and hydrological conditions of the Cuatien River, the gabion and riprap models were introduced to investigate the effect of ecological revetment on the slope stability and ecological restoration characteristics. The effect of prevailing climatic indicators, such as temperature, precipitation, sunlight hours, and humidity were investigated to ascertain the characteristics of weather conditions on the subtropical area. On the surface soil layer of the gabion and riprap, the nutrient indicators of soil organic matter (SOM) and available nitrogen (AN) increased in the spring, summer, and winter, but decreased in autumn, and available phosphorus (AP) did not show an obvious change in the four seasons. The biomass growth rate of Vetiver grass on the gabion and riprap revetments was found to be the highest during the summer, at 15.11 and 17.32 g/month, respectively. The root system of Vetiver and other native plants could increase the cohesion of soil. After 6 and 12 months, the shear strength of the soil behind the gabion revetment increased by 59.6% and 162.9%, while the shear strength of the soil under the riprap also increased by 115.6% and 239.1%, respectively. The results also indicated that the gabion and riprap revetments could improve the river water purification effect and increase the ecological diversity in the region. In the current study, 26 floral and 9 faunal species were detected in the riprap revetment, whereas 14 floral and 5 faunal species were detected in the gabion revetment, respectively. Through high sequencing technology, the number of bacterial species in the present study was found to be 198, 332, and 351 in the water, gabion, and riprap samples, respectively.

Keywords: ecological revetment; Vetiver grass; gabion; riprap; microbial diversity

1. Introduction

Since the 1990s, the economy of Vietnam has experienced a very rapid development. As a result, existing construction technologies in the country have not been able to keep up with the demand of a more modern urban infrastructure [1]. Urban river ecosystems and their quality in Vietnam has been severely affected due to industrial activities, rapid urbanization, which has led to several negative effects on human health and the environment [1,2]. Ecological revetment is one of the effective, cost-efficient, sustainable eco-technologies that can be used to restore urban ecological systems. Vinh city is among the rapidly-developing cities in Vietnam [2], where effective urban development planning and more efficient construction technologies are required for flood control and environmental protection.

Conventionally, concrete and rock materials have been widely used in river bank revetment to control flooding and erosion. The strong revetment with stable construction is essential to ensure the safety of people and their property. However, an appropriately designed, hard revetment can negatively affect the ecosystem, including aquatic and amphibian habitats, river water quality, and aesthetic value [3]. In addition, after heavy rainfall events, storm water runoff scours the road surface, which results in riverbed erosion and water pollution related problems [4]. In order to construct a secure bank revetment with fewer negative effects on the river ecosystem, the ecological impact of revetment must be considered. With the consideration of topographical, hydrological, and ecological conditions of the river, ecological revetment must be combined with civil engineering technologies and ecological science to ensure good stability of the bank slope and ecological restoration effects [5]. Ecological revetment (which integrates plants and a more porous structure) can have a positive effect on the ecosystem as it can facilitate ground and river water circulation, and river bank ecological restoration. This is an indispensable part of the concept of "sponge city" [5–7]. Ecological revetment can also be constructed using locally-available limestone materials [6–8]. Ecological revetment with a porous structure can provide a habitat for microbial and plant reproduction, which plays an important role in removing water pollutants. Plant and microbial growth and their diversity can also improve the river water quality.

There have been some recent examples of revetment design with systematic ecological consideration. In 2001, the Soil and Water Conservation Department of Taiwan worked on the development and integration of ecological engineering, and presented some examples of existing ecological engineering methods using stone revetment, boxed gabion revetment, and arc-shape stone streambed sill for bank protection, respectively [6]. In another study, Chen et al. [7] evaluated the ecological restoration capability of the revetment, and proposed hexagonal precast blocks, bamboo, and complex natural material to construct ecological revetment works for the Liudaxian channel bank in China. Gabriel et al. [8] demonstrated that a well designed riprap can have positive effects in increasing the ecological diversity and the amount of fauna, flora, and microbial communities present in the region, and enhancing pollutant removal.

Plant roots and the stones present in the ecological revetment provides a good habitat for microbial reproduction, which can improve the water quality, as well as the stability of river slopes [7,8]. Ramli et al. [9] studied the ability of gabion to collectively deform under aggravated loads, under the influence of soil-hydrostatic pressures, and evaluated the stability of the gabion wall for earth retaining structure in flood-prone areas. Helal et al. [10] showed that plant roots provide the ideal habitat for microbial growth and reproduction that can improve the effect of naturally occurring biodegradation and the adsorption of pollutants from soil and water. Previous studies have proved that Vetiver grass (*Vetiveria zizanioides*) has a high ability to increase the stability of the soil slope. Ali et al. [11] showed that Vetiver grass, due to its unique root characteristics, could increase the shear strength of the soil and improve the stability of the bank slope. Many existing ecological engineering works have proved the feasibility, ecological restoration, and bank slope stable effectiveness of the ecological revetment and Vetiver grass.

Being low-cost and easy to construct, gabion and riprap revetments are ideal for fast growing countries, like Vietnam. Therefore, in this study, riprap and gabion revetments were combined with Vetiver grass to increase the slope stability and the ecological restoration effect. The gabion and riprap models were designed and applied to investigate the ecological revetment structure and performance, by considering the prevailing ecological, topographical, and hydrological conditions of the Cuatien River. The objectives of this study can be stated as follows:

- Investigate the effect of climatic factors, such as temperature, precipitation, sunlight hours, and humidity, which directly affect the growth of Vetiver grass on the ecological revetments.
- Determine the soil organic matter, available nitrogen, and phosphorus contents in order to understand the soil nutrient effects on the growth of Vetiver grass.
- Investigate the soil strength to envisage the effect of plant root system on the soil shear strength values.
- Evaluate the ecological revetments with regard to the faunal, floral, and the microbial diversity and their impact on their overall performance.

2. Materials and Methods

2.1. Experimental Site

Vinh City is located in the Nghe An province, Vietnam and it has subtropical monsoon [2,12]. The current study sites were located near the Tanphuong bridge area (18°42′ latitude and 105°39′ longitude) of the Cuatien River. These sites were located within a straight reach. The width of the riprap and gabion revetments were 8 and 3 m, respectively, along the river bank. The distance between the two experimental plots was approximately 500 m, to ensure that the environmental and hydrological regimes and their influences were nearly similar in the two experimental plots. The experimental plots were constructed at a location on the Cuatien River bank which were eroded by the rapid fluctuations in river water level and flow. Cuatien River has a length of 25 km, a width of 18 to 60 m, and a depth of 0.86 to 2.45 m. It starts from Namdan county and flows through the delta of Hungnguyen county and merges with the Lam River in Vinh city (Figure 1). It is not only an important source of freshwater for the region, but it also plays an important role in the water ecological system of Vinh city. There are residential areas, farm lands, and animal farms on both sides of the Cuatien River catchments, which produce large amounts of domestic sewage and garbage. The lack of proper garbage collection and sewage treatment systems are the key factors that contribute to the deterioration of river water quality.

A water flow meter, fitted with a flow probe (Xiuyan Precision Instrument Factory, Anshan Liaoning, China) was used to determine the velocity of the river flow. The flow meter was fixed to the boat to determine the flow velocity along the river. The velocity of the river flow ranged from 0.05–0.09 m/s during the dry season and 0.31–0.53 m/s during the rainy season. Although the flow velocity is not high in Cuatien River, the hydrological behavior is significantly affected by the exchange of tide regime at the mouth of the Lam River. Therefore, the water level changes frequently, which is the main cause of erosion in the river banks (Figure 2). Hence, the prevailing river hydrological conditions were considered to design and construct the ecological revetments. In addition, on both sides of the river, the soil levee work was constructed to prevent flood that is too easy to convert and construct the ecological revetment work. Sandy clay soil with the following characteristics: cohesion C = 6.8 (kPa), internal friction angle Φ = 23°, and proportion Ƴ = 1.8 (t/m^3), was used as the main construction material to build the levee structures. Since the slope along the river bank is different from the topography, which has a slope rate of 18–63°, a proper design of the ecological revetment to transform the bank structure is extremely important. Riprap is relatively easy to install, is flexible and can be easily repaired, requires low maintenance, and has a natural appearance. Side slopes from 1:3 (18°) to 1:2 (26°) are recommended for riprap stability [13]. The riprap revetment structure was applied in the low slope banks where the slope is lower than 26° (steepness m = 0.5), whereas the gabion revetment structure was applied on a high slope bank where the slope is higher than 26° (steepness m = 0.5) [14].

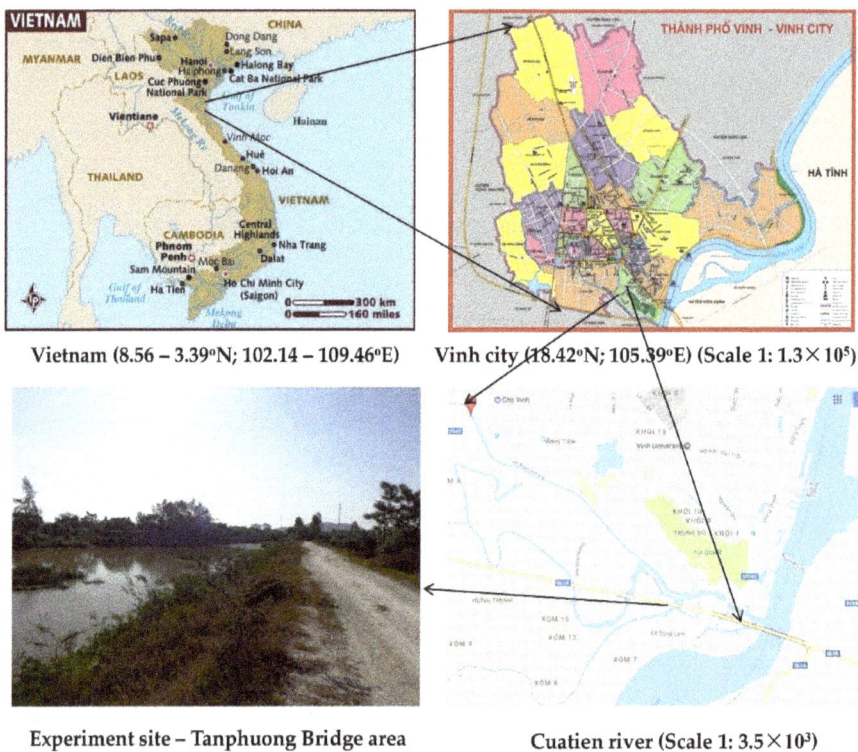

Vietnam (8.56 – 3.39°N; 102.14 – 109.46°E) Vinh city (18.42°N; 105.39°E) (Scale 1: 1.3×10⁵)

Experiment site – Tanphuong Bridge area Cuatien river (Scale 1: 3.5×10³)

Figure 1. Location of the experimental site at Cuatien River.

Figure 2. Experimental site at Cuatien River bank: (**a**) low slope bank; and (**b**) high slope bank (photos were taken on 9 November 2015).

2.2. Construction Materials

Two ecological revetment models (experimental setup) were constructed using natural materials, such as limestone, soil, and Vetiver grass, in conjunction with artificial materials (steel frames, polyethylene nets, geotextile, mesh wire 5 × 5 cm). The construction of riprap and gabion revetments were done in accordance with the design presented in Figure 3.

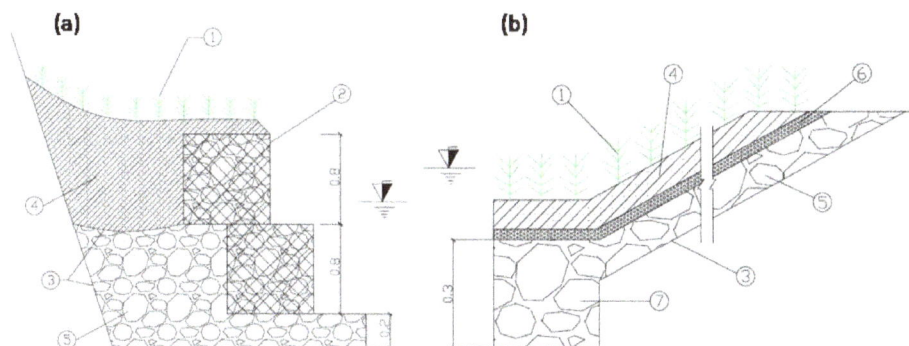

Figure 3. Schematic design of: (**a**) gabion revetment structure; and (**b**) riprap revetment structure (1. Vetiver plants; 2. gabion; 3. geotextile; 4. fill soil; 5. limestone; 6. gravel; and 7. base structure).

The river bank was selected to construct the gabion at a slope of approximate 57° (steepness m = 1.54). The gabion revetment was constructed as shown in Figure 4a. Firstly, the surface soil layer was excavated at a depth of 0.15 m, and then the geo-textiles were covered on the pit, and the stones were used as filling material to form the stable base. Three gabion structures with the following dimension: 0.5 × 0.5 × 0.8 m of length, height, and width were installed to form the gabion revetment. In order to ensure sufficient porosity and stability of the gabion revetment, which is beneficial for the development of plants on the gabion, the diameter of the stones was selected in the range of 8–12 cm (>5 cm) to fill the steel-framed structure [15]. These structures were installed at a stable position and then filled with limestone. In the next step, the soil was used to cover the top of the gabion structure.

Figure 4. Experimental site: (**a**) gabion; and (**b**) riprap revetments (photos were taken on 19 November 2015).

The riprap revetment structure was installed in an area of 4 × 8 m. The sloping bank was selected to construct the riprap revetment, with a slope of approximately 22° (steepness m = 0.42). The riprap revetment was constructed as shown in Figure 4b. In the first step, the surface soil layer was excavated up to a depth of 0.1 m. Then, geo-textiles were used to separate the soil and the limestone layer.

In order to ensure proper porosity and stability of the revetment, the limestone diameter was selected in the range of 0.15–0.2 m, as a heap, for covering the ground surface. Thereafter, the polyethylene net was covered on the surface of the limestone layer to prevent erosion caused due to the water flow [16]. In the bottom of the riprap, the soil layer was excavated to a depth of 0.3 m and filled with stones to increase the stability of the revetment structure [17].

After construction of the revetment structures, Vetiver grass was planted on the top soil layers. The weight of five random Vetiver plants at the time of plantation was 2.31 ± 0.28 g. In order to ensure proper growth of Vetiver grass, the distance between the plants was maintained at 22–25 cm, resulting in a plant density of 25 plants/m^2.

2.3. Climate Data Collection

The data pertaining to temperature, precipitation, sunlight hours, and humidity in Vinh city were collected on a daily basis. The average values of climatic data in different seasons, such as spring, summer, autumn, and winter were calculated.

2.4. Soil Nutrient Test

The mid-point of the soil surface layer of riprap and gabion were selected for the collection of soil samples. A shear soil ring with 80 mm diameter and 20 mm height was used to collect the soil samples from the revetment surface. Soil samples were collected during the initial stages of the experimental work (18 December 2015), at the end of the spring (22 April 2016), the summer (16 July 2016), the autumn (20 October 2016), and the winter (20 January 2016) seasons, respectively. Experiments pertaining to the measurement of soil nutrient tests of the soil samples were performed at the Chemical Laboratory, Vinh University (Vietnam). The available nitrogen (AN) content was determined using the micro-diffusion technique after alkaline hydrolysis of the samples [18]. The available phosphorus (AP) was determined according to the Olsen method [18], and the soil organic matter (SOM) was determined using the K_2CrO_7-H_2SO_4 oxidation method [19], respectively.

2.5. Plant Growth Determination

The growth of Vetiver grass was determined once a month, and five Vetiver plant samples were selected from the center point and at sites adjacent to the four sides of the gabion and riprap surface, respectively. The plant weight was determined by an ES-S electronic balance (Leqi, Tianjin, China). The average weight of five samples values was used to determine the growth rate of the Vetiver grass. In order to determine the differences in weight of the five plants, the standard deviation (SD) was calculated. Accordingly, small values of SD means that the values in the data are close to the averaged mean, and a large SD means that the values in the data are farther from the average.

2.6. Soil Shear Strength Test

The center riprap surface point beneath the riprap rock layer and sites behind the gabion wall were selected as sites to collect the soil samples. Soil layer with a depth of 10 cm was dug out to make the space for experimental sampling. Shear soil ring with 80 mm diameter and 20 mm height was used to collect the soil samples. The soil samples were collected at the initial stage of the experimental work, i.e., after six months and after 12 months, respectively. The soil direct shear strength experiment was performed at the soil mechanics laboratory of Vinh University (Figure 5). The shear strength (τ) of the soil samples was determined by a direct shear strength experimental device EDJ-2 (Yueke Instrument Co., Ltd., Hangzhou, Zhejiang, China) with normal stress (δ) values of 10, 20, 50, and 100 kPa, respectively. The tests were performed according to the ASTM D3080 standards with a shear rate of 1.5 mm/min. The cohesion C and internal friction angle Φ were determined according to the Coulomb formula shown in Equation (1) [20]:

$$\tau = C + \delta tg\Phi \tag{1}$$

where τ is the shear strength (kPa); C is the cohesion (kPa); δ is the normal stress (kPa); and Φ is the internal friction angle.

Figure 5. Experimental setup for determining the direct shear strength of the soil: (**a**) soil sample; and (**b**) experimental device for measuring the direct shear strength (photos were taken on 26 May 2016).

2.7. Determination of Floral and Faunal Diversity

Floral and faunal diversity were investigated in the current work. The studied area was as follows: surface of natural slope (4 × 8 m), gabion (4 × 3 m), and riprap (4 × 8 m). Samples were collected three times per week to investigate the growth of floral and faunal species, as well as for the frequency of the occurrence of the species in the experimental area. This experimental work was carried out for a period of one year.

2.8. Determination of Microbial Diversity

Water and mud samples were collected to determine the characteristics of microorganisms. Samples included: #1 river water near the shore, #2 gabion near the water mud interface, and #3 riprap near the water mud interface. High-throughput deoxyribonucleic acid (DNA) sequencing technology was adopted to explore the influence of spray microbial liquid to improve the microorganism community in river water and sediments [21]. The Miseq high-throughput sequencing process was performed as follows: (1) The extracted genomic DNA was detected with the help of 1% agarose gel electrophoresis; (2) Specific primers with "5" Barcode "3" were synthesized according to the designated sequencing region; (3) All samples were tested according to the standard protocol, wherein each sample measurement was repeated thrice and the polymerase chain reaction (PCR) product of the same samples were purified, mixed, and detected with 2% agarose gel electrophoresis; (4) The PCR products were quantified by Qubit according to the preliminary quantitative result of electrophoresis; and (5) Miseq sequencing on the machine. Adequate data analysis was performed on the experimental data in the form of valid sequence statistics, optimization of sequence statistics, diversity index, and species classification [21].

3. Results and Discussion

3.1. Climate Conditions

The average daily temperature is 24.5 °C, the temperature reached the highest value during the summer period (29.4 °C), and reached the lowest value during the winter period (19.9 °C). The average yearly precipitation is 155 mm, with a lowest average precipitation of 54.2 mm during the spring period. The climate of Vinh city can be divided into the dry and the rainy seasons. The dry season (winter and spring) starts from November to April, where the average rainfall is 54.2 mm (spring) and 111.8 mm (winter). During this season, a longer duration of rainfall prevails; however, with a low rainfall intensity (3.3 mm/h). The rainy season starts from May to October, and the average rainfall is 131.2 mm (spring) and 322.6 mm (autumn), respectively. The duration of rainfall is short, but with a high rainfall intensity of 87.3 mm/h. This often causes severe flooding in the region. The average monthly sunshine hours for the year is 142.5 h/month. The monthly sunshine hour value is high during the summer (224.7 h) and the autumn (171.5 h) period, but it is low during the spring (85.3 h) and the winter (88.5 h) seasons. Additionally, the humidity is always high in Vinh city, wherein the values vary between 81.3% and 87.7%. Olsen et al. [12] carried out a similar study in Nghe An province (Vietnam) and reported that the average temperature was 24.2 °C, annual precipitation was 1610 mm and the average humidity was 84%, respectively. The average values of daily temperature, humidity, and monthly sunshine hours during the four seasons are shown in Table 1.

Table 1. Characteristics of the prevailing climatic conditions of Vinh city, Vietnam.

Season	Spring (Feb.–Apr.)	Summer (May–Jul.)	Autumn (Aug.–Oct.)	Winter (Nov.–Jan.)	Average (yearly)
Daily temperature (°C)	21.3	29.4	27.3	19.9	24.5
Average precipitation (mm/month)	54.2	131.2	322.6	111.8	155
Monthly sunshine hours (hours/month)	85.3	224.7	171.5	88.5	142.5
Humidity (%)	89.3	83.3	81.3	87.7	85.4

3.2. Soil Nutrient Test

The results show that the SOM and AN levels were increasing in the seasons of spring and summer. In the three months of spring, the riprap's SOM and AN indicator values of the surface soil layer increased to 3.74 g/kg and 19.02 mg/kg, while the gabion's SOM and AN indicator values increased to 2.75 g/kg and 19.02 mg/kg, respectively (Figure 6). During the three months of summer, the riprap's SOM and AN indicator values increased to 7.06 g/kg and 84.08 mg/kg, while the gabion's SOM and AN indicator values increased to 3.86 g/kg and 21.95 mg/kg, respectively.

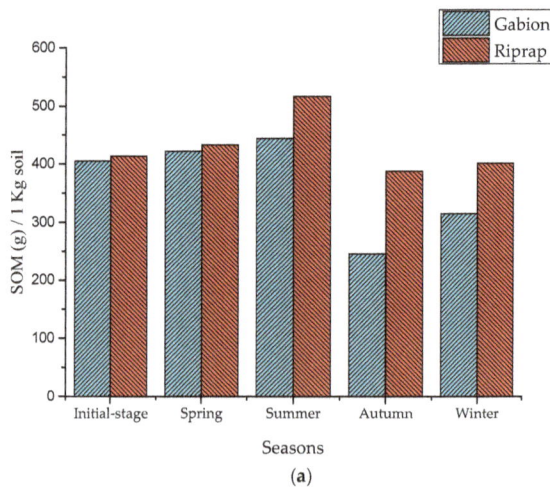

(a)

Figure 6. *Cont.*

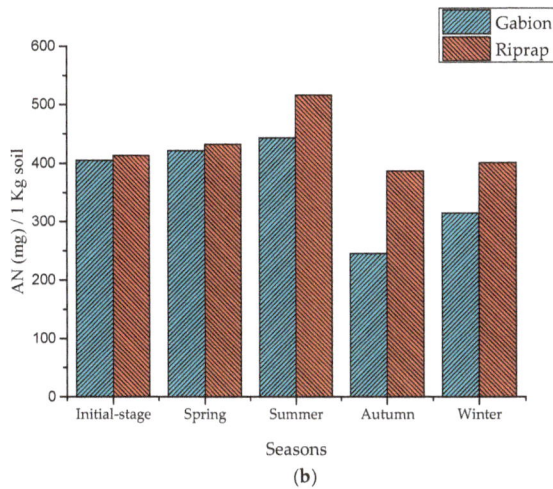

Figure 6. Nutrient indicators in the slope soil layer: (**a**) SOM indicators; and (**b**) AN indicators.

This increase is presumably because, during the spring and summer periods of small and medium rainfall, the pollutants present in the road surface were scoured and transferred to the revetment surface by storm runoff, and the pollutants accumulated in the revetment surface soil layer [4]. The amount of SOM and AN accumulated in the riprap revetment surface soil layer was higher than the gabion revetment. The riprap has a gentle slope and, therefore, the water runoff velocity was slow on the surface soil layer. Hence, the sediments and the pollutants were easily captured by the plant system [22].

During the period of three months in autumn, the SOM and AN indicator values in the riprap revetment decreased to 9.69 g/kg and 129.21 mg/kg, respectively, while the gabion's SOM and AN indicator values decreased to 13.75 g/kg and 198.19 mg/kg, respectively. The autumn season, with the continuous occurrence of heavy storms, caused extreme runoff periods which had diluted the scoured sediments and nutrients on the surface soil layer of the revetment [22,23]. This eventually led to a decrease in the nutrient concentration, subsequently leading to a decrease in the SOM and AN indicator values. Khan et al. [23] reported nutrient losses of 8.46 kg mineral nitrogen (N), 19 kg phosphorus (P), and 220 kg organic matter ha^{-1} due to water erosion from maize crops in the upland sloping soil. The amount of SOM and AN lost in the gabion revetment was higher than the riprap revetment. With its vertical form and porous structure, the nutrients and organic maters easily permeated through the voids of the gabion, and this structural configuration caused the soil layer to easily lose its nutrients. Concerning the riprap, it has a gentle slope that was covered with high floral diversity and distribution. This feature caused the riprap soil to improve its nutrient accumulation capacity, and avoided the nutrient scouring effect from stormwater [22]. Nevertheless, during the winter season, with the decrease in rainfall intensity, the SOM and AN indicator values showed a steep increase. It is noteworthy to mention that, during all the four seasons, AP indicator values were not significant because the P index in the soil was high (85.24–87.63 mg/kg), and the P content in storm water did not show any effect on the AP of the soil. These results clearly prove that the storm runoff has a significant effect on the slope soil nutrient concentration. This was also reflected on the good growth of Vetiver grass on the riprap surface. The SOM and AN indicator values significantly increased during the spring, summer, and winter periods; however, they decreased during the autumn season.

3.3. Plant Growth Conditions

The experimental data obtained for the five samples (S1–S5), the average and SD values for Vetiver biomass grown on the gabion and the riprap revetments are shown in Tables 2 and 3, and Figures 7 and 8, respectively.

Table 2. Vetiver biomass grown on the gabion revetment.

Time	S1	S2	S3	S4	S5	Avg.	SD
Nov.	1.7	1.8	1.9	1.8	1.82	1.814	0.064
Dec.	2.1	2.3	2.4	2.4	2.5	2.34	0.136
Jan.	4.1	4.7	4.8	4.4	4.6	4.52	0.248
Feb.	11.5	12.7	12.3	13.5	13.2	12.64	0.703
Mar.	28.8	30.2	32.8	29.6	31.7	30.62	1.446
Apr.	56.2	48.7	52.4	49.2	52.1	51.72	2.689
May	65.9	59.5	63.3	57.3	64.3	62.06	3.178
Jun.	71.6	80.2	79.5	81.8	75.3	77.68	3.722
Jul.	79.9	82.7	92.4	84.3	87.2	85.3	4.265
Aug.	92.6	104.3	101.8	106.2	96.8	100.34	4.992
Sep.	113.6	105.3	97.6	101.2	108.8	105.3	5.608
Oct.	119.3	102.8	107.6	108.4	113.5	110.32	5.628
Nov.	125.8	119.7	108.8	112.3	115.8	116.48	5.902
Dec.	119.8	129.3	123.8	117.9	111.1	120.38	6.064
Jan.	125.3	132.7	128.5	121.6	113.5	124.3	6.529

Note: S1–S5 are the sample numbers, SD—standard deviation, Avg.—Average value.

Table 3. Vetiver biomass grown on the riprap revetment.

Time	S1	S2	S3	S4	S5	Avg.	SD
Nov.	1.83	1.82	1.91	1.72	1.74	1.804	0.068
Dec.	1.96	2.09	2.12	2.2	1.97	2.068	0.092
Jan.	4.32	3.84	3.94	4.32	4.5	4.184	0.251
Feb.	11.3	9.63	10.9	9.82	10.17	10.364	0.638
Mar.	25.6	28.5	29.4	26.3	28.8	27.72	1.491
Apr.	39.6	41.3	45.4	40.2	39.1	41.12	2.262
May	55.6	57.5	60.3	52.6	54.2	56.04	2.672
Jun.	64.6	58.9	62.3	67.2	59.3	62.46	3.154
Jul.	61.8	62.2	65.9	67.3	70.4	65.5	3.225
Aug.	63.7	71.3	66.1	73.2	68.3	68.5	3.428
Sep.	78.1	68.1	70.2	74.5	70.2	72.22	3.603
Oct.	75.8	80.2	68.9	75.2	77.9	75.6	3.783
Nov.	71.6	82.1	79.7	75.6	73.6	76.52	3.869
Dec.	81.6	72.8	78.2	79.5	71.6	76.74	3.881
Jan.	73.8	83.7	74.5	79.3	73.8	77.02	3.922

Note: S1–S5 are the sample numbers, SD—standard deviation, Avg.—Average value.

At the beginning of the plantation time, on the gabion and riprap revetments, the average weight of Vetiver grass was 2.31 ± 0.28 g. During the spring season (from February to April), with a mild warm climate (21.3 °C), high humidity (89.3%), the Vetiver grass began to develop, and its biomass weight on the gabion and riprap revetments was 10.41 ± 0.52 and 12.13 ± 0.61 g, respectively. The rate of Vetiver biomass growth on the gabion and riprap revetments was 2.70 and 3.27 g/month, respectively. In summer (from May to July), under high rainfall (131.2 mm/month), relatively warm temperature (29.4 °C), and sunshine hour (224.7 h/month) conditions, the tropical flora showed the best growth conditions. In addition, the soil nutrients in terms of SOM and AN reached the highest value during this period. Therefore, the Vetiver grass could strongly grow on the riprap and gabion revetment surfaces and yield higher biomass. The Vetiver grass biomass weighed 55.74 ± 2.79 and 64.1 ± 3.11 g, respectively, in the gabion and riprap revetments. The results from this study shows that the rate of Vetiver biomass growth on the gabion and riprap revetments were 15.11 and 17.32 g/month, respectively. The storm runoff scours containing high nutrient concentrations and pollutants from the road surface caused the amount of SOM and AN to reach its highest values in the gabion and riprap soil layers [4].

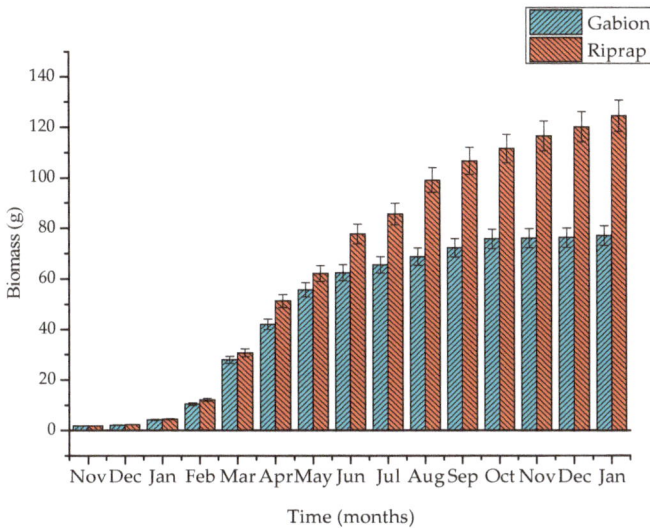

Figure 7. Growth of Vetiver grass on two types of ecological revetments (the error bars represent the standard deviation for the weight of Vetiver grass).

Figure 8. Vetiver growth on: (**a**) gabion; and (**b**) riprap revetments (photos were taken on 23 October 2016).

During the autumn season (from August to October), the growth of Vetiver grass on the gabion and riprap revetments were 68.79 ± 3.43 and 99.14 ± 4.96 g, respectively. The Vetiver biomass growth rate in the gabion and riprap revetments was 4.35 and 11.68 g/month, respectively. During the autumn season, the continuous heavy storms (332.6 mm/month) caused a reduction in the soil nutrient concentrations, leading to low values of SOM and AN. This adversely affected the Vetiver grass biomass on the gabion and riprap revetments. With a vertical form and porous structure, the gabion soil layer lost nutrient and soil particles significantly, causing dissatisfaction of Vetiver grass nutrition.

However, after a planting period of eight months, the riprap's Vetiver grass root pierced through the stone layer and directly absorbed nutrients from beneath the riprap stone layer. Therefore, Vetiver did not primarily depend on the nutrient source from the riprap surface soil layer. Under such conditions, the biomass weight of the gabion's Vetiver grass was significantly lower than the riprap's. During the winter season (from November to January), with lower temperatures (19.9 °C), rainfall (111.8 mm/month), and low sunshine hours (88.5 h/month), Vetiver plant biomass growth in the gabion and riprap revetments significantly declined. The biomass of Vetiver grass on the gabion and riprap revetments were 77.12 ± 3.94 and 124.61 ± 6.23 g, respectively. The Vetiver plant biomass growth rate in gabion and riprap revetments were 2.78 and 8.49 g/month, respectively. Riprap's Vetiver grass showed rapid growth in spring, summer, autumn, and slow growth in winter. On the other hand, the gabion's Vetiver grass showed rapid growth in spring, summer, and slow growth in autumn and winter. Mickovski et al. [24] showed that the dry shoots of 22 Vetiver plants could reach a weight of 70.05 ± 8.04 g with in a period of one year (2012–2013) when grown on different types of ecological revetments. The experimental results from this study revealed that Vetiver grass has good ability for growth and reproduction on the ecological revetments. The results indicated that high values of temperature and humidity, long sunshine hours. And soil nutrient content during the spring and the summer season, will facilitate the rapid development of the Vetiver biomass. However, the Vetiver biomass began to decrease with rainfall, due to the decline of the nutrient content during the autumn season.

3.4. Slope Stability Effect

To investigate the effect of shear strengths (τ_1, τ_2, and τ_3) of the soil samples during 0, 6, and 12 months, the soil samples were subjected to direct shear test under normal strength as $\sigma_1 = 10$ kPa, $\sigma_2 = 20$ kPa, $\sigma_3 = 50$ kPa, and $\sigma_4 = 100$ kPa. The results of the shear strength experiments are shown in Figure 9. The cohesion C, and internal friction angle Φ, calculated through the Coulomb formula (Equation (1)) are presented in Table 4.

The results of direct shear strength tests revealed that the cohesion of soil increased, but the internal friction angle showed a rather slow increase by only 1–3°. After a period of 6 and 12 months, the cohesion of soil samples collected from the gabion revetment increased by 59.6% and 162.9%, respectively. The cohesion in soil samples collected from the riprap revetment after 6 and 12 months increased by 115.6% and 239.1%, respectively. Therefore, the shear strength results proved that the root system of Vetiver grass and native plants could increase the cohesion and internal friction angle of soil, and enhance the bank slope stability.

Ali et al. [11] studied the shear strength of soil containing vegetable roots and noticed that after 6 months, under soil suction-free condition, the root of *Leucaena leucocephala* could increase the cohesion of soil by 116 (0.1 m), 225.0 (0.3 m) and 413.4% (0.5 m), which increases the shear strength of the soil to improve the stability of the bank. In addition, the Vetiver grass root system has high tensile and shear strength, and this unique property would increase the soil shear strength. Noorasiyikin et al. [25] also carried out a tensile strength test of the Vetiver grass root system and showed a significant relationship between the soil shear strength and plant root diameter. Hawke et al. [26] reported that the high moisture condition in the soil can reduce the soil metric tension, and decrease the soil cohesion, leading to a severe instability of the bank slope. As examples, the following two literatures have clearly shown that the Vetiver grass can grow to deeper depths and that their roots are strong enough to provide good reinforcement in the lands where they grow: Truong [27] showed that the Vetiver grass was able to penetrate to a depth of 3–4 m in the soil within a period of one year. In another study, Islam et al. [28] also indicated that Vetiver root grew up to 0.9 m in pure sand soil within three months. This tendency of root growth increased the soil cohesion, but it also absorbed the soil moisture, nutrients, and other pollutants present in the soil layer [24,29]. Current results proved that the roots of the Vetiver grass penetrated up to a depth of 3–4 m in the soil and increased the soil cohesion, enhancing the stability of the river bank.

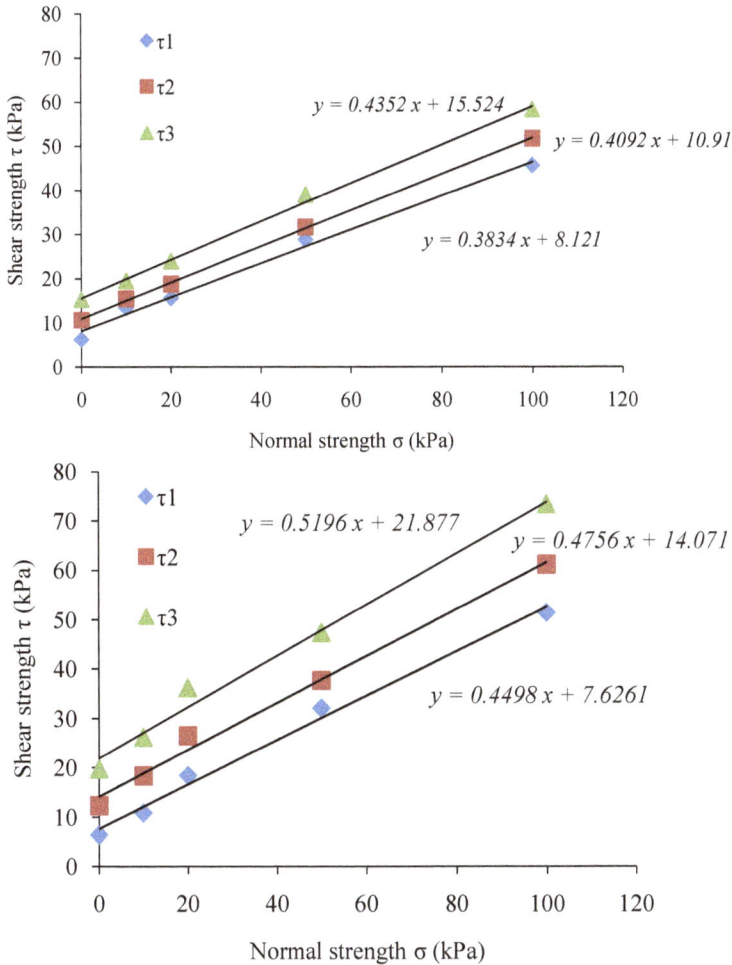

Figure 9. Normal strength and shear strength values for: (**a**) gabion; and (**b**) riprap revetments.

Table 4. Soil shear strength parameters for gabion and riprap revetments.

Revetment Type	Sample	Shear Strength Parameter	Increase in Cohesion (kPa)
	0 month	C = 6.2 kPa; Φ = 20.9°	-
Gabion	After 6 months	C = 9.9 kPa; Φ = 22.3°	3.7
	After 12 months	C = 16.3 kPa; Φ = 23.5°	10.1
	0 month	C = 6.4 kPa; Φ = 24.2°	-
Riprap	After 6 months	C = 13.8 kPa; Φ = 25.0°	7.4
	After 12 months	C = 21.7 kPa; Φ = 27.4°	15.3

During periods of storm runoff scouring, the increase of soil cohesion characteristics by the plant root system significantly reduces the soil layer erosion in slopes. Cuellar et al. [30] simulated the transient erosion caused by a vertical fluid jet impinging on the surface of a granular assembly and indicated that the relative eroded mass decreased when the cohesive bond increased from 0–2.5 N, and the relative eroded mass was zero when the cohesion C = 2.5 N. In another study, Niu et al. [31]

ascertained the role of *Cleistogenes songorica* for the sustainable development and protection of the soil ecosystem and indicated that *Cleistogenes songorica* roots has a major role in increasing the soil aggregates and enhances the resistance to soil erosion under storm-runoff incidents. In conclusion, the current results revealed that the Vetiver grass with a developed root system could increase the soil cohesion, which significantly reduced the loss of the soil layer by storm runoff.

3.5. Improving the Diversity of Flora and Fauna on the Revetments

3.5.1. Floral Diversity

In addition to Vetiver grass, 24, 14, and 26 native plant species were detected on the natural slope, gabion, and riprap revetments, respectively (Table 5). The plant species are as follows: *Ageratum conyzoides, Lantana camara, Mimosa pudica L., Bindens pilosa, Trianthema portulacastrum L., Eclipta alba, Euphorbia thymifolia Burm, Fimbristylis miliacea, Cyperusiria, Cypirus digitatus, Taraxacum mongolicum Hand-Mazz, Sonchus arvensis Linn, Portulaca oleracea L., Sorghum halepense, Alligator Alternanthera Herb, Conyza canadensis, Ageratum conyzoides, Asplenium antiquum, Oxalis corymbosa DC, Centella asiatica, Cyperus tegetiformis, Fimbristylis miliacea, Echinochloa crus-galli, Leptochloa chinensis, Choromolaena odorata,* and *Cenchrus echinatus* L. The gabion's plant species number is lower than that of the natural slope and riprap revetment because of low surface area and lesser amount of soil nutrients. Additionally, the natural plant's species number was lower than the riprap. This might be due to the nature of soil used to fill the riprap revetment. Strayer et al. [32] studied the vegetation of the riprap revetments along the Hudson River (USA) and showed that 11 plant species were detected within a range of 25 m from the shore. The result from this study confirms the fact that the gabion and riprap revetments, with their highly porous structures, could provide an ideal habitat for plant growth and reproduction.

Table 5. The floral and faunal diversity in different revetment types.

Revetment Type	Natural Slope (4×8 m^2)	Gabion Revetment (4×3 m^2)	Riprap Revetment (4×8 m^2)
Floral	24	14	25
Faunal	9	5	9

3.5.2. Faunal Diversity

In the present study nine, five, and nine faunal species were detected on the natural slope, gabion, and riprap revetment, respectively (Table 5). The insects identified on the revetments were as follows: *Acrida chinensis, Gryllidae, Rhopalocera, Mantodea, Pheidole megacephala, Cryllustestaceus walker,* and *Catharsius molossus* L. The results also revealed that the phylum Nemathelminthes and centipede insect species were present in the plant root system. The gabion's faunal species number were found to be lower than the natural slope and riprap, because the gabion surface area and the development of flora were lower than the natural slope and riprap revetment, causing a lack of ideal habitat space for faunal living and reproduction. Gabriel et al. [8] compared the plant and animal species in five adjacent pairs of riprap and natural shorelines and reported that the proportional coverage by aquatic macrophytes at the armored shoreline was obviously higher than the natural site. Additionally, the amount of fish captured by the fyke net in the armored riprap shoreline was lower than that of the natural shoreline. These results clearly reveal that the riprap revetments with a porous structure could provide a good living habitat to aquatic flora and fauna and the fish can take refuge in the riprap porous structure to avoid capture by the fyke net.

3.6. Microbial Community Diversity

In the present study, water and mud samples were collected and analyzed through Miseq high-throughput sequencing [21] to determine the microbial population diversity attached on the surface of gabion and riprap revetments. The sequence information and the diversity index of the three samples are shown in Table 6. The ACE index is also used to estimate the number of OTUs (operational taxonomic units). The total number of species was high based on the Chao and ACE index [21,33,34]. It can be seen from Table 6 that the total number of species in samples determined by Chao index and ACE index is #3 > #2 > # 1, from high to low. The Shannon index and Simpson index were used to estimate the microbial community diversity in the samples. According to Shannon index and Simpson index values, the diversity of microbial community in the samples were ranked as #3 > #2 > #1 from high to low [35,36].

Table 6. Sequence information and the diversity index.

Samples	Reads			Biodiversity Index (3% Cutoff)				
	Unified Sequencing Depth	OTU	Chao Index	Shannon Index	ACE Index	Simpson Index	Goods Coverage
1	50,000	2815	5192	5.1243	7889	0.0218	0.9689
2	50,000	4782	7605	6.2803	8627	0.0102	0.9561
3	50,000	4998	7901	6.4997	9398	0.0087	0.9572

The results showed that the total number and diversity of microorganisms attached on the gabion and riprap revetments were much higher than that of the suspended microorganisms in the water. The results revealed that the total number of microbial communities and their diversity in the riprap revetment was highest, because the riprap revetment is suitable for the growth of plants, it has a rather strong root systems, and the gap between the stones also provide adequate conditions for the growth of microorganisms. It is one of the main reasons for rich microbial diversity in the riprap revetment [33]. Through high sequencing technology, the number of bacteria was estimated at 198, 332, and 351 in water #1, gabion #2 and riprap #3 samples, respectively. The results of species composition, based on its detection and classification at the genus level, is shown in Figure 10.

It can be seen from Figure 10 that the total amount of microbes and the relative content of dominant microorganisms on the two kinds of ecological bank revetments are obviously higher than those in the water body. This clearly indicates that the gaps and plant roots can provide good habitat for microbial attachment and growth.

Microorganisms in the river will gradually attach, reproduce and develop to form the biofilm on the surface of rock and plants root. It plays an important role in transformation and reduction of the pollutants in river water. The amount and diversity of microorganisms attached to the ecological revetments are significantly higher than the water-suspended microbes. Among them, *Polynuclebacter, Menicius, Mssila, Geobacter, Gemmatimonas, Albidiferax, Ramlibacter, Janthino bacterium, Sphaerotilus, Clostridium sensu stricto, pauludibacter, Dechoromonas, Sphingomonas,* and *Dechloromonas* were significantly higher. Previous studies have shown that *Polynuclebacter, Geobacter, Gemmatimonas, Ramlibacter, Clostridium sensu strict, pauludibacter, Dechoromonas,* and *Sphingomonas* have a strong ability to reduce pollutant levels in river water [36]. Its main role is to uptake nutrient in river water to form a biofilm. Earlier studies have classified *Ramlibacter, Gemmatimonas,* and *Sphingomonas* as aerobic denitrifying bacteria [37], and *Pauludibacter* and *Geobacter* as anaerobic denitrifying bacteria [38]. Gabion and riprap revetments, with their highly porous structure, could provide a good habitat for the growth and reproduction of a large number of microorganisms which also enhances the natural purification process of the river.

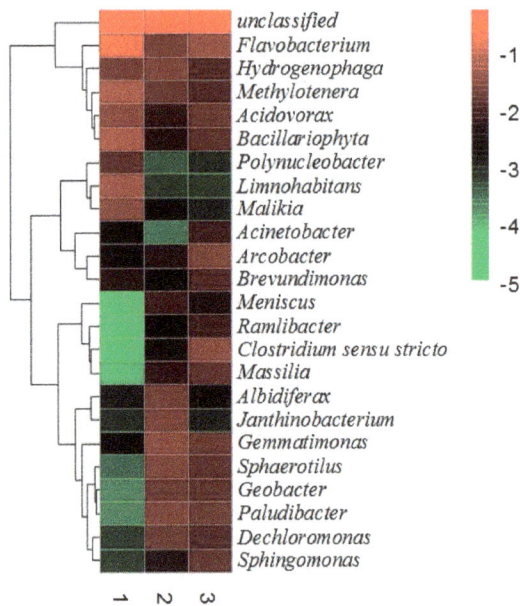

Figure 10. The classification at the genus level.

4. Conclusions

The study area, Vinh city, Vietnam, is located in a subtropical monsoon area. The average daily temperature, monthly precipitation, monthly sunshine hours, and humidity were 24.5 °C, 155 mm/month, 142.5 h/month, and 85.4%, respectively. The soil nutrient levels on the revetments increased during the spring, autumn, and winter seasons, but it decreased during the autumn season due to high rainfall. However, during the summer season, with high temperature and soil nutrient concentrations, the Vetiver grass biomass of riprap and gabion revetments reached its highest growth rate as 15.11 and 17.32 g/month, respectively. With the decrease in temperature and nutrient concentration during the autumn and winter season, the growth rate of Vetiver grass biomass decreased. The Vetiver grass biomass on the riprap was higher than the gabion, which proved that the riprap structure with a gentle slope was not affected severely by the storm runoff erosion. The results revealed that the root system of Vetiver grass and native plants increase the soil cohesion, i.e., it increased by 162.9% in the gabion and 239.1% in the riprap revetment, respectively, after a period of one year and the internal friction angle from 2.6° to 3.2°, leading to an increase in the soil shear strength and good stability of the bank slope. The floral and faunal species detected on the gabion and riprap revetments were similar to the species found on the natural slope. In addition, the rich microbial diversity was also observed in the gabion and riprap revetments. These observations proved the fact that the gabion and riprap revetments provided a good habitat for floral, faunal, and microbial growth, which enhances the ecological restoration capacity of the river bank.

Author Contributions: D.F. and V.T.T. conceived and designed the experiments; V.T.T., T.N.B., and T.T.T.S. performed the experiments at the site in Vietnam; R.P.S. and V.T.T. contributed to the procurement of reagents, materials, monitoring, and data analysis; and R.P.S., V.T.T., and E.R.R. wrote the paper.

Acknowledgments: This study was co-funded by the National Key Technologies R and D Program, China (no. 2015BAL02B05) and the National Natural Science Foundation of China (no. 51650410657).

Conflicts of Interest: The authors declare no conflict of interest.

References

1. Bolay, J.C.; Cartoux, S.; Cunha, A.; Du, T.T.N.; Bass, M. Sustainable development and urban growth: Precarious habitat and water management in Ho Chi Minh City, Vietnam. *Habitat. Int.* **1997**, *21*, 185–197. [CrossRef]
2. Open Letter. Available online: http://vinhcity.gov.vn/?detail=2/open-letter (accessed on 10 March 2008).
3. Böhm, H.R.; Schramm, S.; Bieker, S.; Zeig, C.; Anh, T.H.; Thanh, N.C. The semi-centralized approach to integrated water supply and treatment of solid waste and wastewater—A flexible infrastructure strategy for rapidly growing urban regions: The case of Hanoi/Vietnam. *Clean. Technol. Environ.* **2011**, *13*, 617–623. [CrossRef]
4. Chen, J.; Chang, N.; Fen, C.; Chen, C. Assessing the storm-runoff impact to an urban river ecosystem using estuarine water quality simulation model. *Civ. Eng. Environ. Syst.* **2010**, *21*, 33–49. [CrossRef]
5. Yao, S.; Yue, H.; Li, L. Analysis on current situation and development trend of ecological revetment works in middle and lower reaches of Yangtze river. *Procedia Eng.* **2012**, *28*, 307–313.
6. Wu, H.L.; Feng, Z.Y. Ecological engineering methods for soil and water conservation in Taiwan. *Ecol. Eng.* **2006**, *28*, 333–344. [CrossRef]
7. Chen, Y.; Xu, S.; Jin, Y. Evaluation on ecological restoration capability of revetment in land restricted channel. *KSCE J. Civ. Eng.* **2016**, *20*, 2548–2558. [CrossRef]
8. AGabriel, O.; Bodensteiner, L.R. Impacts of riprap on wetland shorelines, upper Winnebago pool lakes, Wisconsin. *Wetlands* **2012**, *32*, 105–117. [CrossRef]
9. Ramli, M.; Karasu, T.J.R.; Dawood, E.T. The stability of gabion walls for earth retaining structures. *Alexandria Eng. J.* **2013**, *52*, 705–710. [CrossRef]
10. Helal, H.M.; Beck, D.S. Effect of plant roots on carbon metabolism of soil microbial biomass. *J. Plant Nutr. Soil Sci.* **1986**, *149*, 181–188. [CrossRef]
11. Ali, F.H. Shear strength of a soil containing vegetation roots. *Soils Found.* **2008**, *48*, 587–596. [CrossRef]
12. Olsen, A.; Thuan, L.K.; Murrell, K.D.; Dalsgaard, A.; Johansen, M.V.; De, N.V. Cross-sectional parasitological survey for helminth infections among fish farmers in Nghe An province, Vietnam. *Acta Trop.* **2006**, *100*, 199–204. [CrossRef] [PubMed]
13. Yang, G.L.; Liu, Z.Z.; Xu, G.L.; Huang, X.J. Protection technology and application of gabion. In Proceedings of the 13th International Conference on Structural & Geotechnical Engineering, Hangzhou, China, 8–10 September 2009; pp. 915–919.
14. Pagliara, S.; Chiavaccini, P. *Urban Stream Restoration Structures*; Springer Publishing: Dordrecht, The Netherlands, 2004; Volume 43, pp. 239–252.
15. Beikircher, B.; Florineth, F.; Mayr, S. Restoration of rocky slopes based on planted gabions and use of drought-preconditioned woody species. *Ecol. Eng.* **2010**, *36*, 421–426. [CrossRef]
16. Salivcop, V.G. Protection of banks and roadbeds from erosion on river "Nip". *Power Technol. Eng.* **1986**, *20*, 575–580.
17. Crusoe, G.E., Jr.; Cai, Q.; Shu, J.; Han, L.; Barvor, Y.J. Effects of weak layer angle and thickness on the stability of rock slopes. *Int. J. Min. Geo-Eng.* **2016**, *50*, 97–110.
18. Conway, A. *Soil Physical—Chemical Analysis*; Institute of Soil Science: Nanjing, China; Technology Press: Shanghai, China, 1978.
19. Nelson, D.W.; Sommers, L.E.; Sparks, D.L.; Page, A.L.; Helmke, P.A.; Loeppert, R.H.; Soltanpour, P.N.; Tabatabai, M.A.; Johnston, C.T.; Sumner, M.E. *Total Carbon, Organic Carbon, and Organic Matter. Methods of Soil Analysis: Part 3—Chemical Methods*; Soil Science Society of America Book Series; Crop Science Society of America (CSSA): Madison, WI, USA, 1996; pp. 961–1010.
20. Farooq, K.; Rogers, J.D.; Ahmed, M.F. Effect of densification on the shear strength of landslide material: A case study from salt range, Pakistan. *Earth Sci. Res. J.* **2008**, *48*, 587–596. [CrossRef]
21. Bing, W.; Liu, J.; Singh, R.P.; Fu, D. Effect of alternate dry-wet patterns on the performance of bioretention units for nitrogen removal. *Desal. Water Treat.* **2016**, *59*, 295–303. [CrossRef]
22. Peng, T.; Wang, S.J. Effects of land use, land cover and rainfall regimes on the surface runoff and soil loss on karst slopes in southwest China. *Catena* **2012**, *90*, 53–62. [CrossRef]

23. Khan, F.; Bhatti, A.U.; Khattak, R.A. Soil and nutrient losses through sediment and surface runoff under maize monocropping and maize-legumes intercropping from upland sloping field. *Pak. J. Soil Sci.* **2001**, *19*, 32–40.

24. Mickovski, S.B.; Beek, L.P.H.V.; Salin, F. Uprooting of *vetiver* uprooting resistance of *vetiver* grass (*Vetiveria zizanioides*). *Plant Soil* **2005**, *278*, 33–41. [CrossRef]

25. Noorasyikin, M.N.; Zainap, M. A tensile strength of Bermuda grass and Vetiver grass in terms of root reinforcement ability toward soil slope stabilization. *Mater. Sci. Eng.* **2016**, *136*, 12–29. [CrossRef]

26. Hawke, R.; McConchie, J. In situ measurement of soil moisture and pore-water pressures in an 'incipient' landslide: Lake Tutira, New Zealand. *J. Environ. Manag.* **2011**, *92*, 266–274. [CrossRef] [PubMed]

27. Truong, P.; Van, T.T.; Pinners, E. *Vetiver System Applications Technical Reference Manual*; The Vetiver Network International (TVNI): San Antonio, TX, USA, 2008; pp. 1–126.

28. Islam, M.S.; Arif, M.Z.U.; Badhon, F.F.; Mallick, S.; Islam, T. Investigation of vetiver root growth in sandy soil. In Proceedings of the BUET-ANWAR ISPAT 1st Bangladesh Civil Engineering SUMMIT 2016, Dhaka, Bangladesh, 23–26 November 2016.

29. Mathew, M.; Rosary, S.C.; Sebastian, M.; Cherian, S.M. Effectiveness of Vetiver for treatment of wastewater from institutional kichen. *Procedia Technol.* **2016**, *24*, 203–209. [CrossRef]

30. Cuellar, P.; Philippe, P.; Bonelli, S.; Benahmed, N.; Brunier-Coulin, F.; Ngoma, J.; Delenne, J.; Radjaï, F. Micromechanical analysis of the surface erosion of a cohesive soil by means of a coupled LBM-DEM model. In Proceedings of the International Conference on Particles, Barcelona, Spain, 28–30 September 2015.

31. Niu, X.; Nan, Z. Roots of *Cleistogenes songorica* improved soil aggregate cohesion and enhance soil water erosion resistance in rainfall simulation experiments. *Water Air Soil Pollut.* **2017**, *228*, 109. [CrossRef]

32. Strayer, D.L.; Kiviat, E.; Findlay, S.E.G.; Slowik, N. Vegetation of rip rapped revetments along the freshwater tidal Hudson River, New York. *Aquatic Sci.* **2016**, *78*, 605–614. [CrossRef]

33. Xu, M.; Liu, W.; Li, C.; Xiao, C.; Ding, L.; Xu, K.; Geng, J.; Ren, H. Evaluation of the treatment performance and microbial communities of a combined constructed wetland used to treat industrial park wastewater. *Environ. Sci. Pollut. Res. Int.* **2016**, *23*, 10990–11001. [CrossRef] [PubMed]

34. Zhao, J.; Zhang, R.; Xue, C.; Xun, W.; Sun, L.; Xu, Y.; Shen, Q. Pyrosequencing reveals contrasting soil bacterial diversity and community structure of two main winter wheat cropping systems in China. *Microb. Ecol.* **2014**, *67*, 443–453. [CrossRef] [PubMed]

35. Jennifer, B.H.; Jessica, J.H.; Taylor, H.R.; Brendan, J.M.B. Counting the uncountable: Statistical approaches to estimating microbial diversity. *Appl. Environ. Microbiol.* **2001**, *67*, 4399–4406.

36. Telias, A.; White, J.R.; Pahl, D.M.; Ottesen, A.R.; Walsh, C.S. Bacterial community diversity and variation in spray water sources and the tomato fruit surface. *BMC Biotechnol.* **2011**, *11*, 81. [CrossRef] [PubMed]

37. Naoki, T.; Catalan-Sakairi, M.A.B.; Yasushi, S.; Isao, K.; Zhou, Z.; Shoun, H. Aerobic denitrifying bacteria that produce low levels of nitrous oxide. *Appl. Environ. Microbiol.* **2003**, *69*, 3152–3157.

38. Jalil, J.; Alireza, M.; Ramin, N.; Mohammad, H.; Hossein, K.; Amir, H.M. Influence of upflow velocity on performance and biofilm characteristics of anaerobic fluidized bed reactor (AFBR) in treating high-strength wastewater. *J. Environ. Sci. Health* **2014**, *12*, 139.

water

MDPI

Article

The Influences of Sponge City on Property Values in Wuhan, China

Shiying Zhang [1], Chris Zevenbergen [2,*], Paul Rabé [3] and Yong Jiang [4]

[1]   SF Express Holding, Wuhan 430015, China; pattyzsy@hotmail.com
[2]   Department of Water Science Engineering, IHE Delft Institute for Water Education,
      2611 AX Delft, The Netherlands
[3]   Institute for Housing and Urban Development Studies (IHS), Erasmus University Rotterdam,
      3000 DR Rotterdam, The Netherlands; rabe@ihs.nl
[4]   Department of Integrated Water Systems and Governance, IHE Delft Institute for Water Education,
      2611 AX Delft, The Netherlands; y.jiang@un-ihe.org
*   Correspondence: c.zevenbergen@un-ihe.org; Tel.: +31-653-599-654

Received: 7 May 2018; Accepted: 2 June 2018; Published: 12 June 2018

Abstract: Rapid urbanization in China and global climate change have increased urban flood exposure in Wuhan, and the increased flood risk has reduced property values in flood-prone areas. The central government of China is promoting the application of the sponge city concept to reduce urban flood risk and improve the environment in cities. Wuhan is one of the pilot cities of this initiative. A shortage of funds is one of the main obstacles to sponge city construction, as is the lack of a suitable business model. To test residents' willingness to pay for sponge city construction, this research analyzed the impact of sponge city construction on the housing values of areas covered by sponge city interventions. The authors conducted interviews and analyzed secondary data to gauge residents' awareness and perceptions of sponge city interventions. The results show that more than half of residents in Wuhan are willing to pay for sponge city measures, but the amount they are willing to pay is limited. Residents are more willing to pay for improvements of their living environment than for flood reduction measures.

Keywords: urbanization; flood risk; flood risk management; perceptions of flooding; willingness to pay

1. Introduction

During the past 30 years, triggered by economic reforms, China's economy has undergone rapid expansion. Urbanization has been one of the most significant driving forces of this economic expansion. Wuhan is the capital city of Hubei Province and is located in the middle reaches of the Yangtze River. Its history is intertwined with the fight against water and flooding [1]. The Yangtze River and its biggest tributary cross the city's geographic center. This geographic feature means that the water levels of the two rivers have a direct influence on the whole city's safety. Over time, as a result of rapid urbanization, the original lakes and rivers within the city boundaries have been filled in and transformed into impermeable concrete. These man-made transformations have significantly reduced the city's ability to absorb water and have exposed it to regular flooding following heavy rains. Wuhan is one of the top 10 largest cities in China which suffers most from frequent river and pluvial flooding. Wuhan is also a symbol of the uncontrolled, rapid expansion of Chinese cities, resulting in a significant loss of storm water retention capacity and a drainage infrastructure which is unable to keep pace with these changes and growth.

In 2016, two months into the Yangtze River's flooding season, a continual intensive rainfall period affected the catchment area of the Yangtze River (The strongest storm in Wuhan since 1998, Wuhan

Morning News, 2016; Available online: http://hb.qq.com/a/20160703/004392.htm). The high-water level of the Yangtze River, on the one hand, made it hard to discharge urban water into the river, and, on the other hand, the capacity of the city's pumping station was limited. Partly as a result, on the morning of 6 July 2016, a major part of the city was flooded by rainwater. Roads and metro stations were inundated and trains were cancelled. Because of waterlogging within the city, two main metro lines were out of operation and thousands of cars were inundated by water and trapped on the street. This urban flooding event caused an economic loss of approximately CNY 2.2 billion, and 10 million people in the city were severely affected because of the dysfunction of the transportation system and the power supply system [2].

To deal with the challenges and problems resulting from urban flooding, on 12 December 2014, China's central government launched a policy of promoting the construction of sponge cities. On 31 December 2014, China's Ministry of Finance published its "policy of allowance" related to the construction of sponge cities [3]. This policy of allowance gives a detailed description of how the central government will support the local government in the construction of sponge cities.

Shortly after the central government's announcement, Wuhan city was selected as one of the first batches of pilot cities for sponge city construction. The Wuhan pilot started operations in mid-2015 and was planned to finish by the end of 2017. According to the working plan of Wuhan City Government for sponge city development [4], five objectives related to urban environmental conditions and waterlogging were formulated that should be achieved and checked before the end of 2017. These objectives were intended to restore the city's capacity to absorb, infiltrate, store, purify and drain (rain) water. However, the process of constructing the sponge city has fallen behind schedule, mainly due to insufficient funds.

To promote the construction of sponge city in Wuhan, China's central government committed itself to providing CNY 1 billion to the Wuhan government. However, the funding requirement for sponge city construction was estimated at CNY 15 billion for a period of two years, so the subsidy from the central government is insufficient. The funding shortage was assumed to be financed through private-public partnerships (PPPs). However, non-government sources of finance are difficult to mobilize through PPPs, as there are no successful PPP models to collect revenue from sponge cities currently. Thus, the challenge is to develop a business model for financing sponge city projects to address the lack of public funding. Under these circumstances, it is worth investigating whether the construction of sponge cities can increase housing values, as this may support identification and development of a business model and financing instruments to cover the funding gap. This research analyses the impacts of the construction of a sponge city on the housing values of the area that the sponge city construction covers. If the assumption that sponge city construction can improve property values can be verified, an effective business model may be designed accordingly.

2. Background and Context

Over the last two decades, the number of publications on research into the effects of floods and flood hazard on property values has steadily grown. These publications do not (yet) justify a straightforward conclusion: they indicate a range from negative to indifferent effects on property values [5]. These differences are in part a result of different study approaches, data and methods of analysis (e.g., changes to developed residential property values vs. land values) used. One of the conclusions most of these studies share is that an actual flood event seems to have an immediate and greater effect on property values than merely flood risk information provided by, for instance, flood risk maps [5,6]. Flood resistant and resilient retrofitting and repairs made after a flood occurrence generally have a positive effect on property value [7–9].

It is important to note that flood risk is one of many factors affecting property values. Other factors include, chiefly, the demand and the use and location of the property [7]. This, in turn, is typically influenced by legislative differences between countries and regional differences in behaviors and attitudes [5,10]. The levels of flood risk experienced by an individual depends on both exogenous (which are beyond an individual's control) and endogenous (which can be influenced by individuals by taking actions that reduces risk) factors. With respect to the latter, the willingness to pay (WTP) for flood risk reduction is largely influenced by people's subjective assessment of flood risk (referred to as risk perception), rather than a scientifically observed measure. Hence, citizen perceptions of urban flood hazards are often regarded as important factors influencing their WTP to reduce flood risks and to deploy flood resistant and resilient measures [11]. WTP for flood risk reduction may also be influenced by other factors, such as resource limitations, personality (individual characteristics), current risk levels, and acceptability of risks [11].

Very little information is currently available on the public perception and willingness to pay for sponge city initiatives in China. According to [12], education and income have a positive effect on WTP to support sponge city initiatives in two cities (Zibo and Dongying, Shandong Province) of China. A recent comparative study (UK, USA, China and Germany) on the impacts of flooding on property values has revealed that there is generally a lack of information on the impact of flood risk on land and housing markets at present. With the arrival and inter-comparison of more detailed flood impact data of properties, this will likely change and will allow valuation professionals to better reflect risk in their valuations [7]. The contingent valuation method (CVM) is increasingly used to assess the WTP for flood insurance and structural flood resistant and resilient measures (e.g., [13]). CVM involves a survey-based technique in which people are being asked to directly report their WTP. CVM has received wide application in valuing a variety of nonmarket goods such as wildlife, water quality, various risks, and the value of recreational sites [14].

3. Materials and Methods

3.1. Introduction to the Sponge City Concept

The sponge city concept is being promoted by the Chinese government to both reduce urban flood risk and improve the urban environment by mimicking natural processes. The approach aims to enhance a city's ability to "absorb, store, permeate and purify rainwater and to make use of stored water when needed" [15]. General targets have been set with regards to the fraction of annual runoff retained, but not to flood exposure and/or impacts. The central government of China promotes sponge cities as a national, holistic strategy to address urban flooding, because they are designed to offer "a more sustainable, integrative solution to the urban flooding problem than traditional storm water management practices, and one that is more closely related to the eco-city and low-carbon city approach" [16].

3.2. Research Strategy

This research was based on a combination of a survey (based on CVM) and desk research. For the survey, a questionnaire was used as the main method to explore residents' perceptions of flood risk and the sponge city approach, and to test its influence on housing and land values in central districts of Wuhan. The questionnaire consisted of 27 closed-ended questions (see Appendix A). Interviews were conducted with representatives of the public sector—including the city government and planning institutions—the private sector (including developers and professional service firms), and experts on sponge city approaches, in order to analyze residents' perceptions (Representatives of public sector institutions consulted for qualitative information related to the study were from the Wuhan Housing Security and Housing Management Authority, the Wuhan Municipal Development and Reform Committee, and the Wuhan Land Use and Urban Spatial Planning Research Center). In total, 15 in depth interviews were held between January and March 2017. The method of secondary data analysis was used in the strategy of desk research. Land transaction price data and housing transaction prices in the sponge city areas of Wuhan were analyzed to test whether and to what extent the sponge city program would impact property values.

3.3. Data Collection and Analysis

The online platform called Sojump (https://www.sojump.com/) was used to conduct the questionnaire. The questions were put into the platform by the researcher and the platform produced a unique URL link for the researcher. The researcher sent this link to the target respondents through e-mail and other information sharing tools such as Wechat. Participation in the survey was voluntary.

Based on the population of the central districts in Wuhan at the end of 2015 [17] , and using a confidence level of 95% at a confidence interval of 5, a sample size of 384 respondents from the study area(Shows in Figure 1) was established as a representative sample for the survey. Data analysis was undertaken and data were generated using Qualtrics software. To increase the representativeness of the sample, the questionnaire was distributed to residents in all seven districts of central Wuhan. Questionnaires were written in Chinese and list in Appendix A.

Figure 1. Map of Wuhan City, with the study area (central districts) indicated in blue.

Secondary data analysis was used as a supplement to enable the researchers to have a better understanding of the relationship between sponge city construction and housing values. All secondary data were obtained by visiting the website of those institutions or databases. The secondary data collection covered a period of ten years, from 2007 to 2017. Housing transaction price data covered the same seven central districts of Wuhan as the questionnaire.

4. Results

The research findings are presented in three parts. The first part analyses responses to the survey questionnaire. The second part presents qualitative analysis of in-depth interviews of representatives from different sectors. The final part examines collected secondary data of prices for land use rights and houses.

4.1. Survey Responses

4.1.1. Demographic Characteristics of Respondents

The survey collected 452 sets of responses. Twenty-nine of the respondents did not live in one of the seven central districts of Wuhan and thus were not included in the analysis, resulting in a total sample size of 423 in this study.

The demographic characteristics of the respondents are shown in Table 1. The primary age range of the respondents was between 18 and 28 years, accounting for 47.5% of the total samples. Among the respondents there were more females than males. The majority had a college or university degree. The respondents came from seven central districts of Wuhan city, with the relative percentages in the sample equal to at least 10%, implying the sample covered all the central districts of Wuhan (Table 2).

Table 1. Demographic Characteristics of Respondents.

Variable	Category	Number of Respondents	Percentage of Respondents
	Below 18	1	0.2
	18–30	201	47.5
Age	31–45	88	20.8
	46–55	51	12.1
	56–65	80	18.9
	older than 65	2	0.5
Gender	Female	250	59.1
	Male	173	40.9
Education level	Senior school or below	87	20.6
	University or college degree	224	53
	Post-graduate qualification	112	26.5

Table 2. Demographic data of central districts in year-end of 2015.

District	Number of Residents (Unit: Thousand)	Percentage of City Residents by District	Percentage of Samples by District
Jiangan District	719.5	8.7%	15%
Jianhan District	486.4	5.9%	10%
Qiaokou District	526.5	6.3%	11%
Hanyang District	585.4	7.1%	12%
Wuchang District	1056.1	12.7%	22%
Qingshan District	433.7	5.2%	9%
Hongshan District	948.8	11.4%	20%
Total	4756.4	57.3	100

4.1.2. The Respondents' Exposure to Urban Flooding

How residents perceive risk is fundamental to how they respond to flooding and take actions [18]. Flood risk perception is influenced by numerous factors related to the *perceived* probability of flood occurrence on the one hand and to the *perceived* consequences on the other hand. Several studies have shown that these factors are closely related to the proximity (distance) to the flood hazard, past experience (flood frequency), and the ability to take preparedness actions [19]. Based on these findings, five indicators were used in our survey to map the flood risk perception of the respondents. These indicators were: flood exposure, flood frequency, flood impact perception, flood protection measures (taken by residents), and perception of floods on property value.

Flood Exposure

Table 3 shows the extent to which the respondents had been exposed to urban floods over the past twenty years. The survey reveals that 96.7% of the respondents had been exposed to urban flooding in Wuhan in either a direct or an indirect way. This means that urban residents' exposure level in Wuhan is very high. This high level of exposure to floods indicates the importance for the city to take action and to promote sponge city development in Wuhan.

Flood Frequency

Table 4 shows that about 66.5% of the respondents experienced flooding at least once a year over the past ten years.

Table 3. Respondents' exposure to urban floods.

Extent of Exposure	Frequency	Percentage	Cumulative Percentage
Heavy exposure	186	44	44
Limited exposure	144	34	78
Observed other's exposure	79	18.7	96.7
No exposure	14	3.3	100
Total	423	100	

Table 4. Frequency of respondents' exposure to urban floods.

Number of Flood Exposure	Frequency	Percentage
More than once a year	109	25.8
Once a year	172	40.7
Once per two years	55	13
Once per three to five years	59	13.9
Once per six to ten years	10	2.4
Less than once per ten years	4	0.9
Total	409	96.7
Missing value	14	3.3
Total	423	100

Perception of Flood Impacts

Respondents' perceptions of the consequences of flooding differed depending on their lifestyle (for example, their reliance on public transport) and their proximity to previously flooded areas. Figure 2 suggests that 73% of the respondents ranked the disruption of public transportation caused by flooding as the most harmful impact, followed by flooding of their neighborhoods (49% of respondents). In contrast, only 16% and 23% of the respondents experienced direct impacts from flooding of their houses and working places, respectively. The damage to private (movable) property was considered less harmful, as only 5% of the respondents reported property damage and loss. This might be attributed to the fact that flood damage to private property can be reduced as, unlike infrastructure and public buildings, they can be moved to dry places, and are thus less vulnerable to flooding. Respondents also mentioned that floods not only have hygiene and safety impacts but also negatively affect the image of a city.

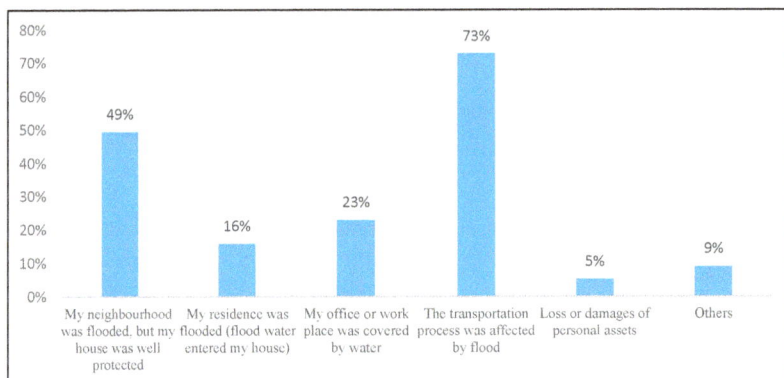

Figure 2. Impacts of urban floods as perceived by survey respondents.

Flood Protection Measures Taken by Residents

Measures that residents took when responding to urban floods are another important indicator of local perceptions of urban floods. They reflect how local people reacted to flooding and demonstrate their ability to reduce flood impact.

Figure 3 demonstrates that only 4% of respondents took no action to cope with floods, implying that the majority (96%) of the respondents took actions to protect themselves against flooding. The protective measures were dominated by recovery actions (taken after flooding), which accounted for 66% of the respondents. About 19% of the respondents claimed to have taken preventive actions to protect their houses and properties from flood water. Measures such as flood barriers, sandbags, and removing valuable assets to dry places were usually deployed, albeit that they were not always perceived as effective. The remaining 11% of the respondents adopted passive actions, such as temporarily leaving the flooded area.

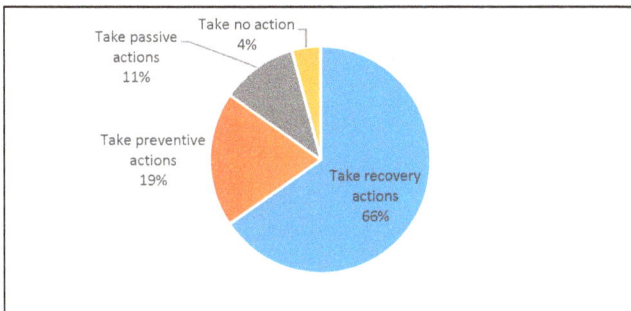

Figure 3. Types of flood protection measures taken by respondents.

Perceptions of Floods and Property Value

Most respondents considered floods to be annoying (59% of the respondents), as shown in Figure 4. About 14% of the respondents regarded their flood experience as traumatic and considered evacuating to a dry place as a preferred option. Those who felt neutral or indifferent about floods accounted for 24% of responses. During the in-depth interviews, two respondents indicated having a neutral attitude toward floods because they got used to them as floods happened every year.

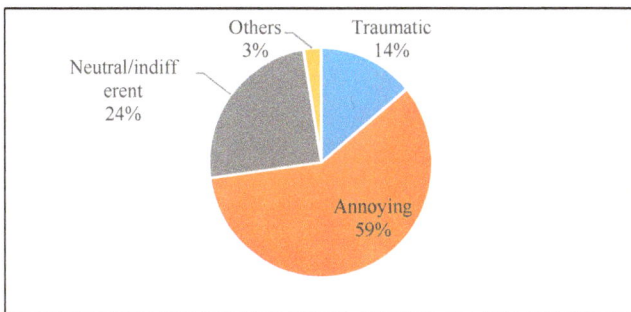

Figure 4. Resident's perceptions of floods.

Figure 5 shows respondents' perceptions of the impact of floods on residential property values. According to the figure, 86% of the respondents whose properties were exposed to flooding during the flooding season suffered from a value decrease compared to residential properties outside the flooded areas. However, 48% of the respondents expected that the value of the properties in the flooded areas would recover to the level before the flooding if adequate flood protection measures would be implemented. Only 25% of the respondents indicated that the value of their properties that were flooded would not return to this pre-flood level even after flood protection measures were implemented.

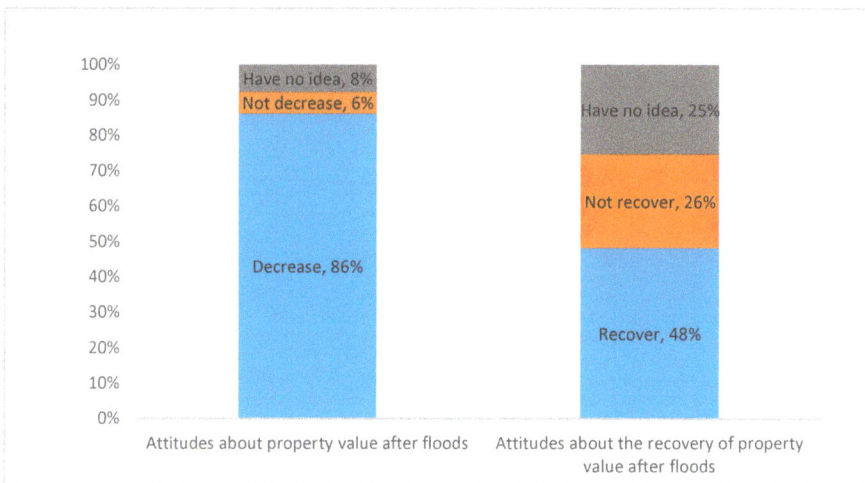

Figure 5. Residents' perceptions of property values after flooding.

4.1.3. Respondents' Perceptions of Sponge City and Its Functions

Respondents' Perceptions of Sponge City

Although the sponge city concept has only recently been introduced, 34% of the respondents in the central area of Wuhan were to some extent familiar with the concept, and 32% had heard about it but had no understanding of its meaning (Table 5). As shown in Table 5, 34% of the respondents did not understand the sponge city concept. As many as 44% of respondents were aware of the fact that Wuhan is currently implementing the sponge city concept through construction activities associated with urban development.

Table 5. Respondents' perceptions about sponge city.

Knowledge about Sponge City	Perception	Frequency	Percent	Cumulative Percent
Concept of sponge city	Know	144	34	34
	Do not know	143	34	68
	Heard, but have no idea what it means	136	32	100
Awareness of sponge city development in Wuhan	Yes	186	44	44
	No	237	56	100

Respondents' Perceptions of the Functions of Sponge City

During the survey process, the concept of sponge city and its main functions and benefits to cities were introduced to the interviewees. More than two-thirds (68.1%) of respondents believed that the construction of sponge city could effectively reduce the negative impact of floods on residential properties. Only 5.7% of respondents did not believe in the effectiveness or usefulness of the sponge city concept. The remaining 26.2% of the respondents had little idea about the objectives and effects of sponge city construction and shared the opinion that the effectiveness of sponge city projects should be demonstrated first (a "wait-and-see attitude").

The survey also assessed residents' appreciation of the multi-functional benefits of the sponge city concept, including the increase or improvement of the coverage of public green space (amenities) and flood safety (Figure 6). About 80% of the respondents ranked improvement of green space coverage as the most important benefit, closely followed by flood safety (74%) and the improvement of the coverage of public space (73%). Hence, most respondents thought that the multiple benefits of the sponge city concept were of high relevance and importance.

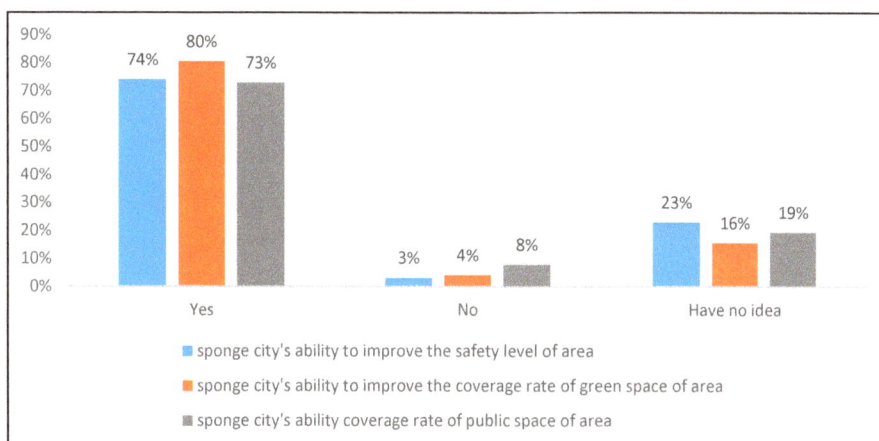

Figure 6. Residence's perceptions of different functions of sponge city.

4.1.4. Respondents' Willingness to Pay for Sponge City

Public willingness to pay is a subjective matter that varies from person to person. The heterogeneous nature of WTP is reflected in different perceptions, which in turn are impacted by differences in socio-economic and cultural backgrounds. It is assumed in this study that an important factor affecting residents' WTP for sponge city construction is residents' perceptions of the relationship between property values and the functional benefits attached to property brought about by sponge city construction. In this study, the above-mentioned assumption can be verified. The results show that 17% and 12% of the respondents believed sponge city construction would not increase their house value and land value, respectively. However, as shown in Figure 7, most respondents believed sponge city construction would increase their property value: 83% of the respondents assumed that their house value would increase, of which 50% expected an increase between 2% and 11% of the original house value. The percentages for land value increase are at a lower level than those for house values. In China, residential land is owned by collectives or by the state, and residents do not have the ability to participate in land transactions related to renting land or buying land user-rights. In other words, the perception of land value is biased by Chinese property laws and does not necessarily align with the perception of house values. As a result, 21% of respondents have no opinion about the impact of the sponge city concept on land values.

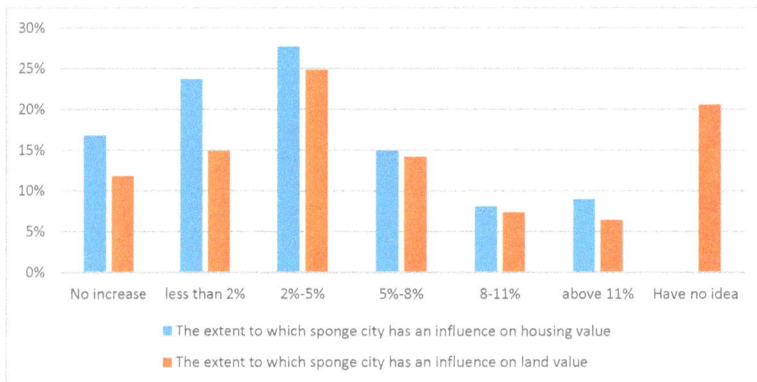

Figure 7. The extent to which sponge city influences housing and land values.

In summary, the data show that respondents were sensitive to flood risk exposure and property values, and were likely to recognize the value of sponge city as they expected their property value to increase. About 83% of respondents assumed that sponge city construction would increase the house value and 67% of the respondents expected their land value to increase in areas that the sponge city construction will cover. Most respondents who assumed that sponge city construction would increase their property value expected an increase in property values of less than 5%.

4.2. Research Findings of in-Depth Interviews

In-depth interviews were intended to better understand the reasons behind respondents' answers, and to gather professional views on the issues addressed in this research related to sponge city construction. From the interviews, it was found that flood risk exposure in Wuhan varied widely across the city. Most interviewees indicated that urban flooding is influencing their daily life either in a direct or an indirect way. The observed high frequency of flooding and the general notion that the current flood mitigation measures are not sufficient contributed to a certain level of tolerance towards urban flooding among the residents of Wuhan. Residents of Wuhan cared more about water management facilities and the services provided in the city and the multiple benefits emerging from sponge city construction. Only during the rainy season did residents value the benefits from flood safety improvements. During spring and winter time, the environmental effects of green infrastructure associated with sponge city construction were valued most, as green infrastructure was perceived to contribute to reducing air-pollution problems of the city.

Urban flood management approaches have significantly influenced urban design and construction policies and practices in recent years (e.g., [20]). In the past, citizens' involvement in flood management was very limited due mainly to a lack of public participation mechanisms [16]. Nowadays, partly because of new communication technologies that enable citizens to be engaged in the processes of urban design and planning, there are opportunities to provide citizens with a new role in decision-making [21]. The increasing exposure to the impacts of urban floods also provides incentives to planners, engineers and architects to integrate water in the urban design process. However, it is hard for all stakeholders involved (including residents) to recognize the values of such an integrated water management approach. The benefits are often remote and to some extent uncertain, while the extra costs are upfront. As a result, residents take it for granted that flood management has been appropriately considered by the government. Components of the sponge city concept have been applied in many new construction projects in Wuhan, but residents do not recognize these interventions as being part of this concept. For example, permeable parking lots, green roofs and small gardens have already been used in real estate developments. However, residents and some developers are not always aware of their functions and their ability to potentially reduce flood risk. In practice, the design of sponge city construction should be integrated into the urban landscape while acknowledging the environmental benefits. Developers tend to spend more on those two aspects than on flood safety alone. It is therefore crucial for the wider uptake of the sponge city construction to take all these multiple benefits into account.

To answer the question of "who should bear the cost for sponge city?", the interviews yielded the following insights. From the developers' perspective, the extra costs should be transferred to potential homebuyers regardless of their WTP. Most experts interviewed admitted that under the current booming real estate market conditions, developers should be able to transfer the costs of sponge city construction to homebuyers. However, they also shared the opinion that the construction costs of collective, local-level flood protection infrastructure, such as flood embankments, pumps and ponds, should be covered by the municipality through local taxes.

4.3. Research Findings from Secondary Data Analysis

4.3.1. The Relationship between Flooding Events and House Prices

Wuhan has the features of a subtropical monsoon climate, with the summer season being from June to September. During those months, the amount of rainfall usually varies between 200 and 300 mm per month, and this amount of rainfall is larger than the total rainfall in the remaining months of the year [22]. Figure 8 illustrates that, in 2016, monthly rainfall was significantly higher in the months from June to August than in the other months.

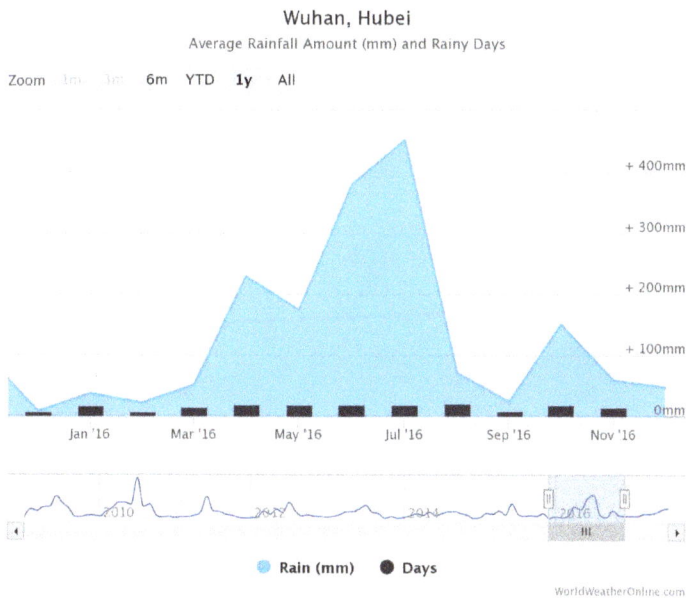

Figure 8. Monthly rainfall in Wuhan (2016). Data source: World Weather Online.

The heavy rainfall that hit Wuhan in July 2016 (see Section 1) caused waterlogging of many neighborhoods across the city. The data in Table 6 show that the residential house transaction prices in most districts were lower than usual that July. Only two districts experienced residential house transaction price increases. Both districts were sponge city construction areas. It seems plausible to assume that the flood event reduced the value of homes that were subject to serious flooding. As a result, the district's monthly average house transaction price dropped in July 2016.

Table 6. Monthly percentage change of transaction prices for housing in central districts of Wuhan in 2016.

Districts	Change of Transaction Prices for Residential Houses Across Districts by Month, %											
	Jan.	Feb.	Mar.	Apr.	May	Jun.	Jul.	Aug.	Sep.	Oct.	Nov.	Dec.
Qiaokou	s	−10	0	−5	0	9	−3	3	2	2	−3	−3
Wuchang	3	2	6	2	2	3	−4	1	5	−36	−4	6
Jiangan	−1	6	2	−2	6	2	−8	9	39	−16	−8	−5
Jianghan	11	−2	−5	16	−10	−8	−5	15	3	−8	−5	−31
Hongshan	−1	1	7	1	3	10	−10	−7	26	−8	−10	−12
Qingshan *	16	−9	−2	5	1	10	6	−5	13	3	6	−15
Hanyang *	−2	0	2	7	0	0	4	3	4	9	4	11

* Refers to districts that were sponge city pilots. Source: China Index Academy, 2017.

4.3.2. The Comparison between Sponge City Area and Non-Sponge City Area

Figure 9 shows the trend of the annual average transaction price for houses in Wuhan central areas over the past six years. Quantitative data on housing transaction prices in Wuhan (both sponge city areas and non-sponge city areas) are from the China Index Academy (accessed in 2017). Quantitative data on land transaction prices in Wuhan (both sponge city areas and non-sponge city areas) are from the Wuhan Land Transaction Center (accessed in 2017). The non-sponge city areas covered five districts: Jianghan, Jiangan, Hongshan, Wuchang and Qiaokou. The sponge city areas included two districts: Hanyang and Qingshan. From 2012 to 2014, both trend lines are very similar, indicating no difference between the average unit house transaction prices of non-sponge city and sponge city areas. In 2015, the trend line representing the sponge city area dropped by more than 10% to −1%. A plausible explanation for this abrupt change is the sudden increase of the housing supply in Qingshan District. In the past five years, the average housing supply was around 2300 houses per year, but the supply in 2015 reached a level of 6900 houses. This increased supply caused the average unit house transaction price of Qingshan District to decrease from 9520 yuan per square meter to 8743 yuan per square meter. In 2016 as well as the first half of 2017, the trend line of the sponge city areas was above that of the non-sponge city areas, suggesting that the average unit house transaction prices in sponge city areas were at a level higher than those of the non-sponge city areas. It seems, therefore, reasonable to speculate that sponge city construction may have had a positive influence on housing transaction prices in this period. This is a field of research that needs more attention in the future.

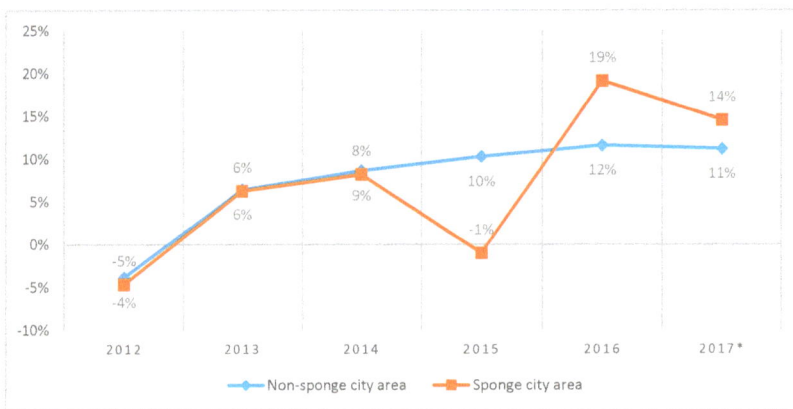

Figure 9. Annual average unit housing transaction prices in Wuhan central areas in the period 2012–2017. Source: China Index Academy, 2017.

5. Conclusions

5.1. Research Conclusion

5.1.1. Flood Risk Perceptions of Residents in Wuhan

A very large proportion (96.7%) of the residents of Wuhan has been exposed to flooding in either a direct or an indirect way. Apart from the scale of exposure, the flood frequency was also high, as 69% of residents claimed to suffer from urban flooding at least once a year. Disruption to the public transportation system and roads caused by flooding were perceived to be the most stressful impacts of urban flooding on the residents of Wuhan. Flooding of residential properties ranked as the second-most stressful impact. The most important type of interventions to manage flood risk in the study area were considered to be recovery actions. Preventive actions at household level had been taken by 19% of the residents, even though most households valued the effectiveness of preventive actions as low in terms of reducing flood risk. Compared with the large-scale exposure to, and the high frequency nature of, urban floods in Wuhan, residents' attitudes toward flood risk seemed to be relatively modest. Only 14% of respondents considered their flood experiences to be "traumatic". Most residents classified them merely as "annoying". The results of this research show that flooding had a negative impact on property values, as 86% of respondents thought that the value of their residential property would decrease if it had been flooded during the recent rainfall events that occurred in Wuhan. However, 48% of respondents believed that the value of residential properties could recover after the floods when adequate flood proofing measurements had been taken.

5.1.2. Residents' Perceptions of Sponge City and Its Functions

Although the term "sponge city" is a relatively academic concept in urban water management, more than 66% of the residents were familiar with the concept. Half of those residents had merely heard of the concept and had a limited understanding of its meaning. Past flood experiences and the level of understanding of the sponge city concept were the two factors that had an influence on residents' perceptions and valuing of the sponge city concept. A reduction of flood risk and an increase of the coverage of green space and public space are the three main benefits that are being delivered by sponge city construction. More than 70% of respondents acknowledged these functions of the sponge city. Most of the respondents expected that the sponge city concept would improve the livability of their neighborhood. They perceived an increase of the coverage of green space to be the most important factor that contributes to a better living environment. The sponge city's ability to improve the safety level of the area was acknowledged by most of the residents. This was particularly true for those who suffered frequently from flooding. However, one fifth of respondents still had reservations about its actual effectiveness in terms of reducing the city's flood risk.

5.1.3. Residents' Willingness to Pay for Sponge City

Residents' exposure to floods had an influence on their perception of the impact of sponge city construction on property values. About 83% of respondents believed that sponge city construction would increase the value of those houses located in the area covered by the construction, and 67% of respondents thought that sponge city construction would increase the land value of the area that the construction covers. Most respondents assumed that the range of increase would be less than 5% of the property's original value.

Residents' willingness to pay for sponge city construction was affected by the expected multiple benefits of the sponge city. Although 76% of respondents were willing to pay for the sponge city through an increase of their housing prices, the degree of their WTP was related to the real estate market in Wuhan. The current booming housing market in Wuhan reduces residents' bargaining power. Therefore, their WTP is relatively strong. It is highly plausible that residents' WTP would decrease if the real estate market became less volatile in the future. The actual transaction data supports the assumption that flood events have a negative influence on housing prices. From the actual transaction data, it is reasonable to infer that sponge city construction may contribute to an increment of housing transaction prices, but the degree of the price increase is hard to estimate.

In summary, most of the residents were willing to pay in the form of increased housing prices for sponge city construction, but to a limited extent only. The drivers of their WTP were a better living environment and lower flood risk. Under these conditions, low impact developments might seem to be the most suitable way to meet residents' requirements.

5.1.4. Sponge City Concepts and Residents' Perceptions of Housing Values

The research results support the theory that sponge city constructions are one of the factors that influence property values and prices in urban areas. Based on the research data, 83% of residents interviewed believed that sponge city construction in Wuhan would increase the value of houses located in the area covered by sponge city. The sponge city concept has an impact on residents' perceptions of housing values in two ways. The first relates to residents' perceptions of flood risk. The data analysis reveals that the flood exposure level of residents had an influence on residents' perceptions of flood risk. For those residents who have been exposed to flooding, property values were strongly impacted by the perceived flood risk level. The higher the level of flood exposure (in terms of frequency and/or impact), the more the residents believed that housing values will be positively affected by sponge city construction. The second relates to the residents' recognition of the multiple benefits provided by sponge city construction. Past experiences and degrees of exposure to flooding, and understanding of the sponge city concept, were considered to be the main factors influencing residents' confidence in the sponge city concept. The achievement of an improved living environment was valued as the most significant benefit by respondents in Wuhan.

This study reveals that, in the context of Wuhan, the sponge city concept impacts housing values. Most residents believed that sponge city construction has increased housing values by a maximum of 5% of the original housing value. Finally, both the results of the questionnaires and the in-depth interviews indicate a wide degree of support for the view that the costs of sponge city construction should be shared between the private sector and the government. This view holds that it is the government's responsibility to invest in basic infrastructure such as collective flood protection infrastructure including embankments, drainage systems and retention ponds, and that it is the private sector's responsibility to invest in multiple benefit interventions that enhance the livability of urban neighborhoods.

5.2. Recommendations

Although more than half of the residents were willing to pay for the sponge city concept, they valued equally—or even more importantly—the multiple benefits such as improved livability and a better living environment associated with the concept, instead of flood safety gains alone. Therefore, the first recommendation is that designers, architects and engineers of sponge city constructions should be fully aware of the opportunities the concept may offer to create sponge city interventions, which maximize multi-functional uses. This study reveals that residents are willing to pay for these additional benefits.

The second recommendation relates to public–private partnerships. PPPs are generally considered to be an important potential funding source of the sponge city program. However, in the current pilot period, the government is playing a leading role in the sponge city construction process, including finance, but private engagement is virtually absent. This study shows that there are opportunities for PPPs to engage in the process of sponge city construction. Promotion of the PPP model and the participation of social capital require market-oriented construction and an understanding and valuing (monetizing) of the multiple benefits associated with sponge city construction.

Finally, it is important to note here that care must be taken to consider the outcomes of this study for other Chinese cities, as local contextual differences may have an impact on citizens' perceptions.

Although this study provides useful information for other Chinese cities, it is recommended to validate the outcomes if applied to another city using a similar type of survey and field research.

Author Contributions: Conceptualization, Methodology, Formal analysis, Investigation, Data Curation and Writing- Original Draft preparation, S.Z.; Conceptualization, formal analysis, and writing (both original draft preparation and review and editing), C.Z., P.R.; Formal analysis, Y.J.

Acknowledgments: The authors wish to thank the Wuhan Housing Security and Housing Management Authority, the Wuhan Municipal Development and Reform Committee, and the Wuhan Land Use and Urban Spatial Planning Research Center for their technical assistance provided during the research.

Conflicts of Interest: The authors declare no conflict of interest.

Appendix A. Questionnaire Used in This Study

Personal information

1. What's your age currently?

   - Below 18
   - 18–30
   - 31–45
   - 46–55
   - 56–65
   - Above 65

2. What's your gender?

   - female
   - male

3. Which is your education level?

   - Senior school or below
   - University or college degree
   - Post-graduate qualification

4. Which of district that your current residence locates in?

   - Jianghan
   - Jiangan
   - Qiaokou
   - Qingshan
   - Wuchang
   - Hongshan
   - Hanyang
   - Donghu Hi-tech development zone
   - Wuhan economic development zone
   - Others

   Exposure to urban floods

5. Do you ever have exposure to urban floods in the past twenty years in Wuhan?

   - Heavy exposure (once a year or more)
   - Limited exposure (less than once a year)
   - Observed other's exposure

- No exposure

6. Which kinds of affects the urban floods had played on you and your asset?

    - My neighbourhood was flooded, but my house was well protected
    - My residence was flooded (flood water entered my house)
    - My office or work place was covered by water
    - The transportation process was affected by flood
    - Loss or damages of personal assets
    - Others affects made by floods

7. How do you feel about the experience of these floods?

    - Traumatic (e.g., if possible would change the residence for a safe and flood free one)
    - Annoying
    - Neutral/indifferent

8. What's the frequency that your house has been affected by urban floods (in the past 10 years)?

    - More than once a year
    - Once a year
    - Once per two years
    - Once per three to five years
    - Once per six to ten years
    - Less than once per ten years

9. Which kind of measurements below that do you prefer to take when responded to flooding?

    - Take recovery actions: flood resilient repair/replace/ cleaning/
    - Take preventive actions to protect from flooding water entering the house (such as flood boards to close the door(s), sand bags, remove valuable assets to dry place)
    - Leave the house
    - Take no action (Except for cleaning the house after the flood)

10. Do you think the property value of areas that heavily affected by urban floods will decrease after the urban floods?

    - Yes
    - No
    - I don't know

11. Do you think the government had taken effective measurements to reduce the effects played by the flood?

    - Yes
    - No
    - I don't know

    General perception of the sponge city program

12. Do you know the concept of the sponge city?

    - Yes
    - No
    - Heard, but have no idea of what does it mean

'The general objectives of the concept entail 'restore' the city's capacity to absorb, infiltrate, store, purify, drain and manage rainwater and 'regulate' the water cycle as much as possible to mimic the natural hydrological cycle.' [23].

13. Do you know that Wuhan government has conducted the construction of the sponge city?

- Yes
- No

14. Do you think that the construction of the sponge city can effectively reduce the negative effects made by floods on your property/house?

- Yes
- No
- I don't know

15. To what extent you think a reduction of flood risk (change of occurrence), which is the result of the construction of the sponge city, will increase the residential value of areas that the construction covers?

- Increased less than 2% of the house value
- Increased the house value from 2% to 4.9%
- Increased the house value from 5% to 7.9%
- Increased the house value from 8% to 10.9%
- Increased more than 11% of the house value
- Make no effect on house value

16. To what extent you think a reduction of flood risk (change of occurrence), which is the result of the construction of the sponge city, will increase the land value of areas that the construction covers?

- Increased less than 2% of the land value
- Increased the land value from 2% to 4.9%
- Increased the land value from 5% to 7.9%
- Increased the land value from 8% to 10.9%
- Increased more than 11% of the land value
- Make no effect on land value
- I have no idea on this issue

Perception of the positive effects brought by the sponge city program

17. Do you think that the construction of the sponge city can effectively improve the safety level of the area that the construction covers?

- Yes
- No
- I don't know

18. To what extent you think the increasing level of safety (reduction of impact/consequences) at the household level, which is the result of the construction of the sponge city, will increase the residential value of areas that the construction covers?

- Increased less than 2% of the house value
- Increased the house value from 2% to 4.9%
- Increased the house value from 5% to 7.9%

- Increased the house value from 8% to 10.9%
- Increased more than 11% of the house value
- Make no effect on house value

19. To what extent you think the increasing level of safety (reduction of impact/consequences) at the household level, which is the result of the construction of the sponge city, will increase the land value of areas that the construction covers?

- Increased less than 2% of the land value
- Increased the land value from 2% to 4.9%
- Increased the land value from 5% to 7.9%
- Increased the land value from 8% to 10.9%
- Increased more than 11% of the land value
- Make no effect on land value
- I have no idea on this issue

20. Do you think that the construction of the sponge city can effectively improve the coverage rate of green space of area that the construction covers?

- Yes
- No
- I don't know

21. To what extent you think the improvement of the coverage rate of green space, which is the result of the construction of the sponge city, will increase the residential value of areas that the construction covers?

- Increased less than 2% of the house value
- Increased the house value from 2% to 4.9%
- Increased the house value from 5% to 7.9%
- Increased the house value from 8% to 10.9%
- Increased more than 11% of the house value
- Make no effect on house value

22. To what extent you think the improvement of the coverage rate of green space, which is the result of the construction of the sponge city, will increase the land value of areas that the construction covers?

- Increased less than 2% of the land value
- Increased the land value from 2% to 4.9%
- Increased the land value from 5% to 7.9%
- Increased the land value from 8% to 10.9%
- Increased more than 11% of the land value
- Make no effect on land value
- I have no idea on this issue

23. Do you think that the construction of the sponge city can effectively improve the coverage rate of public space of area that the construction covers?

- Yes
- No
- I don't know

24. To what extent you think the improvement of the coverage rate of public space, which is the result of the construction of the sponge city, will increase the residential value of areas that the construction covers?

    - Increased less than 2% of the house value
    - Increased the house value from 2% to 4.9%
    - Increased the house value from 5% to 7.9%
    - Increased the house value from 8% to 10.9%
    - Increased more than 11% of the house value
    - Make no effect on house value

25. To what extent you think the improvement of the coverage rate of public space, which is the result of the construction of the sponge city, will increase the land value of areas that the construction covers?

    - Increased less than 2% of the land value
    - Increased the land value from 2% to 4.9%
    - Increased the land value from 5% to 7.9%
    - Increased the land value from 8% to 10.9%
    - Increased more than 11% of the land value
    - Make no effect on land value
    - I have no idea on this issue

Summary

26. Compared with houses that do not construct under the sponge city concept, whether you will prefer to buy the houses that constructed under the sponge city concept? (Under the assumption that other factors are similar?)

    - Yes
    - No
    - I don't know

27. To what extent you are willing to pay an extra cost for the construction of the sponge city in the area that your targeting house locates?

    - Not pay any more for the sponge city
    - Pay less than the 2%
    - Pay the extra cost from 2% to 4.9%
    - Pay the extra cost from 5% to 7.9%
    - Pay the extra cost from 8% to 10.9%
    - Pay more than 11% of the extra cost

References

1. Wang, Q.L. Effects of changing of water-route in Ming and Qing Dynasty. *Jianghan Acad.* **2013**, *13*, 108–113.
2. Today's Sydney. The Most Destructive Floods in Wuhan for 21th Century, 2016. Available online: http://www.gzhphb.com/article/25/257574.html (accessed on 1 March 2016).
3. Ministry of Finance of the People's Republic of China. A Noticed of the Scheme of Subsidies on the Pilot Project of Sponge City Construction by Finance Department of Central Government, 2014. Available online: http://jjs.mof.gov.cn/zhengwuxinxi/tongzhigonggao/201501/t20150115_1180280.html (accessed on 1 March 2017).
4. The Office of Wuhan City Government. The Program of Functional Improvement of Waterfront Area in Wuhan, 2014. Available online: http://www.wh.gov.cn/hbgovinfo_47/szfggxxml/zcfg/bgtwj/201611/t20161110_93233.html (accessed on 1 March 2017).

5. Yeo, S. Effects of disclosure of flood-liability on residential property values. *Aust. J. Emerg. Manag.* **2003**, *18*, 35–44.

6. Montz, B.E. Hazard area disclosure in New Zealand: The impacts on residential property values in two communities. *Appl. Geogr.* **1993**, *13*, 225–242. [CrossRef]

7. Lamond, J.; Bhattacharya-Mis, N. Risk perception and vulnerability of value: A study in the context of commercial property sector. *Int. J. Strateg. Prop. Manag.* **2016**, *20*, 252–264.

8. Eves, C.; Wilkinson, S. Assessing the immediate and short-term impact of flooding on residential property participant behaviour. *Nat. Hazards* **2014**, *71*, 1519–1536. [CrossRef]

9. Tobin, G.A.; Montz, B.E. The flood hazard and dynamics of the urban residential land market. *Water Resour. Bull.* **1994**, *30*, 673–685. [CrossRef]

10. Troy, A.; Romm, J. Assessing the Price Effects of Flood Hazard Disclosure Under the California Natural Hazard Disclosure Law (AB 1195). *J. Environ. Plan. Manag.* **2003**, *47*, 1–35. [CrossRef]

11. Botzen, W.J.W.; van den Bergh, J.C.J.M. Risk attitudes to low-probability climate change risks: WTP for flood insurance. *J. Econ. Behav. Organ.* **2012**, *82*, 151–166. [CrossRef]

12. Wang, Y.; Sun, M.; Song, B. Public perceptions of and willingness to pay for sponge city initiatives in China. *Resour. Conserv. Recycl.* **2017**, *122*, 11–20. [CrossRef]

13. Daun, M.C.; Clark, D. *Flood Risk and Contingent Valuation Willingness to Pay Studies: A Methodological Review and Applied Analysis*; Technical Report #6; Institute for Urban Environmental Risk Management, Marquette University: Milwaukee, WI, USA, 2000.

14. McLean, D.; Mundy, B. The Addition of Contingent Valuation and Conjoint Analysis to the Required Body of Knowledge for the Estimation of Environmental Damages to Real Property. *J. Real Estate Pract. Educ.* **1998**, *1*, 1–19.

15. Shao, W.; Zhang, H.; Liu, J.; Yang, G.; Chen, X.; Yang, Z.; Huang, H. Data integration and its application in the sponge city construction of China. *Procedia Eng.* **2016**, *154*, 779–786. [CrossRef]

16. Yang, Y.; Lin, G. A review on sponge city. *South Archit.* **2015**, *3*, 59–64.

17. *Wuhan Statistics Yearbook 2016*; No. 29; China Statistics Press: Beijing, China, 2017.

18. Renn, O. Individual and social perception of risk. In *Ökologisches Handeln als Sozialer Prozess. Themenhefte (Schwerpunktprogramm Umwelt/Programme Prioritaire Environnement/Priority Programme Environment)*; Fuhrer, U., Ed.; Birkhäuser: Basel, Switzerland, 1995; pp. 27–50.

19. Birkholz, S.; Muro, M.; Jeffrey, P.; Smith, H.M. Rethinking the relationship between flood risk perception and flood management. *Sci. Total Environ.* **2014**, *478*, 12–20. [CrossRef] [PubMed]

20. Wang, M.; Sweetapple, C.; Fu, G.; Farmani, R.; Butler, D. A framework to support decision making in the selection of sustainable drainage system design alternatives. *J. Environ. Manag.* **2017**, *201*, 145–152. [CrossRef] [PubMed]

21. Wehn, U.; Rusca, M.; Evers, J.; Lanfranchi, V. Participation in flood risk management and the potential of citizen observatories: A governance analysis. *Environ. Sci. Policy* **2015**, *48*, 225–236. [CrossRef]

22. Monthly Rainfall in Wuhan (2016). World Weather Online. Available online: https://www.worldweatheronline.com/wuhan-weather-averages/hubei/cn.aspx (accessed on 5 July 2017).

23. Zevenbergen, C.; Boogaard, F. *Sponge City Scoping Report*; UNESCO-IHE: Delft, The Netherlands, 2016; 22p.

MDPI

St. Alban-Anlage 66

4052 Basel

Switzerland

Tel. +41 61 683 77 34

Fax +41 61 302 89 18

www.mdpi.com

Water Editorial Office

E-mail: water@mdpi.com

www.mdpi.com/journal/water

www.ingramcontent.com/pod-product-compliance
Lightning Source LLC
Chambersburg PA
CBHW051703210326
41597CB00032B/5355